Interactions of Photons and Neutrons with Matter

Second Edition

Interactions of Photons and Neutrons with Matter

Second Edition

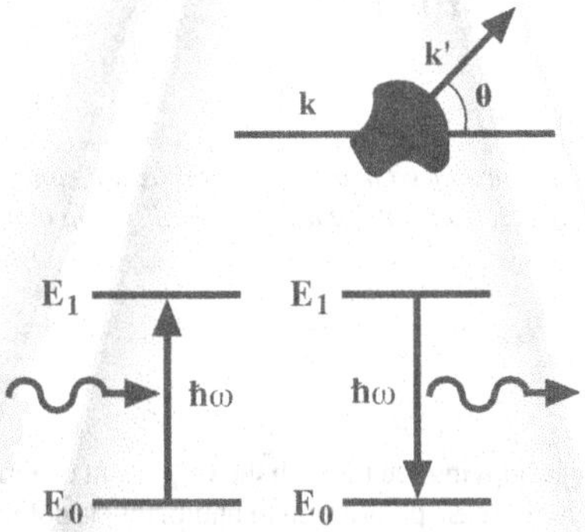

Sow-Hsin Chen
Massachusetts Institute of Technology

Michael Kotlarchyk
Rochester Institute of Technology

World Scientific

NEW JERSEY • LONDON • SINGAPORE • BEIJING • SHANGHAI • HONG KONG • TAIPEI • CHENNAI

Published by

World Scientific Publishing Co. Pte. Ltd.
5 Toh Tuck Link, Singapore 596224
USA office: 27 Warren Street, Suite 401-402, Hackensack, NJ 07601
UK office: 57 Shelton Street, Covent Garden, London WC2H 9HE

British Library Cataloguing-in-Publication Data
A catalogue record for this book is available from the British Library.

INTERACTIONS OF PHOTONS AND NEUTRONS WITH MATTER
Second Edition

ISBN-13 978-981-02-4214-5
ISBN-10 981-02-4214-X

Printed in Singapore.

To our wives, Ching-chih and Jill.

"I would rather have my ignorance than another man's knowledge, because I have got so much more of it."

Mark Twain
(Samuel Langhorne Clemens)

Preface

This is the second edition of a book that is an outgrowth from a set of lecture notes developed for a one-semester graduate course on the non-relativistic quantum theory of radiation. One of the authors (SHC) has been teaching this course for the past thirty-five years to engineering students preparing to enter doctoral research in the area of Nuclear Science and Technology (NST) in the Department of Nuclear Science and Engineering at the Massachusetts Institute of Technology. The course, entitled *Interaction of Radiation with Matter*, has been required for all students in NST prior to their taking of the Ph.D. qualifying examination. The aim has been to teach graduate students in engineering the basic principles underlying the interaction of electromagnetic and neutron radiations with atomic and molecular systems. The emphasis is on applications to modern-day materials research.

The primary audience for this book is first and second-year graduate students having a BS degree in science or engineering from a standard four-year university. It is assumed that the student has taken a year of general physics and chemistry, advanced calculus, differential equations, and linear algebra. In addition, it is extremely helpful if the student has experienced a one-semester course in modern physics in which basic quantum theory and introductory nuclear physics are taught. The basic lecture notes developed by SHC were designed for a one-semester course, however, this book reflects a highly expanded presentation of the original material, in terms of both content and exposition, developed by MK. Depending on the choice of topics covered by the instructor, this book could serve as the text for either a one-semester or a two-semester course. In either case, most of the first nine chapters should be covered. A two-semester course allows for the study of extended applications appearing in Chapters 10–13.

The aim of the first four chapters of this book is to rapidly elevate the reader's knowledge of fundamental physics to a more advanced level. The student is introduced to the Lagrangian and Hamiltonian formulations of classical mechanics, important elements of classical electrodynamics (including Maxwell's equations, electromagnetic potentials, and the wave equation), and standard nonrelativistic quantum mechanics based on the Dirac formalism. In addition, the treatment of electrodynamics (Chapter 4) includes substantial material on the theory of classical light scattering.

Chapters 5–7 focus on the quantization of radiation and lattice-displacement fields, time-dependent perturbation theory and transition probabilities, as well as the equilibrium and time-evolution properties of the density operator in statistical mechanics. Chapter 6 also contains a section on the formulation of the double-differential cross-section for the scattering of photons and thermal neutrons (the latter based on the Fermi pseudo-potential). At this point, the student should be prepared to understand basic aspects of the quantum theory of radiation. Chapters 8 and 9 present the topics of photon emission, absorption, and scattering.

The next three chapters illustrate how the theories learned earlier on can be used to understand a number of useful experimental techniques. Examples are chosen from the areas of nuclear magnetic resonance, photon correlation spectroscopy, and thermal neutron scattering. These choices reflect both the expertise of the authors and the interests of students in the M.I.T. Department of Nuclear Science and Engineering.

The final chapter presents, in a concise way, the modern view of the general relationship between theory and experiment in terms of equilibrium correlation functions.

The authors feel that it is of paramount importance to strike a balance between the amount of information one would like to convey in a course like this and the practical limits set by what a capable student can reasonably learn and digest during a four-month time period. The material in no way constitutes a complete coverage of the quantum theory of radiation interaction with matter. Rather, it is our intent that the material presented in this book be somewhat different from the standard material covered in, for example, an equivalent graduate course offered by a physics department. Our main goal is not to try to be complete in coverage, but instead to select topics that we feel are useful to students learning how to comprehend and analyze important types of laboratory experiments. In this regard, it is our opinion that the relativistic treatment, which is so standard in a course of this nature, can essentially be omitted.

In order to illustrate the principles presented in each chapter, numerous examples are included that have been worked out in detail. It is our opinion that the value of a new textbook is largely determined by the selection of non-trivial examples it provides to the reader. Furthermore, in this new second edition of the text, the most significant change is the addition of fifty-six substantial end-of-chapter problems. These are not just brief exercises requiring only a simple rehashing of the material found in each chapter. Our philosophy, instead, has been to design problems that serve a pedagogical function, challenging the student to extend his or her understanding of the material, and aimed toward deriving some significant results not already developed within the text. In our opinion, this added feature of the new edition greatly enhances the usefulness of the book as a course text.

The other substantive addition to the book is a discussion of high-resolution inelastic x-ray scattering appearing in Chapter 9. The specific application of this technique to investigating the dynamics of low-temperature liquid water is presented in Example 9.3.

Other, more minor, changes for this edition are the following:

- Example 4.3 on the Hamiltonian formulation for the electromagnetic field plus charged particles has been revised and corrected.
- Various equations throughout the text have been corrected.
- Both the end-of-chapter and end-of-book reference lists have been updated.

We would like to take this opportunity to thank the many graduate students who have taken this course over the years. Their comments about the usefulness of the course have provided us with encouragement for completing this project. Many students have commented that the material presented has helped them significantly in their comprehension of the wide range of literature available to them.

M. Kotlarchyk is appreciative of the sabbatical-leave opportunity provided to him by both the M.I.T. Department of Nuclear Science and Engineering and the R.I.T. Department of Physics during the Spring of 1994 when the first edition of this book was being written. During that time, he was able to complete two chapters of the manuscript. Later that year, the writing of a third chapter was made possible through the support of an R.I.T. College of Science Dean's Summer Research Fellowship. We are also grateful to the Bernard M. Gordon M.I.T. School of Engineering Curriculum Development Fund which helped us in the final production of the camera-ready manuscript, to Sung-Min Choi for contributions to Chapter 10, and to John Chen for his assistance with some of the illustrations. Finally, the authors are extremely grateful to our patient family members who have persevered through the long process associated with the preparation of both the first and second editions of this book.

Sow-Hsin Chen
Michael Kotlarchyk

Cambridge, Massachusetts, USA
January 2007

Contents

List of Examples

Chapter 1
INTRODUCTION

The exchange of energy between a photon or neutron and a material medium is a fundamentally quantum-mechanical phenomenon. For scientists and engineers who need to deal with the application of radiation probes to condensed matter research, learning to calculate experimentally relevant quantities using the principles of quantum mechanics is a skill that is indispensable. This book is intended to meet this need. It is presumed that any prior training or coursework the reader may have in the formal aspects of standard theoretical physics is minimal.

The most relevant quantity in connection with interpreting data derived from modern laboratory experiments involving radiation-matter interactions is the so-called *double-differential scattering cross-section*; this quantity basically gives the probability that a particular radiation-probe exchanges a given amount of energy and momentum with a target material. In order to calculate this cross-section, one first needs to become familiar with the quantum-mechanical concept of a *transition probability per unit time* between two states. Consequently, the most direct route to an understanding of the interaction between photons and neutrons with matter is to develop some of the basic time-dependent quantum-mechanical perturbation theory leading to the calculation of these transition probabilities.

Even though the foundations of quantum mechanics were formulated early in the twentieth century, many aspects of the theory make use of the much older classical concept of conjugate pairs of canonical mechanical variables and the *Hamiltonian function* of a physical system. It is important, therefore, that the reader have some understanding of the formalism of classical mechanics before exploring elements of the quantum theory.

Another branch of classical physics relevant to the interaction of photons with matter is the theory of electromagnetic (EM) fields and waves. For the purpose of our present study, it is essential that the reader also develop a firm grounding in Maxwell's classical theory of EM radiation because, ultimately, the concept of a photon (i.e., the treatment of the EM field as a mechanical entity) depends on being able to quantize the classical field.

In order to make connections to real physical systems, and hence to actual experimental results, one must also learn some of the basic principles of classical and quantum statistical mechanics. This allows one to perform calculations of various statistical averages of mechanical variables for a system with many degrees of freedom at a specified temperature.

Figure 1.1 is a flowchart mapping the connections between the various topics treated within this book. In Chapters 2–4 we present, in sequence (and in what we feel are sufficient detail), the cornerstone subjects—classical mechanics, quantum mechanics, and classical electrodynamics. Since these subject areas form the basis for all further discussion in the book, they appear as boxes at the top of the flowchart and, in all probability, should not be skipped by the reader. The presentation is designed to be at a sufficiently elementary level, which allows one to read and understand it with only a minimum of guidance provided by the instructor.

The second level of the flowchart includes the subjects of Chapters 5–7, namely, quantization of the field, the calculation of transition probabilities using the time-dependent perturbation theory, and the role of the density operator in the calculation of statistical averages at a given temperature. Essentially, this level represents the most important part of the book. The subject of "linear response theory" also appears at this level in the chart, but a discussion of this topic is postponed until the final chapter. The authors feel that this is appropriate because, by doing so, we are able to present more examples that are applicable to experiments discussed in the later chapters.

The third level consists of a series of practical applications including photon correlation spectroscopy, nuclear magnetic resonance, and thermal neutron scattering. As mentioned in the preface, these topics should be considered optional for a one-semester course, and are indicated so by the use of dashed boxes in the flowchart. Chapters 8 and 9 on first and second-order radiation processes are considered to be applications at the most fundamental level. Hence, these two topics appear in boxes that are not dashed, and should not be skipped in a course of any length.

The examples presented in each chapter have been carefully selected to illustrate the various principles and techniques introduced. The authors feel that the working principles of quantum mechanics, which are based on rather abstract concepts, are best learned through the use of specific illustrative examples. The reader should pay special attention to these examples, since they show how the various principles can be applied in a concrete manner. Sufficient details are supplied so that the reader can follow along without the assistance of an instructor.

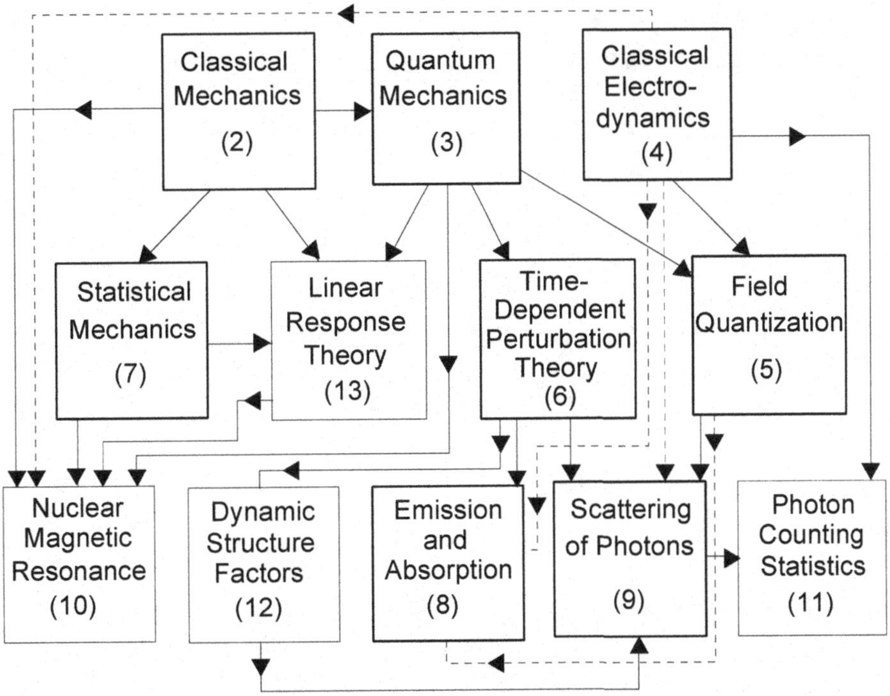

Figure 1.1 Connections between the topics treated in the various chapters of this book. (The numbers in parentheses indicate the relevant chapters.)

Chapter 2
AN OVERVIEW OF CLASSICAL MECHANICS

The primary motivation for this chapter is to lay the groundwork necessary for making connections between the classical and quantum-mechanical description of a physical system. In particular, the dynamic description of mechanical systems in terms of Lagrangian and Hamiltonian mechanics is presented, with a special emphasis on oscillators. It is only by introducing the so-called *canonical coordinates* and by representing the total energy in Hamilton's form that one can ultimately make the transition from a classical to a quantum mechanical description of both atomic systems and the interacting electromagnetic radiation field.

2.1 The Lagrangian Formulation

In many simple mechanical systems, the most natural way to obtain the equations of motion is to apply Newton's Second Law to each particle, namely

$$\dot{\mathbf{p}}_i = \mathbf{F}_i, \tag{2.1}$$

which relates the force $\mathbf{F}_i$ acting on the ith particle in the system to the rate of change* of the linear momentum $\mathbf{p}_i = m_i\mathbf{v}_i$ of that particle, where $\mathbf{v}_i$ is the velocity of particle i. In most cases, the mass m_i of each particle is fixed and Newton's Second Law reduces to the more familiar form

$$\mathbf{F}_i = m\mathbf{a}_i, \tag{2.2}$$

where $\mathbf{a}_i = \dot{\mathbf{v}}_i$ is the acceleration of a given particle in the system.

Consider, for example, the simple case of a single particle under the influence of a linear restoring force along the x-axis. The equation of motion is that for a simple harmonic oscillator, i.e.,

$$m\ddot{x} = -Kx, \tag{2.3}$$

where K is a positive (spring) constant. Alternatively, this can be written as

$$\frac{d}{dt}\left(\frac{1}{2}m\dot{x}^2 + \frac{1}{2}Kx^2\right) = 0, \tag{2.4}$$

*Each dot above a letter indicates a time derivative.

which means that the expression in the parentheses remains fixed (or conserved) as the system evolves in time. This *constant of the motion* can be defined as E, the total energy of the system:

$$E = \frac{1}{2}m\dot{x}^2 + \frac{1}{2}Kx^2. \tag{2.5}$$

Generally, Eq. 2.5 is written as

$$E = T(\dot{x}) + V(x), \tag{2.6}$$

where T and V are the kinetic and potential energy functions of the system. In general, the condition for energy conservation is that the forces performing work in the system can be set equal to the negative gradient of the potential function, i.e.,

$$\mathbf{F} = -\boldsymbol{\nabla} V. \tag{2.7}$$

A more general and powerful method for obtaining the equations of motion is through the Lagrangian formulation of mechanics. Here, one introduces the so-called *Lagrangian* of the system, which is a function of a set of *generalized coordinates* and *generalized velocities* suitable for describing the system. For example, x and $\dot{x}$ are the generalized coordinate and velocity for the simple harmonic oscillator. The Lagrangian $\mathcal{L}$ is defined as

$$\mathcal{L}(x,\dot{x}) = T(\dot{x}) - V(x) = \frac{1}{2}m\dot{x}^2 - \frac{1}{2}Kx^2. \tag{2.8}$$

The equation of motion, Eq. 2.3, can then be obtained from *Lagrange's equation*, which has the form

$$\frac{d}{dt}\left(\frac{\partial \mathcal{L}}{\partial \dot{x}}\right) - \frac{\partial \mathcal{L}}{\partial x} = 0. \tag{2.9}$$

Lagrange's equation can be derived from the *principle of least action* which states that the system evolves in such a way that the *action*, S, given by the integral

$$S = \int_{t_1}^{t_2} \mathcal{L}(x,\dot{x},t)dt, \tag{2.10}$$

takes on an extremum. Equivalently, with the two end-points held fixed (see Fig. 2.1), one requires that a variation in the action, δS, vanishes, i.e.,

$$\delta S = \int_{t_1}^{t_2} \left(\frac{\partial \mathcal{L}}{\partial x}\delta x + \frac{\partial \mathcal{L}}{\partial \dot{x}}\delta \dot{x}\right) dt = 0. \tag{2.11}$$

Using the fact that $\delta\dot{x} = d(\delta x)/dt$ and integrating the second term by parts gives

$$\delta S = \frac{\partial \mathcal{L}}{\partial \dot{x}}\delta x \,|_{t_1}^{t_2} + \int_{t_1}^{t_2} \left[\frac{\partial \mathcal{L}}{\partial x}\delta x - \frac{d}{dt}\left(\frac{\partial \mathcal{L}}{\partial \dot{x}}\right)\delta x\right] dt. \tag{2.12}$$

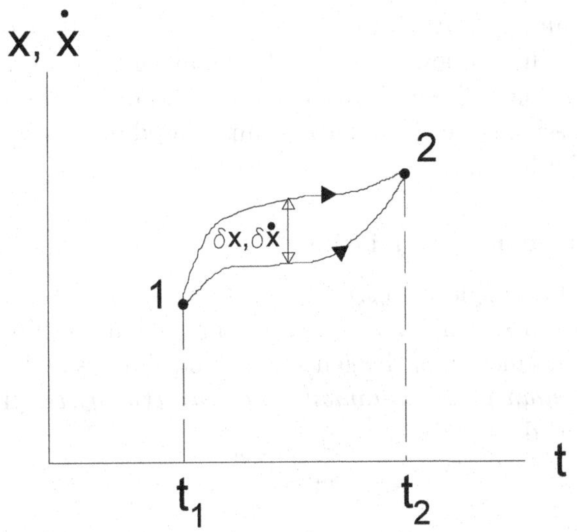

Figure 2.1 The variation of x or $\dot{x}$ in the principle of least action.

The first term vanishes because the two end-points are fixed and the variation, δx, must therefore be zero at these points. Since Eq. 2.12 must hold for arbitrary δx, the integrand must vanish, and Eq. 2.9 follows.

A major advantage of the Lagrangian formalism for doing mechanics is that it can naturally incorporate constraints that exist between various particles. In particular, even though a system may contain M particles, requiring $3M$ coordinates to specify the particle positions, constraints between the particles may reduce the number of degrees of freedom to a number significantly less than $3M$. For example, in a rigid body, all particles of the body are constrained to be at fixed positions relative to each other. Then, even though the number of particle coordinates $3M$ can be huge, the actual number of independent coordinates is only six. Hence, the Lagrangian is a function of only six generalized coordinates and their associated velocities. The generalized coordinates are the three Cartesian coordinates of the center-of-mass and the three Eulerian angles specifying the orientation of the rigid body. In general, a many-particle system with N degrees of freedom can be characterized by a set of generalized coordinates $\{q_1, q_2, ..., q_k, ..., q_N\}$ and generalized velocities $\{\dot{q}_1, \dot{q}_2, ..., \dot{q}_k, ..., \dot{q}_N\}$. For a conservative system (i.e., one with constant total energy), one can define the Lagrangian generally by

$$\mathcal{L}(\dot{q}_k, q_k) = T(\dot{q}_k) - V(q_k, \dot{q}_k) \tag{2.13}$$

and Lagrange's equations can be generalized to

$$\frac{d}{dt}\left(\frac{\partial \mathcal{L}}{\partial \dot{q}_k}\right) - \frac{\partial \mathcal{L}}{\partial q_k} = 0 \qquad (k = 1, 2, ..., N). \tag{2.14}$$

There is one equation for each degree of freedom.

There are further advantages to working with Lagrange's equations. Later in this chapter it will be shown how the formulation can naturally be extended to continuous systems (i.e., those with an infinite number of degrees of freedom). In Chapter 4, we will see how the formulation can be used in conjunction with a velocity-dependent potential function.

2.2 The Hamiltonian Formulation

By definition, the Lagrangian $\mathcal{L}(q_k, \dot{q}_k, t)$ is a function of the generalized coordinates and generalized velocities. In contrast, the Hamiltonian formulation of classical mechanics introduces a function of the generalized coordinates and *generalized momenta* called the *Hamiltonian* or *Hamiltonian function* $H(p_k, q_k, t)$. The generalized momenta, p_k, are defined as

$$p_k = \frac{\partial \mathcal{L}}{\partial \dot{q}_k}, \tag{2.15}$$

and from Eq. 2.14, we also have

$$\dot{p}_k = \frac{\partial \mathcal{L}}{\partial q_k}. \tag{2.16}$$

The Hamiltonian of the system is defined as†

$$H(p_k, q_k, t) = \sum_k \dot{q}_k p_k - \mathcal{L}(q_k, \dot{q}_k, t). \tag{2.17}$$

If the potential energy function of the system is independent of the $\dot{q}_k$'s, then the Hamiltonian becomes

$$H = \sum_k \dot{q}_k \frac{\partial T}{\partial \dot{q}_k} - T + V. \tag{2.18}$$

Because the kinetic energy T is a purely quadratic function of the generalized velocities, the summation term simply reduces to $2T$ and the resulting Hamiltonian reduces to the total energy of the system.

The equations of motion for the system can now be formulated in terms of Hamilton's equations. These can be derived from Lagrange's equations by first writing down the total differential of Eq. 2.17 and applying Eqs. 2.15 and 2.16:

$$\begin{aligned} dH &= \sum_k \dot{q}_k\, dp_k + \sum_k p_k\, d\dot{q}_k - \sum_k (\partial \mathcal{L}/\partial q_k)\, dq_k - \sum_k (\partial \mathcal{L}/\partial \dot{q}_k)\, d\dot{q}_k - (\partial \mathcal{L}/\partial t)\, dt \\ &= \sum_k \dot{q}_k\, dp_k + \sum_k p_k\, d\dot{q}_k - \sum_k \dot{p}_k\, dq_k - \sum_k p_k\, d\dot{q}_k - (\partial \mathcal{L}/\partial t)\, dt \\ &= \sum_k \dot{q}_k\, dp_k - \sum_k \dot{p}_k\, dq_k - (\partial \mathcal{L}/\partial t)\, dt. \end{aligned} \tag{2.19}$$

†This is called a "Legendre transformation" from the set of variables $\{q_k, \dot{q}_k, t\}$ to the set $\{q_k, p_k, t\}$.

A comparison of the latter total differential to the form

$$dH = \sum_k \frac{\partial H}{\partial p_k} dp_k + \sum_k \frac{\partial H}{\partial q_k} dq_k + \frac{\partial H}{\partial t} dt \tag{2.20}$$

requires that

$$\dot{q}_k = \frac{\partial H}{\partial p_k}, \qquad \dot{p}_k = -\frac{\partial H}{\partial q_k}, \tag{2.21}$$

and

$$\frac{\partial H}{\partial t} = -\frac{\partial \mathcal{L}}{\partial t}. \tag{2.22}$$

Equations 2.21 are Hamilton's equations of motion, also known as the *canonical equations*. For a system with N degrees of freedom, these form a set of $2N$ first-order differential equations for the so-called *canonical coordinates* and *conjugate momenta*, replacing the N second-order differential equations in the Lagrangian formulation for the generalized coordinates and velocities. Equation 2.22 shows that when $\mathcal{L}$ does not contain the time explicitly, neither does H.

The Hamiltonian for the harmonic oscillator is a function of the single pair of canonically conjugate coordinates x and p, where

$$p = \frac{\partial L}{\partial \dot{x}} = m\dot{x}. \tag{2.23}$$

From Eq. 2.17, the Hamiltonian becomes

$$H = \dot{x}p - \frac{1}{2}m\dot{x}^2 + \frac{1}{2}Kx^2. \tag{2.24}$$

Replacing $\dot{x}$ with p/m gives

$$H(x,p) = \frac{p^2}{2m} + \frac{1}{2}Kx^2. \tag{2.25}$$

Applying the canonical equations results in

$$\dot{x} = \frac{p}{m} \qquad \text{and} \qquad \dot{p} = -Kx. \tag{2.26}$$

The first of these is the definition of p and the latter is Newton's Second Law.

For a general system, it is important to notice two characteristic features of the Hamiltonian. First, observe that any time the Hamiltonian does not explicitly contain a particular canonical coordinate, q_k, we have $\dot{p}_k = 0$, and the corresponding conjugate momentum is a constant of the motion. Such missing coordinates in the Hamiltonian are called *cyclic* or *ignorable* coordinates.

Secondly, the total time derivative of the Hamiltonian is

$$\dot{H} = \sum_k \frac{\partial H}{\partial p_k}\dot{p}_k + \sum_k \frac{\partial H}{\partial q_k}\dot{q}_k + \frac{\partial H}{\partial t}. \tag{2.27}$$

Because of Eqs. 2.21, the two summation terms cancel, resulting in

$$\dot{H} = \frac{\partial H}{\partial t}. \tag{2.28}$$

This means that as long as the Hamiltonian is not explicitly a function of time, H is a constant of the motion, i.e., energy is conserved.

Example 2.1 A Particle in a Central Potential. In situations where two particles (masses m_1 and m_2) interact, it is convenient to separately treat the motion of the center-of-mass and the motion of the particles relative to the center-of-mass (see Fig. 2.2). This latter motion can be viewed as equivalent to a single particle of reduced mass $m = m_1 m_2/(m_1 + m_2)$ moving about a fixed point at the origin. The position vector $\mathbf{r}$ then corresponds to the relative position $\mathbf{r}_1 - \mathbf{r}_2$ between the two particles. In many important situations, the interparticle potential is only a function of the separation, r, between the two particles, i.e., $V(\mathbf{r}_1 - \mathbf{r}_2) = V(|\mathbf{r}_1 - \mathbf{r}_2|) = V(r)$, and the two-body problem reduces to a central potential problem. Examples of important situations falling into this category are two-body gravitational problems, the electron-proton system of the hydrogen atom, diatomic molecules, the deuteron problem, and the scattering of one particle by the field of another.

The spherical symmetry of the central potential function suggests choosing the spherical coordinates r, θ, and ϕ as the generalized coordinates for the three degrees of freedom of the reduced-mass particle, where θ is the polar angle from the z-axis and ϕ is the azimuthal angle about the z-axis (see Fig. 2.3). The transformations between the Cartesian and spherical coordinates are

$$x = r\sin\theta\cos\phi, \qquad y = r\sin\theta\sin\phi, \qquad z = r\cos\theta, \tag{2.29}$$

allowing us to write the Lagrangian for the system:

$$\begin{aligned}\mathcal{L} &= T - V = \frac{1}{2}m(\dot{x}^2 + \dot{y}^2 + \dot{z}^2) - V(r) \\ &= \frac{1}{2}m(\dot{r}^2 + r^2\dot{\theta}^2 + r^2\dot{\phi}^2 \sin^2\theta) - V(r).\end{aligned} \tag{2.30}$$

To obtain the Hamiltonian, we first determine the generalized momenta p_r, p_θ, and p_ϕ conjugate to the generalized coordinates, i.e.,

$$\begin{aligned}p_r &= \partial\mathcal{L}/\partial\dot{r} = m\dot{r} \\ p_\theta &= \partial\mathcal{L}/\partial\dot{\theta} = mr^2\dot{\theta} \\ p_\phi &= \partial\mathcal{L}/\partial\dot{\phi} = mr^2\dot{\phi}\sin^2\theta.\end{aligned} \tag{2.31}$$

Then,

$$H = \dot{r}p_r + \dot{\theta}p_\theta + \dot{\phi}p_\phi - \mathcal{L} \tag{2.32}$$

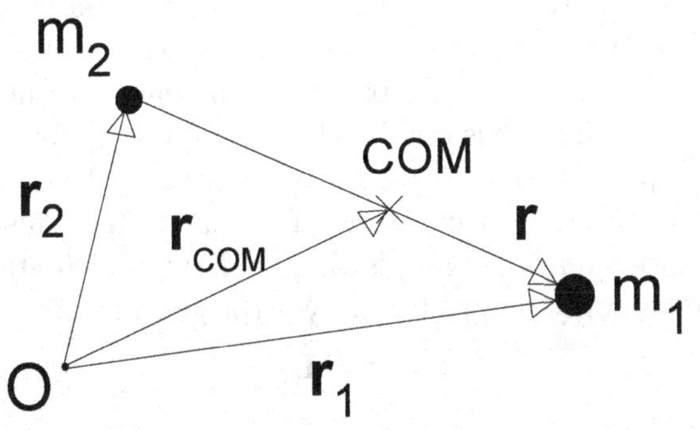

Figure 2.2 Center-of-mass and relative coordinates in the two-body problem.

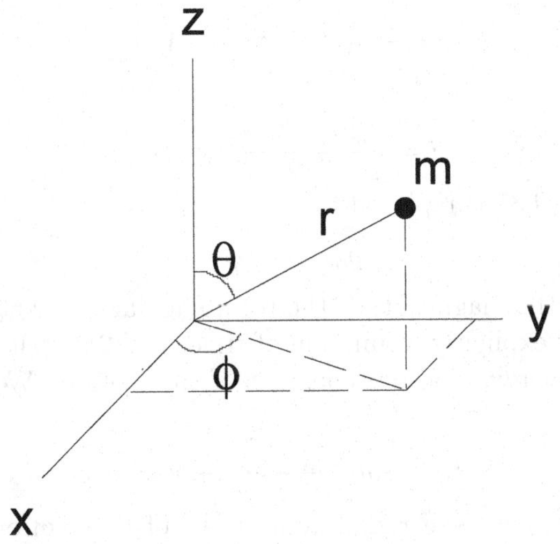

Figure 2.3 Spherical coordinate system for central potential problem.

or

$$H = \frac{p_r^2}{2m} + \frac{p_\theta^2}{2mr^2} + \frac{p_\phi^2}{2mr^2 \sin^2\theta} + V(r). \tag{2.33}$$

Note that the coordinate ϕ does not appear in the Hamiltonian, hence it is a cyclic coordinate and the conjugate momentum, p_ϕ, is a constant of the motion. p_ϕ corresponds to the z-component of the particle's angular momentum about the origin. Since the orientation of the z-axis can be chosen arbitrarily for a central potential, it must be that all components of the angular momentum are constant, i.e., the angular momentum vector $\mathbf{L}$ must be a constant of the motion. This result is consistent with the rotational version of Newton's Second Law, which states that an angular momentum change can only be produced by a torque, $\boldsymbol{\tau}$:

$$\boldsymbol{\tau} = \dot{\mathbf{L}}. \tag{2.34}$$

The torque is more basically defined as

$$\boldsymbol{\tau} = \mathbf{r} \times \mathbf{F}. \tag{2.35}$$

Since the force vector in a central potential is directed purely radially, i.e., $\mathbf{F} = -(\partial V/\partial r)\mathbf{e}_r$, the torque produced by the central field is zero.

Because the orientation of the angular momentum vector remains unchanged during the motion, the particle must confine its motion to a plane. By specifying the motion as being in the x-y plane, one eliminates one degree of freedom, i.e., the generalized coordinate θ is fixed at $\pi/2$ and its conjugate momentum p_θ vanishes. Equations 2.30 and 2.33 get replaced by

$$\mathcal{L} = \frac{1}{2}m(\dot{r}^2 + r^2\dot{\phi}^2) - V(r) \tag{2.36}$$

and

$$H = \frac{p_r^2}{2m} + \frac{p_\phi^2}{2mr^2} + V(r), \tag{2.37}$$

while the last of Eqs. 2.31 is replaced by

$$p_\phi = mr^2\dot{\phi}. \tag{2.38}$$

p_ϕ now represents L, the magnitude of the total angular momentum of the system.

Since H is not explicitly a function of time, Eq. 2.28 tells us that the Hamiltonian, or the total energy, is also a constant of the motion. We can now write the total energy as

$$E = \frac{1}{2}m\dot{r}^2 + \frac{L^2}{2mr^2} + V(r). \tag{2.39}$$

This expression shows that the radial component of the motion is analogous to a one-dimensional motion as long as one replaces the actual central potential $V(r)$ by a so-called *effective potential*

$$V_{eff}(r) = V(r) + \frac{L^2}{2mr^2}, \tag{2.40}$$

where $L^2/2mr^2$ is called the *centrifugal potential*.

Example 2.2 A Relativistic Particle. For a particle moving at speeds that are not negligible relative to the speed of light c, Einstein's theory of special relativity requires that the linear momentum of a particle be redefined to be

$$\mathbf{p} = \gamma m \mathbf{v}, \tag{2.41}$$

where $\gamma = (1-\beta^2)^{-\frac{1}{2}}$ and $\beta^2 = (\dot{x}^2 + \dot{y}^2 + \dot{z}^2)/c^2$. This form for the momentum guarantees that the law of conservation of momentum is obeyed from the point of view of all inertial (non-accelerating) frames of reference. The corresponding expression for the kinetic energy of the particle can be obtained from the work-energy theorem, which states that the change in kinetic energy of a particle is identical to the work performed on that particle, i.e.,

$$T - T_0 = \int \mathbf{F}\cdot d\mathbf{r}, \tag{2.42}$$

where the integral is performed over the path of the particle. Here, the force on the particle must be consistent with the most general form of Newton's Second Law as given by Eq. 2.1, i.e.,

$$\mathbf{F} = \dot{\mathbf{p}} = \frac{d}{dt}(\gamma m \mathbf{v}). \tag{2.43}$$

It is important to note that, in general, $\mathbf{F}$ is not proportional to the acceleration, so the form $\mathbf{F} = m\mathbf{a}$ is not obeyed by a particle moving at relativistic speeds. Returning to Eq. 2.42, one can, for simplicity, calculate the kinetic energy change for a particle moving along the x-axis under the influence of forces directed along this axis. If the particle starts from rest, then the change in and hence the final kinetic energy follows from

$$T = \int_{x_0}^{x} F\,dx = \int_{x_0}^{x} \frac{dp}{dt}dx = \int_0^p \frac{dx}{dt}dp = \int_0^p v\,dp = pv - \int_0^v p\,dv, \tag{2.44}$$

where the last step is an integration by parts. Inserting Eq. 2.41 for the relativistic momentum, and integrating, one obtains

$$T = (\gamma - 1)mc^2. \tag{2.45}$$

The quantity mc^2 is called the *rest energy* of the particle. For speeds much less than the speed of light ($\beta \ll 1$), the latter reduces to the classical expression for the kinetic energy:

$$T = (1 + \frac{1}{2}\beta^2 + ...)mc^2 - mc^2 \cong \frac{1}{2}mv^2. \tag{2.46}$$

It is tempting to form the Lagrangian by using the classical $\mathcal{L} = T - V$. However, this would mean that for a particle acted on by conservative forces independent of velocity, we have for the generalized momenta $\partial L/\partial \dot{q}_k = \gamma^3 m \dot{q}_k$, and inserting into Lagrange's equations gives $d(\gamma^3 m \dot{q}_k)/dt = -(\partial V/\partial q_k)$. The right-hand side in the latter equation is just the kth force component, but the left-hand side is not the corresponding rate of change of the momentum, as it should be. We must, therefore, redefine the Lagrangian to be something other than $T - V$, namely

$$\mathcal{L} = -\frac{mc^2}{\gamma} - V. \tag{2.47}$$

Now the generalized momenta are $\partial\mathcal{L}/\partial\dot{q}_k = \gamma m\dot{q}_k$, and the corresponding Lagrange's equations become $d(\gamma m\dot{q}_k)/dt = -(\partial V/\partial q_k)$, as expected. For $\beta \ll 1$, the Lagrangian given by Eq. 2.47 becomes

$$\mathcal{L} = -\,mc^2(1 - \frac{1}{2}\beta^2 + ...) - V \cong -mc^2 + \frac{1}{2}mv^2 - V, \tag{2.48}$$

which is different from the classical Lagrangian by the additive constant $-mc^2$. However, since Lagrange's equations only involve derivatives of $\mathcal{L}$, this has no effect on the classical equations of motion.

The Hamiltonian can be found from Eq. 2.17, i.e.,

$$\begin{aligned} H &= \sum_k \dot{q}_k(\gamma m\dot{q}_k) + mc^2/\gamma + V \\ &= \gamma m\left(\beta^2c^2 + c^2/\gamma^2\right) + V \\ &= \gamma mc^2 + V. \end{aligned} \tag{2.49}$$

From Eq. 2.45, we see that the term γmc^2 is equal to the sum of the particle's kinetic and rest energies. Equation 2.49 is not yet the correct form for the Hamiltonian, since it must ultimately be expressed in terms of the particle's momentum. However, this can be accomplished by observing the following energy-momentum relation:

$$(\gamma mc^2)^2 - (pc)^2 = (\gamma mc^2)^2 - (\gamma m\beta c^2)^2 = (mc^2)^2. \tag{2.50}$$

In the limit when $T \ll mc^2$, this reduces to the classical energy-momentum relation, namely, $T = p^2/2m$. The Hamiltonian then becomes

$$H = \sqrt{p^2c^2 + m^2c^4} + V, \tag{2.51}$$

where $p^2 = \sum_k p_k^2$.

2.3 Trajectories in Phase Space

The Hamiltonian function determines the equations of motion, and hence, completely defines the time evolution of a classical system once the initial values of the canonical variables are specified. This can be visualized by picturing a $2N$-dimensional *phase space* with coordinate axes that correspond to each of the N generalized coordinates and N conjugate momenta. At each instant, the "state" of the system is defined by specifying the coordinates of a *phase point* in this space, and the evolution of the system in time is represented by a trajectory in the phase space. Hamilton's equations dictate the velocity of the phase point along the trajectory. The fact that the Hamiltonian governs the behavior of the system (both in the past and in the future) in a completely deterministic way requires that trajectories never intersect in phase space. For example, in the case of the harmonic oscillator, since the Hamiltonian is the constant total energy of the motion, Eq. 2.25 defines an elliptical trajectory in the two-dimensional p-x phase plane (see Fig. 2.4). The initial conditions, $x(t=0)$ and $p(t=0)$, determine the starting point in phase space, and hence, which ellipse (or energy) the oscillator follows. The subsequent motion around the ellipse is found by solving the equation(s) of motion, giving the following parametric relations:

$$\begin{aligned} x(t) &= \sqrt{2E/K}\cos(\omega_0 t + \phi) \\ p(t) &= \sqrt{2mE}\sin(\omega_0 t + \phi). \end{aligned} \tag{2.52}$$

$\omega_0 = \sqrt{K/m}$ is the natural vibration frequency of the oscillator.

Instead of following the course of an individual point in phase space, it is often useful to follow the development of a region, or cloud, of phase points. This construct is a fundamental starting point for classical statistical mechanics, where such a cloud of phase points represents a large *ensemble* of physical systems that are identical to one another, but with different initial conditions. The most fundamental equation of classical statistical mechanics is *Liouville's theorem*, which states that the density of points representing an ensemble of systems does not change as a function of time. A derivation of Liouville's theorem follows:

Let $\rho(p,q,t)$ represent the density of phase points in an ensemble. For notational simplicity, we denote the set of N pairs of canonical variables, i.e., the q_k's and p_k's, by q and p, respectively. Then, at any time t, the number of members of the ensemble in the phase volume $dp\,dq$ is given by $\rho(p,q,t)\,dp\,dq$. Since the total number of systems in the ensemble is fixed, the function ρ must satisfy the continuity equation consistent with conservation of particles (i.e., phase points):

$$\frac{\partial \rho}{\partial t} = -\boldsymbol{\nabla}\cdot(\rho \mathbf{v}). \tag{2.53}$$

Here, $\mathbf{v}$ represents the "velocity" of a point in the ensemble along its trajectory through phase space, and the divergence is taken with respect to the $2N$ canonical variables. Evaluating the right-hand side produces

$$\frac{\partial \rho}{\partial t} = -\mathbf{v}\cdot\boldsymbol{\nabla}\rho - \rho\boldsymbol{\nabla}\cdot\mathbf{v}. \tag{2.54}$$

The second term on the right-hand side, however, vanishes since

$$\boldsymbol{\nabla}\cdot\mathbf{v} = \sum_{i=1}^{N}\left(\frac{\partial \dot{q}_i}{\partial q_i} + \frac{\partial \dot{p}_i}{\partial p_i}\right) = \sum_{i=1}^{N}\left[\frac{\partial}{\partial q_i}\left(\frac{\partial H}{\partial p_i}\right) - \frac{\partial}{\partial p_i}\left(\frac{\partial H}{\partial q_i}\right)\right] = 0. \tag{2.55}$$

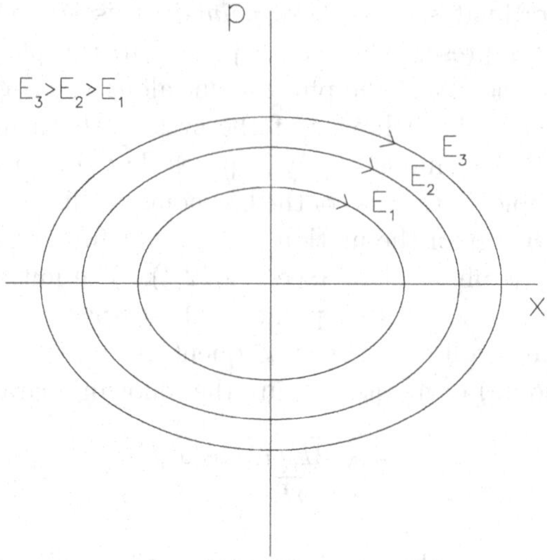

Figure 2.4 Phase portrait for harmonic oscillator.

This gives the result

$$\frac{\partial \rho}{\partial t} = -\sum_{i=1}^{N} \left(\frac{\partial \rho}{\partial q_i} \dot{q}_i + \frac{\partial \rho}{\partial p_i} \dot{p}_i \right) . \tag{2.56}$$

Equation 2.56 is Liouville's theorem. Notice that since $\rho = \rho(p, q, t)$, the total time derivative of ρ is

$$\dot{\rho} = \frac{\partial \rho}{\partial t} + \sum_{i=1}^{N} \left(\frac{\partial \rho}{\partial q_i} \dot{q}_i + \frac{\partial \rho}{\partial p_i} \dot{p}_i \right) . \tag{2.57}$$

Then, by Eq. 2.56,

$$\dot{\rho} = 0, \tag{2.58}$$

and the density of phase points of the ensemble is constant in time. Since the number of points in the ensemble is fixed, it follows that although the shape of a given volume of phase points may change as it wanders through phase space, its volume is preserved. The motion of phase points resembles that of an incompressible fluid.

Liouville's theorem is a consequence of the fact that $\nabla \cdot \mathbf{v}$, the divergence of the velocity field in phase space, vanishes. One should take note, however, that Liouville's theorem is not obeyed by macroscopic systems involving dissipative, or frictional, forces, for which $\nabla \cdot \mathbf{v} < 0$. For such systems, Liouville's theorem would be replaced by

$$\dot{\rho} = -\rho \nabla \cdot \mathbf{v}. \tag{2.59}$$

In these cases, the density of phase points, and hence the phase volume, changes as time proceeds. This occurs because such systems cannot be represented by a

Hamiltonian function. Therefore, the substitution of Hamilton's equations performed in Eq. 2.55 is not permitted. For example, one can revise Eqs. 2.26, the equations of motion for the simple harmonic oscillator, to include dissipation in the form of a damping force that is oppositely directed, but proportional to, the particle's velocity. The equations of motion for the damped oscillator take the form

$$\dot{x} = \frac{p}{m} \qquad \text{and} \qquad \dot{p} = -Kx - b\dot{x} = -Kx - \frac{b}{m}p, \tag{2.60}$$

where b is a *friction constant*. For this situation, we have

$$\boldsymbol{\nabla}\cdot\mathbf{v} = \frac{\partial \dot{x}}{\partial x} + \frac{\partial \dot{p}}{\partial p} = -\frac{b}{m}, \tag{2.61}$$

so, from Eq. 2.59,

$$\dot{\rho} = \frac{b}{m}\rho. \tag{2.62}$$

This means that the density of phase points increases, or the phase volume decreases, exponentially in time. This is a consequence of the non-conservative nature of the system.

2.4 Variations on the Pendulum

We now illustrate some of the concepts outlined up to this point by analyzing various pendulum configurations. We begin by considering the free, undamped, spherical pendulum. Here, the pendulum bob, of mass m, is attached to a weightless rigid rod of length ℓ that is free to swing in any direction about one end fixed at the origin. The bob moves under the influence of its weight ($m\mathbf{g}$) while being constrained to be at a distance ℓ from the origin. The system has two degrees of freedom, which can be specified by the angles θ and ϕ in a spherical coordinate system. The polar (z) axis is chosen to be oriented vertically downward, as shown in Fig. 2.5. The Lagrangian can then be obtained from Eq. 2.30, with the radial coordinate fixed at $r = \ell$ and with $V(r)$ replaced by the gravitational potential energy $V(\theta) = mg\ell(1 - \cos\theta)$:

$$\mathcal{L} = \frac{1}{2}m\ell^2(\dot{\theta}^2 + \dot{\phi}^2 \sin^2\theta) - mg\ell(1 - \cos\theta). \tag{2.63}$$

Here it is assumed that the potential energy vanishes at the bottom of the sphere of motion. Aided by Eq. 2.33, with $p_r = 0$ and $V(r)$ replaced by $V(\theta)$, the Hamiltonian becomes

$$H = \frac{p_\theta^2}{2m\ell^2} + \frac{p_\phi^2}{2m\ell^2 \sin^2\theta} + mg\ell(1 - \cos\theta). \tag{2.64}$$

The conjugate momenta are

$$p_\theta = m\ell^2\dot{\theta} \tag{2.65}$$

and

$$p_\phi = m\ell^2\dot{\phi}\sin^2\theta. \tag{2.66}$$

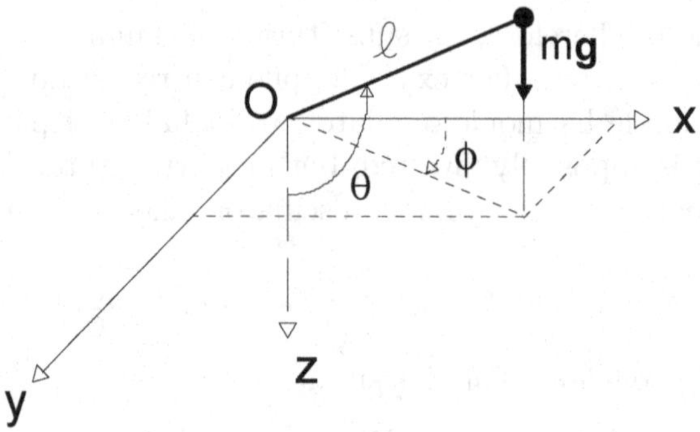

Figure 2.5 Coordinate system for spherical pendulum.

The coordinate ϕ is cyclic, hence p_ϕ, the z-component of the particle's angular momentum, is conserved. The basic equation of motion can be found from Lagrange's equation

$$\frac{d}{dt}\left(\frac{\partial \mathcal{L}}{\partial \dot{\theta}}\right) - \frac{\partial \mathcal{L}}{\partial \theta} = 0. \tag{2.67}$$

The result is

$$\ddot{\theta} + \frac{g}{\ell}\sin\theta - \dot{\phi}^2 \sin\theta\cos\theta = 0, \tag{2.68}$$

or, using Eq. 2.66,

$$\ddot{\theta} + \frac{g}{\ell}\sin\theta - \frac{p_\phi^2}{m^2\ell^4}\frac{\cos\theta}{\sin^3\theta} = 0. \tag{2.69}$$

We now consider the equation of motion for two special cases, namely, the conical pendulum and the simple (planar) pendulum.

In the case of the conical pendulum, the bob traces out a horizontal circle specified by $\theta = \theta_0 =$ constant. Then $\ddot{\theta} = \dot{\theta} = 0$, and using Eq. 2.68 one has the following requirement for $\dot{\phi}$, the angular speed of the bob:

$$\dot{\phi}^2 = \frac{g}{\ell}\sec\theta_0. \tag{2.70}$$

The bob exhibits uniform circular motion, with an angular speed that is an increasing function of the cone angle θ_0. The two-dimensional phase trajectory for a given θ_0 simply follows a horizontal line in the p_ϕ-ϕ plane.‡ The speed of the phase point along its horizontal path changes linearly as one jumps from one trajectory to the next. This means that an initially rectangular region of phase points develops into a parallelogram that becomes more and more skewed as time goes on. However, since

‡Since ϕ must remain bounded, one imposes periodic boundary conditions on ϕ, as depicted in Fig. 2.6. One interprets these boundaries on the phase space as follows: If the pendulum bob goes around in the positive-ϕ direction, its phase point disappears off the right edge of the picture and immediately reappears on the left side.

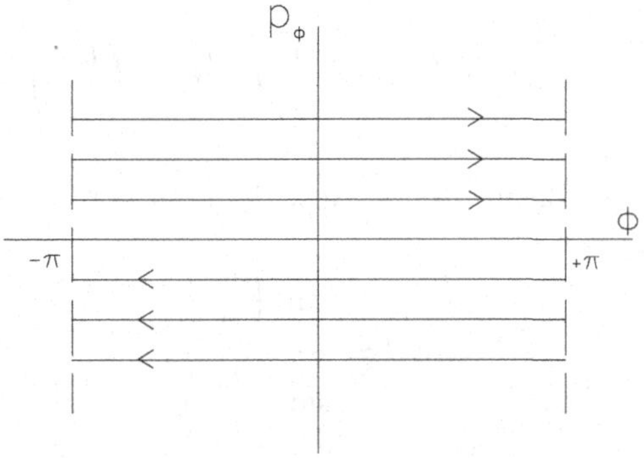

Figure 2.6 Phase portrait for conical pendulum.

the size of the base and height remain fixed, the area of the developing region does not change, as expected from Liouville's theorem.

In the case of the simple pendulum, the motion takes place in a fixed vertical plane, i.e., $\phi = \phi_0 =$ constant. With $p_\phi = \dot{\phi} = 0$, the equation of motion reduces to

$$\ddot{\theta} + \omega_0^2 \sin\theta = 0, \tag{2.71}$$

where $\omega_0^2 = g/\ell$. For small amplitude oscillations, $\sin\theta \approx \theta$, and the equation of motion takes the simple form

$$\ddot{\theta} + \omega_0^2 \theta = 0. \tag{2.72}$$

This is identical to the equation of motion for the simple harmonic oscillator. The solution is

$$\theta(t) = \theta_m \cos(\omega_0 t + \varphi), \tag{2.73}$$

with period $T = 2\pi/\omega_0 = 2\pi\sqrt{\ell/g}$. The amplitude θ_m and phase φ are integration constants that depend on the initial conditions. To analyze the period for arbitrary amplitude (when the motion is oscillatory), it is convenient to start with the expression for the total energy of the pendulum:

$$E = \frac{1}{2} m\ell^2 \dot{\theta}^2 + mg\ell(1 - \cos\theta). \tag{2.74}$$

This equation can be integrated directly to give

$$\omega_0 t = \frac{1}{\sqrt{2}} \int_{\theta_0}^{\theta} \left(\frac{E}{mg\ell} + \cos\theta - 1 \right)^{-1/2} d\theta. \tag{2.75}$$

The total energy is identical to the potential energy when θ reaches its maximum value of θ_m, i.e., $E = mg\ell(1 - \cos\theta_m)$. Equation 2.75 can be rewritten as

$$\omega_0 t = \frac{1}{\sqrt{2}} \int_{\theta_0}^{\theta} (\cos\theta - \cos\theta_m)^{-1/2}\, d\theta = \frac{1}{2} \int_{\theta_0}^{\theta} \left[\sin^2\left(\frac{\theta_m}{2}\right) - \sin^2\left(\frac{\theta}{2}\right) \right]^{-1/2} d\theta, \tag{2.76}$$

where we used $\cos\theta = 1 - 2\sin^2(\theta/2)$. Now define a new variable β by

$$\sin\beta = \frac{\sin\left(\frac{\theta}{2}\right)}{\sin\left(\frac{\theta_m}{2}\right)}. \tag{2.77}$$

Upon substitution, we obtain a standard elliptic integral

$$\omega_0 t = \int_{\beta_0}^{\beta} \left[1 - \sin^2\left(\frac{\theta_m}{2}\right) \sin^2\beta \right]^{-1/2} d\beta. \tag{2.78}$$

The integrand can be expanded in powers of $\sin^2(\theta_m/2)$:

$$\omega_0 t = \int_{\beta_0}^{\beta} d\beta \left[1 + \frac{1}{2} \sin^2\left(\frac{\theta_m}{2}\right) \sin^2\beta + \ldots \right]. \tag{2.79}$$

The time t represents one-half of the period if the limits of the above integral are set at $\beta_0 = -\pi/2$ and $\beta = +\pi/2$ (i.e., $\theta_0 = -\theta_m$, $\theta = +\theta_m$). Doing so results in the following expression for the period:

$$T = \frac{2\pi}{\omega_0} \left[1 + \frac{1}{4} \sin^2\left(\frac{\theta_m}{2}\right) + \ldots \right]. \tag{2.80}$$

This shows that as the oscillation amplitude increases, the period becomes only slightly larger than for small oscillations.

The phase portrait for the simple pendulum is shown in Fig. 2.7. For small oscillations, where the period is independent of the amplitude, the curves of constant energy take the form of circles centered at the origin. For larger amplitude oscillations, the shapes of the phase trajectories begin to change. Once the energy of the motion exceeds $2mg\ell$, the oscillatory motion is replaced by a continuous rotation of the bob. The curves that separate the regions of oscillatory and rotational behavior are called *separatrices*; they exhibit discontinuities in their slopes as the bob reaches unstable equilibrium points at $\theta = \pm\pi$, $p_\theta = 0$.

The effect of a damping force on the simple pendulum is found by adding a term into the equation of motion that is proportional, but oppositely directed, to the velocity of the bob, i.e.,

$$\ddot{\theta} + \gamma\dot{\theta} + \omega_0^2 \sin\theta = 0, \tag{2.81}$$

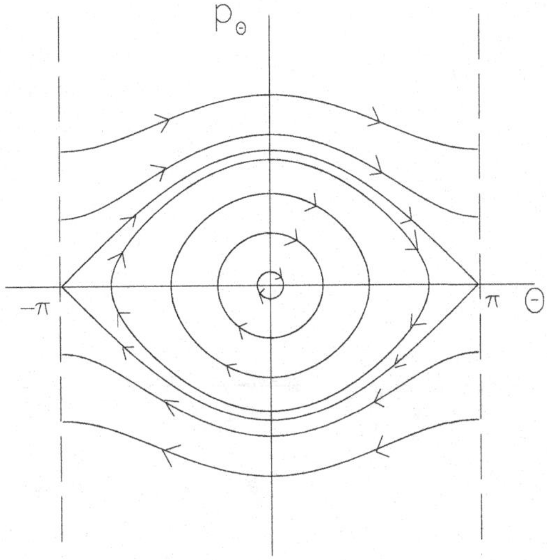

Figure 2.7 Phase portrait for simple pendulum.

where γ is a positive damping constant. For small amplitude oscillations ($\sin\theta \approx \theta$), the equation is solved by assuming a trial solution of the form $\theta(t) = \exp(\alpha t)$, where α is complex. There are three types of solutions, depending on the value of $\gamma^2 - 4\omega_0^2$, namely

$$\begin{array}{lll} \text{underdamped:} & \theta(t) = Ae^{-\gamma t/2}\cos(\omega_d t - \varphi), & (\gamma^2 - 4\omega_0^2 < 0) \\ \text{critically damped:} & \theta(t) = (At + B)e^{-\gamma t/2}, & (\gamma^2 - 4\omega_0^2 = 0) \\ \text{overdamped:} & \theta(t) = Ae^{-(\gamma-\delta)t/2} + Be^{-(\gamma+\delta)t/2}, & (\gamma^2 - 4\omega_0^2 > 0), \end{array} \tag{2.82}$$

where $\omega_d^2 = -\delta^2 = \omega_0^2 - (\gamma/2)^2$, and the constants A, B, and φ depend on the initial conditions. The angular displacement as a function of time is shown in Fig. 2.8 for all three cases. The motion is oscillatory only when the damping is relatively low, (underdamped case).

The damped pendulum is a dissipative system, i.e., the energy of the system decreases with time. This can easily be seen by considering the rate of change of the total energy, i.e.,

$$\begin{aligned} \dot{E} &= \frac{d}{dt}\left[\frac{1}{2}m\ell^2\dot{\theta}^2 + mg\ell(1-\cos\theta)\right] \\ &= m\ell^2\dot{\theta}\left(\ddot{\theta} + \omega_0^2\sin\theta\right) = -m\ell^2\gamma\dot{\theta}^2, \end{aligned} \tag{2.83}$$

where use of Eq. 2.81 was made in the last step. Since γ and $\dot{\theta}^2$ are both positive definite, $\dot{E} < 0$ and the energy of the pendulum continuously decreases. As a result, the phase trajectories for all three types of damping decay toward the equilibrium

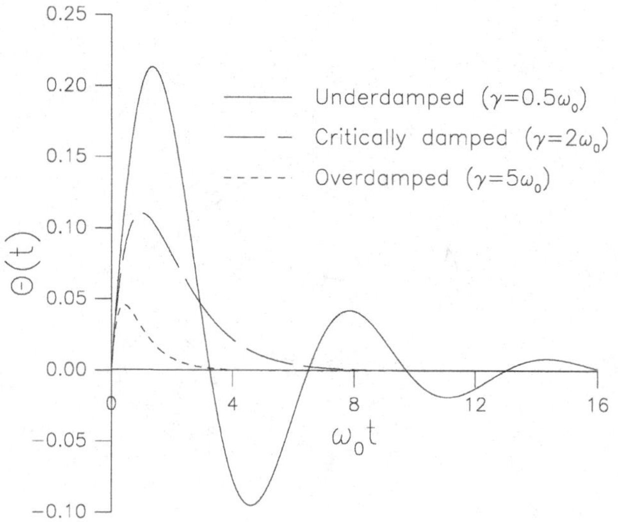

Figure 2.8 Angular displacement vs. time for damped pendulum.

point at the origin of phase space, as shown in Fig. 2.9. This point is also known as an *attractor* (having a geometric dimension of zero).

The following pendulum equation of motion includes the addition of a sinusoidally varying driving force:

$$\ddot{\theta} + \gamma\dot{\theta} + \omega_0^2 \sin\theta = G\cos\Omega t. \tag{2.84}$$

G and Ω represent the amplitude and frequency of the drive, respectively. For small oscillations and drive amplitude, the motion reaches a periodic steady state once the transients have died away, as given by the particular solution of Eq. 2.84, namely,

$$\theta(t) = A(\Omega)\cos\left[\Omega t - \varphi(\Omega)\right], \tag{2.85}$$

where

$$A(\Omega) = \frac{G}{\sqrt{(\omega_0^2 - \Omega^2)^2 + \gamma^2\Omega^2}} \quad \text{and} \quad \tan\varphi = \frac{\gamma\Omega}{\omega_0^2 - \Omega^2}. \tag{2.86}$$

$A(\Omega)$, the functional dependence of the steady-state oscillation amplitude on the drive frequency, is known as the *resonance curve*. It has the form shown in Fig. 2.10. The amplitude of the motion is a maximum when the pendulum is driven at the *resonant frequency* of $\Omega = \sqrt{\omega_0^2 - (\gamma^2/2)}$. The width of the resonance curve is proportional to the damping constant. The phase portrait for a given set of initial conditions (see Fig. 2.11) follows a trajectory that asymptotically approaches a closed curve representing the periodic steady-state of the pendulum. This curve is a one-dimensional attractor, also known as a *limit cycle*. Even for large-scale pendulum motions, as long as the steady state is periodic, a limit cycle will be approached.

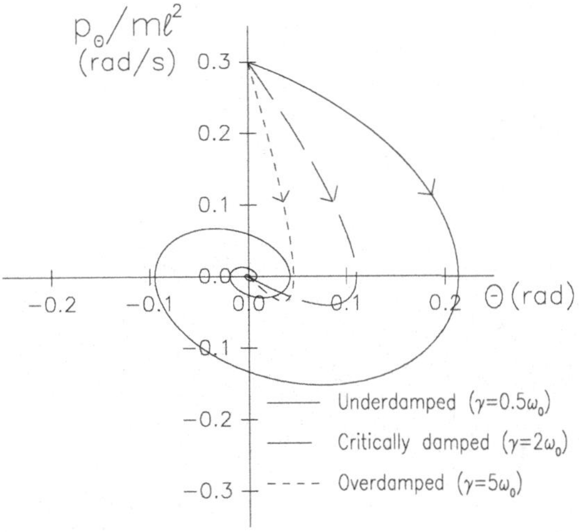

Figure 2.9 Phase trajectories for damped pendulum.

Example 2.3 The Possibility for Chaotic Behavior in the Driven Pendulum. The most interesting types of motion exhibited by the driven pendulum are not periodic at all. It turns out that for certain parameter sets, $\{\omega_0, \gamma, \Omega, G\}$, the driven pendulum exhibits *chaotic behavior.*[§] In these cases, it becomes impossible in practice to predict the configuration of the system (i.e., know the values of θ and $\dot{\theta}$) with any reasonable degree of accuracy at future instants in time. In principle, once the initial conditions are specified, the equation of motion still permits one to trace the time-evolution of the system through phase space in a completely deterministic way. However, when chaotic behavior is present, there is extreme *sensitivity to the initial conditions.* This means that if two identical pendula are set into motion with almost identical initial conditions $(\theta_0, \dot{\theta}_0)$ and $(\theta_0 + \delta\theta, \dot{\theta}_0 + \delta\dot{\theta})$, the ensuing trajectories through phase space will be such that the separation between the two phase points will, on the average, increase exponentially with time. Numerical solutions of Eq. 2.84 on a small computer illustrate this point. For example, Fig. 2.12 shows the pronounced effect of sensitivity to initial conditions on the development of phase trajectories for neighboring initial points. Unless knowledge of the initial conditions is perfect, one quickly loses the ability to track the state of the system in time.

The simple pendulum is only one example among many physical systems capable of exhibiting chaotic behavior. In general, the necessary conditions for possible chaos are (a) the equations of motion can be written as a set of autonomous,[¶] first-

[§]See Baker and Golub [1] for a good introductory description of chaotic dynamics that focuses exclusively on the damped, driven pendulum.

[¶]This means that the equations are not explicit functions of time.

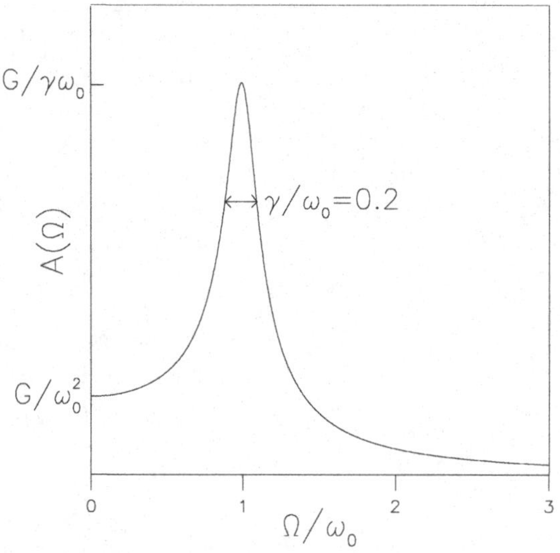

Figure 2.10 Resonance curve for weakly driven pendulum—dependence of steady-state oscillation amplitude A on driving frequency Ω.

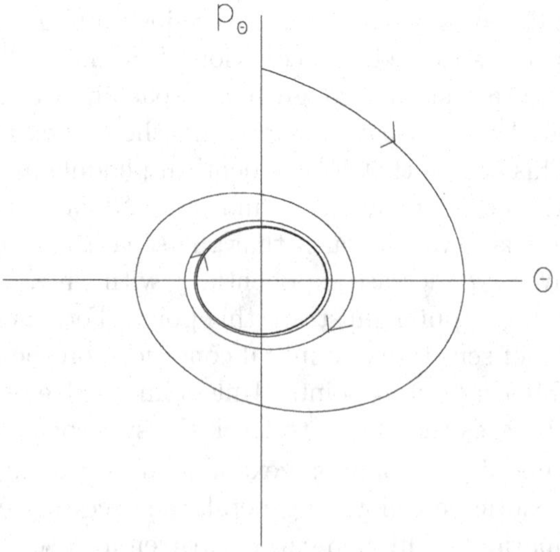

Figure 2.11 Limit cycle for damped, driven pendulum.

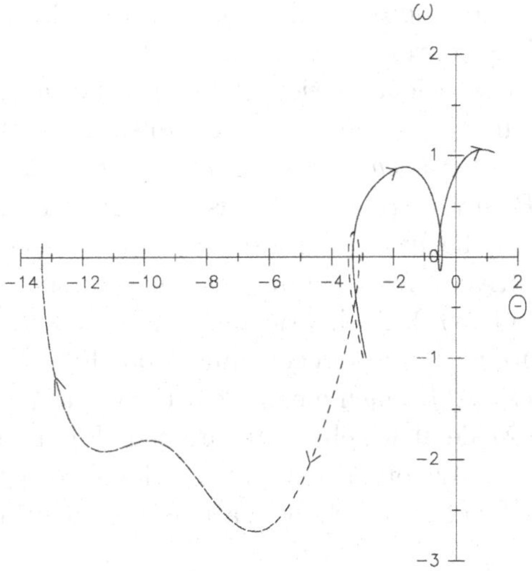

Figure 2.12 Illustration of sensitivity to initial conditions in the driven pendulum with $G/\omega_0^2 = 1.05$, $\Omega/\omega_0 = 0.625$, and $\gamma/\omega_0 = 0.25$.

order differential equations of the form

$$\begin{aligned} \dot{x}_1 &= f_1(x_1, x_2, ..., x_D) \\ \dot{x}_2 &= f_2(x_1, x_2, ..., x_D) \\ \vdots \quad & \quad \vdots \quad \vdots \\ \dot{x}_D &= f_3(x_1, x_2, ..., x_D), \end{aligned} \tag{2.87}$$

where D must be at least 3, and (b) the equations contain at least one non-linear coupling term between some of the variables. To see that the driven pendulum meets these requirements, one can identify $x_1 = \omega$, $x_2 = \theta$, and $x_3 = \vartheta$ as the three variables, and rewrite Eq. 2.84 as the following set:

$$\begin{aligned} \dot{\omega} &= -\gamma\omega - \omega_0^2 \sin\theta + G\cos\vartheta \\ \dot{\theta} &= \omega \\ \dot{\vartheta} &= \Omega. \end{aligned} \tag{2.88}$$

According to Eq. 2.62 for a simple damped harmonic oscillator (with $\gamma = b/m$), the phase-space distribution function is given by $\dot{\rho} = \gamma\rho$, and the density of phase points representing an ensemble of identical pendula increases as time goes on; i.e., the system is dissipative. While the volume occupied by the set of phase points shrinks, sensitivity to initial conditions causes the points to quickly move apart. These two effects can take place simultaneously if there is a stretching of the set of phase points along certain preferred directions, while there is an accompanying contraction along other directions, as illustrated in Fig. 2.13. As the system evolves, these preferred directions continuously change.

Once any initial transient motion of the pendulum has died out, the phase points find themselves moving about a very complex geometric object in a bounded region of phase space called a *chaotic attractor*. The attractor is most clearly displayed as a so-called *Poincaré section*. This is a set of phase points that is generated by sampling the state of the system once during each period of the drive, at exactly the same phase in the drive. That is to say, one takes a "snapshot" of the phase point at equal time intervals of $2\pi/\Omega$. A chaotic attractor is an example of a *fractal*, a complex geometric structure with non-integer dimension. Fractals are characterized by a property called *self-similarity*, which means that the basic structure of the object does not depend on the size-scale at which it is examined. Figure 2.14 shows two different magnifications of the same chaotic attractor and the resulting self-similar structures. Because of the unusual nature of chaotic attractors, they are often referred to as *strange attractors*.

The apparent striated structure of the attractor comes about because of a process called *stretching and folding*. As the phase trajectory develops, initially neighboring phase points must, on the average, move apart at an exponential rate while remaining confined to the localized region defined by the attractor. At the same time, the fact that phase trajectories emanating from different initial points can never cross still holds true for chaotic systems. To meet these apparently conflicting criteria, a given region of phase points must undergo a two-step process: First, the region must be rapidly stretched in some direction in order to pull the phase points apart. Secondly, the resulting set of points must be folded back onto itself so as to remain in the bounded region defined by the attractor. By repeating this two-step sequence over and over, the system develops an infinitely-layered attractor structure. Furthermore, this way of envisioning how the attractor develops also provides for an understanding of how chaos leads to the loss of predictability in these systems. While the stretching process causes initially adjacent phase points to diverge, the basic result of the folding process is to continually "shuffle" the relative locations of the phase points. Without perfect knowledge of the initial conditions, there is no way to even estimate how far apart the phase points of two systems will be even a relatively short time after the pendulum motions are started.

2.5 Coupled Oscillations

The coupled harmonic oscillator problem (including that of the simple pendulum in the limit of small oscillation amplitude) can be solved in closed analytic form. The solution is of general interest since numerous physical systems in both the microscopic and macroscopic world are well-approximated by such a model. Imagine two identical oscillators of mass m characterized by the same spring constant K that are coupled to each other through a spring constant κ, as shown in Fig. 2.15. The displacements along the x-direction, x_1 and x_2, of the individual oscillators relative to their equilib-

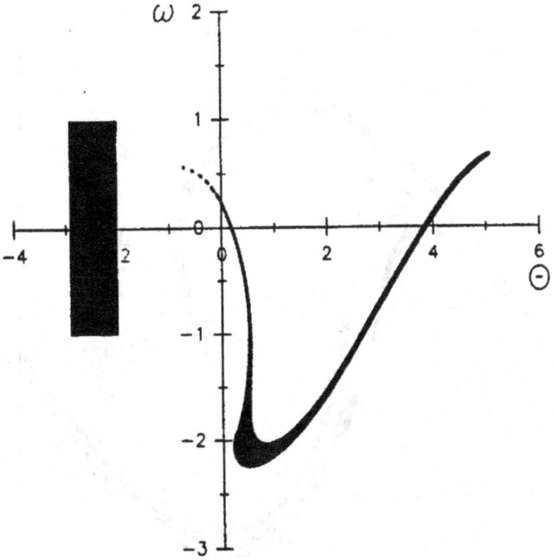

Figure 2.13 Illustration of the stretching of points in phase space for the driven pendulum with $G/\omega_0^2 = 1.05$, $\Omega/\omega_0 = 0.625$, and $\gamma/\omega_0 = 0.25$. Each point in the rectangular block on the left represents a set of initial conditions. After one-half of the drive period, the phase points are mapped into the locus of points on the right.

rium positions, represent the two degrees of freedom of the system.[‖] The Lagrangian is

$$\mathcal{L} = \frac{1}{2}m(\dot{x}_1^2 + \dot{x}_2^2) - \frac{1}{2}K(x_1^2 + x_2^2) - \frac{1}{2}\kappa(x_1 - x_2)^2. \tag{2.89}$$

The third term represents the potential energy due to coupling. Lagrange's equations then take the form

$$\begin{aligned} m\ddot{x}_1 + Kx_1 + \kappa(x_1 - x_2) &= 0 \\ m\ddot{x}_2 + Kx_2 - \kappa(x_1 - x_2) &= 0. \end{aligned} \tag{2.90}$$

Note that adding or subtracting the above equations shows that the linear combinations $\xi_a = x_1 + x_2$ and $\xi_b = x_1 - x_2$ obey the equations

$$\begin{aligned} m\ddot{\xi}_a + K\xi_a &= 0 \\ m\ddot{\xi}_b + (K + 2\kappa)\xi_b &= 0. \end{aligned} \tag{2.91}$$

ξ_a and ξ_b are called *normal coordinates*—each independently exhibits simple harmonic motion, i.e.,

$$\begin{aligned} \xi_a(t) &= A\cos(\omega_a t + \varphi_a) \\ \xi_b(t) &= B\cos(\omega_b t + \varphi_b), \end{aligned} \tag{2.92}$$

[‖]The discussion also applies to two simple pendula (length ℓ, generalized coordinates θ_1, θ_2) coupled by a spring, in the limit of small oscillation amplitude. Then, the spring constant K represents mg/ℓ, and the coordinates x_1 and x_2 correspond to $\ell\theta_1$ and $\ell\theta_2$, respectively.

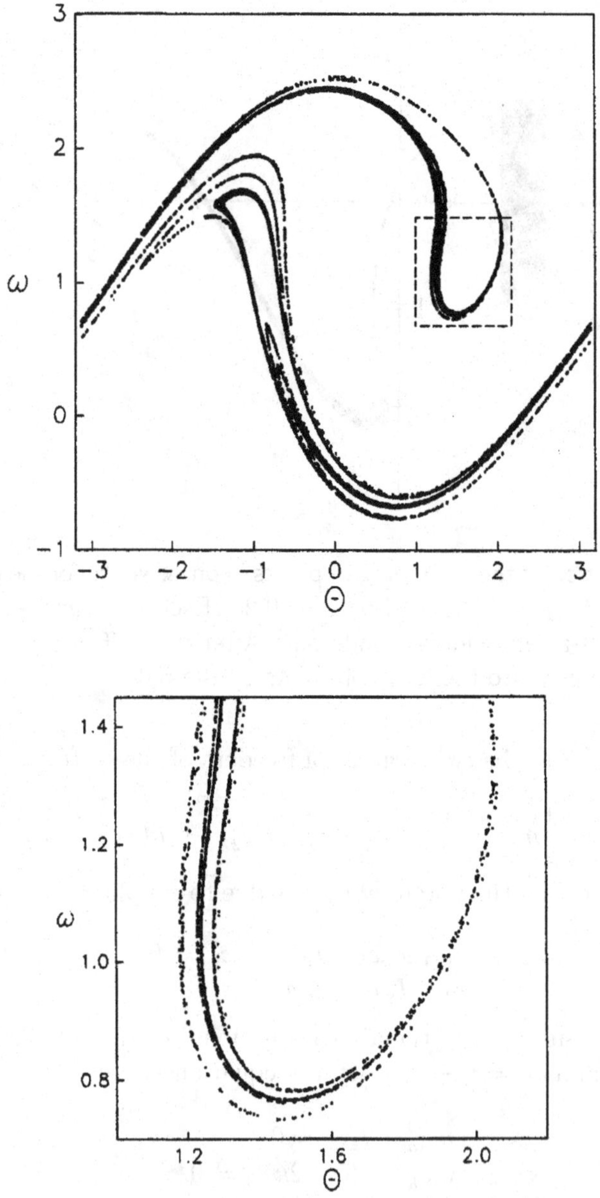

Figure 2.14 The top figure is a Poincaré section for the driven pendulum with $G/\omega_0^2 = 1.05$, $\Omega/\omega_0 = 0.625$, and $\gamma/\omega_0 = 0.25$. A region of the strange attractor (dashed box) then appears magnified in the bottom figure.

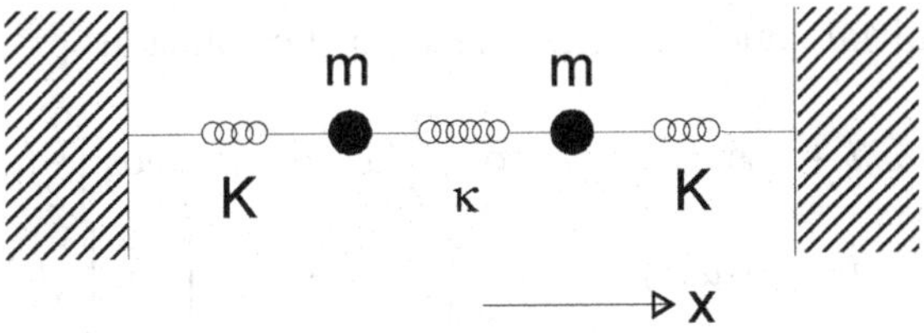

Figure 2.15 Coupled harmonic oscillator.

each with its own characteristic frequency

$$\omega_a = \sqrt{\frac{K}{m}} \quad \text{and} \quad \omega_b = \sqrt{\frac{K+2\kappa}{m}}. \tag{2.93}$$

In general, the motion of the coupled oscillator is a superposition of these solutions, i.e.,

$$\begin{aligned} x_1(t) &= \frac{1}{2}[\xi_a(t) + \xi_b(t)] \\ x_2(t) &= \frac{1}{2}[\xi_a(t) - \xi_b(t)]. \end{aligned} \tag{2.94}$$

At $t = 0$, suppose the two oscillators are released from rest after being displaced by an equal amount in the same direction. Then, the initial conditions, $x_1(0) = x_2(0) = x_0$ and $\dot{x}_1(0) = \dot{x}_2(0) = 0$, allow one to find the undetermined amplitudes and phases in Eqs. 2.92. The motions of the two oscillators can then be found from Eqs. 2.94 to be

$$x_1(t) = x_2(t) = x_0 \cos \omega_a t. \tag{2.95}$$

Alternatively, if the two oscillators are initially displaced by the same amount in opposite directions, and then released from rest, the initial conditions are $x_1(0) = -x_2(0) = x_0$ and $\dot{x}_1(0) = \dot{x}_2(0) = 0$. The resulting motions are then

$$x_1(t) = -x_2(t) = x_0 \cos \omega_b t. \tag{2.96}$$

The motions represented by Eqs. 2.95 and 2.96 are called *normal modes*. The first is a *symmetric mode* and the second is an *antisymmetric mode*. In general, for an oscillating system with N degrees of freedom, there are N normal modes, each representing a very special motion in that all of the generalized coordinates oscillate with one and the same frequency characteristic of that mode.

For an arbitrary set of initial conditions, the solution is generally a superposition of normal mode oscillations. For example, suppose the two oscillators are initially at rest, and we start the motion by applying a displacement x_0 to the first oscillator. These initial conditions then lead to the solution

$$\begin{aligned} x_1(t) &= \frac{x_0}{2}(\cos\omega_a t + \cos\omega_b t) = x_0 \cos\left[\frac{1}{2}(\omega_a+\omega_b)t\right] \cos\left[\frac{1}{2}(\omega_b-\omega_a)t\right] \\ x_2(t) &= \frac{x_0}{2}(\cos\omega_a t - \cos\omega_b t) = x_0 \sin\left[\frac{1}{2}(\omega_a+\omega_b)t\right] \sin\left[\frac{1}{2}(\omega_b-\omega_a)t\right]. \end{aligned} \tag{2.97}$$

In the limit of weak coupling ($\kappa \ll K$), we find that the main oscillator frequency, given by $(\omega_a + \omega_b)/2 \simeq \omega_a = \sqrt{K/m}$, is almost identical to that of an uncoupled oscillator. The difference frequency (or *beat frequency*), $(\omega_b - \omega_a)/2 \simeq (\kappa/K)\sqrt{K/m}$, represents a very slow modulation of the motion associated with the gradual transfer of energy between the two oscillators.

Example 2.4 Continuum Limit of Longitudinal Vibrations in a Linear Mass Chain. Consider the vibration of a linear chain of point masses (each of mass m) along the x-axis that are connected by a series of springs having the same equilibrium length a and stiffness constant K (see Fig. 2.16). The linear mass-density of the chain is $\mu = m/a$. Let the generalized coordinates η_i denote the displacement along the x-direction of each point mass from its equilibrium position. Then the Lagrangian of this system takes the form

$$\begin{aligned} \mathcal{L} &= \sum_i \frac{1}{2} m\dot{\eta}_i^2 - \sum_i \frac{1}{2} K(\eta_{i+1} - \eta_i)^2 \\ &= \frac{1}{2}\sum_i a(\mu\dot{\eta}_i^2 - Ka\epsilon_i^2), \end{aligned} \tag{2.98}$$

where we have introduced the displacement (or elongation) per unit length, $\epsilon_i = (\eta_{i+1} - \eta_i)/a$.

We now go to the continuum limit of an elastic rod characterized by a Young's modulus Y. The rod obeys Hooke's law

$$F = Y\epsilon. \tag{2.99}$$

Here, F is the force (or tension) required to produce the elongation ϵ in one of the springs, i.e.,

$$F = K(\eta_{i+1} - \eta_i) = Ka\epsilon. \tag{2.100}$$

Comparison of Eqs. 2.99 and 2.100 shows that Ka plays the role of Young's modulus. Using this fact, we can now replace the discrete Lagrangian, Eq. 2.98, with a continuous one by letting $a \to 0$. Then the summation over index i is replaced with an integral over a continuous position coordinate x, and the elongation is replaced using

$$\epsilon_i = \lim_{a\to 0}\left(\frac{\eta_{i+1} - \eta_i}{a}\right) \Rightarrow \lim_{a\to 0}\left[\frac{\eta(x+a) - \eta(x)}{a}\right] = \frac{\partial\eta}{\partial x}. \tag{2.101}$$

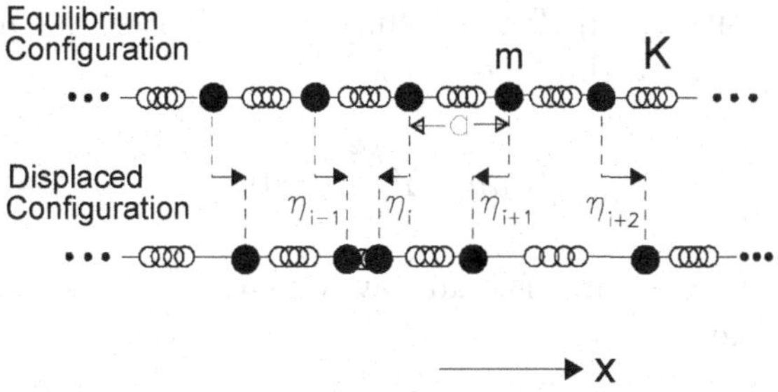

Figure 2.16 Linear mass chain vibrating longitudinally.

$\partial\eta/\partial x$ denotes the strain in the rod. With dx playing the role of a, the Lagrangian for the elastic rod becomes

$$\mathcal{L} = \int \mathbb{L}\, dx \tag{2.102}$$

where $\mathbb{L}$, known as the *Lagrangian density*, is

$$\mathbb{L} = \frac{1}{2}\left[\mu\dot{\eta}^2 - Y\left(\frac{\partial\eta}{\partial x}\right)^2\right]. \tag{2.103}$$

Because our system has an infinite number of degrees of freedom, i.e., a different possible longitudinal displacement η at each position coordinate x in the rod, we must somehow derive a tractable version of Lagrange's equations useful for this type of continuous system. This can be accomplished by returning to the concept of an action integral, S, introduced in Sect. 2.1 to derive Lagrange's equations. In a one-dimensional continuous medium, the action takes on the following more general form:

$$S = \int_{t_1}^{t_2} dt \int \mathbb{L}\left(\eta, \dot{\eta}, \frac{\partial\eta}{\partial x}, t\right) dx. \tag{2.104}$$

As before, the principle of least action requires that a variation in the action, δS, vanishes, while the end-points remain fixed:

$$\delta S = \int_{t_1}^{t_2} dt \int \left[\frac{\partial\mathbb{L}}{\partial\eta}\delta\eta + \frac{\partial\mathbb{L}}{\partial\dot{\eta}}\delta\dot{\eta} + \frac{\partial\mathbb{L}}{\partial(\partial\eta/\partial x)}\delta(\partial\eta/\partial x)\right] dx = 0. \tag{2.105}$$

But $\delta(\partial\eta/\partial x) = \partial(\delta\eta)/\partial x$ and $\delta\dot{\eta} = \partial(\delta\eta)\partial t$, allowing one to integrate the last two terms by parts. This leads to

$$\int_{t_1}^{t_2} dt \int \left\{\frac{\partial\mathbb{L}}{\partial\eta}\delta\eta - \frac{d}{dt}\left(\frac{\partial\mathbb{L}}{\partial\dot{\eta}}\right)\delta\eta - \frac{d}{dx}\left[\frac{\partial\mathbb{L}}{\partial(\partial\eta/\partial x)}\right]\delta\eta\right\} dx = 0. \tag{2.106}$$

For an arbitrary variation, $\delta\eta$, the integrand itself must vanish. This produces the appropriate Lagrange's equation:

$$\frac{d}{dt}\left(\frac{\partial\mathbb{L}}{\partial\dot{\eta}}\right) + \frac{d}{dx}\left[\frac{\partial\mathbb{L}}{\partial(\partial\eta/\partial x)}\right] - \frac{\partial\mathbb{L}}{\partial\eta} = 0. \tag{2.107}$$

The Lagrangian density of Eq. 2.103 results in the following equation of motion for longitudinal vibrations in a thin elastic rod:

$$\mu\ddot{\eta} - Y\frac{\partial^2\eta}{\partial x^2} = 0. \tag{2.108}$$

This is in the form of a one-dimensional wave equation with solutions having the form of travelling waves, i.e.,

$$\eta(x,t) = f(x \pm vt), \tag{2.109}$$

where the upper sign corresponds to a disturbance propagating in the negative x-direction with a speed $v = \sqrt{Y/\mu}$, and the lower sign corresponds to a disturbance propagating with the same speed in the positive x-direction. Although f can be any differentiable function, the most fundamental travelling wave is one having a harmonic form $A\sin[2\pi(x \pm vt)/\lambda]$, where A and λ denote the amplitude and wavelength of the displacement, respectively. A superposition of oppositely directed harmonic waves, having the same A and λ, results in a standing wave of the form

$$\eta(x,t) = 2A\sin(kx)\cos(\omega t), \tag{2.110}$$

where $k = 2\pi/\lambda$ is the *wavenumber* and the angular frequency is given by $\omega = vk$. This represents a situation where the longitudinal displacements at all positions along the rod have the same vibration frequency, while the vibration amplitude varies sinusoidally with position. In addition, the vibration at any point is either in phase or π-radians out of phase with any other point. This motion represents a normal mode of the system. Since the system has an infinite number of degrees of freedom, there are an infinite number of normal modes available to the system, i.e., one at any chosen λ (and hence ω). If one imposes boundary conditions at the ends of the rod, one is no longer free to choose any arbitrary λ. For example, suppose one clamps a rod of length ℓ between two immovable walls. This demands that the vibration amplitudes at the end-points vanish, hence only allowing integral multiples of half of a wavelength to exist within the rod, i.e., $\lambda = 2\ell/n$. However, notice that since an infinite number of degrees of freedom still exist, there are still an infinite number of oscillation modes. The difference here is that the possible λ's, and hence normal mode frequencies, are discrete.

2.6 Theory of Small Oscillations

A system is considered to be in *equilibrium* when no net forces are experienced by the particles. Furthermore, for a system in *stable equilibrium*, small displacements from equilibrium are countered by restoring forces that tend to return the system back to the equilibrium configuration. A general conservative system can exhibit simple harmonic motion about a stable equilibrium point, provided that the oscillations are sufficiently small.

For a system with N degrees of freedom, one can expand the potential energy function in a Taylor series about a stable equilibrium point postulated to be at the coordinates $q_1, \ldots, q_N = 0$:

$$V(q_1, \ldots, q_N) = V_0 + \sum_{i=1}^{N} q_i \left(\frac{\partial V}{\partial q_i}\right)_{q_1,\ldots,q_N=0} + \frac{1}{2} \sum_{i,j=1}^{N} q_i q_j \left(\frac{\partial^2 V}{\partial q_i \partial q_j}\right)_{q_1,\ldots,q_N=0} + \ldots. \tag{2.111}$$

Without loss of generality, one may chose the potential energy V_0 at equilibrium to be zero. In addition, since the $\partial V/\partial q_i$'s represent the ith force components, the definition of an equilibrium point requires that the terms in the first summation vanish. For small oscillations, we do not consider terms past the second order, so that

$$V(q_1, \ldots, q_N) = \frac{1}{2} \sum_{i,j=1}^{N} K_{ij} q_i q_j, \qquad \text{where } K_{ij} = K_{ji} = \left(\frac{\partial^2 V}{\partial q_i \partial q_j}\right)_{q_1,\ldots,q_N=0}. \tag{2.112}$$

The stability condition requires that this quadratic form for the potential energy be positive definite. For small displacements, the kinetic energy can be written as a homogeneous quadratic function of the generalized velocities, i.e.,

$$T(\dot{q}_1, \ldots, \dot{q}_N) = \frac{1}{2} \sum_{i,j=1}^{N} m_{ij} \dot{q}_i \dot{q}_j, \tag{2.113}$$

where the coefficients m_{ij} are constants taking on the values at the equilibrium point. Then, the general form of the Lagrangian for a system in the vicinity of a stable equilibrium point becomes

$$\mathcal{L} = \frac{1}{2} \sum_{i,j=1}^{N} (m_{ij} \dot{q}_i \dot{q}_j - K_{ij} q_i q_j). \tag{2.114}$$

From Eqs. 2.14, the equations of motion are found to be

$$\sum_{i=1}^{N} (m_{ij} \ddot{q}_i + K_{ij} q_i) = 0 \qquad (j = 1, 2, \ldots, N). \tag{2.115}$$

One can solve for the normal modes of the system by assuming solutions of the form

$$q_i = A_i \cos \omega t. \tag{2.116}$$

This produces the following set of linear homogeneous equations for the A_i's:

$$\sum_{i=1}^{N} (K_{ij} - m_{ij} \omega^2) A_i = 0 \qquad (j = 1, 2, \ldots, N). \tag{2.117}$$

The existence of non-trivial solutions requires that the determinant of the coefficients of the A_i's vanishes, in other words,

$$\left|K_{ij} - m_{ij}\omega^2\right| = 0. \tag{2.118}$$

The result is an Nth-degree equation for the normal mode frequencies. Once these are found, Eqs. 2.117 can be used to find the relations between the corresponding A_i's, thus completely specifying the normal mode motion for each characteristic frequency.

Example 2.5 Vibrations of the CO_2 Molecule. At equilibrium, the three atoms of the carbon dioxide molecule are collinear, as shown in Fig. 2.17. The carbon atom (mass m) is situated midway between the two oxygen atoms (mass M). The equilibrium separation between the carbon atom and each of the oxygen atoms is denoted by a. In all, there are nine degrees of freedom associated with this system, i.e., three for each atom. Let z_1, z_2, and z_3 denote the longitudinal displacements of the three atoms from their equilibrium positions, where subscript 2 is associated with the carbon atom. x_1 and y_1 denote the displacements perpendicular to the z-axis for the first atom, and likewise for the transverse displacements of atoms 2 and 3. We will assume two force constants: K is the force constant along the z-direction between the carbon and each of the oxygen atoms and κ is the force constant in the transverse direction.

In actuality, there are only four degrees of freedom associated with vibrations of the molecule, since three degrees of freedom correspond to translation of the center-of-mass and two correspond to rotations about the center-of-mass. We can eliminate the latter translational degrees of freedom by fixing the equilibrium position of the carbon atom at the origin so that $\mathbf{r}_{\text{com}} = \sum m_i \mathbf{r}_i = 0$. The coordinates for the carbon atom can then be replaced with

$$x_2 = -\mu(x_1 + x_3), \qquad y_2 = -\mu(y_1 + y_3), \qquad z_2 = -\mu(z_1 + z_3), \tag{2.119}$$

where $\mu = M/m$. By also requiring that the total angular momentum ($\mathbf{L}$) about the center-of-mass be zero, we can eliminate the rotational degrees of freedom. Using the fact that $\mathbf{L} = \mathbf{r}\times\mathbf{p}$, we have

$$L_x = aM(\dot{y}_1 - \dot{y}_3) = 0 \qquad \text{and} \qquad L_y = aM(\dot{x}_3 - \dot{x}_1) = 0, \tag{2.120}$$

where the angular momentum of the carbon atom is assumed to be negligible for small vibrations. We then have

$$\begin{aligned} \dot{y}_1 &= \dot{y}_3, & \dot{x}_1 &= \dot{x}_3 \\ y_1 &= y_3, & x_1 &= x_3. \end{aligned} \tag{2.121}$$

This leaves four independent coordinates to specify the vibrational motion.

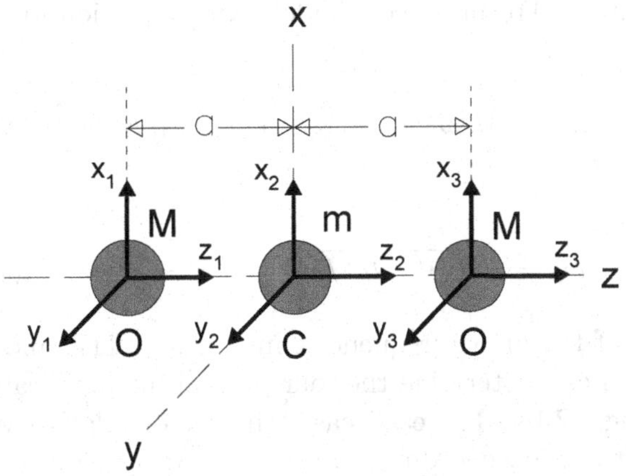

Figure 2.17 Coordinate system for CO_2 molecule.

With the aid of Eqs. 2.119 and 2.121, the vibrational kinetic energy is

$$\begin{aligned} T &= \frac{1}{2}M(\dot{x}_1^2 + \dot{y}_1^2 + \dot{z}_1^2) + \frac{1}{2}m(\dot{x}_2^2 + \dot{y}_2^2 + \dot{z}_2^2) + \frac{1}{2}M(\dot{x}_3^2 + \dot{y}_3^2 + \dot{z}_3^2) \\ &= \frac{1}{2}M(2+4\mu)(\dot{x}_1^2 + \dot{y}_1^2) + \frac{1}{2}M(1+\mu)(\dot{z}_1^2 + \dot{z}_3^2) + \mu M \dot{z}_1 \dot{z}_3, \end{aligned} \tag{2.122}$$

and the potential energy is

$$\begin{aligned} V &= \frac{1}{2}K[(z_1 - z_2)^2 + (z_3 - z_2)^2] \\ &\quad + \frac{1}{2}\kappa[(x_1 - x_2)^2 + (y_1 - y_2)^2 + (x_3 - x_2)^2 + (y_3 - y_2)^2] \\ &= \frac{1}{2}K[(1 + 2\mu + 2\mu^2)(z_1^2 + z_3^2) + 4\mu(1+\mu)z_1 z_3] \\ &\quad + \frac{1}{2}\kappa[2(1+2\mu)^2(x_1^2 + y_1^2)]. \end{aligned} \tag{2.123}$$

Comparing these expressions to the forms of Eqs. 2.112 and 2.113, with the identification $q_1 = x_1$, $q_2 = y_1$, $q_3 = z_1$, $q_4 = z_3$, produces the following $\mathbf{K}$ and $\mathbf{m}$ matrices:

$$\mathbf{K} = \begin{pmatrix} 2\kappa(1+2\mu)^2 & 0 & 0 & 0 \\ 0 & 2\kappa(1+2\mu)^2 & 0 & 0 \\ 0 & 0 & K(1+2\mu+2\mu^2) & 2K\mu(1+\mu) \\ 0 & 0 & 2K\mu(1+\mu) & K(1+2\mu+2\mu^2) \end{pmatrix} \tag{2.124}$$

$$\mathbf{m} = \begin{pmatrix} 2M(1+2\mu) & 0 & 0 & 0 \\ 0 & 2M(1+2\mu) & 0 & 0 \\ 0 & 0 & M(1+\mu) & M\mu \\ 0 & 0 & M\mu & M(1+\mu) \end{pmatrix}. \tag{2.125}$$

Substituting into Eq. 2.118 produces the following equation for ω, the normal mode frequencies:

$$[\kappa(1+2\mu) - M\omega^2]^2[M^2\omega^4 - 2KM(1+\mu)\omega^2 + K^2(1+2\mu)] = 0. \tag{2.126}$$

The solutions are

$$\omega_a = \sqrt{K/M}, \qquad \omega_b = \sqrt{K(1+2\mu)/M}, \qquad \omega_c = \omega_d = \sqrt{\kappa(1+2\mu)/M}. \tag{2.127}$$

By inserting each of the above frequencies into Eqs. 2.117, and making use of Eqs. 2.119 and 2.121, we can determine the four normal mode vibration patterns. These are depicted in Fig. 2.18. For each case, the set of relative vibration amplitudes corresponding to the nine coordinates, (x_1, y_1, z_1; x_2, y_2, z_2; x_3, y_3, z_3), are as follows:

$$\begin{aligned} \omega_a &: \ (0,0,A;0,0,0;0,0,-A) \\ \omega_b &: \ (0,0,A;0,0,-2\mu A;0,0,A) \\ \omega_c &: \ (A,0,0;-2\mu A,0,0;A,0,0) \\ \omega_d &: \ (0,A,0;0,-2\mu A,0;0,A,0). \end{aligned} \tag{2.128}$$

The first two modes represent longitudinal oscillations of the molecule, while the last two are transverse. The transverse modes are *degenerate* since they vibrate with the same frequency.

Using infrared spectroscopy, one finds the following measured vibrational frequencies for CO_2 [2]: $\omega_a = 2.62\times10^{14}$ s^{-1}, $\omega_b = 4.43\times10^{14}$ s^{-1}, $\omega_c = \omega_d = 1.26\times10^{14}$ s^{-1}. Since $\mu \cong 16/12$, the calculated frequency ratio, $\omega_b/\omega_a = \sqrt{1+2\mu}$, is about 1.9 using the above classical considerations, as compared to the measured value of about 1.7. One can also estimate the ratio of the molecular force constants. Depending on whether one uses $K/\kappa = (\omega_b/\omega_c)^2$ or $(1+2\mu)(\omega_a/\omega_c)^2$, we find that κ/K is either $\sim$ 12 or 16 for the carbon dioxide molecule.

2.7 Poisson Brackets

We shall shortly find that by writing the equations of motion in the so-called *Poisson bracket* form, the transition from classical to quantum mechanics becomes rather natural. We define the Poisson bracket of two functions of the canonical variables $u(q_k, p_k, t)$ and $v(q_k, p_k, t)$ by

$$\{u, v\} = \sum_k \left(\frac{\partial u}{\partial q_k}\frac{\partial v}{\partial p_k} - \frac{\partial u}{\partial p_k}\frac{\partial v}{\partial q_k} \right). \tag{2.129}$$

For example, we find that

$$\{q_i, p_j\} = \sum_k \left(\frac{\partial q_i}{\partial q_k}\frac{\partial p_j}{\partial p_k} - \frac{\partial q_i}{\partial p_k}\frac{\partial p_j}{\partial q_k} \right) = \delta_{ij} \tag{2.130}$$

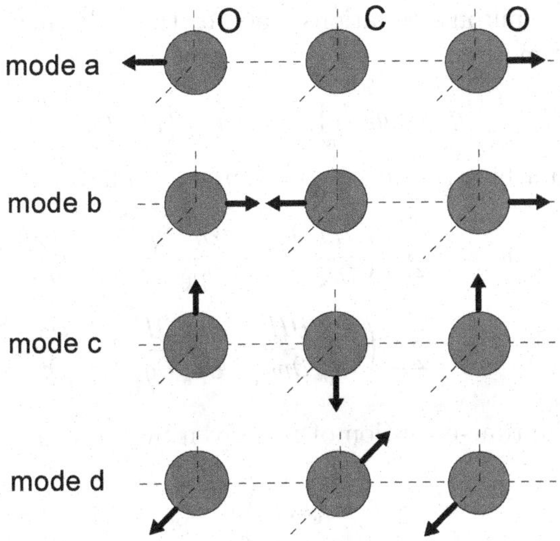

Figure 2.18 Normal mode vibration patterns of the CO_2 molecule.

and

$$\{p_i, p_j\} = \sum_k \left(\frac{\partial p_i}{\partial q_k} \frac{\partial p_j}{\partial p_k} - \frac{\partial p_i}{\partial p_k} \frac{\partial p_j}{\partial q_k} \right) = 0. \tag{2.131}$$

Some algebraic properties of Poisson brackets are

$$\{u, u\} = 0 \tag{2.132}$$

$$\{u, \text{constant}\} = 0 \tag{2.133}$$

$$\{u, v\} = -\{v, u\} \tag{2.134}$$

$$\{u + v, w\} = \{u, w\} + \{v, w\} \tag{2.135}$$

$$\{u, vw\} = \{u, v\} w + v \{u, w\}. \tag{2.136}$$

The most useful property of Poisson brackets is that one is able to write equations of motion in terms of them. Consider the Poisson bracket of the q's and p's with the Hamiltonian function, i.e.,

$$\{q_i, H\} = \sum_k \left(\frac{\partial q_i}{\partial q_k} \frac{\partial H}{\partial p_k} - \frac{\partial q_i}{\partial p_k} \frac{\partial H}{\partial q_k} \right) = \frac{\partial H}{\partial p_i} \tag{2.137}$$

and

$$\{p_i, H\} = \sum_k \left(\frac{\partial p_i}{\partial q_k} \frac{\partial H}{\partial p_k} - \frac{\partial p_i}{\partial p_k} \frac{\partial H}{\partial q_k} \right) = -\frac{\partial H}{\partial q_i}. \tag{2.138}$$

Comparison with Hamilton's equations leads to the following Poisson bracket form for the equations of motion:

$$\dot{q}_i = \{q_i, H\}, \qquad \dot{p}_i = \{p_i, H\}. \tag{2.139}$$

Furthermore, for an arbitrary function $u(q_k, p_k, t)$, we have

$$\begin{aligned} \dot{u} &= \sum_k \left(\frac{\partial u}{\partial q_k}\frac{\partial q_k}{\partial t} + \frac{\partial u}{\partial p_k}\frac{\partial p_k}{\partial t} \right) + \frac{\partial u}{\partial t} \\ &= \sum_k \left(\frac{\partial u}{\partial q_k}\frac{\partial H}{\partial p_k} - \frac{\partial u}{\partial p_k}\frac{\partial H}{\partial q_k} \right) + \frac{\partial u}{\partial t}, \end{aligned} \tag{2.140}$$

which shows that the time-evolution of u is given by

$$\dot{u} = \{u, H\} + \frac{\partial u}{\partial t}. \tag{2.141}$$

Therefore, if the function u is not explicitly a function of time, it is a constant of the motion when $\{u, H\} = 0$. In particular, if u is the Hamiltonian itself, then $\dot{H} = \{H, H\} = 0$ when H is not an explicit function of time, and the Hamiltonian represents the constant energy of the system.

An immediate consequence of Eq. 2.141 involves Liouville's theorem in statistical mechanics, which states that the density of points in phase space is a constant of the motion, i.e., $\dot{\rho} = 0$. Thus, letting u be the function ρ in Eq. 2.141, we have

$$\frac{\partial \rho}{\partial t} = -\{\rho, H\}. \tag{2.142}$$

If a collection of phase points is chosen to represent an equilibrium ensemble, then the density, ρ, at any fixed spot in phase space should be independent of time, i.e., $\partial\rho/\partial t = 0$, and $\{\rho, H\} = 0$. This shows that a general way to construct an ensemble representing an equilibrium situation is to choose ρ to be a function only of the system's constants of motion—this guarantees that the latter Poisson bracket vanish. For example, a common choice for an equilibrium ensemble in the statistical mechanics of conservative systems is the *microcanonical ensemble*. In this case, ρ is chosen to be a function of the constant total energy of the system.

Example 2.6 Angular Momentum Conservation and Poisson Brackets. The angular momentum of a particle about some point O is defined as

$$\mathbf{L} = \mathbf{r} \times \mathbf{p}, \tag{2.143}$$

where $\mathbf{r}$ is the position vector of the particle relative to O, and $\mathbf{p}$ is the particle's linear momentum. The three Cartesian components of $\mathbf{L}$ are

$$L_x = yp_z - zp_y, \qquad L_y = zp_x - xp_z, \qquad L_z = xp_y - yp_x, \tag{2.144}$$

and $L^2 = L_x^2 + L_y^2 + L_z^2$. In Example 2.1 we showed that the angular momentum vector $\mathbf{L}$ is conserved for motion in a central force field specified by a potential $V(r)$. This result can also be demonstrated using Poisson brackets. To show that $\mathbf{L}$ is a constant of the motion, we need to prove that $\dot{L}_x = \dot{L}_y = \dot{L}_z = 0$. According to Eq. 2.141, this occurs if $\{L_x, H\} = \{L_y, H\} = \{L_z, H\} = 0$.

Let us consider $\{L_x, H\}$, i.e.,

$$\{L_x, H\} = \left\{L_x, \frac{p^2}{2m} + V(r)\right\} = \frac{1}{2m}\{L_x, p^2\} + \{L_x, V(r)\}, \tag{2.145}$$

where $p^2 = p_x^2 + p_y^2 + p_z^2$. The first bracket on the right-hand side is

$$\begin{aligned}\{L_x, p^2\} &= \left(\frac{\partial L_x}{\partial x}\frac{\partial p^2}{\partial p_x} - \frac{\partial L_x}{\partial p_x}\frac{\partial p^2}{\partial x}\right) + \left(\frac{\partial L_x}{\partial y}\frac{\partial p^2}{\partial p_y} - \frac{\partial L_x}{\partial p_y}\frac{\partial p^2}{\partial y}\right) \\ &\quad + \left(\frac{\partial L_x}{\partial z}\frac{\partial p^2}{\partial p_z} - \frac{\partial L_x}{\partial p_z}\frac{\partial p^2}{\partial z}\right).\end{aligned} \tag{2.146}$$

Only two terms survive, but they cancel, giving a vanishing result:

$$\{L_x, p^2\} = \frac{\partial L_x}{\partial y}\frac{\partial p^2}{\partial p_y} + \frac{\partial L_x}{\partial z}\frac{\partial p^2}{\partial p_z} = (p_z)(2p_y) + (-p_y)(2p_z) = 0. \tag{2.147}$$

The second bracket in Eq. 2.145 is

$$\begin{aligned}\{L_x, V(r)\} &= \left(\frac{\partial L_x}{\partial x}\frac{\partial V(r)}{\partial p_x} - \frac{\partial L_x}{\partial p_x}\frac{\partial V(r)}{\partial x}\right) + \left(\frac{\partial L_x}{\partial y}\frac{\partial V(r)}{\partial p_y} - \frac{\partial L_x}{\partial p_y}\frac{\partial V(r)}{\partial y}\right) \\ &\quad + \left(\frac{\partial L_x}{\partial z}\frac{\partial V(r)}{\partial p_z} - \frac{\partial L_x}{\partial p_z}\frac{\partial V(r)}{\partial z}\right).\end{aligned} \tag{2.148}$$

Since $V(r)$ is independent of momentum, the $\partial V(r)/\partial p_i$'s all vanish. Also, $-\partial V/\partial x = -F_x$, and similarly for the other spatial derivatives. And finally, using Eqs. 2.144 to find the derivatives of the angular momentum components, we obtain

$$\{L_x, V(r)\} = yF_z - zF_y. \tag{2.149}$$

But, from Eq. 2.35, the right-hand side is just τ_x, the x-component of the torque exerted on the system. The fact that a central field produces no torque means that

$$\{L_x, V(r)\} = 0. \tag{2.150}$$

So, we now have the expected result for one component of the angular momentum, i.e., $\dot{L}_x = \{L_x, H\} = 0$.

Calculations of the other two brackets, $\{L_y, H\}$ and $\{L_z, H\}$, just produce the other two torque components, τ_y and τ_z, respectively. Since these also vanish, we have

$$\{L_x, H\} = \{L_y, H\} = \{L_z, H\} = 0, \tag{2.151}$$

and **L** is conserved. Clearly, this means that the magnitude of **L** is a constant of the motion as well, so

$$\{L^2, H\} = 0. \tag{2.152}$$

Suggested References

Standard references on the general formalism of classical mechanics are

[a] H. Goldstein, *Classical Mechanics*, 3rd ed. (Addison-Wesley, Reading, MA, 2002).

[b] L. D. Landau and E. M. Lifshitz, *Mechanics*, 2nd ed. (Pergamon Press, Oxford, 1969).

[c] K. R. Symon, *Mechanics*, 3rd ed. (Addison-Wesley, Reading, MA, 1971).

An excellent general reference on phase space and Liouville's theorem is

[d] D. A. McQuarrie, *Statistical Mechanics*, 2nd ed. (University Science Books, Sausalito, CA, 2000).

The following are excellent overviews of chaotic behavior in physical systems:

[e] F. C. Moon, *Chaotic and Fractal Dynamics: An Introduction for Applied Scientists and Engineers* (John Wiley and Sons, New York, 1992).

[f] S. H. Strogatz, *Nonlinear Dynamics and Chaos: With Applications to Physics, Biology, Chemistry, and Engineering* (Perseus Books Publishing, Cambridge, MA, 1994).

Problems

1. Consider a system of N particles interacting among themselves via pair potentials $V(|\mathbf{r}_i - \mathbf{r}_j|)$.

   (a) Write down the Lagrangian and Hamiltonian of the system. Find the equations of motion in each case and compare the results with Newton's equations of the same system.

   (b) Consider the case of a two-particle system. Explicitly write out Newton's equations of motion for the two particles. By direct calculation, work out the relation between the two interparticle forces, $\mathbf{F}_{12}$ and $\mathbf{F}_{21}$. Does the result agree with Newton's Third Law?

(c) Continuing with the two-particle system, transform to the center-of-mass and rewrite the equations of motion in terms of the center-of-mass position vector, **R**, and the relative position vector, **r**.

2. Consider the longitudinal vibrations of a linear chain of point particles, each of mass m, positioned along the x-axis and connected by a series of springs having equilibrium length a and and stiffness constant K. Denote the equilibrium position of particle s by x_s and the displacement from the equilibrium position by η_s. The Lagrangian function is then given by Eq. 2.98, i.e.,

$$\mathcal{L} = \sum_s \frac{1}{2} m\dot{\eta}_s^2 - \sum_s \frac{1}{2} K(\eta_{s+1} - \eta_s)^2.$$

(a) Derive Lagrange's equation for the system.

(b) Assume that the solution to the equation of motion can be written in a so-called "normal mode expansion," i.e., as a summation over modes

$$\eta_s(x_s, t) = \eta_k(t)\, e^{ikx_s}$$

corresponding to various "wavevectors" k (where k is identical to the wavenumber, except here it can be positive or negative). Show that each normal mode $\eta_k(t)$ satisfies a simple harmonic oscillator equation with a frequency ω that depends on the wavevector k. Specifically, derive the dispersion relation

$$\omega = 2\sqrt{\frac{K}{m}}\left|\sin\left(\frac{ka}{2}\right)\right|.$$

3. A linear chain of point particles connected by identical springs (stiffness constant K, equilibrium length a)contains two types of particles. All the even particle-sites (indexed by $2s$) are occupied by a mass M and all the odd sites (with index $2s+1$) by a mass m. Denote the displacement of a given mass from its equilibrium position by η_{2s} if the value of the mass is M, or by η_{2s+1} if the mass is m. This is a model for a one-dimensional diatomic lattice.

(a) Write down the Lagrangian of the system and derive the equations of motion for the two types of masses.

(b) Solve the equations by assuming the following normal-mode forms:

$$\eta_{2s} = Ae^{-i\omega t}e^{i2ska}$$
$$\eta_{2s+1} = Be^{-i\omega t}e^{i(2s+1)ka}.$$

As in the previous problem, ω and k denote the mode frequency and wavevector. By demanding non-trivial solutions for the mode amplitudes A and B, derive the dispersion relation for the system, namely,

$$\omega^2 = \frac{K}{M}(1+\mu)\left[1 \pm \sqrt{1 - \frac{4\mu}{(1+\mu)^2}\sin^2 ka}\right],$$

where we define the mass ratio $\mu \equiv M/m$.

(c) Plot the dispersion relation (ω vs. ka) for the range $-\pi/2 < ka < +\pi/2$ for various values of μ, showing the two dispersion curves. The lower and upper curves correspond, respectively, to the so-called *acoustic* (frequency $\omega_{\rm a}$) and *optical* (frequency $\omega_{\rm o}$) lattice vibration (or *phonon*) branches. Find the values of $\omega_{\rm a}$ and $\omega_{\rm o}$ when (i) $k = 0$ and (ii) $k = \pm\pi/2a$. Show that the gap between the two branches widens as the mass ratio μ increases.

(d) Plot B/A as a function of k. For $k = 0$, show that $B/A = 1$ for the acoustic branch and $B/A = -\mu$ for the optical branch. Discuss the nature of the lattice vibrations for each branch.

4. Consider a symmetrical linear triatomic molecule, ABA, lying along the x-axis. Number the atoms from left to right by indices 1, 2, and 3.

(a) First examine the case of longitudinal vibrations, i.e., vibrations along the x-direction. Denoting the displacements from the equilibrium positions by x_1, x_2, and x_3, the Lagrangian function is

$$\mathcal{L} = \frac{1}{2}m_A\left(\dot{x}_1^2 + \dot{x}_3^2\right) + \frac{1}{2}m_B\dot{x}_2^2 - \frac{1}{2}K\left[(x_1 - x_2)^2 + (x_3 - x_2)^2\right].$$

Find the normal coordinates, Q_a and Q_s, corresponding to the antisymmetric and symmetric vibrations. **You will need to use the condition that the center-of-mass of the molecule is stationary during the vibrations.** For each vibration pattern, determine the vibration frequency and draw the corresponding displacement pattern.

(b) Now examine the case of transverse vibration along the y-direction. The Lagrangian is

$$\mathcal{L} = \frac{1}{2}m_A\left(\dot{y}_1^2 + \dot{y}_3^2\right) + \frac{1}{2}m_B\dot{y}_2^2 - \frac{1}{2}\kappa\ell^2\delta^2,$$

where ℓ is the equilibrium bond-length in the molecule and

$$\delta = \frac{1}{\ell}\left[(y_1 - y_2) + (y_3 - y_2)\right]$$

is the deviation of the angle ABA from the value π. The variable δ is the most physically relevant quantity associated with the transverse vibration and can be chosen as the normal coordinate. Express the Lagrangian in terms of this generalized coordinate. **Again, you will need to use the condition that the center-of-mass is stationary. In addition, you will need to use the condition that the angular momentum about the center-of-mass is zero.** Determine the vibration frequency and draw the displacement pattern.

5. Consider a long, thin strip of elastic material having thickness h, width w, density ρ, and Young's modulus Y. It undergoes very small flexural (i.e., transverse) motions along its length (x-axis). Let the variable $\eta(x,t)$ represent the transverse displacement of the strip at position x and time t. **Assume that the motions are small enough such that dilational (i.e., tensile) stresses, not shear stresses, are associated with the deformation of the medium.**

   (a) Let R represent the radius of curvature of the medial plane for a small section of the elastic strip. For a plane a distance z from the medial plane, show that the dilational strain (along the length of the strip) is given by $\delta(z) = z/R$. (This assumes that the medial plane remains unstrained.)

   (b) Show that the potential energy density (i.e., potential energy per unit length) of a plane within the elastic strip is given by $\frac{1}{2}hwY\delta^2$. Then integrate the potential energy density over the thickness of the strip.

   (c) Using the fact that the radius of curvature can be written as $R = \left(\partial^2\eta/\partial x^2\right)^{-1}$, write the Lagrangian density $\mathbb{L}$ of the elastic strip.

   (d) Show that the equation of motion is

$$\frac{\partial^2\eta}{\partial t^2} + \frac{h^2 Y}{12\rho}\frac{\partial^4\eta}{\partial x^4} = 0.$$

Chapter 3
THE TRANSITION TO QUANTUM MECHANICS

In classical mechanics, all the mechanical properties of a system are functions of the canonical coordinates q_i and conjugate momenta p_i. At a particular instant in time, the state of the system is specified by giving the values of all the q_i's and p_i's. Furthermore, it is assumed that any of the classical variables can be measured to within an arbitrary degree of precision without affecting the measurement of any of the other variables.

In contrast, the formalism of quantum mechanics requires the replacement of each classical variable by a mathematical *operator*. These, in turn, act on *state vectors* which specify the state of the system. In quantum mechanics, certain pairs of variables cannot be measured simultaneously to within arbitrary precision.

The most general way of formulating quantum mechanics is through the use of abstract linear vector spaces. The theory of linear vector spaces was fully developed by mathematicians early in the twentieth century, but a new vehicle called *bra-ket notation* was introduced by P.A.M. Dirac specifically for the new quantum mechanics in the 1930's. This chapter provides a brief overview of the basic Dirac formulation of quantum mechanics, with particular emphasis on the simple harmonic oscillator. We will consider time evolution from two different viewpoints, or *pictures*, namely, the *Heisenberg picture* and the *Schrödinger picture*. The transition from classical mechanics to quantum mechanics is most naturally accomplished by way of the Heisenberg picture. Therefore, this picture shall be introduced first, followed by a transformation to the Schrödinger picture. A third viewpoint, referred to as the *interaction picture*, will be introduced in Chapter 6 in the context of time-dependent perturbation theory. The elegant Dirac formulation naturally facilitates transformations between the different pictures. The chapter closes with a discussion of angular momentum and quantum mechanics in three-dimensions.

3.1 Basic Dirac Formulation

When one is introduced to quantum phenomena for the first time, the theory is usually cast in the form of *wave mechanics*. The Dirac formulation is a more fundamental approach that elucidates the very underpinnings of quantum mechanics. The following sections attempt to provide an overview of the basic elements of this formalism.

3.1.1 The State Vector: Kets, Bras, and Inner Products

The wave-mechanical treatment of quantum mechanics is centered around the concept of a complex *wavefunction*, or *probability amplitude*, representing the state of a system. For a particle in one dimension, the wavefunction is denoted by $\psi(x,t)$. Although $\psi(x,t)$ cannot be measured, the quantity $|\psi(x,t)|^2$, called the *probability density*, can be accessed by experiment. $|\psi(x,t)|^2\, dx$ represents the probability that a measurement of the particle's position at time t produces an outcome between x and $x + dx$. Wave mechanics is extremely powerful when one is interested in visualizing spatial probabilities, however, it is limited in that it only offers a partial glimpse of the total quantum-mechanical picture. A more abstract, but general, way to specify the state of a system is through the concept of a *state function* or *state vector*, denoted by $|\psi\rangle$.

To understand the idea of a state vector, consider the following: For some fixed time t, expand the wavefunction in terms of a set of basis functions $f_n(x)$, i.e.,

$$\psi(x) = \sum_n c_n f_n(x), \tag{3.1}$$

where the c_n's are expansion coefficients which, in general, may be complex. Mathematically, the only requirements for the expansion of an arbitrary $\psi(x)$ are that the set of functions $\{f_n\}$ be *orthogonal* and *complete*. If one further requires that the functions be *normalized*, we have the following *orthonormality* condition which must hold over the interval of the wavefunction:

$$\int f_n^*(x)\, f_m(x)\, dx = \delta_{nm}. \tag{3.2}$$

The symbol δ_{nm} is called the *Kronecker delta* and is defined by

$$\delta_{nm} = \begin{cases} 0, & n \neq m \\ 1, & n = m. \end{cases} \tag{3.3}$$

The completeness criterion means that the set $\{f_n\}$ contains a sufficient number of functions to represent an arbitrary $\psi(x)$. Now imagine that the state of the system ψ corresponds to a vector in a multi-dimensional abstract vector space. We represent this state by a state vector, or *ket*, $|\psi\rangle$. One can now think of the functions f_n as playing the role of the basis vectors for the space, represented by basis kets $|f_n\rangle$. The c_n's correspond to projections of the state vector $|\psi\rangle$ onto each of the basis kets. In Dirac notation, Eq. 3.1 becomes

$$|\psi\rangle = \sum_n c_n \,|f_n\rangle\,. \tag{3.4}$$

There is a very strong parallel between the kets of quantum mechanics and the more familiar ordinary vectors in three-dimensional space. For example, a vector in three-dimensions, say $\mathbf{V}$, can be expanded in terms of the three basis vectors $\mathbf{e}_1$, $\mathbf{e}_2$, $\mathbf{e}_3$ as

$$\mathbf{V} = \sum_{n=1}^{3} (\mathbf{e}_n \cdot \mathbf{V})\, \mathbf{e}_n, \tag{3.5}$$

where the inner products (or scalar products) represent components of the vector $\mathbf{V}$ in the chosen basis, i.e., they are projections of the vector $\mathbf{V}$ onto each of the unit basis vectors. Because any arbitrary vector can be expanded in this fashion, we say that the basis vectors form a *complete set*, spanning the entire three-dimensional space. Eq. 3.5 is completely analogous to Eq. 3.4, with the components $\mathbf{e}_n \cdot \mathbf{V}$ playing the role of the expansion coefficients c_n. Furthermore, the basis vectors are normalized (i.e., they are unit vectors) and they are mutually orthogonal. The orthonormality condition for the basis vectors takes the form

$$\mathbf{e}_n \cdot \mathbf{e}_m = \delta_{nm}. \tag{3.6}$$

As in ket space, the particular choice for the basis vectors is completely arbitrary, as long as they form a complete orthogonal set. Just as the values of the c_n's depend on the choice of basis functions (kets), the resulting components for a given vector in three-dimensional space depend on the particular choice of basis vectors.

Continuing this parallel, one would also like to be able to form inner products in our quantum-mechanical space. This is accomplished by defining *bra space*, which is a *dual vector space* to ket space. That is to say, for each ket vector $|\psi\rangle$ there is a corresponding bra vector $\langle\psi|$. The connection between bra vectors and ket vectors is made through the definition of the inner product. We denote the inner product of a bra $\langle\varphi|$ and a ket $|\psi\rangle$ as a *bra-ket* $\langle\varphi \mid \psi\rangle$. The result is a complex number such that

$$\langle\varphi \mid \psi\rangle = \langle\psi \mid \varphi\rangle^*. \tag{3.7}$$

In keeping with the standard interpretation for the inner product, one can interpret $\langle\varphi \mid \psi\rangle$ as the projection of state $|\psi\rangle$ onto state $|\varphi\rangle$. This means that one can represent a component of some vector $|\psi\rangle$ in the $|f_n\rangle$-basis simply as $c_n = \langle f_n \mid \psi\rangle$. Equation 3.4 can now be written as

$$|\psi\rangle = \sum_n \langle f_n \mid \psi\rangle\, |f_n\rangle = \sum_n |f_n\rangle\, \langle f_n \mid \psi\rangle. \tag{3.8}$$

This is a statement of the completeness of the basis states in Dirac notation. The orthonormality condition for the basis kets mimics Eq. 3.6, i.e.,

$$\langle f_n \mid f_m\rangle = \delta_{nm}. \tag{3.9}$$

3.1.2 Operators

In quantum mechanics, mathematical *operators* have the effect of transforming bra and ket state vectors into new state vectors in the same space. In three-dimensions, this would be equivalent to a rotation and/or change in length of a vector. Operators are marked with a "hat," for example $\hat{A}$. They only attain meaning when they act on state vectors. When an operator acts on a ket from the left, it produces a new ket, i.e.,

$$\hat{A}\,|\psi\rangle = |\varphi\rangle\,, \tag{3.10}$$

and when it acts on a bra from the right, it produces a new bra

$$\langle\psi|\,\hat{A} = \langle\chi|\,. \tag{3.11}$$

As indicated, the action of an operator on a corresponding bra $\langle\psi|$ and ket $|\psi\rangle$ does not in general produce corresponding bras and kets. However, $\hat{A}\,|\psi\rangle$ has a dual correspondence defined as

$$\hat{A}\,|\psi\rangle \quad \underleftrightarrow{\text{dual correspondence}} \quad \langle\psi|\,\hat{A}^\dagger, \tag{3.12}$$

where $\hat{A}^\dagger$ is called the *Hermitian adjoint* of the operator $\hat{A}$. One can then take the complex conjugate of an inner product containing an operator using

$$\langle\psi \mid \hat{A} \mid \varphi\rangle^* = \left[\langle\psi \mid \left(\hat{A}\,|\varphi\rangle\right)\right]^* = \left(\langle\varphi|\,\hat{A}^\dagger\right)|\psi\rangle \tag{3.13}$$

or

$$\langle\psi \mid \hat{A} \mid \varphi\rangle^* = \langle\varphi \mid \hat{A}^\dagger \mid \psi\rangle. \tag{3.14}$$

If Eq. 3.14 holds for all ψ and φ, it is, in fact, the defining relation between an operator and its adjoint. In special cases when $\hat{A}^\dagger = \hat{A}$, one says that $\hat{A}$ is a *self-adjoint* or *Hermitian operator*, and we have $\langle\psi \mid \hat{A} \mid \varphi\rangle = \langle\varphi \mid \hat{A} \mid \psi\rangle^*$.

A few particularly useful relations involving operators and their Hermitian adjoints follow:

- Consider the special case where the operator $\hat{A}$ corresponds to a complex constant c. We then have

  $$\langle\psi \mid c \mid \varphi\rangle^* = c^*\langle\psi \mid \varphi\rangle^* = c^*\langle\varphi \mid \psi\rangle = \langle\varphi \mid c^* \mid \psi\rangle. \tag{3.15}$$

  From Eq. 3.14, since $\langle\psi \mid c \mid \varphi\rangle^* = \langle\varphi \mid c^\dagger \mid \psi\rangle$, the operator corresponding to the Hermitian adjoint of a complex number is just its complex conjugate:

  $$c^\dagger = c^*. \tag{3.16}$$

  Operators that are real constants are Hermitian since $c^* = c$.

- Here, we prove that $\hat{A}^{\dagger\dagger} = \hat{A}$. In Eq. 3.14, replace $\hat{A}$ with the operator $\hat{A}^{\dagger\dagger}$ giving

  $$\langle\psi \mid \hat{A}^\dagger \mid \varphi\rangle^* = \langle\varphi \mid \hat{A}^{\dagger\dagger} \mid \psi\rangle. \tag{3.17}$$

Now take the complex conjugate of Eq. 3.14, after reversing the placement of φ and ψ:

$$\langle \varphi \mid \hat{A} \mid \psi \rangle = \langle \psi \mid \hat{A}^\dagger \mid \varphi \rangle^*. \tag{3.18}$$

Comparing Eqs. 3.17 and 3.18 produces

$$\langle \varphi \mid \hat{A}^{\dagger\dagger} \mid \psi \rangle = \langle \varphi \mid \hat{A} \mid \psi \rangle, \tag{3.19}$$

yielding the relation

$$\hat{A}^{\dagger\dagger} = \hat{A}. \tag{3.20}$$

- Consider the operator $\left(\hat{A}\hat{B}\right)^\dagger$. We can write the following:

$$\langle \psi \mid \hat{A}\hat{B} \mid \varphi \rangle^* = \left(\langle \varphi \mid \hat{B}^\dagger\right)\left(\hat{A}^\dagger \mid \psi \rangle\right) = \langle \varphi \mid \hat{B}^\dagger \hat{A}^\dagger \mid \psi \rangle. \tag{3.21}$$

From Eq. 3.14, we also have

$$\langle \psi \mid \hat{A}\hat{B} \mid \varphi \rangle^* = \langle \varphi \mid \left(\hat{A}\hat{B}\right)^\dagger \mid \psi \rangle. \tag{3.22}$$

Comparing the last two equations produces the useful result

$$\left(\hat{A}\hat{B}\right)^\dagger = \hat{B}^\dagger \hat{A}^\dagger. \tag{3.23}$$

In general, the order of operators cannot be interchanged, i.e., $\hat{A}\hat{B} \neq \hat{B}\hat{A}$. One defines the *commutator* of two operators by

$$\left[\hat{A}, \hat{B}\right] = \hat{A}\hat{B} - \hat{B}\hat{A}. \tag{3.24}$$

If $\left[\hat{A}, \hat{B}\right] = 0$, one says that the operators $\hat{A}$ and $\hat{B}$ *commute.* A commutator is, in itself, another operator. Some general commutator relations are

$$\left[\hat{A} + \hat{B}, \hat{C}\right] = \left[\hat{A}, \hat{C}\right] + \left[\hat{B}, \hat{C}\right] \tag{3.25}$$

$$\left[\hat{A}, \hat{B}\hat{C}\right] = \left[\hat{A}, \hat{B}\right]\hat{C} + \hat{B}\left[\hat{A}, \hat{C}\right]. \tag{3.26}$$

Since quantum-mechanical operators are, for the most part *linear,** commutators involving a (complex) constant c become very simple:

$$\left[\hat{A}, c\right] = 0 \tag{3.27}$$

*An operator $\hat{A}$ is linear if $\hat{A}\left(c_1 |\psi\rangle + c_2 |\varphi\rangle\right) = c_1 \hat{A} |\psi\rangle + c_2 \hat{A} |\varphi\rangle$.

$$\left[\hat{A}, c\hat{B}\right] = \left[c\hat{A}, \hat{B}\right] = c\left[\hat{A}, \hat{B}\right]. \tag{3.28}$$

Of particular importance is the fact that any two functions of an operator $\hat{A}$ commute, i.e.,

$$\left[f(\hat{A}), g(\hat{A})\right] = 0. \tag{3.29}$$

This is true because a function of an operator is interpreted by expanding it in a Taylor series

$$f(\hat{A}) = \sum_{n=0}^{\infty} \frac{f^{(n)}(0)}{n!} \hat{A}^n, \tag{3.30}$$

where $f^{(n)}$ is the nth derivative of the function and $\hat{A}^n = \hat{A}........\hat{A}$ (n times). Equation 3.29 then follows since any power of $\hat{A}$ in the function f commutes with any power of $\hat{A}$ in the function g.

One can define the *outer product* between a bra and ket by

$$\hat{\Lambda} =| \psi\rangle\langle\varphi | . \tag{3.31}$$

$\hat{\Lambda}$ is an operator because $\hat{\Lambda}\,|\gamma\rangle =| \psi\rangle\langle\varphi \mid \gamma\rangle$ is a constant times a ket. A commonly used outer product is a *projection operator* defined by

$$\hat{P}_n = |f_n\rangle \langle f_n| , \tag{3.32}$$

where $|f_n\rangle$ is one of the basis kets. It has the effect of extracting from a vector that component which is parallel to $|f_n\rangle$. Now observe the following: Each term in the summation of Eq. 3.8 just produces one of the components of an arbitrary state vector $|\psi\rangle$, i.e.,

$$|\psi\rangle = \sum_n |f_n\rangle \langle f_n \mid \psi\rangle = \sum_n \hat{P}_n |\psi\rangle . \tag{3.33}$$

This just says that a state vector $|\psi\rangle$ can be assembled from its individual components. One can consider $\sum_n \hat{P}_n$ to be the *identity operator* $\hat{I}$; it does not change the state vector, it just re-expresses it in terms of components:

$$\hat{I} = \sum_n \hat{P}_n = \sum_n |f_n\rangle \langle f_n| . \tag{3.34}$$

This is called the *completeness* or *closure relation.*

3.1.3 Matrix Representations

Finally, here we briefly show how to cast the abstract state vectors and operators of the Dirac formalism into the more concrete representations of *matrix mechanics.* This step allows one to perform numerical calculations in quantum mechanics . To achieve this transformation, one first decides to work in the context of a particular set of basis states. This is analogous to choosing a particular coordinate system when working in the ordinary space of three dimensions. Once the basis states are chosen, say the set $\{|f_n\rangle\}$, any ket is represented by a column vector:

$$|\psi\rangle \to \begin{pmatrix} c_1 \\ c_2 \\ \vdots \\ c_n \\ \vdots \end{pmatrix} = \begin{pmatrix} \langle f_1 \mid \psi\rangle \\ \langle f_2 \mid \psi\rangle \\ \vdots \\ \langle f_n \mid \psi\rangle \\ \vdots \end{pmatrix}. \tag{3.35}$$

The elements of the column vector correspond to components (projections) of $|\psi\rangle$ in the $|f_n\rangle$ basis. Clearly, a change of basis results in a column vector with different elements, i.e., a different *matrix representation.* In order to be able to form inner products, bras are represented by row vectors:

$$\langle\varphi| \to \begin{pmatrix} d_1^*, & d_2^*, & \cdots, & d_n^*, & \cdots \end{pmatrix} = \begin{pmatrix} \langle\varphi \mid f_1\rangle, & \langle\varphi \mid f_2\rangle, & \cdots, & \langle\varphi \mid f_n\rangle, & \cdots \end{pmatrix}. \tag{3.36}$$

The row vector that represents a bra is the complex conjugate and transpose of the column vector for the corresponding ket. Now one can use the usual rules for matrix multiplication to form an inner product, i.e.,

$$\langle\varphi \mid \psi\rangle = \langle\varphi \mid f_1\rangle\langle f_1 \mid \psi\rangle + \langle\varphi \mid f_2\rangle\langle f_2 \mid \psi\rangle + \cdots + \langle\varphi \mid f_n\rangle\langle f_n \mid \psi\rangle + \cdots. \tag{3.37}$$

Note that the right-hand side can be re-written as $\langle\varphi \mid (\sum_n |f_n\rangle\langle f_n|)\,|\psi\rangle$. Since the quantity in parentheses is just the identity operator $\hat{I}$, the result is not surprising.

To see how operators are handled in a particular representation, consider the general action of an operator $\hat{A}$ on an arbitrary ket, as given by Eq. 3.10. Let's also insert the identity operator $\hat{I} = \sum_n |f_n\rangle\langle f_n|$ into the equation:

$$\hat{A}\hat{I}\,|\psi\rangle = |\varphi\rangle\,. \tag{3.38}$$

Suppose we now project both sides of this equation onto each of the basis kets. If the basis is N-dimensional, then this will produce N equations, as follows:

$$\begin{gathered}
\langle f_1 \mid \hat{A} \mid f_1\rangle\langle f_1 \mid \psi\rangle + \langle f_1 \mid \hat{A} \mid f_2\rangle\langle f_2 \mid \psi\rangle + \cdots + \langle f_1 \mid \hat{A} \mid f_N\rangle\langle f_N \mid \psi\rangle = \langle f_1 \mid \varphi\rangle \\
\langle f_2 \mid \hat{A} \mid f_1\rangle\langle f_1 \mid \psi\rangle + \langle f_2 \mid \hat{A} \mid f_2\rangle\langle f_2 \mid \psi\rangle + \cdots + \langle f_2 \mid \hat{A} \mid f_N\rangle\langle f_N \mid \psi\rangle = \langle f_2 \mid \varphi\rangle \\
\vdots \\
\langle f_N \mid \hat{A} \mid f_1\rangle\langle f_1 \mid \psi\rangle + \langle f_N \mid \hat{A} \mid f_2\rangle\langle f_2 \mid \psi\rangle + \cdots + \langle f_N \mid \hat{A} \mid f_N\rangle\langle f_N \mid \psi\rangle = \langle f_N \mid \varphi\rangle.
\end{gathered} \tag{3.39}$$

These can be expressed as a single matrix equation:

$$\begin{pmatrix} \langle f_1 \mid \hat{A} \mid f_1\rangle & \langle f_1 \mid \hat{A} \mid f_2\rangle & \cdots & \langle f_1 \mid \hat{A} \mid f_N\rangle \\ \langle f_2 \mid \hat{A} \mid f_1\rangle & \langle f_2 \mid \hat{A} \mid f_2\rangle & \cdots & \langle f_2 \mid \hat{A} \mid f_N\rangle \\ \vdots & \vdots & & \vdots \\ \langle f_N \mid \hat{A} \mid f_1\rangle & \langle f_N \mid \hat{A} \mid f_2\rangle & \cdots & \langle f_N \mid \hat{A} \mid f_N\rangle \end{pmatrix} \begin{pmatrix} \langle f_1 \mid \psi\rangle \\ \langle f_2 \mid \psi\rangle \\ \vdots \\ \langle f_N \mid \psi\rangle \end{pmatrix} = \begin{pmatrix} \langle f_1 \mid \varphi\rangle \\ \langle f_2 \mid \varphi\rangle \\ \vdots \\ \langle f_N \mid \varphi\rangle \end{pmatrix}. \tag{3.40}$$

This shows that an operator $\hat{A}$ is represented by an $N \times N$ matrix

$$\hat{A} \to \begin{pmatrix} \langle f_1 \mid \hat{A} \mid f_1 \rangle & \langle f_1 \mid \hat{A} \mid f_2 \rangle & \cdots \\ \langle f_2 \mid \hat{A} \mid f_1 \rangle & \langle f_2 \mid \hat{A} \mid f_2 \rangle & \cdots \\ \vdots & \vdots & \ddots \end{pmatrix} = \begin{pmatrix} A_{11} & A_{12} & \cdots \\ A_{21} & A_{22} & \cdots \\ \vdots & \vdots & \ddots \end{pmatrix}, \tag{3.41}$$

i.e., the matrix elements are given by

$$A_{ij} = \langle f_i \mid \hat{A} \mid f_j \rangle. \tag{3.42}$$

Equation 3.14 shows that the elements for the Hermitian adjoint matrix are obtained by taking the complex conjugate and then the transpose of $\hat{A}$, i.e.,

$$\left(A^{\dagger}\right)_{ij} = A^{*}_{ji}. \tag{3.43}$$

It then follows that a matrix is Hermitian if $A_{ij} = A^{*}_{ji}$. Once all the matrix elements for a particular representation are known, one can determine the effect of the operator on any state vector. In particular, Eq. 3.40 takes the form

$$\sum_j A_{ij} c_j = d_i. \tag{3.44}$$

3.2 The Quantum Postulates

With the Dirac formulation of complex vector spaces in hand, we now attach physical meaning to the operators and state vectors. This is accomplished by introducing the central postulates of quantum mechanics. Different references present these postulates in various forms. One version is presented here:

3.2.1 Observables, Operators, and Measurement

Any quantity that can be measured by experiment is called an *observable*. The first postulate attaches a mathematical operator to each observable:

- **Postulate 1**

  Corresponding to every physical observable, A, there is a Hermitian operator, $\hat{A}$, in an abstract linear vector space. Any measurement of A will produce a result, a_n, that corresponds to an eigenvalue of $\hat{A}$. In other words, each a_n represents a solution to the eigenvalue equation

  $$\hat{A} \left| a_n \right\rangle = a_n \left| a_n \right\rangle, \tag{3.45}$$

  where $\left| a_n \right\rangle$ is the eigenvector in the linear vector space corresponding to the eigenvalue a_n. Furthermore, immediately upon producing a particular measurement result a_n, the system is left in the state characterized by the eigenvector $\left| a_n \right\rangle$, sometimes called an eigenstate or eigenket of the system.

For example, the total energy of a system is represented by a particular Hamiltonian operator, $\hat{H}$. For a system in an arbitrary state $|\psi\rangle$, the possible results of an energy measurement, E_n, must correspond to one of the energy eigenstates, $|E_n\rangle$, satisfying the eigenvalue equation

$$\hat{H}\,|E_n\rangle = E_n\,|E_n\rangle\,. \tag{3.46}$$

Depending on the form of the Hamiltonian, the eigenvalues might form a discrete spectrum, a continuous spectrum, or possibly a combination of the two. Initially the system does not have to be in a state of definite energy. However, upon making an energy measurement, the state of the system collapses into one of the eigenstates, $|E_n\rangle$.

The fact that observables are represented by Hermitian operators is very important. To see this, lets first establish a few general properties about the eigenvectors and eigenvalues of Hermitian operators. Let the sets $\{|a_n\rangle\}$ and $\{a_n\}$ represent the eigenstates and eigenvalues for the operator $\hat{A}$, respectively. Then, from Eq. 3.45, observe that

$$\langle a_n \mid \hat{A} \mid a_m\rangle = a_m\langle a_n \mid a_m\rangle. \tag{3.47}$$

Furthermore, if $\hat{A}$ is a Hermitian operator, then

$$\langle a_n \mid \hat{A} \mid a_m\rangle = \langle a_m \mid \hat{A} \mid a_n\rangle^* = a_n^*\langle a_n \mid a_m\rangle. \tag{3.48}$$

Subtracting the latter two equations gives

$$(a_m - a_n^*)\langle a_n \mid a_m\rangle = 0. \tag{3.49}$$

Letting $a_n = a_m$, we must have $a_n = a_n^*$. This demands that *the eigenvalues of a Hermitian operator are real.* For $a_n \neq a_m$, we have $\langle a_n \mid a_m\rangle = 0$, that is, *the eigenstates corresponding to different eigenvalues are orthogonal for a Hermitian operator.*

Since the eigenvalues for quantum-mechanical operators represent actual measurement results from real experiments, it only makes sense that the a_n's be real. In addition, we see that the eigenkets of an operator $\hat{A}$ corresponding to any physical observable can be chosen to act as the basis kets for our abstract quantum-mechanical vector space. This is true because if we demand that the eigenkets are normalized, they form an orthonormal set, i.e.,

$$\langle a_n \mid a_m\rangle = \delta_{nm}. \tag{3.50}$$

If we further assume that they form a complete set, as was the case in Eq. 3.8, then any arbitrary ket in the vector space can be expanded as a superposition of basis eigenkets

$$|\psi\rangle = \sum_n c_n\,|a_n\rangle = \sum_n \mid a_n\rangle\langle a_n \mid \psi\rangle \tag{3.51}$$

and the closure relation for the identity operator $\hat{I}$ can be represented in terms of eigenkets:

$$\hat{I} = \sum_n | a_n\rangle\langle a_n | . \tag{3.52}$$

If the eigenvalues form a continuous spectrum, then the summations over the discrete indices in the latter relations need to be changed to integrals over the continuum of eigenvalues. As a result, Eqs. 3.51 and 3.52 get replaced by

$$|\psi\rangle = \int da \; | a\rangle\langle a | \psi\rangle \tag{3.53}$$

$$\hat{I} = \int da \; | a\rangle\langle a | . \tag{3.54}$$

The orthonormality condition, Eq. 3.50, is replaced by

$$\langle a | a'\rangle = \delta(a - a'), \tag{3.55}$$

where $\delta(a - a')$ is the *Dirac delta function.*[†]

3.2.2 Probabilities and Expectation Values

The second postulate relates the state vector to the probabilistic nature of measurement in quantum mechanics:

- **Postulate 2**

  The state vector contains all information about the state of the system. Given a system represented by the state vector $|\psi\rangle$, measurements of a physical observable A will produce an expectation value given by

$$\langle A\rangle = \langle\psi | \hat{A} | \psi\rangle. \tag{3.56}$$

  This represents the average value for a set of measurements performed on an ensemble of identical systems, all prepared in the state $|\psi\rangle$.

[†]The Dirac delta function is defined by the following two properties:

$$\int_{-\infty}^{+\infty} f(x)\delta(x - x')\,dx = f(x')$$

$$\delta(x - x') = 0 \qquad (\text{for } x \neq x').$$

Let's see what this implies. By expanding $|\psi\rangle$ in a complete set of eigenstates (Eq. 3.51), one can find the expectation value of an observable A:

$$\begin{aligned}
\langle A \rangle &= \langle \psi \mid \hat{A} \mid \psi \rangle \\
&= \sum_{n,m} \langle \psi \mid a_m \rangle \langle a_m \mid \hat{A} \mid a_n \rangle \langle a_n \mid \psi \rangle \\
&= \sum_{n,m} \langle \psi \mid a_m \rangle a_n \delta_{mn} \langle a_n \mid \psi \rangle \\
&= \sum_{n} |\langle a_n \mid \psi \rangle|^2 a_n. \qquad (3.57)
\end{aligned}$$

This means that the expansion coefficients $c_n = \langle a_n \mid \psi \rangle$ are directly linked to the possible outcomes for an experiment. The c_n's, which are complex numbers, are generally referred to as *probability amplitudes*. More specifically, $\langle a_n \mid \psi \rangle$ is referred to as "the amplitude for state $|\psi\rangle$ to be in state $|a_n\rangle$". The amplitudes themselves are not directly measurable. However, if a system is in state $|\psi\rangle$, the probability that a measurement of A will produce the result a_n is just $|\langle a_n \mid \psi \rangle|^2$, i.e.,

$$P(a_n) = |\langle a_n \mid \psi \rangle|^2. \qquad (3.58)$$

Attaching this probabilistic interpretation to the expansion coefficients leads to the requirement that all state vectors be normalized. To see this, observe that

$$\langle \psi \mid \psi \rangle = \sum_n \langle \psi \mid a_n \rangle \langle a_n \mid \psi \rangle = \sum_n |\langle a_n \mid \psi \rangle|^2 = \sum_n P(a_n). \qquad (3.59)$$

The last expression represents the total probability of obtaining any possible measurement outcome, i.e., it is just unity. The result is the *normalization condition*:

$$\langle \psi \mid \psi \rangle = 1. \qquad (3.60)$$

In cases when the eigenvalues form a continuum, Eqs. 3.57 and 3.58 are replaced by the following:

$$\langle A \rangle = \int da \, |\langle a \mid \psi \rangle|^2 a \qquad (3.61)$$

$$P(a)\, da = |\langle a \mid \psi \rangle|^2 \, da. \qquad (3.62)$$

Here, $P(a)\, da$ represents the probability that a measurement of the observable A will produce an outcome between a and $a + da$.

3.2.3 Classical Correspondence and the Role of Commutators

The formal connection between classical mechanics and quantum mechanics is made in the third postulate:

- **Postulate 3**

  In making the transition from classical mechanics to quantum mechanics, one replaces the Poisson bracket of two dynamical variables with a commutator, i.e.,

$$\{u, v\} \to \frac{1}{i\hbar}\left[\hat{u}, \hat{v}\right]. \tag{3.63}$$

  $\hat{u}$ and $\hat{v}$ are the quantum-mechanical operators that correspond to the classical variables u and v.

$\hbar$ stands for $h/2\pi$, where h is *Planck's constant* given by $h = 6.626 \times 10^{-27}$erg·s.

From this postulate and Eq. 2.141, the classical equation of motion for a dynamical variable A is replaced by the operator equation of motion:

$$\frac{d\hat{A}}{dt} = \frac{1}{i\hbar}\left[\hat{A}, \hat{H}\right] + \frac{\partial \hat{A}}{\partial t}. \tag{3.64}$$

This is called *Heisenberg's equation of motion* for the operator $\hat{A}$. It is the governing equation for time evolution in the Heisenberg picture. In this picture, operators are time dependent, while state vectors are not. An immediate consequence is that one can determine the time dependence of expectation values since it obviously follows that

$$\frac{d\langle A\rangle}{dt} = \frac{1}{i\hbar}\langle\psi \mid \left[\hat{A}, \hat{H}\right] \mid \psi\rangle + \langle\psi \mid \frac{\partial \hat{A}}{\partial t} \mid \psi\rangle. \tag{3.65}$$

This shows that as long as an operator $\hat{A}$ is not explicitly a function of time (i.e., $\partial\hat{A}/\partial t = 0$), the expectation value of the variable A corresponding to that operator is a constant of the motion whenever the operator commutes with $\hat{H}$. Comparing this result to the observations concerning Eq. 2.141, we see that when a classical variable A obeys a conservation law, then so does the expectation value $\langle A\rangle$ of the corresponding observable in quantum mechanics:

$$\begin{array}{ccc} \underline{\text{classical mechanics}} & & \underline{\text{quantum mechanics}} \\ \dfrac{dA}{dt} = 0 & \longleftrightarrow & \dfrac{d\langle A\rangle}{dt} = 0. \end{array} \tag{3.66}$$

Be aware, however, that conservation laws in quantum mechanics are statistical in nature, i.e., they apply to ensemble averages of identically prepared systems.

A more general connection between classical and quantum mechanics is a statement of *Ehrenfest's principle*:

> *Expectation values in quantum mechanics obey the laws of classical mechanics.*

This is easily demonstrated. Using Eq. 3.65, $\langle q_i \rangle$ and $\langle p_i \rangle$ obey the equations of motion

$$\frac{d\langle q_i \rangle}{dt} = \frac{1}{i\hbar}\langle \psi \mid \left[\hat{q}_i, \hat{H}\right] \mid \psi \rangle \quad \text{and} \quad \frac{d\langle p_i \rangle}{dt} = \frac{1}{i\hbar}\langle \psi \mid \left[\hat{p}_i, \hat{H}\right] \mid \psi \rangle. \tag{3.67}$$

But from the third postulate, the commutators can be obtained from the corresponding Poisson brackets. With the aid of Eqs. 2.137 and 2.138, we have

$$\begin{aligned} \left[\hat{q}_i, \hat{H}\right] &= i\hbar \{q_i, H\} = i\hbar \frac{\partial H}{\partial p_i} = i\hbar \left(\frac{p_i}{m}\right) \\ \left[\hat{p}_i, \hat{H}\right] &= i\hbar \{p_i, H\} = -i\hbar \frac{\partial H}{\partial q_i} = -i\hbar \left(\frac{\partial V}{\partial q_i}\right). \end{aligned} \tag{3.68}$$

The last step in each case follows because the Hamiltonian is of the form

$$H = \frac{1}{2m}\sum_k p_k^2 + V(q_k). \tag{3.69}$$

The result is Ehrenfest's principle, stated as

$$\frac{d\langle q_i \rangle}{dt} = \frac{\langle p_i \rangle}{m} \tag{3.70}$$

$$\frac{d\langle p_i \rangle}{dt} = -\left\langle \frac{\partial V}{\partial q_i} \right\rangle. \tag{3.71}$$

Equation 3.70 resembles the classical relationship between velocity and momentum. Since $-\partial V/\partial q_i$ represents the ith force component in classical mechanics, Eq. 3.71 has the same form as Newton's Second Law, $d\langle p_i \rangle/dt = \langle F_i \rangle$.

Also, by relating the commutators of quantum mechanics to the Poisson brackets calculated in classical mechanics, the third postulate allows us to determine which pairs of operators commute. For example, from Eqs. 2.130 and 2.131 we can immediately write the following commutator relations:

$$[\hat{q}_i, \hat{p}_j] = i\hbar\delta_{ij} \tag{3.72}$$

$$[\hat{q}_i, \hat{q}_j] = [\hat{p}_i, \hat{p}_j] = 0 \tag{3.73}$$

To see why relations such as these are so important, we first state the following *commutator theorem*:

Operators that commute have a common (simultaneous) set of eigenvectors.

The proof is as follows:‡ Consider the operator $\hat{A}$ and its standard eigenvalue problem, Eq. 3.45. Now operate on both sides with another operator, $\hat{B}$:

$$\hat{B}\hat{A}\,|a_n\rangle = a_n\hat{B}\,|a_n\rangle\,. \tag{3.74}$$

If the operators $\hat{A}$ and $\hat{B}$ commute, we can write

$$\hat{A}\left(\hat{B}\,|a_n\rangle\right) = a_n\left(\hat{B}\,|a_n\rangle\right). \tag{3.75}$$

This means that the kets $\hat{B}\,|a_n\rangle$ are the eigenvectors of $\hat{A}$, at least to within a multiplicative constant, i.e.,

$$\hat{B}\,|a_n\rangle = (\text{constant})\,|a_n\rangle\,. \tag{3.76}$$

This proves that the states $|a_n\rangle$ are also eigenvectors of the operator $\hat{B}$.

When $\hat{A}$ and $\hat{B}$ commute, one says that the observables A and B are *compatible*. A simultaneous measurement of A and B can be performed without any inherent conflict. According to the first postulate, the measurement process will force the system into one of the eigenstates common to the operators for the two observables. The outcomes obtained for A and B will just be the corresponding eigenvalues associated with this eigenstate. Both measurements can be made precisely.

On the other hand, if $\hat{A}$ and $\hat{B}$ do not commute, the eigenstates of the two operators are different, and it becomes impossible to simultaneously measure A and B to within arbitrary precision, i.e., a measurement of A interferes with ones ability to measure B. For a given state $|\psi\rangle$, the uncertainties $\triangle A$ and $\triangle B$ associated with the simultaneous measurement of two observables, are linked by a general *uncertainty principle* stated as§

$$\triangle A\,\triangle B \geq \frac{1}{2}\left|\langle\psi \mid \left[\hat{A},\hat{B}\right] \mid \psi\rangle\right|. \tag{3.77}$$

For example, the important commutator relation of Eq. 3.72 gives the famous Heisenberg position-momentum uncertainty relation

$$\triangle q_i\,\triangle p_i \geq \hbar/2. \tag{3.78}$$

Here, q_i and p_i are a conjugate pair, for example, x and p_x. Note that no uncertainty relation exists between non-conjugate coordinate-momentum pairs, such as x and p_y, since they commute.

Example 3.1 Quantization of the Harmonic Oscillator. The procedure for quantizing the one-dimensional simple harmonic oscillator is based on regarding q and p as non-commuting operators as demanded by the third quantum postulate, i.e.,

$$[\hat{q},\hat{p}] = i\hbar. \tag{3.79}$$

‡The proof given here assumes that the eigenvalues are non-degenerate, i.e., corresponding to each eigenvalue, a_n, there is only one eigenvector, $|a_n\rangle$.

§See, for example, Merzbacher [3].

The Hamiltonian is then represented by the operator

$$\hat{H}(\hat{p},\hat{q}) = \frac{\hat{p}^2}{2m} + \frac{1}{2}m\omega^2\hat{q}^2, \tag{3.80}$$

where $\omega = \sqrt{k/m}$ is the natural frequency of the oscillator. It becomes convenient to introduce a non-Hermitian operator, $\hat{a}$, and its adjoint defined as follows:

$$\hat{a} = \sqrt{\frac{m\omega}{2\hbar}}\left(\hat{q} + \frac{i}{m\omega}\hat{p}\right) \quad \text{and} \quad \hat{a}^\dagger = \sqrt{\frac{m\omega}{2\hbar}}\left(\hat{q} - \frac{i}{m\omega}\hat{p}\right). \tag{3.81}$$

Alternatively, one can write

$$\hat{p} = -i\sqrt{\frac{m\hbar\omega}{2}}\left(\hat{a} - \hat{a}^\dagger\right) \quad \text{and} \quad \hat{q} = \sqrt{\frac{\hbar}{2m\omega}}\left(\hat{a} + \hat{a}^\dagger\right). \tag{3.82}$$

We then find

$$\left[\hat{a},\hat{a}^\dagger\right] = \frac{m\omega}{2\hbar}\left[\hat{q} + \frac{i}{m\omega}\hat{p}, \hat{q} - \frac{i}{m\omega}\hat{p}\right] = \frac{m\omega}{2\hbar}\left\{-\frac{2i}{m\omega}\left[\hat{q},\hat{p}\right]\right\} = 1. \tag{3.83}$$

The last step follows from Eq. 3.79. The Hamiltonian can now be rewritten as

$$\begin{aligned} \hat{H} &= \frac{1}{2m}\left(-\frac{m\hbar\omega}{2}\right)\left(\hat{a} - \hat{a}^\dagger\right)^2 + \frac{1}{2}m\omega^2\left(\frac{\hbar}{2m\omega}\right)\left(\hat{a} + \hat{a}^\dagger\right)^2 \\ &= \frac{1}{2}\hbar\omega\left(\hat{a}\hat{a}^\dagger + \hat{a}^\dagger\hat{a}\right) \\ &= \hbar\omega\left(\hat{a}^\dagger\hat{a} + \frac{1}{2}\right), \end{aligned} \tag{3.84}$$

where the last step follows from Eq. 3.83. The operator $\hat{N} = \hat{a}^\dagger\hat{a}$ commutes with the Hamiltonian, i.e., $\left[\hat{a}^\dagger\hat{a}, \hat{H}\right] = 0$. This means that $\hat{N}$ and $\hat{H}$ have a common set of eigenvectors, and it is only necessary to solve the eigenvalue problem for $\hat{N}$, namely,

$$\hat{N}\left|n\right\rangle = n\left|n\right\rangle. \tag{3.85}$$

Using Eq. 3.23, note that

$$\hat{N}^\dagger = \left(\hat{a}^\dagger\hat{a}\right)^\dagger = \hat{a}^\dagger\hat{a} = \hat{N}. \tag{3.86}$$

Therefore, $\hat{N}$ is a Hermitian operator and the eigenvalues must be real.

The task at hand is to find the eigenvalues and eigenvectors of $\hat{N}$. This is best accomplished by considering the following commutators (see Eq. 3.26):

$$\begin{aligned} \left[\hat{a},\hat{N}\right] &= \left[\hat{a},\hat{a}^\dagger\hat{a}\right] = \left[\hat{a},\hat{a}^\dagger\right]\hat{a} + \hat{a}^\dagger\left[\hat{a},\hat{a}\right] = \hat{a} \\ \left[\hat{a}^\dagger,\hat{N}\right] &= \left[\hat{a}^\dagger,\hat{a}^\dagger\hat{a}\right] = \left[\hat{a}^\dagger,\hat{a}^\dagger\right]\hat{a} + \hat{a}^\dagger\left[\hat{a}^\dagger,\hat{a}\right] = -\hat{a}^\dagger. \end{aligned} \tag{3.87}$$

We thus have

$$\begin{aligned}\hat{N}(\hat{a}|n\rangle) &= \hat{N}\hat{a}|n\rangle = \left(\hat{a}\hat{N} - \hat{a}\right)|n\rangle = (n-1)(\hat{a}|n\rangle) \\ \hat{N}(\hat{a}^\dagger|n\rangle) &= \hat{N}\hat{a}^\dagger|n\rangle = \left(\hat{a}^\dagger\hat{N} + \hat{a}^\dagger\right)|n\rangle = (n+1)(\hat{a}^\dagger|n\rangle).\end{aligned} \tag{3.88}$$

Equations 3.88 show that both $\hat{a}|n\rangle$ and $\hat{a}^\dagger|n\rangle$ are eigenvectors of the operator $\hat{N}$, with eigenvalues $(n-1)$ and $(n+1)$, respectively. Thus, except for a multiplicative constant, $\hat{a}$ has the effect of transforming the state $|n\rangle$ to the state $|n-1\rangle$. The operator $\hat{a}$ goes by the name *step-down*, *lowering*, or *annihilation operator*. On the other hand, $\hat{a}^\dagger$ transforms the state $|n\rangle$ to the state $|n+1\rangle$, giving it the name *step-up*, *raising*, or *creation operator*. In summary,

$$\hat{a}|n\rangle = c_-|n-1\rangle \qquad \text{and} \qquad \hat{a}^\dagger|n\rangle = c_+|n+1\rangle. \tag{3.89}$$

The constants c_- and c_+ can be found (to within an undetermined phase factor) by taking inner products:

$$\begin{aligned}|c_-|^2 &= \langle n-1|c_-^*c_-|n-1\rangle = \langle n|\hat{a}^\dagger\hat{a}|n\rangle = \langle n \mid \hat{N} \mid n\rangle = n \\ |c_+|^2 &= \langle n+1|c_+^*c_+|n+1\rangle = \langle n|\hat{a}\hat{a}^\dagger|n\rangle = \langle n \mid \left(\hat{N}+1\right) \mid n\rangle = n+1.\end{aligned} \tag{3.90}$$

Therefore $c_- = \sqrt{n}$ and $c_+ = \sqrt{n+1}$. We now have the relations

$$\hat{a}|n\rangle = \sqrt{n}\,|n-1\rangle \qquad \text{and} \qquad \hat{a}^\dagger|n\rangle = \sqrt{n+1}\,|n+1\rangle. \tag{3.91}$$

In addition,

$$n = \langle n \mid \hat{N} \mid n\rangle = \langle n \mid \hat{a}^\dagger\hat{a} \mid n\rangle \geq 0. \tag{3.92}$$

The last step follows by observing that $\hat{a} \mid n\rangle$ can just be represented by some ket $|\psi\rangle$, so $\langle n \mid \hat{a}^\dagger\hat{a} \mid n\rangle = \langle\psi \mid \psi\rangle \geq 0$. This property, taken together with the properties of the step-up and step-down operators, demands that the eigenvalues of $\hat{N}$ be limited to

$$n = 0, 1, 2, \ldots. \tag{3.93}$$

Hence, $\hat{N}$ is called the *number operator*. It is now a simple matter to find the energy eigenvalues, E_n, which must satisfy

$$\hat{H}|n\rangle = E_n|n\rangle. \tag{3.94}$$

Since $\hat{H}$ and $\hat{N} = \hat{a}^\dagger\hat{a}$ are characterized by the common set of eigenvectors, $|n\rangle$, replacing $\hat{H}$ with the form from Eq. 3.84 immediately gives the result

$$E_n = \hbar\omega\left(n + \frac{1}{2}\right) \qquad (n = 0, 1, 2, \ldots). \tag{3.95}$$

As shown in Fig. 3.1, there is a spacing of $\hbar\omega$ between successive energies in the spectrum, and the oscillator has a minimum possible energy of $\hbar\omega/2$—this is called

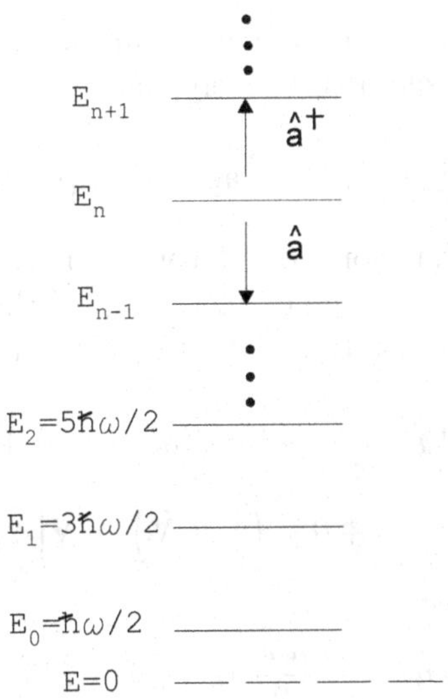

Figure 3.1 Energy levels of one-dimensional harmonic oscillator.

the *ground state* or *zero-point energy*. Since a change in n by one unit corresponds to an energy change of $\hbar\omega$, one can think of the oscillator as containing n *quanta* of energy $\hbar\omega$.

The state $|n\rangle$ can be generated from the ground state through the use of the step-up operator. Observe the following pattern:

$$\begin{aligned}
&\hat{a}^\dagger |0\rangle = \sqrt{1}\,|1\rangle \\
&\left(\hat{a}^\dagger\right)^2 |0\rangle = \hat{a}^\dagger\sqrt{1}\,|1\rangle = \sqrt{2}\sqrt{1}\,|2\rangle \\
&\left(\hat{a}^\dagger\right)^3 |0\rangle = \hat{a}^\dagger\sqrt{2}\sqrt{1}\,|2\rangle = \sqrt{3}\sqrt{2}\sqrt{1}\,|3\rangle \\
&\vdots \\
&\left(\hat{a}^\dagger\right)^n |0\rangle = \sqrt{n}\sqrt{n-1}...\sqrt{2}\sqrt{1}\,|n\rangle\,.
\end{aligned} \tag{3.96}$$

We see that

$$|n\rangle = \frac{1}{\sqrt{n!}}\left(\hat{a}^\dagger\right)^n |0\rangle\,. \tag{3.97}$$

Also recall that the eigenstates are orthonormal and complete:

$$\langle n \mid n'\rangle = \delta_{nn'} \tag{3.98}$$

$$\hat{I} = \sum_{n=0}^{\infty} \mid n\rangle\langle n \mid\,. \tag{3.99}$$

The ground-state of the harmonic oscillator represents a minimum uncertainty state, i.e., $\triangle q\triangle p = \hbar/2$. This can be shown by assuming that the uncertainties, $\triangle q$

and Δp, correspond to one standard deviation in the measured results for the position and momentum of the particle, respectively, i.e.,

$$\Delta q = \sqrt{\langle q^2 \rangle - \langle q \rangle^2} \qquad \text{and} \qquad \Delta p = \sqrt{\langle p^2 \rangle - \langle p \rangle^2}. \tag{3.100}$$

For a given state $|n\rangle$ of the harmonic oscillator, the required expectation values can be computed without knowledge of the explicit form of the wavefunctions. Instead, one only needs Eqs. 3.82, along with the properties of the raising and lowering operators:

$$\begin{aligned} \langle q^2 \rangle &= \langle n \mid \hat{q}^2 \mid n \rangle = \frac{\hbar}{2m\omega} \langle n \mid \left(\hat{a}\hat{a} + \hat{a}^\dagger \hat{a}^\dagger + \hat{a}\hat{a}^\dagger + \hat{a}^\dagger \hat{a} \right) \mid n \rangle \\ &= \frac{\hbar}{2m\omega} \langle n \mid \left[0 + 0 + \left(1 + \hat{N}\right) + \hat{N} \right] \mid n \rangle = \frac{\hbar}{2m\omega} (1 + 2n) \end{aligned} \tag{3.101}$$

$$\begin{aligned} \langle p^2 \rangle &= \langle n \mid \hat{p}^2 \mid n \rangle = -\frac{m\hbar\omega}{2} \langle n \mid \left(\hat{a}\hat{a} + \hat{a}^\dagger \hat{a}^\dagger - \hat{a}\hat{a}^\dagger - \hat{a}^\dagger \hat{a} \right) \mid n \rangle \\ &= -\frac{m\hbar\omega}{2} \langle n \mid \left[0 + 0 - \left(1 + \hat{N}\right) - \hat{N} \right] \mid n \rangle = \frac{m\hbar\omega}{2} (1 + 2n)\,. \end{aligned} \tag{3.102}$$

$\langle q \rangle$ and $\langle p \rangle$ vanish because $\langle n \mid \hat{a} \mid n \rangle \sim \langle n \mid n-1 \rangle = 0$ and $\langle n \mid \hat{a}^\dagger \mid n \rangle \sim \langle n \mid n+1 \rangle = 0$ due to the orthogonality of the eigenstates. So we have

$$\begin{aligned} \Delta q &= \sqrt{\langle q^2 \rangle} = \sqrt{\frac{\hbar}{2m\omega}(2n+1)} \\ \Delta p &= \sqrt{\langle p^2 \rangle} = \sqrt{\frac{m\hbar\omega}{2}(2n+1)}\,, \end{aligned} \tag{3.103}$$

which satisfies the Heisenberg uncertainty principle, i.e.,

$$\Delta q \Delta p = (2n+1)\frac{\hbar}{2} \geq \frac{\hbar}{2}. \tag{3.104}$$

For $n = 0$ the equality holds, and we have a minimum uncertainty state.

Finally, we can find the time dependence of the various operators in the Heisenberg picture by writing down Heisenberg's equations of motion, Eq. 3.64. For the step-up and step-down operators, the time dependence is not explicit (i.e., $\partial \hat{a}/\partial t = \partial \hat{a}^\dagger/\partial t = 0$), so

$$\begin{aligned} \frac{d\hat{a}}{dt} &= \frac{1}{i\hbar}\left[\hat{a}, \hbar\omega\left(\hat{a}^\dagger\hat{a} + \tfrac{1}{2}\right)\right] = \frac{1}{i\hbar}\hbar\omega\left[\hat{a}, \hat{a}^\dagger\hat{a}\right] = -i\omega\hat{a} \\ \frac{d\hat{a}^\dagger}{dt} &= \frac{1}{i\hbar}\left[\hat{a}^\dagger, \hbar\omega\left(\hat{a}^\dagger\hat{a} + \tfrac{1}{2}\right)\right] = \frac{1}{i\hbar}\hbar\omega\left[\hat{a}^\dagger, \hat{a}^\dagger\hat{a}\right] = +i\omega\hat{a}^\dagger, \end{aligned} \tag{3.105}$$

where the commutator identity, Eq. 3.26, was used in the last step in each case. Solving the two first-order equations of motion gives the time dependence

$$\hat{a}(t) = \hat{a}(0)\, e^{-i\omega t} \qquad \text{and} \qquad \hat{a}^\dagger(t) = \hat{a}^\dagger(0)\, e^{+i\omega t}. \tag{3.106}$$

Inserting this result into Eqs. 3.82 produces the time dependence of $\hat{q}$ and $\hat{p}$, namely

$$\begin{aligned}\hat{q}(t) &= \hat{q}(0)\cos\omega t + \frac{\hat{p}(0)}{m\omega}\sin\omega t \\ \hat{p}(t) &= \hat{p}(0)\cos\omega t - m\omega\hat{q}(0)\sin\omega t.\end{aligned} \tag{3.107}$$

These equations resemble the classical solution (Eqs. 2.52), except the classical variables are replaced by operators.

3.3 Transformation to the Schrödinger Picture

In the Heisenberg picture, we have time-dependent operators obeying Eq. 3.64, and it is assumed that the state vector, $|\psi\rangle$, does not evolve with time. In the Schrödinger picture, by contrast, the situation is reversed. Here, the state vector $|\psi_s(t)\rangle$ changes with time, while the operators, denoted by $\hat{A}_s$, are assumed fixed, except, possibly, for an explicit time dependence. Since the expectation value $\langle A\rangle$ of a physical observable should be independent of what picture one happens to choose, we can make the following *unitary transformation* between the two pictures:

$$\langle\psi \mid \hat{A}(t) \mid \psi\rangle = \langle\psi \mid \hat{U}^{\dagger}\hat{U}\hat{A}(t)\hat{U}^{\dagger}\hat{U} \mid \psi\rangle = \langle\psi_s(t) \mid \hat{A}_s \mid \psi_s(t)\rangle. \tag{3.108}$$

$\hat{U}(t)$ is an example of a *unitary operator*, which means that it satisfies

$$\hat{U}^{\dagger}\hat{U} = \hat{U}\hat{U}^{\dagger} = \hat{I}. \tag{3.109}$$

As a result, one can switch between the two pictures via the transformation equations

$$|\psi_s(t)\rangle = \hat{U}(t)\,|\psi\rangle \qquad \text{or} \qquad |\psi\rangle = \hat{U}^{\dagger}(t)\,|\psi_s(t)\rangle \tag{3.110}$$

and

$$\hat{A}_s = \hat{U}(t)\hat{A}(t)\hat{U}^{\dagger}(t) \qquad \text{or} \qquad \hat{A}(t) = \hat{U}^{\dagger}(t)\hat{A}_s\hat{U}(t). \tag{3.111}$$

The state vectors in the two pictures are identical at $t = 0$, i.e., $|\psi\rangle = |\psi_s(t = 0)\rangle$. Thus, Eq. 3.110 provides an interpretation for the operator $\hat{U}(t)$. By operating on the state vector at $t = 0$ in the Schrödinger picture, it has the effect of evolving the state forward to time t. It goes by the name *time-evolution operator*.

In order to deduce the form of $\hat{U}(t)$, we take a time derivative of Eq. 3.111:

$$\frac{d\hat{A}_s}{dt} = \frac{d\hat{U}}{dt}\hat{A}\hat{U}^{\dagger} + \hat{U}\frac{d\hat{A}}{dt}\hat{U}^{\dagger} + \hat{U}\hat{A}\frac{d\hat{U}^{\dagger}}{dt} = \frac{\partial\hat{A}_s}{\partial t}. \tag{3.112}$$

The last equality occurs because the only time dependence carried by operators in the Schrödinger are explicit. Using Heisenberg's equation of motion (Eq. 3.64) for $d\hat{A}/dt$, we have

$$\frac{d\hat{U}}{dt}\hat{A}\hat{U}^{\dagger} + \frac{1}{i\hbar}\hat{U}\left[\hat{A},\hat{H}\right]\hat{U}^{\dagger} + \hat{U}\frac{\partial\hat{A}}{\partial t}\hat{U}^{\dagger} + \hat{U}\hat{A}\frac{d\hat{U}^{\dagger}}{dt} = \frac{\partial\hat{A}_s}{\partial t}. \tag{3.113}$$

Each operator in this equation can be transformed to the Schrödinger picture by using Eq. 3.111. This produces

$$\left(\frac{d\hat{U}}{dt}\hat{U}^\dagger - \frac{1}{i\hbar}\hat{H}_s\right)\hat{A}_s + \hat{A}_s\left(\hat{U}\frac{d\hat{U}^\dagger}{dt} + \frac{1}{i\hbar}\hat{H}_s\right) = 0. \quad (3.114)$$

In order for this equation to hold for any arbitrary $\hat{A}_s$, the two expressions enclosed in parentheses must separately vanish. However, note that these expressions are just Hermitian adjoints of each other. Thus, the result is a single differential equation for the time-evolution operator, namely,

$$i\hbar\frac{d}{dt}\hat{U} = \hat{H}_s\hat{U}. \quad (3.115)$$

At this time note that Eq. 3.115 suggests the following identification for the energy operator:

$$\hat{E} \to i\hbar\frac{d}{dt}. \quad (3.116)$$

This is useful if one considers the commutator $\left[\hat{E}, t\right]$. The latter can be evaluated by allowing it to operate on an arbitrary state:

$$\begin{aligned}\left[\hat{E}, t\right]|\psi_s(t)\rangle &= i\hbar\frac{d}{dt}t\,|\psi_s(t)\rangle - i\hbar t\frac{d}{dt}|\psi_s(t)\rangle \\ &= i\hbar t\frac{d}{dt}|\psi_s(t)\rangle + i\hbar\,|\psi_s(t)\rangle - i\hbar t\frac{d}{dt}|\psi_s(t)\rangle \\ &= i\hbar\,|\psi_s(t)\rangle, \end{aligned} \quad (3.117)$$

or

$$\left[\hat{E}, t\right] = i\hbar. \quad (3.118)$$

Equation 3.77 now allows us to state the time-energy uncertainty principle

$$\Delta E\,\Delta t \geq \hbar/2. \quad (3.119)$$

This relationship becomes important when discussing the lifetimes and linewidths of states.

Now suppose we act on an arbitrary state $|\psi\rangle$ in the Heisenberg picture with Eq. 3.115. Then, with the aid of Eq. 3.110 and its time-derivative, we produce

$$i\hbar\frac{d}{dt}|\psi_s(t)\rangle = \hat{H}_s\,|\psi_s(t)\rangle, \quad (3.120)$$

which is the dynamical equation for the state vector in the Schrödinger picture, known as the *(time-dependent) Schrödinger equation.*

Here, let's consider the solution to the Schrödinger equation for the case of conservative systems, where the Hamiltonian is not an explicit function of time. Then, Eq. 3.115 can be integrated, giving the following explicit form for the time-evolution operator:

$$\hat{U}(t) = e^{-i\hat{H}_s t/\hbar} \qquad \text{or} \qquad \hat{U}^\dagger(t) = e^{+i\hat{H}_s t/\hbar}. \tag{3.121}$$

In this case, we then see that the Hamiltonian operator is actually the same in the two pictures, i.e.,

$$\begin{aligned} \hat{H} &= \hat{U}^\dagger \hat{H}_s \hat{U} = e^{i\hat{H}_s t/\hbar} \hat{H}_s e^{-i\hat{H}_s t/\hbar} \\ &= \left(1 + i\hat{H}_s t/\hbar + ...\right) \hat{H}_s \left(1 - i\hat{H}_s t/\hbar + ...\right) = \hat{H}_s, \end{aligned} \tag{3.122}$$

since various powers of $\hat{H}_s$ commute. It is common, therefore, to denote the Hamiltonian operator in either picture by $\hat{H}$. From Eq. 3.110, the solution for the state vector is

$$|\psi_s(t)\rangle = e^{-i\hat{H}t/\hbar} |\psi\rangle. \tag{3.123}$$

Let us now suppose that the state of a system is specified at $t = 0$. This initial state an be expanded as a superposition of energy eigenstates, i.e.,

$$|\psi_s(t=0)\rangle = |\psi\rangle = \sum_n c_n |E_n\rangle, \tag{3.124}$$

where $c_n = \langle E_n \mid \psi_s(t=0)\rangle$. The $|E_n\rangle$'s are the eigenkets satisfying Eq. 3.46, the energy eigenvalue problem for the Hamiltonian operator. This equation is commonly referred to as the *time-independent Schrödinger equation.* We therefore say that the energy eigenstates and associated energy eigenvalues form the solutions to the time-independent Schrödinger equation. Once this equation is solved, it is straightforward to write the solution to the time-dependent Schrödinger equation for $|\psi_s(t)\rangle$. From Eq. 3.123, we have

$$\begin{aligned} |\psi_s(t)\rangle &= \sum_n c_n e^{-i\hat{H}t/\hbar} |E_n\rangle \\ &= \sum_n c_n \left(1 - i\hat{H}t/\hbar + ...\right) |E_n\rangle = \sum_n c_n \left(1 - iE_n t/\hbar + ...\right) |E_n\rangle \\ &= \sum_n c_n e^{-i\omega_n t} |E_n\rangle, \end{aligned} \tag{3.125}$$

where $\omega_n = E_n/\hbar$ is the characteristic frequency contributed by eigenstate $|E_n\rangle$. Thus, in the Schrödinger picture of quantum mechanics, the Hamiltonian operator determines the time evolution of the state vector, given the initial state vector, whereas in classical mechanics, the Hamiltonian function determines the time evolution of the canonical coordinates and conjugate momenta, given the initial coordinates and momenta.

Of special note is a situation where the system is prepared in one of the energy eigenstates at $t = 0$. The time dependence for such a state appears as a multiplicative phase factor, i.e., $|\psi_s(t)\rangle = e^{-i\omega_n t}|E_n\rangle$. These are called *stationary states* because they are characterized by the following important property: For any physical observable A, the probability of obtaining any particular measurement result, a_n, does not vary with time (as long as $\partial\hat{A}/\partial t = 0$), i.e.,

$$P(a_n) = |\langle a_n \mid \psi_s(t)\rangle|^2 = \left|e^{-i\omega_m t}\langle a_n \mid E_m\rangle\right|^2 = |\langle a_n \mid E_m\rangle|^2 \tag{3.126}$$

is independent of time. As a consequence, it then also follows that the expectation value of any physical observable is also independent of time since $\langle A\rangle = \sum_n a_n P(a_n)$.

Example 3.2 Classical Correspondence and the Oscillator. According to Ehrenfest's principle discussed in Section 3.2.3, expectation values in quantum mechanics satisfy the classical equations. For the harmonic oscillator, Eqs. 3.70 and 3.71 become

$$\begin{aligned} \frac{d\langle q\rangle}{dt} &= \frac{\langle p\rangle}{m} \\ \frac{d\langle p\rangle}{dt} &= -\left\langle \frac{\partial}{\partial q}\left(\frac{1}{2}m\omega^2 q^2\right)\right\rangle = -m\omega^2\langle q\rangle\,. \end{aligned} \tag{3.127}$$

Combining these gives the equation of motion for the expectation value $\langle q\rangle$ as

$$\frac{d^2\langle q\rangle}{dt^2} + \omega^2\langle q\rangle = 0 \tag{3.128}$$

with the solution

$$\langle q\rangle = A\cos(\omega t + \phi). \tag{3.129}$$

This expression resembles the classical solution, Eq. 2.52. However, if the correspondence is to be exact, we must have $A \rightarrow \sqrt{2E/m\omega^2}$. When is this in fact the case? To answer this question, one needs to realize that the oscillating behavior predicted by Eq. 3.129 only occurs when the system is in a superposition of energy eigenstates. Recall that time dependence is never observed for a pure stationary state. Therefore, let's calculate $\langle q\rangle$ for a completely general superposition state. Choosing to work in the Schrödinger picture (remember that expectation values do not depend on the choice of picture used), we write

$$|\psi_s(t)\rangle = \sum_{n=0}^{\infty} c_n\, e^{-i\omega_n t}\,|n\rangle\,. \tag{3.130}$$

Since $\omega_n = E_n/\hbar = \left(n + \frac{1}{2}\right)\omega$, this is equivalent to

$$|\psi_s(t)\rangle = e^{-i\omega t/2}\sum_{n=0}^{\infty} c_n\, e^{-in\omega t}\,|n\rangle\,. \tag{3.131}$$

The expectation value of q is then

$$\langle q\rangle = \langle\psi_s(t) \mid \hat{q} \mid \psi_s(t)\rangle = \sum_{n,n'=0}^{\infty} c_n^* \, c_{n'} \, e^{i(n-n')\omega t} \langle n \mid \hat{q} \mid n'\rangle. \tag{3.132}$$

The matrix elements are

$$\begin{aligned}\langle n \mid \hat{q} \mid n'\rangle &= \sqrt{\frac{\hbar}{2m\omega}} \, \langle n \mid (\hat{a} + \hat{a}^\dagger) \mid n'\rangle \\ &= \sqrt{\frac{\hbar}{2m\omega}} \left[\sqrt{n+1}\langle n+1 \mid n'\rangle + \sqrt{n}\langle n-1 \mid n'\rangle\right] \\ &= \sqrt{\frac{\hbar}{2m\omega}} \left[\sqrt{n+1}\, \delta_{n+1,n'} + \sqrt{n}\, \delta_{n-1,n'}\right].\end{aligned} \tag{3.133}$$

Therefore, after some manipulation,

$$\langle q\rangle = \sqrt{\frac{\hbar}{2m\omega}} \sum_{n=1}^{\infty} \sqrt{n} \left(c_{n-1}^* \, c_n \, e^{-i\omega t} + c_n^* \, c_{n-1} \, e^{+i\omega t}\right). \tag{3.134}$$

Letting $c_n = |c_n| \, e^{i\phi}$ gives the result

$$\langle q\rangle = \sqrt{\frac{2}{m\omega^2}} \sum_{n=0}^{\infty} \sqrt{n\hbar\omega} \; |c_n| \, |c_{n-1}| \cos(\omega t + \phi_{n-1} - \phi_n). \tag{3.135}$$

The only way to make this precisely match the classical expression is to demand that n become very large so that $n-1 \to n$ and $n\hbar\omega \to E_n$, in which case

$$\langle q\rangle = \sqrt{\frac{2}{m\omega^2}} \left[\sum_{n=0}^{\infty} |c_n|^2 \sqrt{E_n}\right] \cos(\omega t + \phi). \tag{3.136}$$

The quantity in square brackets is just $\left\langle\sqrt{E_n}\right\rangle$, thus

$$\langle q\rangle = \left\langle\sqrt{\frac{2E}{m\omega^2}}\right\rangle \cos(\omega t + \phi), \tag{3.137}$$

in agreement with the classical result. The observation that the quantum-mechanical behavior of physical systems reduces to the classical behavior for large n is known as the *correspondence principle*.

3.4 Representations in Position Space

The connection between Dirac's abstract vector space and the formalism of wave mechanics is made by expressing all state vectors and operators in the *position representation*. This is also called the *Schrödinger representation* (which is not to be confused with the Schrödinger picture). Here, the basis vectors are chosen to be the eigenkets of the position operator $\hat{q}$. These form a continuum of basis kets $|q\rangle$ satisfying the position eigenvalue equation

$$\hat{q}\,|q\rangle = q\,|q\rangle\,. \tag{3.138}$$

For the time being, we are working in one spatial dimension. The eigenkets must form an orthonormal set

$$\langle q \mid q'\rangle = \delta(q - q') \tag{3.139}$$

and they must be complete

$$\hat{I} = \int_{-\infty}^{+\infty} dq\ \mid q\rangle\langle q \mid . \tag{3.140}$$

Any state vector $|\psi\rangle$ can now be expanded in terms of this basis set, i.e.,

$$|\psi\rangle = \int_{-\infty}^{+\infty} dq\ \mid q\rangle\langle q \mid \psi\rangle. \tag{3.141}$$

The amplitude $\langle q \mid \psi\rangle$, written as $\psi(q)$, is the projection of the state vector onto an eigenstate in position space. In this context, $|\psi\rangle$ can be thought of as infinite-dimensional vector having elements given by the various amplitudes. The position representation of the state vector in the Schrödinger picture is, in fact, the usual wavefunction $\psi(q,t)$ of wave mechanics, i.e.,

$$\psi(q,t) = \langle q \mid \psi_s(t)\rangle. \tag{3.142}$$

By using Eq. 3.140, any inner product can be expressed as an overlap integral involving wavefunctions in the Schrödinger picture:

$$\langle \psi \mid \varphi\rangle = \int_{-\infty}^{+\infty} dq\,\langle \psi \mid q\rangle\langle q \mid \varphi\rangle = \int_{-\infty}^{+\infty} dq\,\psi^*(q)\,\varphi(q). \tag{3.143}$$

It then follows that the normalization condition $\langle \psi \mid \psi\rangle = 1$ can be written as

$$\int_{-\infty}^{+\infty} dq\ |\psi(q)|^2 = 1. \tag{3.144}$$

Consider the eigenstates and eigenvalues of some operator $\hat{A}$, and expand an arbitrary state $|\psi\rangle$, as was previously done in Eq. 3.51. Taking an inner product with $\langle q|$ gives

$$\langle q \mid \psi\rangle = \sum_n \langle q \mid a_n\rangle\langle a_n \mid \psi\rangle. \tag{3.145}$$

The result can be re-expressed using the language of wavefunctions:

$$\psi(q) = \sum_n c_n f_n(q). \tag{3.146}$$

Here, $f_n(q) = \langle q \mid a_n \rangle$ is the nth position-space *eigenfunction* of the operator $\hat{A}$, and the c_n's are the expansion coefficients $\langle a_n \mid \psi \rangle$. Equation 3.146 parallels Eq. 3.1 for expanding an arbitrary function in terms of a set of complete, orthogonal basis functions. What is the position-space eigenfunction that represents the eigenket $|q'\rangle$? It is just $\langle q \mid q' \rangle$. But from Eq. 3.139, we see that this is simply the Dirac delta function, i.e., $\langle q \mid q' \rangle = \delta(q - q')$.

What are the Schrödinger representations for the operators $\hat{q}$ and $\hat{p}$? This question is answered by finding the various matrix elements $\langle q \mid \hat{q} \mid q' \rangle$ and $\langle q \mid \hat{p} \mid q' \rangle$. For the position operator, we immediately have

$$\langle q \mid \hat{q} \mid q' \rangle = q\, \delta(q - q'). \tag{3.147}$$

We now argue that the momentum operator in the position basis is given by

$$\langle q \mid \hat{p} \mid q' \rangle = -i\hbar \frac{\partial}{\partial q} \delta(q - q'). \tag{3.148}$$

In order for this to be the case, we must show that this representation is consistent with the fundamental commutator relation of Eq. 3.72. Let's examine what happens if $[\hat{q}, \hat{p}]$ operates on an arbitrary state $|\psi\rangle$, with the result projected onto position space:

$$\langle q \mid [\hat{q}, \hat{p}] \mid \psi \rangle = i\hbar \langle q \mid \psi \rangle. \tag{3.149}$$

Rewriting, we have

$$q\langle q \mid \hat{p} \mid \psi \rangle - \langle q \mid \hat{p}\hat{q} \mid \psi \rangle = i\hbar\psi(q). \tag{3.150}$$

Now evaluate the bra-ket $\langle q \mid \hat{p} \mid \psi \rangle$ on the left-hand side:

$$\begin{aligned}
\langle q \mid \hat{p} \mid \psi \rangle &= \langle \psi \mid \hat{I}\hat{p} \mid q \rangle^* = \int_{-\infty}^{+\infty} dq'\, \langle \psi \mid q' \rangle^* \langle q' \mid \hat{p} \mid q \rangle^* \\
&= i\hbar \int_{-\infty}^{+\infty} dq'\, \psi(q') \frac{\partial}{\partial q'} \delta(q' - q) = -i\hbar \int_{-\infty}^{+\infty} dq'\, \frac{\partial \psi(q')}{\partial q'} \delta(q' - q) \\
&= -i\hbar \frac{\partial}{\partial q} \psi(q).
\end{aligned} \tag{3.151}$$

In a similar fashion, the second bra-ket on the left of Eq. 3.150 can be shown to be

$$\langle q \mid \hat{p}\hat{q} \mid \psi \rangle = -i\hbar \left[q \frac{\partial \psi(q)}{\partial q} + \psi(q) \right]. \tag{3.152}$$

Using these last two identities, we see that Eq. 3.150 is indeed satisfied, and the expression in Eq. 3.148 is the position representation for $\hat{p}$.

A fundamental result in itself is Eq. 3.151. It shows that, in the Schrödinger representation, one makes the following replacement for the momentum operator:

$$\hat{p} \rightarrow -i\hbar\frac{\partial}{\partial q}. \tag{3.153}$$

Likewise, since

$$\langle q \mid \hat{q} \mid \psi\rangle = q\langle q \mid \psi\rangle = q\,\psi(q), \tag{3.154}$$

one has the following trivial replacement for the position operator:

$$\hat{q} \rightarrow q. \tag{3.155}$$

It is now left as an exercise for the reader to show that if $A(\hat{q}, \hat{p})$ is some general function of the operators $\hat{p}$ and $\hat{q}$, then one has the relations stated below:

$$\langle q \mid A(\hat{q}, \hat{p}) \mid q'\rangle = A\left(q, -i\hbar\frac{\partial}{\partial q}\right)\delta(q - q') \tag{3.156}$$

$$\begin{gathered}\langle q \mid A(\hat{q}, \hat{p}) \mid \psi\rangle = A\left(q, -i\hbar\frac{\partial}{\partial q}\right)\psi(q) \\ \left[\text{i.e., } A(\hat{q}, \hat{p}) \rightarrow A\left(q, -i\hbar\frac{\partial}{\partial q}\right)\right]\end{gathered} \tag{3.157}$$

$$\langle \varphi \mid A(\hat{q}, \hat{p}) \mid \psi\rangle = \int_{-\infty}^{+\infty} dq\,\varphi^*(q)\,A\left(q, -i\hbar\frac{\partial}{\partial q}\right)\psi(q). \tag{3.158}$$

In the Schrödinger representation, the Schrödinger equation takes on a more familiar form. Taking an inner product of Eq. 3.120 with $\langle q|$ gives

$$i\hbar\frac{\partial}{\partial t}\langle q \mid \psi_s(t)\rangle = \langle q \mid \hat{H} \mid \psi_s(t)\rangle, \tag{3.159}$$

or, replacing the Hamiltonian operator with operators for the canonical coordinate and conjugate momentum, we have

$$i\hbar\frac{\partial}{\partial t}\psi(q,t) = \langle q \mid \left[\frac{\hat{p}^2}{2m} + V(\hat{q})\right] \mid \psi_s(t)\rangle. \tag{3.160}$$

Using Eq. 3.157, we now have the time-dependent Schrödinger equation in one-dimension:

$$i\hbar\frac{\partial}{\partial t}\psi(q,t) = \left[-\frac{\hbar^2}{2m}\frac{\partial^2}{\partial q^2} + V(q)\right]\psi(q,t). \tag{3.161}$$

In a similar manner, the time-independent Schrödinger equation, Eq. 3.46, can be shown to have the following representation in position space:

$$\left[-\frac{\hbar^2}{2m}\frac{d^2}{dq^2} + V(q)\right]\psi_n(q) = E_n\psi_n(q). \tag{3.162}$$

Here, the $\psi_n(q)$'s are the stationary state wavefunctions corresponding to the energy eigenkets, i.e., $\psi_n(q) = \langle q \mid E_n \rangle$.¶ It is relatively straightforward to extend the derivations to three spatial dimensions. The time-dependent and time-independent Schrödinger equations become

$$i\hbar\frac{\partial}{\partial t}\psi(\mathbf{r},t) = \left[-\frac{\hbar^2}{2m}\nabla^2 + V(\mathbf{r})\right]\psi(\mathbf{r},t) \tag{3.163}$$

and

$$\left[-\frac{\hbar^2}{2m}\nabla^2 + V(\mathbf{r})\right]\psi_n(\mathbf{r}) = E_n\psi_n(\mathbf{r}). \tag{3.164}$$

When the Hamiltonian is not an explicit function of time, the general solution to the time-dependent Schrödinger equation is found by taking an inner product of Eq. 3.125 with $\langle q|$, and then extending to three dimensions, i.e.,

$$\psi(\mathbf{r},t) = \sum_n c_n e^{-i\omega_n t}\psi_n(\mathbf{r}), \tag{3.165}$$

where, as before, $\omega_n = E_n/\hbar$.

Example 3.3 The Harmonic Oscillator Eigenfunctions. We now generate the stationary state wavefunctions for the one-dimensional harmonic oscillator, i.e., the projections of the $|n\rangle$ states onto coordinate space:

$$\psi_n(q) = \langle q \mid n \rangle. \tag{3.166}$$

Clearly, these can be found by solving Eq. 3.162, however the properties of the operators $\hat{a}$ and $\hat{a}^\dagger$ allow one to generate the wavefunctions in a more elegant fashion. Since stepping down the state $|0\rangle$ results in no state at all, we have

$$\hat{a}\,|0\rangle = 0. \tag{3.167}$$

Now use the first of Eqs. 3.81 to write

$$\left(\hat{q} + \frac{i}{m\omega}\hat{p}\right)|0\rangle = 0. \tag{3.168}$$

Taking an inner product with $\langle q|$ produces the differential equation for the ground-state wavefunction

$$\left(q + \frac{\hbar}{m\omega}\frac{d}{dq}\right)\psi_0(q) = 0, \tag{3.169}$$

with the solution

$$\psi_0(q) = Ce^{-m\omega q^2/2\hbar}. \tag{3.170}$$

¶Keep in mind that there may, in fact, be a continuum of solutions to the time-independent Schrödinger equation, despite the choice to label them with the discrete index n.

The constant C is determined from the normalization condition

$$\int_{-\infty}^{+\infty} dq \, |\psi_0(q)|^2 = 1. \tag{3.171}$$

We then have the ground-state wavefunction

$$\psi_0(q) = \langle q \mid 0 \rangle = \left(\frac{m\omega}{\pi\hbar}\right)^{1/4} e^{-m\omega q^2/2\hbar}. \tag{3.172}$$

To obtain the wavefunction for the first excited state, $|1\rangle$, we can apply the step-up operator to the ground state, i.e.,

$$\psi_1(q) = \langle q \mid 1 \rangle = \langle q \mid \hat{a}^\dagger \mid 0 \rangle = \langle q \mid \sqrt{\frac{m\omega}{2\hbar}} \left(\hat{q} - \frac{i}{m\omega}\hat{p}\right) \mid 0 \rangle, \tag{3.173}$$

to obtain

$$\begin{aligned} \psi_1(q) &= \sqrt{\frac{m\omega}{2\hbar}} \left(q - \frac{\hbar}{m\omega}\frac{d}{dq}\right) \psi_0(q) \\ &= \left[\frac{4}{\pi}\left(\frac{m\omega}{\hbar}\right)^3\right]^{1/4} q\, e^{-m\omega q^2/2\hbar}. \end{aligned} \tag{3.174}$$

Similarly, the wavefunction for state $|n\rangle$ is

$$\psi_n(q) = \langle q \mid n \rangle = \frac{1}{\sqrt{n!}} \left(\sqrt{\frac{m\omega}{2\hbar}}\right)^n \left(q - \frac{\hbar}{m\omega}\frac{d}{dq}\right)^n \left(\frac{m\omega}{\pi\hbar}\right)^{1/4} e^{-m\omega q^2/2\hbar}. \tag{3.175}$$

The normalized wavefunctions have the following functional form:

$$\psi_n(q) = \sqrt{\frac{\beta}{\pi^{1/2}\, 2^n\, n!}}\, e^{-\beta^2 q^2/2}\, H_n(\beta q). \tag{3.176}$$

Here, $\beta = \sqrt{m\omega/\hbar}$ and the H_n's are *Hermite polynomials* of order n. The wavefunctions and the corresponding position-space probability densities, $|\psi_n(q)|^2 = |\langle q \mid n \rangle|^2$, are displayed in Fig. 3.2. Notice that the number of nodes possessed by the wavefunction is equal to n. Also, the stationary state wavefunctions have either even or odd parity. The general definition for the parity of a function is

$$\psi(-\mathbf{r}) = \begin{cases} +\psi(\mathbf{r}) & \text{(even parity)} \\ -\psi(\mathbf{r}) & \text{(odd parity).} \end{cases} \tag{3.177}$$

Definite parity of ψ is guaranteed whenever the potential function, $V(\mathbf{r})$, has even parity $[V(\mathbf{r}) = V(-\mathbf{r})]$. This is because a symmetric potential function demands that the observed spatial probability density also be symmetric, i.e., $|\psi(-\mathbf{r})|^2 = |\psi(\mathbf{r})|^2$.

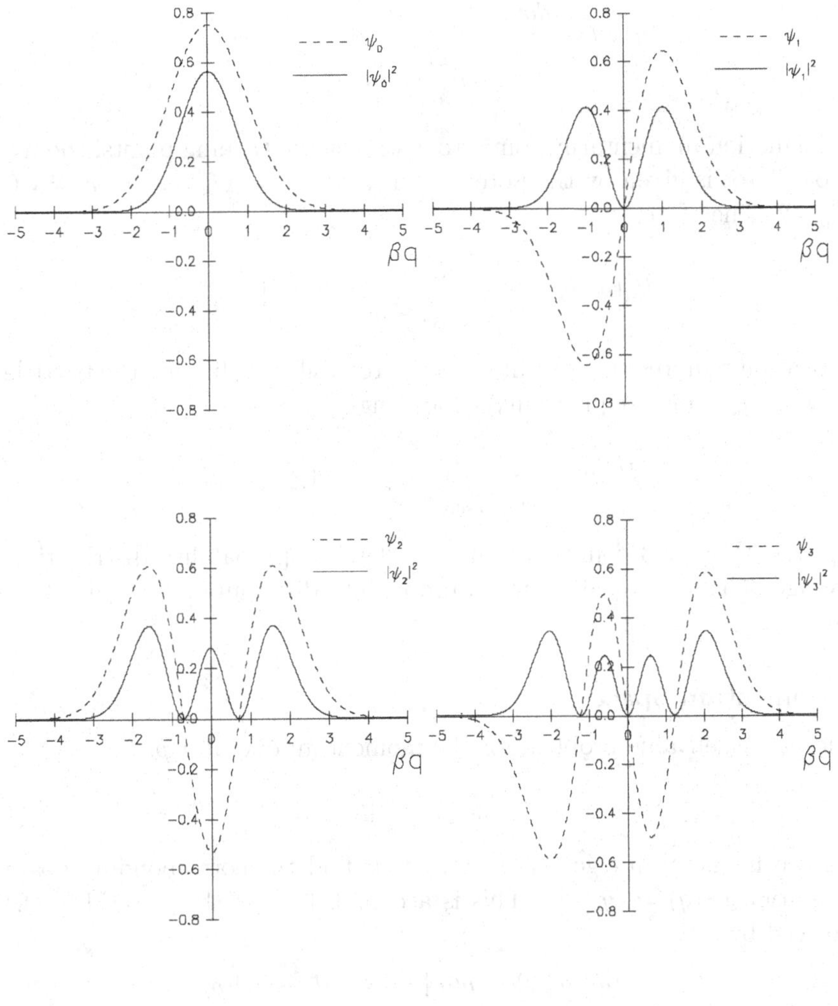

Figure 3.2 First four harmonic oscillator wavefunctions and probability densities.

As a final observation, let's show that the behavior of $\psi_n(q)$ for large n approaches that of the classical harmonic oscillator in accord with the correspondence principle. Figure 3.3 is a sketch of the quantum probability density $|\psi_n(q)|^2$ for large n, along with the corresponding classical distribution function, $P_{cl}(q)$. For locations between the classical turning points of $\pm q_{max}$, the quantity $P_{cl}(q)\,dq$ is proportional to the amount of time the oscillating mass spends between q and $q + dq$, i.e.,

$$P_{cl}(q)\,dq \propto \frac{dq}{\dot{q}} = \frac{dq}{\sqrt{\dfrac{2E}{m} - \omega^2 q^2}} \qquad (|q| < q_{max})\,. \tag{3.178}$$

Since the motion momentarily comes to a stop at the turning points, the total energy of the oscillator is given by the potential energy at $q = \pm q_{max}$, i.e., $E = V(\pm q_{max}) = m\omega^2 q_{max}^2/2$. Therefore,

$$P_{cl}(q)\,dq \propto \frac{dq}{\omega\sqrt{q_{max}^2 - q^2}} \qquad (|q| < q_{max})\,. \tag{3.179}$$

Normalization requires that the integrated probability of finding the particle between $-q_{max}$ and $+q_{max}$ must equal unity, producing

$$P_{cl}(q) = \frac{1}{\pi\sqrt{q_{max}^2 - q^2}} \qquad (|q| < q_{max})\,. \tag{3.180}$$

As expected, Fig. 3.3 confirms that this classical probability distribution matches the average behavior of the quantum probability distribution $|\psi_n(q)|^2$ in the limit of large n.

3.5 Momentum Space

Consider the eigenvalue problem for the momentum operator $\hat{p}$:

$$\hat{p}\,|p\rangle = p\,|p\rangle\,. \tag{3.181}$$

For a given momentum eigenstate, $|p\rangle$, let us find the corresponding position-space representation $\psi_p(q) = \langle q \mid p\rangle$. This is accomplished by the use of Eq. 3.151, with $|\psi\rangle$ replaced by $|p\rangle$:

$$\langle q \mid \hat{p} \mid p\rangle = p\langle q \mid p\rangle = -i\hbar\frac{\partial}{\partial q}\langle q \mid p\rangle. \tag{3.182}$$

Integrating this equation produces the desired position-space wavefunction:

$$\psi_p(q) = \langle q \mid p\rangle = Ce^{ipq/\hbar}. \tag{3.183}$$

Alternatively, one can write

$$\psi_k(q) = \langle q \mid k\rangle = C' e^{ikq}, \tag{3.184}$$

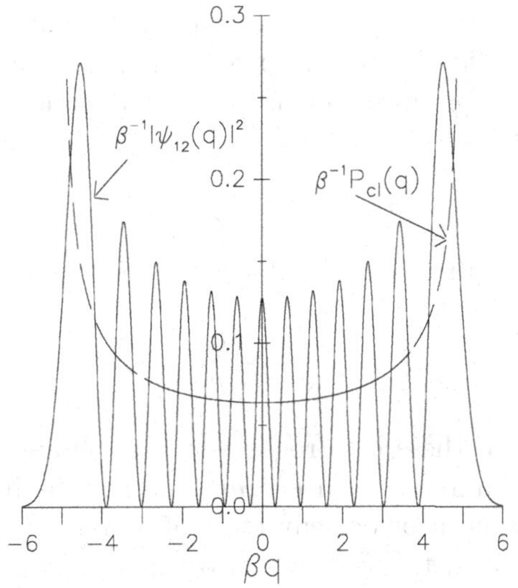

Figure 3.3 Comparison between the classical and quantum ($n = 12$) probability densities for the 1-D oscillator.

where we define $k = p/\hbar$. Here, $|k\rangle$ denotes the momentum eigenstate that produces the momentum eigenvalue $p_k = \hbar k$, i.e.,

$$p\,|k\rangle = p_k\,|k\rangle = \hbar k\,|k\rangle\,. \tag{3.185}$$

The $|k\rangle$'s form a continuum of orthonormal and complete states:

$$\langle k \mid k' \rangle = \delta(k - k') \tag{3.186}$$

$$\hat{I} = \int_{-\infty}^{+\infty} dk\ \mid k\rangle\langle k\mid . \tag{3.187}$$

The normalization constant C' can be obtained by observing that

$$\delta(q - q') = \langle q \mid q' \rangle = \int_{-\infty}^{+\infty} dk\,\langle q \mid k\rangle\langle k \mid q'\rangle = |C'|^2 \int_{-\infty}^{+\infty} dk\, e^{ik(q-q')}. \tag{3.188}$$

Comparison of this result to the well-known delta-function representation

$$\delta(q - q') = \frac{1}{2\pi}\int_{-\infty}^{+\infty} dk\, e^{ik(q-q')} \tag{3.189}$$

shows that the normalization constant is $C' = 1/\sqrt{2\pi}$. Therefore,

$$\psi_k(q) = \frac{1}{\sqrt{2\pi}}\, e^{ikq}. \tag{3.190}$$

The momentum eigenfunctions correspond to states of definite momentum. They are spatially periodic with wavelength $\lambda = 2\pi/k$, since shifting q by any integral multiple of this amount produces no change in the value of the function. We thus identify k as the *wavenumber*

$$k = \frac{2\pi}{\lambda}. \tag{3.191}$$

Since $k = p/\hbar$, this is equivalent to the famous *de Broglie relation*, namely

$$\lambda = \frac{h}{p}. \tag{3.192}$$

The time dependence of these de Broglie waves is obtained by observing that the momentum operator commutes with $\hat{p}^2/2m$, which is the free-particle Hamiltonian. Thus, the $|k\rangle$'s are simultaneous eigenvectors of $\hat{p}$ and $\hat{H}_{\text{free particle}}$. As well as being momentum eigenfunctions, the ψ_k's are also the stationary states that satisfy the time-independent Schrödinger equation for the free particle. The time dependence of the ψ_k's is the same as that of the stationary states. Suppose $|k_0\rangle$ is an eigenstate with a positive momentum $\hbar k_0$. Then note that

$$\hat{H}_{\text{free particle}} |\pm k_0\rangle = \frac{\hat{p}^2}{2m} |\pm k_0\rangle = \frac{(\pm\hbar k_0)^2}{2m} |\pm k_0\rangle = \frac{\hbar^2 k_0^2}{2m} |\pm k_0\rangle = E_0 |\pm k_0\rangle . \tag{3.193}$$

Each energy eigenvalue for the free particle, $E_0 = \hbar^2 k_0^2/2m$, has two eigenstates associated with it. We say that the energy states exhibit a two-fold *degeneracy*. The time dependence of both states is given by the same factor $e^{-i\omega_0 t}$, where $\omega_0 = E_0/\hbar = \hbar k_0^2/2m$. Thus, the two momentum states associated with energy E_0 are

$$\begin{aligned} \psi_{+k_0}(q,t) = \psi_{+k_0}(q)\, e^{-i\omega_0 t} &= \frac{1}{\sqrt{2\pi}}\, e^{+ik_0 q}\, e^{-i\omega_0 t} = \frac{1}{\sqrt{2\pi}}\, e^{+ik_0(q-v_0 t)} \\ \psi_{-k_0}(q,t) = \psi_{-k_0}(q)\, e^{-i\omega_0 t} &= \frac{1}{\sqrt{2\pi}}\, e^{-ik_0 q}\, e^{-i\omega_0 t} = \frac{1}{\sqrt{2\pi}}\, e^{-ik_0(q+v_0 t)}, \end{aligned} \tag{3.194}$$

where $v_0 = \omega_0/k_0 = \hbar k_0/2m$ is the phase velocity of the de Broglie wave.‖ Because they are of the general form $f(q \pm vt)$, the ψ_k's are travelling waves. ψ_{+k_0} propagates with speed v_0 in the $+q$-direction, which is in the same sense as its momentum, and ψ_{-k_0} propagates the opposite way with the same speed.

In a manner similar to the development of the position representation in the last section, one can also speak of the *momentum representation*. Here, the basis vectors are chosen to be the eigenkets $|k\rangle$ of the momentum operator. Upon projecting

‖Note that the wave travels at a speed of $\hbar k_0/2m = p_0/2m$, which is one-half of the speed possessed by a classical particle with the same momentum. However, by superimposing momentum eigenstates, one can construct a "wavepacket" having a group velocity of p_0/m representative of the classical particle motion.

an arbitrary state $|\psi\rangle$ onto this basis, one produces the set of amplitudes $\langle k \mid \psi\rangle$ corresponding to the *momentum space wavefunction*:

$$\phi(k) = \langle k \mid \psi\rangle. \tag{3.195}$$

$|\phi(k)|^2\, dk$ represents the probability that a measurement of the particle's momentum produces an outcome between $\hbar k$ and $\hbar(k + dk)$. The normalization condition in the momentum representation takes the form

$$\langle \psi \mid \psi\rangle = \int_{-\infty}^{+\infty} dk\, |\phi(k)|^2 = 1. \tag{3.196}$$

Now examine the two identities

$$\begin{aligned} \langle q \mid \psi\rangle &= \int_{-\infty}^{+\infty} dk\, \langle q \mid k\rangle\langle k \mid \psi\rangle \\ \langle k \mid \psi\rangle &= \int_{-\infty}^{+\infty} dq\, \langle k \mid q\rangle\langle q \mid \psi\rangle. \end{aligned} \tag{3.197}$$

These can be rewritten as

$$\psi(q) = \frac{1}{\sqrt{2\pi}} \int_{-\infty}^{+\infty} dk\, e^{ikq}\phi(k) \tag{3.198}$$

$$\phi(k) = \frac{1}{\sqrt{2\pi}} \int_{-\infty}^{+\infty} dq\, e^{-ikq}\psi(q). \tag{3.199}$$

This shows that the position-space and momentum-space wavefunctions constitute a Fourier transform pair. It is important to note that, in general, as the characteristic width of a function decreases, the width of its Fourier transform increases. In this context, one now has a physical interpretation for our result—Eq. 3.198 basically expresses $\psi(q)$ as a superposition of the various momentum eigenfunctions given by $e^{ikq}/\sqrt{2\pi}$. As one becomes more certain of the outcome of a momentum measurement, the width of the function $\phi(k)$ decreases. As a result, the width of the Fourier transform function, $\psi(q)$, increases and one becomes less certain of the outcome of a position measurement. This is just an alternate statement of the position-momentum uncertainty principle. For example, suppose a particle is in a momentum eigenstate $|k'\rangle$. In position space, the state is represented by

$$\psi(q) = \psi_{k'}(q) = \frac{1}{\sqrt{2\pi}} e^{ik'q}. \tag{3.200}$$

The particle is completely unlocalized since $|\psi(q)|^2 = \text{constant}$ for all q, i.e., $\Delta q \to \infty$. The corresponding momentum-space wavefunction, as found from Eq. 3.199, is

$$\phi(k) = \phi_{k'}(k) = \frac{1}{2\pi} \int_{-\infty}^{+\infty} dq\, e^{-iq(k-k')} = \delta(k - k'). \tag{3.201}$$

Actually, this should be no surprise since, by definition, $\phi_{k'}(k) = \langle k \mid k' \rangle$ and, from the orthonormality condition of Eq. 3.186, $\phi_{k'}(k) = \delta(k-k')$. The probability density for measuring various momenta, $|\phi(k)|^2$, vanishes for all k except $k = k'$. Thus, as expected, the particle has a definite momentum of $\hbar k'$, and $\Delta p \to 0$.

What are the momentum-space representations for the operators $\hat{q}$ and $\hat{p}$? In order to answer this question, let's first calculate the quantity $\langle k \mid \hat{q} \mid \psi \rangle$:

$$\begin{aligned}
\langle k \mid \hat{q} \mid \psi \rangle &= \int_{-\infty}^{+\infty} dq \, \langle k \mid \hat{q} \mid q \rangle \langle q \mid \psi \rangle = \int_{-\infty}^{+\infty} dq \, \langle q \mid k \rangle^* q \langle q \mid \psi \rangle \\
&= \int_{-\infty}^{+\infty} dq \left(\frac{1}{\sqrt{2\pi}} e^{-ikq} \right) q \, \psi(q) = \frac{1}{\sqrt{2\pi}} \int_{-\infty}^{+\infty} dq \left[i \frac{\partial}{\partial k} \left(e^{-ikq} \right) \right] \psi(q) \\
&= i \frac{\partial}{\partial k} \left[\frac{1}{\sqrt{2\pi}} \int_{-\infty}^{+\infty} dq \, e^{-ikq} \psi(q) \right] = i \frac{\partial}{\partial k} \phi(k). \qquad (3.202)
\end{aligned}$$

This means that, in momentum space, we have

$$\hat{q} \to i \frac{\partial}{\partial k}. \qquad (3.203)$$

If $|\psi\rangle$ is replaced with $|k'\rangle$ in Eq. 3.202, we find the matrix elements of the position operator:

$$\langle k \mid \hat{q} \mid k' \rangle = i \frac{\partial}{\partial k} \delta(k - k'). \qquad (3.204)$$

The representation for the momentum operator is trivial. Since

$$\langle k \mid \hat{p} \mid \psi \rangle = \hbar k \, \langle k \mid \psi \rangle = \hbar k \, \phi(k), \qquad (3.205)$$

we have

$$\hat{p} \to \hbar k. \qquad (3.206)$$

The reader can verify that if $A(\hat{q}, \hat{p})$ is an arbitrary function of the operators $\hat{q}$ and $\hat{p}$, then

$$\begin{gathered}
\langle k \mid A(\hat{q}, \hat{p}) \mid \psi \rangle = A\left(i \frac{\partial}{\partial k}, \hbar k \right) \phi(k) \\
\left[\text{i.e., } A(\hat{q}, \hat{p}) \to A\left(i \frac{\partial}{\partial k}, \hbar k \right) \right].
\end{gathered} \qquad (3.207)$$

Example 3.4 Harmonic Oscillator in Momentum Space. In position space, we previously showed that the energy eigenstates for the simple harmonic oscillator have the form given by Eq. 3.176. Here, we determine the momentum representation of these energy eigenstates. Luckily, the procedure is rather simple. This is due to the fact that the position-space and momentum-space representations of the harmonic-oscillator Hamiltonian are completely symmetric in form.

The energy eigenvalue problem is

$$\left(\frac{\hat{p}^2}{2m} + \frac{1}{2}m\omega^2\hat{q}^2\right)|n\rangle = E_n|n\rangle. \tag{3.208}$$

In the Schrödinger representation, this becomes the time-independent Schrödinger equation in position space:

$$\left(-\frac{\hbar^2}{2m}\frac{d^2}{dq^2} + \frac{1}{2}m\omega^2q^2\right)\psi_n(q) = E_n\psi_n(q). \tag{3.209}$$

We can also represent Eq. 3.208 in momentum space by making the replacements $\hat{q} \to i\dfrac{d}{dk}$, $\hat{p} \to \hbar k$, and $|n\rangle \to \phi_n(k)$, producing

$$\left(\frac{\hbar^2k^2}{2m} - \frac{1}{2}m\omega^2\frac{d^2}{dk^2}\right)\phi_n(k) = E_n\phi_n(k). \tag{3.210}$$

This is the time-independent Schrödinger equation in momentum space. Recalling that $E_n = \left(n + \frac{1}{2}\right)\hbar\omega$ and $\beta = \sqrt{m\omega/\hbar}$, Eqs. 3.209 and 3.210 can be rearranged and expressed in dimensionless fashion:

$$\left(\frac{1}{\beta^2}\frac{d^2}{dq^2} - \beta^2q^2\right)\psi_n(q) = -2\left(n + \frac{1}{2}\right)\psi_n(q) \tag{3.211}$$

$$\left(\beta^2\frac{d^2}{dk^2} - \frac{1}{\beta^2}k^2\right)\phi_n(k) = -2\left(n + \frac{1}{2}\right)\phi_n(k). \tag{3.212}$$

Inspection shows that the momentum-space equation is identical to the position-space equation, except that β is replaced by $1/\beta$. Thus, from the form of the position-space solutions (Eq. 3.176), the normalized energy eigenfunctions in momentum space are

$$\phi_n(k) = \sqrt{\frac{1}{\pi^{1/2}\,2^n\,n!\,\beta}}\,e^{-k^2/2\beta^2}\,H_n(k/\beta). \tag{3.213}$$

Except for scaling considerations, sketches of the $\phi_n(k)$'s are the same as the ψ's in Fig. 3.2.

3.6 Angular Momentum and Quantum Mechanics in Three Dimensions

Understanding quantum mechanics in three-dimensions requires that we take a new look at the concept of angular momentum. Classically, the angular momentum of a particle was given in Chapter 2 as

$$\mathbf{L} = \mathbf{r}\times\mathbf{p}. \tag{3.214}$$

Equation 2.34 shows that introducing a torque has the effect of changing angular momentum. By choosing the torque properly, one can cause the angular momentum to take on any value desired, i.e., in classical physics there are no inherent restrictions on the magnitude or direction of **L**. Furthermore, it is possible to measure any component of **L** without limiting the precision with which one is able to measure the other components. In quantum mechanics, on the other hand, we will find that both the magnitude and orientation of **L** are quantized. In addition, the measurement of one angular momentum component affects the measurement of other components.

3.6.1 Angular Momentum Operators and Commutator Relations

To make the transition to quantum mechanics, we replace the angular momentum components obtained from Eq. 3.214 with corresponding angular momentum operators:

$$\hat{L}_x = \hat{y}\hat{p}_z - \hat{z}\hat{p}_y, \qquad \hat{L}_y = \hat{z}\hat{p}_x - \hat{x}\hat{p}_z, \qquad \hat{L}_z = \hat{x}\hat{p}_y - \hat{y}\hat{p}_x. \tag{3.215}$$

The commutators for the angular momentum operators can now be formed with the aid of Eqs. 3.72 and 3.73 for the position-momentum commutators, with $(\hat{q}_1, \hat{q}_2, \hat{q}_3) \rightarrow (\hat{x}, \hat{y}, \hat{z})$ and $(\hat{p}_1, \hat{p}_2, \hat{p}_3) \rightarrow (\hat{p}_x, \hat{p}_y, \hat{p}_z)$. The basic commutator properties are found to be

$$\left[\hat{L}_x, \hat{L}_y\right] = i\hbar\hat{L}_z, \qquad \left[\hat{L}_y, \hat{L}_z\right] = i\hbar\hat{L}_x, \qquad \left[\hat{L}_z, \hat{L}_x\right] = i\hbar\hat{L}_y. \tag{3.216}$$

As was demonstrated in the case of the simple harmonic oscillator, the commutation relations are all that one needs to quantize the system. In fact, Eqs. 3.216 are more fundamental than Eqs. 3.215 which were based on the classical definition of angular momentum. We shall find shortly that not all angular momentum in quantum mechanics has a classical analogue. To make this clear, we shall reserve the operators $\hat{L}_x$, $\hat{L}_y$, and $\hat{L}_z$ to represent the components of *orbital angular momentum* **L**, i.e., angular momentum with a classical analogue. We choose the triplet $\hat{S}_x$, $\hat{S}_y$, and $\hat{S}_z$ to be the components of *spin angular momentum* **S**, for which there is no classical analogue. It is tempting to think that **S** corresponds to the type of angular momentum associated with a classical rigid body "spinning" about an axis, like a spinning top or the daily rotation of the earth. This is incorrect, however, since the angular momentum of such a motion really corresponds to an orbital angular momentum that is the sum of the $\mathbf{L}_i$'s about the rotation axis for the individual particles that form the rigid body, i.e., $\mathbf{L} = \sum_i \mathbf{L}_i$. The existence of spin angular momentum **S** is a purely quantum phenomenon that has no counterpart in the classical world. We will see that both types of angular momentum, **L** and **S**, are derived from the same set of commutator relations. Hence, we restate Eqs. 3.216 more generally as

$$\left[\hat{J}_x, \hat{J}_y\right] = i\hbar\hat{J}_z, \qquad \left[\hat{J}_y, \hat{J}_z\right] = i\hbar\hat{J}_x, \qquad \left[\hat{J}_z, \hat{J}_x\right] = i\hbar\hat{J}_y, \tag{3.217}$$

where $\hat{J}_x$, $\hat{J}_y$, and $\hat{J}_z$ are the components of some general angular momentum vector **J**, which may represent orbital angular momentum, spin angular momentum, or a combination of the two.

An immediate consequence of these commutator relations is the existence of an uncertainty relation between any two angular momentum components. Using Eq. 3.77, we have

$$\triangle J_x \triangle J_y \geq \frac{\hbar}{2} \left|\langle J_z\rangle\right|, \qquad \triangle J_y \triangle J_z \geq \frac{\hbar}{2} \left|\langle J_x\rangle\right|, \qquad \triangle J_z \triangle J_x \geq \frac{\hbar}{2} \left|\langle J_y\rangle\right|. \tag{3.218}$$

This reflects the fact that no two angular momentum components have simultaneous eigenvectors. On the other hand, consider the operator $\hat{J}^2$ that corresponds to the square of the angular momentum:

$$\hat{J}^2 = \hat{J}_x^2 + \hat{J}_y^2 + \hat{J}_z^2. \tag{3.219}$$

We see that

$$\begin{aligned}
\left[\hat{J}_z, \hat{J}^2\right] &= \left[\hat{J}_z, \hat{J}_x^2 + \hat{J}_y^2 + \hat{J}_z^2\right] = \left[\hat{J}_z, \hat{J}_x^2\right] + \left[\hat{J}_z, \hat{J}_y^2\right] + 0 \\
&= \hat{J}_x \left[\hat{J}_z, \hat{J}_x\right] + \left[\hat{J}_z, \hat{J}_x\right] \hat{J}_x + \hat{J}_y \left[\hat{J}_z, \hat{J}_y\right] + \left[\hat{J}_z, \hat{J}_y\right] \hat{J}_y \\
&= i\hbar \left(\hat{J}_x \hat{J}_y + \hat{J}_y \hat{J}_x - \hat{J}_y \hat{J}_x - \hat{J}_x \hat{J}_y\right) \\
&= 0,
\end{aligned} \tag{3.220}$$

and similarly,

$$\left[\hat{J}_x, \hat{J}^2\right] = \left[\hat{J}_y, \hat{J}^2\right] = \left[\hat{J}_z, \hat{J}^2\right] = 0. \tag{3.221}$$

This shows that there are simultaneous eigenvectors for $\hat{J}^2$ and the operator for any of the individual angular momentum components. Suppose we consider the common eigenvectors for $\hat{J}^2$ and $\hat{J}_z$. Recalling that $\hbar$ has the dimensions of angular momentum, we choose to write the eigenvalues for $\hat{J}^2$ and $\hat{J}_z$ as $\alpha\hbar^2$ and $m\hbar$, respectively. The common eigenvectors can then be indexed as $|\alpha, m\rangle$ and the eigenvalue equations for the two operators become

$$\begin{aligned}
\hat{J}^2 |\alpha, m\rangle &= \alpha\hbar^2 |\alpha, m\rangle \\
\hat{J}_z |\alpha, m\rangle &= m\hbar |\alpha, m\rangle .
\end{aligned} \tag{3.222}$$

Because $\hat{J}^2$ and $\hat{J}_z$ are Hermitian, the eigenvectors are orthogonal (and postulated to be normalized):

$$\langle \alpha, m \mid \alpha', m' \rangle = \delta_{\alpha\alpha'} \delta_{mm'}. \tag{3.223}$$

The quantization procedure that follows will determine the permissible values of the dimensionless numbers α and m.

3.6.2 Quantization of Angular Momentum

We now introduce the non-Hermitian operators $\hat{J}_+$ and $\hat{J}_-$:

$$\hat{J}_+ = \hat{J}_x + i\hat{J}_y \qquad \hat{J}_- = \hat{J}_x - i\hat{J}_y. \tag{3.224}$$

Note that $\hat{J}_- = \hat{J}_+^\dagger$. It becomes straightforward to show the following commutator relations:

$$\left[\hat{J}^2, \hat{J}_\pm\right] = 0 \tag{3.225}$$

$$\left[\hat{J}_z, \hat{J}_\pm\right] = \pm\hbar\hat{J}_\pm. \tag{3.226}$$

From the definitions of $\hat{J}_+$ and $\hat{J}_-$ (Eq. 3.224), along with Eqs. 3.217 and 3.219, we also have

$$\hat{J}_+\hat{J}_- = \hat{J}^2 - \hat{J}_z^2 + \hbar\hat{J}_z \qquad \text{and} \qquad \hat{J}_-\hat{J}_+ = \hat{J}^2 - \hat{J}_z^2 - \hbar\hat{J}_z. \tag{3.227}$$

Now, from Eqs. 3.222, 3.225, and 3.226, observe the following:

$$\begin{aligned} \hat{J}^2\left(\hat{J}_\pm\left|\alpha, m\right\rangle\right) &= \hat{J}_\pm\hat{J}^2\left|\alpha, m\right\rangle = \alpha\hbar^2\left(\hat{J}_\pm\left|\alpha, m\right\rangle\right) \\ \hat{J}_z\left(\hat{J}_\pm\left|\alpha, m\right\rangle\right) &= \left(\hat{J}_\pm\hat{J}_z \pm \hbar\hat{J}_\pm\right)\left|\alpha, m\right\rangle = (m \pm 1)\hbar\left(\hat{J}_\pm\left|\alpha, m\right\rangle\right). \end{aligned} \tag{3.228}$$

Given that $|\alpha, m\rangle$ is an eigenket of $\hat{J}^2$ and $\hat{J}_z$, these two relations identify $\hat{J}_+|\alpha, m\rangle$ and $\hat{J}_-|\alpha, m\rangle$ as newly generated eigenkets of these same operators. For the operator $\hat{J}_z$, the eigenket $\hat{J}_+|\alpha, m\rangle$ has the eigenvalue $(m+1)\hbar$, while $\hat{J}_-|\alpha, m\rangle$ has the eigenvalue $(m-1)\hbar$. For the operator $\hat{J}^2$, both $\hat{J}_+|\alpha, m\rangle$ and $\hat{J}_-|\alpha, m\rangle$ have the eigenvalue $\alpha\hbar^2$, which is the same as the eigenvalue corresponding to $|\alpha, m\rangle$. These observations can be summarized by

$$\hat{J}_+\left|\alpha, m\right\rangle = b_+\left|\alpha, m+1\right\rangle \qquad \text{and} \qquad \hat{J}_-\left|\alpha, m\right\rangle = b_-\left|\alpha, m-1\right\rangle. \tag{3.229}$$

Each application of $\hat{J}_+$ causes the eigenvalue of $\hat{J}_z$ to increment by one unit of $\hbar$, as if climbing up a ladder with this spacing between rungs. $\hat{J}_-$, on the other hand, causes these eigenvalues to climb down the ladder. Hence, $\hat{J}_+$ and $\hat{J}_-$ are called *ladder operators*.

The constant b_+ in Eq. 3.229 is found from

$$\begin{aligned} |b_+|^2 &= \langle\alpha, m+1 \mid b_+^* b_+ \mid \alpha, m+1\rangle \\ &= \langle\alpha, m \mid \hat{J}_+^\dagger\hat{J}_+ \mid \alpha, m\rangle = \langle\alpha, m \mid \hat{J}_-\hat{J}_+ \mid \alpha, m\rangle \\ &= \langle\alpha, m \mid \left(\hat{J}^2 - \hat{J}_z^2 - \hbar\hat{J}_z\right) \mid \alpha, m\rangle. \end{aligned} \tag{3.230}$$

Evaluating the last expression gives

$$b_+ = \sqrt{\alpha - m(m+1)}\hbar. \tag{3.231}$$

Similarly, b_- is

$$b_- = \sqrt{\alpha - m(m-1)}\hbar. \tag{3.232}$$

Next, we realize that the value of m is bounded. Clearly, the z-component of the angular momentum vector $\mathbf{J}$ cannot exceed the magnitude of that vector, i.e., $m^2\hbar^2 \le \alpha\hbar^2$. This means that, for a fixed α, the value of m must lie between some m_{min} and m_{max}. To investigate these bounds, note that the ladder operator $\hat{J}_+$ cannot raise $| \ \alpha, m\rangle$ to a state any higher than $|\alpha, m_{\text{max}}\rangle$, while $\hat{J}_-$ cannot lower it below $|\alpha, m_{\text{min}}\rangle$. Thus, we have the following two relations:

$$\hat{J}_+ |\alpha, m_{\text{max}}\rangle = 0 \tag{3.233}$$

$$\hat{J}_- |\alpha, m_{\text{min}}\rangle = 0. \tag{3.234}$$

Operating on Eqs. 3.233 and 3.234 with $\hat{J}_-$ and $\hat{J}_+$, respectively, we have, with the aid of Eqs. 3.227,

$$\begin{aligned} \hat{J}_-\hat{J}_+ |\alpha, m_{\text{max}}\rangle &= \left(\hat{J}^2 - \hat{J}_z^2 - \hbar\hat{J}_z\right) |\alpha, m_{\text{max}}\rangle = 0 \\ \hat{J}_+\hat{J}_- |\alpha, m_{\text{min}}\rangle &= \left(\hat{J}^2 - \hat{J}_z^2 + \hbar\hat{J}_z\right) |\alpha, m_{\text{min}}\rangle = 0. \end{aligned} \tag{3.235}$$

Evaluating these produces two constraining equations, namely,

$$\alpha - m_{\text{max}}(m_{\text{max}} + 1) = 0 \qquad \text{and} \qquad \alpha - m_{\text{min}}(m_{\text{min}} - 1) = 0. \tag{3.236}$$

These can be solved for m_{min} in terms of m_{max}, producing the two solutions

$$m_{\text{min}} = -m_{\text{max}}, \ m_{\text{max}} + 1. \tag{3.237}$$

The second solution is obviously inconsistent with the definitions of m_{min} and m_{max}, so we only keep the first. Renaming m_{max} as the quantum number j, we therefore have

$$m_{\text{max}} = j \qquad \text{and} \qquad m_{\text{min}} = -j. \tag{3.238}$$

From either of Eqs. 3.236, we also have

$$\alpha = j(j+1). \tag{3.239}$$

The value of j is also restricted. This is because the state $|\alpha, m_{\text{max}}\rangle$ can be generated form the state $|\alpha, m_{\text{min}}\rangle$ by $2j$ successive applications of the operator $\hat{J}_+$. Therefore, $2j$ can only be 0, 1, 2, ..., etc., so j can only take on the values $j = 0, \ \frac{1}{2}, \ 1, \ \frac{3}{2}, \ 2,$ ..., etc.

Our findings can be summarized by relabeling the states $|\alpha, m\rangle$ as $|j, m\rangle$, and replacing Eqs. 3.222 with the equations below:

$$\begin{aligned} \hat{J}^2 |j, m\rangle &= j(j+1)\hbar^2 |j, m\rangle \\ \hat{J}_z |j, m\rangle &= m\hbar |j, m\rangle . \end{aligned} \tag{3.240}$$

The eigenvalues are limited to discrete values according to the restrictions

$$\begin{aligned} j &= 0,\ \tfrac{1}{2},\ 1,\ \tfrac{3}{2},\ 2,\ \ldots \\ m &= -j,\ -(j-1),\ \ldots,\ 0,\ \ldots,\ +(j-1),\ +j. \end{aligned} \tag{3.241}$$

j and m are called the *angular momentum quantum number* and *magnetic quantum number*, respectively. Associated with each value of j, are $(2j+1)$ values of m. Finally, Eqs. 3.229 can be rewritten as

$$\hat{J}_{\pm}\left|j,m\right\rangle = \sqrt{j(j+1) - m(m\pm 1)}\hbar\left|j, m\pm 1\right\rangle. \tag{3.242}$$

A semiclassical picture of the results can be represented through the so-called *vector model*. It provides one with a partial grasp of the angular momentum concepts in quantum mechanics. The permitted magnitudes and orientations of $\mathbf{J}$ are displayed on a diagram like the one in Fig. 3.4. For example, suppose $j = 2$. Then the corresponding classical angular momentum vector would have a magnitude of $J = \sqrt{2(2+1)}\hbar = \sqrt{6}\hbar$. The orientation of this vector with respect to the z-axis is restricted by the five allowable values of J_z, namely, $J_z = -2\hbar,\ -\hbar,\ 0,\ +\hbar,\ +2\hbar$. Associated with each of these orientations is a cone of uncertainty due to the absence of information about J_x and J_y. Although this model is somewhat useful for representing the angular momentum in a pictorial way, one should be careful not to take its meaning too literally.

3.6.3 Orbital Angular Momentum Eigenfunctions

Later in this section it will be shown that the integral values of j correspond to the case of orbital angular momentum, $\mathbf{J} = \mathbf{L}$, while the half-integral values of j represent states of spin angular momentum, $\mathbf{J} = \mathbf{S}$. Only the orbital angular momentum states have a corresponding spatial representation in the form of a wavefunction. Below, we illustrate a procedure for obtaining the position-space eigenfunctions of the orbital angular momentum operators $\hat{L}^2$ and $\hat{L}_z$. In this case, we replace j by the orbital angular momentum quantum number l and its associated allowed values:

$$l = 0,\ 1,\ 2,\ \ldots. \tag{3.243}$$

In order to generate the eigenfunctions, we first need to express the various orbital angular momentum operators in terms of the Schrödinger (position) representation. We can use Eqs. 3.153 and 3.155 to replace the angular momentum operators of Eq. 3.215 with the Cartesian coordinate representation:

$$\begin{aligned} \hat{L}_x &\to -i\hbar\left(y\frac{\partial}{\partial z} - z\frac{\partial}{\partial y}\right) \\ \hat{L}_y &\to -i\hbar\left(z\frac{\partial}{\partial x} - x\frac{\partial}{\partial z}\right) \\ \hat{L}_z &\to -i\hbar\left(x\frac{\partial}{\partial y} - y\frac{\partial}{\partial x}\right). \end{aligned} \tag{3.244}$$

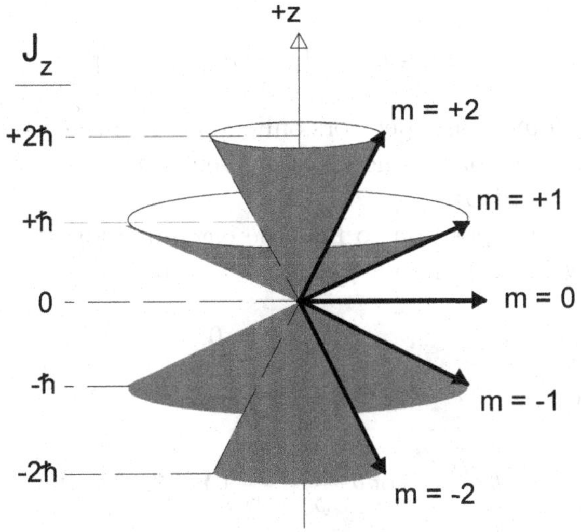

Figure 3.4 Vector model of angular momentum for $j = 2$.

We will find that the eigenfunctions are most conveniently expressed in spherical coordinates. To this end, we note that

$$\frac{\partial}{\partial x} = \frac{\partial \theta}{\partial x}\frac{\partial}{\partial \theta} + \frac{\partial \phi}{\partial x}\frac{\partial}{\partial \phi} + \frac{\partial r}{\partial x}\frac{\partial}{\partial r}, \tag{3.245}$$

and similarly for the other partial derivative operators. Then, using the coordinate transformation equations

$$r^2 = x^2 + y^2 + z^2 \qquad \cos\theta = \frac{z}{r} \qquad \tan\phi = \frac{y}{x}, \tag{3.246}$$

one can find the needed partial derivatives. For example,

$$\frac{\partial \theta}{\partial x} = \frac{\cos\phi\cos\theta}{r}, \qquad \frac{\partial \phi}{\partial x} = -\frac{y}{x^2}\cos^2\phi, \qquad \frac{\partial r}{\partial x} = \frac{x}{r}. \tag{3.247}$$

This procedure, along with some rather involved algebra, produces the desired spherical coordinate representations:

$$\begin{aligned} \hat{L}_x &\to i\hbar\left(\sin\phi\frac{\partial}{\partial\theta} + \cot\theta\cos\phi\frac{\partial}{\partial\phi}\right) \\ \hat{L}_y &\to i\hbar\left(-\cos\phi\frac{\partial}{\partial\theta} + \cot\theta\sin\phi\frac{\partial}{\partial\phi}\right) \\ \hat{L}_z &\to -i\hbar\frac{\partial}{\partial\phi}. \end{aligned} \tag{3.248}$$

These lead to

$$\hat{L}^2 \to -\hbar^2\left[\frac{1}{\sin\theta}\frac{\partial}{\partial\theta}\left(\sin\theta\frac{\partial}{\partial\theta}\right) + \frac{1}{\sin^2\theta}\frac{\partial^2}{\partial\phi^2}\right] \tag{3.249}$$

and

$$\hat{L}_{\pm} \to \hbar\, e^{\pm i\phi} \left(i \cot\theta \frac{\partial}{\partial\phi} \pm \frac{\partial}{\partial\theta} \right). \tag{3.250}$$

Since the angular momentum operators only depend on the coordinates θ and ϕ, and not r, the eigenfunction corresponding to a given state $|l, m\rangle$ can be written as some unknown function $Y_{l,m}(\theta, \phi)$.

We are now in a position to find the eigenfunction for the state $|l, l\rangle$. Start with Eq. 3.233 rewritten using our present notation:

$$\hat{L}_+ \, |l, l\rangle = 0. \tag{3.251}$$

In position space this is

$$\hbar\, e^{i\phi} \left(i \cot\theta \frac{\partial}{\partial\phi} + \frac{\partial}{\partial\theta} \right) Y_{l,l}(\theta, \phi) = 0, \tag{3.252}$$

which is a partial differential equation for the eigenfunction $Y_{l,l}(\theta, \phi)$. It can be solved by the standard technique of separation of variables, where one assumes a solution of the form

$$Y_{l,l}(\theta, \phi) = \Phi(\phi)\, \Theta(\theta). \tag{3.253}$$

Substituting into Eq. 3.252 produces

$$\frac{1}{\Phi}\frac{\partial\Phi}{\partial\phi} = i \tan\theta \frac{1}{\Theta}\frac{\partial\Theta}{\partial\theta}. \tag{3.254}$$

Since the left-hand side is purely a function of ϕ and the right-hand side is purely a function of θ, the two sides can be decoupled and set equal to the same constant, say α. The resulting equations are

$$\frac{\partial\Phi}{\partial\phi} = \alpha\Phi \qquad \text{and} \qquad \frac{\partial\Theta}{\partial\theta} = -i\alpha\,(\cot\theta)\,\Theta. \tag{3.255}$$

The Φ-equation has a solution of the form $e^{\alpha\phi}$. The constant α can be determined by substituting $Y_{l,l}(\theta, \phi)$ into the eigenvalue equation for $\hat{L}_z$ in the position representation, i.e.,

$$-i\hbar \frac{\partial}{\partial\phi} \left[e^{\alpha\phi}\Theta(\theta) \right] = l\hbar \left[e^{\alpha\phi}\Theta(\theta) \right]. \tag{3.256}$$

This can only be satisfied if $\alpha = il$, so

$$\Phi(\phi) \propto e^{il\phi}. \tag{3.257}$$

At this point, we can explain why the l's are forbidden from taking on half-integral values if they correspond to orbital angular momentum states. Since $Y_{l,l}(\theta, \phi)$ is to represent a spatial probability amplitude, it must be single valued at any location in space. In particular, we must have $\Phi(\phi) = \Phi(\phi + 2\pi)$. This condition is only met if l is an integer.

We now return to the Θ-equation. It can be integrated, and the solution is

$$\Theta(\theta) \propto \sin^l\theta. \tag{3.258}$$

Furthermore, the $Y_{l,m}(\theta, \phi)$'s must satisfy the normalization condition

$$\int_0^{2\pi} d\phi \int_0^{\pi} \sin\theta \, d\theta \, |Y_{l,m}(\theta, \phi)|^2 = 1. \tag{3.259}$$

Combining our results, the eigenfunction $Y_{l,l}(\theta, \phi)$ then has the following form:

$$Y_{l,l}(\theta, \phi) = \frac{(-1)^l}{2^l \, l!} \sqrt{\frac{(2l+1)!}{4\pi}} \, e^{il\phi} \sin^l\theta. \tag{3.260}$$

It now becomes straightforward to generate the position representation for the eigenstate $|l, l-1\rangle$ just by stepping down $|l, l\rangle$ with the $\hat{L}_-$ operator:

$$\hat{L}_- \, |l, l\rangle = \sqrt{l(l+1) - l(l-1)} \hbar \, |l, l-1\rangle \,. \tag{3.261}$$

In position space this equation becomes

$$\hbar e^{-i\phi} \left(i \cot\theta \frac{\partial}{\partial\phi} - \frac{\partial}{\partial\theta} \right) Y_{l,l}(\theta, \phi) = \sqrt{2l} \hbar Y_{l,l-1}(\theta, \phi). \tag{3.262}$$

Thus, $Y_{l,l-1}$ is generated from $Y_{l,l}$ simply by carrying out appropriate differentiations. The result is

$$Y_{l,l-1}(\theta, \phi) = -l \frac{(-1)^l}{2^l \, l!} \sqrt{\frac{2l+1}{4\pi}(2l-1)!} \, e^{i(l-1)\phi} \sin^{l-1}\theta \cos\theta. \tag{3.263}$$

Successive applications of the $\hat{L}_-$ operator show the other eigenfunctions to be

$$Y_{l,m}(\theta, \phi) = \sigma_- \frac{(-1)^l}{2^l \, l!} \sqrt{\frac{2l+1}{4\pi} \frac{(l+|m|)!}{(l-|m|)!}} \, e^{im\phi} \frac{1}{\sin^{|m|}\theta} \frac{d^{l-|m|}}{d(\cos\theta)^{l-|m|}} \sin^{2l}\theta, \tag{3.264}$$

where $\sigma_- = (-1)^m$ for $m < 0$ and $\sigma_- = 1$ otherwise. The $Y_{l,m}$'s are called *spherical harmonics*. They are more commonly expressed in the alternative form

$$Y_{l,m}(\theta, \phi) = \sigma_+ \sqrt{\frac{2l+1}{4\pi} \frac{(l+|m|)!}{(l-|m|)!}} \, P_l^m(\cos\theta) \, e^{im\phi}, \tag{3.265}$$

where $\sigma_+ = (-1)^m$ for $m > 0$ and $\sigma_+ = 1$ otherwise. The $P_l^m(\cos\theta)$'s are the *associated Legendre polynomials.* The first few spherical harmonics are listed in Table 3.1.

Table 3.1 The first few normalized spherical harmonics $Y_{l,m}(\theta,\phi)$.

l	m	$Y_{l,m}(\theta,\phi)$
0	0	$\sqrt{\frac{1}{4\pi}}$
1	1	$-\sqrt{\frac{3}{8\pi}}\sin\theta\, e^{i\phi}$
1	0	$\sqrt{\frac{3}{4\pi}}\cos\theta$
1	-1	$\sqrt{\frac{3}{8\pi}}\sin\theta\, e^{-i\phi}$
2	2	$\sqrt{\frac{15}{32\pi}}\sin^2\theta\, e^{2i\phi}$
2	1	$-\sqrt{\frac{15}{8\pi}}\cos\theta\sin\theta\, e^{i\phi}$
2	0	$\sqrt{\frac{5}{16\pi}}(3\cos^2\theta-1)$
2	-1	$\sqrt{\frac{15}{8\pi}}\cos\theta\sin\theta\, e^{-i\phi}$
2	-2	$\sqrt{\frac{15}{32\pi}}\sin^2\theta\, e^{-2i\phi}$

The probability densities $|Y_{l,m}(\theta,\phi)|^2$ for $l = 0$, 1, 2, and 3 are displayed in Fig. 3.5. One interprets $|Y_{l,m}(\theta,\phi)|^2\, d\Omega = |Y_{l,m}(\theta,\phi)|^2 \sin\theta\, d\theta\, d\phi$ as the probability of a position measurement producing an outcome within the solid angle $d\Omega$ about the angular position (θ,ϕ). Since the ϕ-dependence of the $Y_{l,m}$'s only occurs via the phase factor $e^{im\phi}$, the probability densities exhibit no azimuthal dependence. The $l = 0$ state exhibits spherical symmetry—it is called an s-state. The states $l = 1$, 2, and 3 go by the names p, d, and f states, respectively, and exhibit the distinct polar profiles shown in Fig. 3.5.

Example 3.5 The Rigid Rotator and Molecular Rotations. Consider two particles, each with mass M, attached by a massless, rigid rod of length R (see Fig. 3.6). The system is free to rotate about the midpoint of the rod, which is fixed in space. The energy states of this *rigid rotator* are found by solving the energy eigenvalue problem for the Hamiltonian

$$\hat{H} = \frac{\hat{L}^2}{2I}, \tag{3.266}$$

where $I = MR^2/2$ is the moment of inertia of the system. The solution is at once apparent since it has already been determined that the operator $\hat{L}^2$ has eigenvalues $l\,(l+1)\,\hbar^2$ and eigenstates $|l,m\rangle$. Immediately, one is able to write down the energy states for the rotator:

$$E_l = \frac{l\,(l+1)\,\hbar^2}{2I}. \tag{3.267}$$

The energy levels are indexed by the angular momentum quantum number l. Since there are $2l+1$ values of the magnetic quantum number m belonging to each l, it

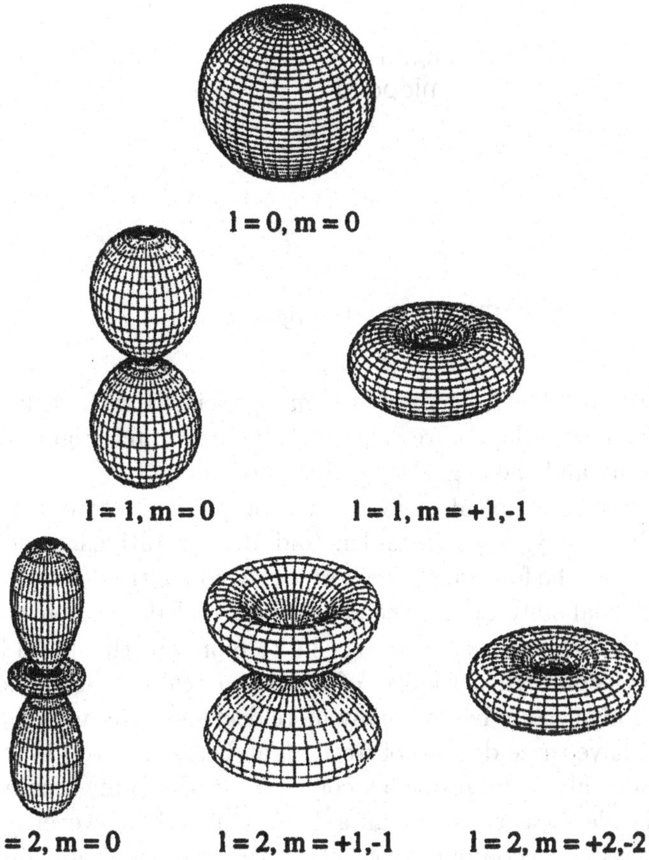

Figure 3.5 Probability densities $|Y_{l,m}(\theta, \phi)|^2$ for the $l = 0$, 1, and 2 spherical harmonics.

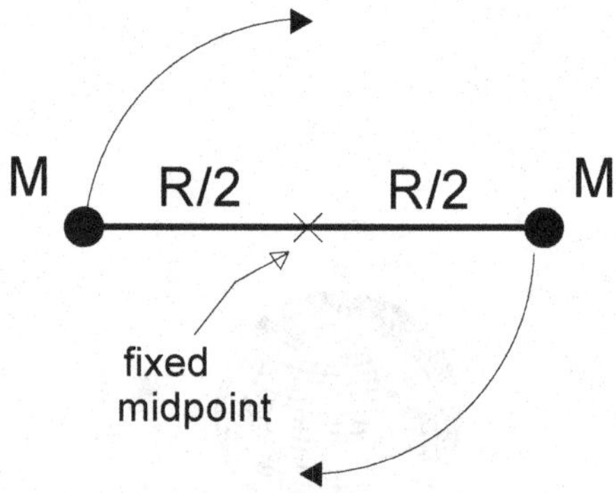

Figure 3.6 The rigid rotator.

follows that there are $2l+1$ eigenstates $|l,m\rangle$ associated with each allowed energy. One says that the energy levels are $(2l+1)$-*fold degenerate*. Figure 3.7 displays the energy-level spacing and the degeneracy of the rotator.

The quantum-mechanical characteristics of the rigid rotator can be carried over to explain, in a basic way, what we understand about rotations in molecules. To put things in perspective, the following is a short digression outlining how the interactions governing the internal state of a molecule can be handled quantum-mechanically:**

Consider the two masses M in Fig. 3.6 to represent the nuclei bound together in a diatomic molecule. In actuality, of course, a real diatomic molecule poses a complex, many-body system having many more degrees of freedom. Specifically, each of the two nuclei have three degrees of freedom and each electron in the molecule has the same. At face value, the problems connected with finding the energy states in even a light molecule appear insurmountable. Fortunately, however, a clever method exists that enables one to compute molecular energy states; it is known as the *Born-Oppenheimer approximation* [5]. Here, one takes advantage of the fact that nuclei are much more massive than electrons—thus, they move much more slowly than the electrons in a molecule. As a first approximation, therefore, the nuclei can be regarded as having a fixed separation (R) in space. For each such separation, one then solves the problem for the eigenvalues of the total electronic energy ε of the molecule, i.e., there is an ε vs. R curve associated with each possible electronic energy level. A central assertion of the Born-Oppenheimer approximation is that, for each electronic level, these same ε vs. R curves also play the role of the internuclear

**A superb general reference explaining molecular energy levels is Herzberg [4].

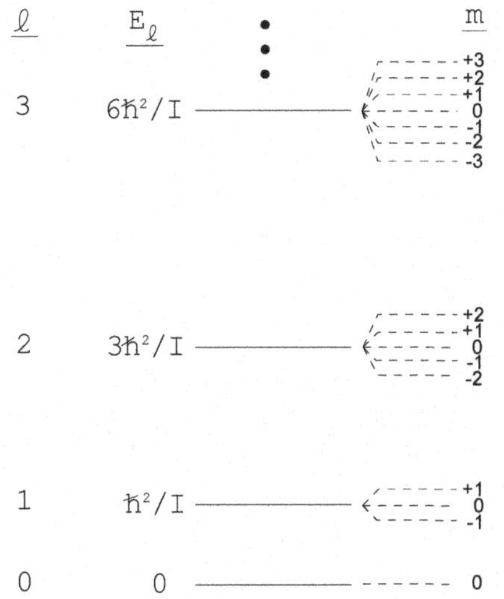

Figure 3.7 Energy-level diagram for rigid rotator.

potential-energy function, $V(R)$. Typically, each $V(R)$ curve has a rather sharp minimum, as illustrated in Fig. 3.8. As a consequence, the two nuclei have a highly-preferred separation at some R_0. The constant $V(R_0)$ corresponds to the binding energy of the molecule when it is in the electronic state under consideration. As far as the basic structure of the molecule is concerned, the nuclear motions may be neglected. However, for the purpose of analyzing molecular spectra, it is important to consider the nuclear degrees of freedom; these give rise to closely-spaced vibrational and rotational states that accompany each electronic level.

For a given electronic state, the characteristics of the associated vibrational energy levels is obtained by expanding the nuclear potential function in a Taylor series about the minimum at $R = R_0$:

$$V(r) = V(R_0) + \frac{1}{2}(R - R_0)^2 \left(\frac{\partial^2 V}{\partial R^2}\right)_{R=R_0} + \dots . \tag{3.268}$$

Superimposed on the electronic energy of $V(R_0)$ is a quadratic potential characteristic of a harmonic oscillator with restoring constant

$$M\omega^2 = \left(\frac{\partial^2 V}{\partial R^2}\right)_{R=R_0} . \tag{3.269}$$

Thus, the vibrational levels are equally spaced, with separation $\hbar\omega$. Let us now estimate how the spacing of the vibrational levels compares with that of the electronic levels. Suppose that the size of the molecule is on the order of a. Then, from the uncertainty principle, the electronic energy is on the order of $\varepsilon \simeq \hbar^2/2m_e a^2$,

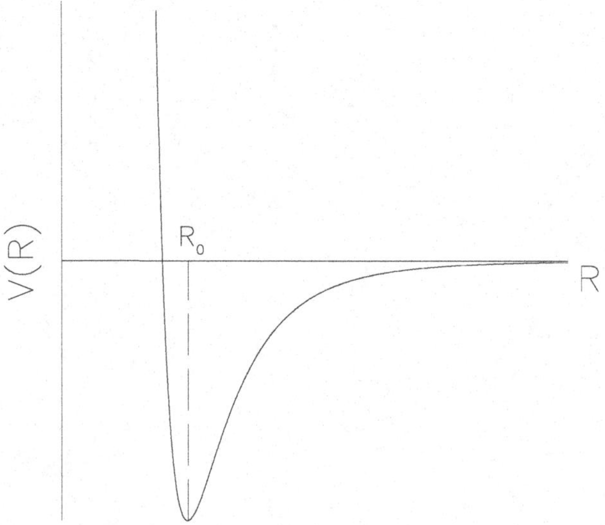

Figure 3.8 Internuclear potential function, $V(R)$.

where m_e is the electron mass. Meanwhile, a basic dimensional argument gives $(\partial^2 V/\partial R^2)_{R=R_0} \simeq \hbar^2/m_e a^4$. Substituting into Eq. 3.269 produces the estimated spacing of vibrational levels:

$$\hbar\omega \simeq \frac{\hbar^2}{m_e a^2}\sqrt{\frac{m_e}{M}} \simeq \varepsilon\sqrt{\frac{m_e}{M}}. \tag{3.270}$$

This shows that the spacing of the vibrational levels is on the order of one-hundred times smaller than that of the electronic levels. Furthermore, for low-lying vibrational states, the molecular vibration amplitude can be estimated from Eq. 3.103 for the position uncertainty associated with the harmonic oscillator states:

$$\triangle R \simeq \sqrt{\frac{\hbar}{2M\omega}} \simeq \left(\frac{m_e}{M}\right)^{1/4} a. \tag{3.271}$$

The vibrational motion is only on the order of one-tenth of the equilibrium separation. This shows that, in fact, the molecule is rather rigid. As a consequence, one is justified in treating the rotational motion of the molecule separately.

To a first approximation, the rotational energy levels essentially correspond to that of the rigid rotator previously described. The level spacings are on the order of

$$\hbar^2/I \simeq \hbar^2/Ma^2 \simeq \varepsilon\left(\frac{m_e}{M}\right) \simeq \hbar\omega\sqrt{\frac{m_e}{M}}. \tag{3.272}$$

This means that the typical spacing between rotational energy levels is on the order of one-hundredth of the vibrational level spacing.

3.6.4 Stationary States for Particle in a Central Potential

In this section, we illustrate how to solve the energy eigenvalue problem for the stationary states when a particle moves under the influence of a spherically symmetric, or central, potential. In Section 3.4, we showed that in the position representation this is equivalent to finding the solutions of the time-independent Schrödinger equation. For a central potential it is most natural to work in spherical coordinates since $V(\mathbf{r}) = V(r)$, so we write the Schrödinger equation as

$$\left[-\frac{\hbar^2}{2\mu}\nabla^2 + V(r)\right]\psi_n(r,\theta,\phi) = E_n\psi_n(r,\theta,\phi), \tag{3.273}$$

where μ denotes the mass of the particle (to avoid confusion with the quantum number m). In spherical coordinates the Laplacian operator takes the form

$$\nabla^2 = \frac{1}{r^2}\frac{\partial}{\partial r}\left(r^2\frac{\partial}{\partial r}\right) + \frac{1}{r^2 \sin\theta}\frac{\partial}{\partial\theta}\left(\sin\theta\frac{\partial}{\partial\theta}\right) + \frac{1}{r^2 \sin^2\theta}\frac{\partial^2}{\partial\phi^2}. \tag{3.274}$$

The stationary states representing the solutions to Eq. 3.273 may turn out to form a discrete or continuous set, depending on the particular $V(r)$ and the energy range in question.

It turns out that the central potential problem can be solved using the separation of variables technique. One assumes the solution is a product of a radial wavefunction $R(r)$ and an angular wavefunction $G(\theta, \phi)$:

$$\psi_n(r,\theta,\phi) = R(r)\,G(\theta,\phi). \tag{3.275}$$

Substituting into Eq. 3.273 yields

$$\frac{\hbar^2}{R}\frac{\partial}{\partial r}\left(r^2\frac{\partial R}{\partial r}\right) + 2\mu r^2\left[E_n - V(r)\right] = -\frac{\hbar^2}{G}\left[\frac{1}{\sin\theta}\frac{\partial}{\partial\theta}\left(\sin\theta\frac{\partial G}{\partial\theta}\right) + \frac{1}{\sin^2\theta}\frac{\partial^2 G}{\partial\phi^2}\right]. \tag{3.276}$$

The only way the pure function of r on the left-hand side can be equal to the function of angular coordinates on the right is for each function to be equal to the same constant, which we shall call γ. The resulting angular equation is then

$$\left\{-\hbar^2\left[\frac{1}{\sin\theta}\frac{\partial}{\partial\theta}\left(\sin\theta\frac{\partial}{\partial\theta}\right) + \frac{1}{\sin^2\theta}\frac{\partial^2}{\partial\phi^2}\right]\right\} G(\theta,\phi) = \gamma\, G(\theta,\phi). \tag{3.277}$$

But now note the following: The expression in the curly brackets is just the operator $\hat{L}^2$ represented in spherical coordinates, i.e., Eq. 3.277 is nothing more than the position-space representation of the $\hat{L}^2$ eigenvalue problem! The solutions can immediately be identified as the spherical harmonics

$$G(\theta,\phi) \equiv G_{l,m}(\theta,\phi) = Y_{l,m}(\theta,\phi) \tag{3.278}$$

with the corresponding eigenvalues

$$\gamma = l(l+1)\hbar^2. \tag{3.279}$$

For a fixed value of l (again, $l = 0,\ 1,\ 2,\ ...$), one can now show that the equation for the radial wavefunction becomes

$$\left[-\frac{\hbar^2}{2\mu}\frac{d^2}{dr^2} + V(r) + \frac{l(l+1)\hbar^2}{2\mu r^2}\right] u_{n,l}(r) = E_{n,l}\, u_{n.l}(r), \tag{3.280}$$

where we define

$$u_{n,l}(r) = r\, R_{n,l}(r). \tag{3.281}$$

This is called the *radial wave equation*. For each orbital angular momentum quantum number l, one solves the radial equation for the energy eigenvalues $E_{n,l}$ and the associated radial wavefunction $R_{n,l}(r)$. Thus, the stationary-state eigenfunctions are best indexed as

$$\psi_{n,l,m}(r,\theta,\phi) = R_{n,l}(r)\, Y_{l,m}(\theta,\phi). \tag{3.282}$$

For a central potential, the stationary states are both states of definite energy and definite angular momentum. That is to say, the $\psi_{n,l,m}$'s (or more generally, the $|n,l,m\rangle$'s) are simultaneous eigenstates of the operators $\hat{H}$, $\hat{L}^2$, and $\hat{L}_z$. At the most basic level, this is because any pair of operators from among these three commute, i.e.,

$$\left[\hat{L}^2, \hat{L}_z\right] = \left[\hat{H}, \hat{L}^2\right] = \left[\hat{H}, \hat{L}_z\right] = 0. \tag{3.283}$$

Equation 3.283 follows because the corresponding classical Poisson brackets vanish, as shown in Chapter 2 (see Eqs. 2.151 and 2.152). Since the angular momentum operators commute with the Hamiltonian, they correspond to constants of the motion.

Now examine the radial equation, Eq. 3.280, and compare it to the one-dimensional time-independent Schrödinger equation, Eq. 3.162. The two forms are identical except for the following observations:

- The radial equation is not an equation for the radial wavefunction $R_{n,l}(r)$. Rather, one solves for the function $u_{n,l}(r) = r\, R_{n,l}(r)$.

- By definition, the coordinate r in the radial equation cannot take on negative values. Furthermore, because $R_{n,l}(r)$ must be finite for $r \geq 0$, the function $u_{n,l}(r)$ must vanish at the origin, i.e., $u_{n,l}(0) = 0$. These two statements are equivalent to imposing boundary conditions at the origin that are equivalent to placing an impenetrable wall ($V \to \infty$) there.

- In the radial equation, the actual potential $V(r)$ is replaced by an effective potential

$$V_{eff}(r) = V(r) + \frac{l(l+1)\hbar^2}{2\mu r^2}. \tag{3.284}$$

This is completely analogous to the effective potential observed for a classical particle in a central force field (see Section 2.2). The term $l(l+1)\hbar^2/2\mu r^2$ acts as a centrifugal potential barrier, where $l(l+1)\hbar^2$ plays the role of L^2, the square of the classical angular momentum.

Finding the stationary states for a particle in a central potential is reduced to solving the radial equation corresponding to each value of the orbital angular momentum quantum number l.

Example 3.6 Isotropic Harmonic Oscillator. Consider a particle of mass μ moving in the potential

$$V(r) = \frac{1}{2}kr^2 = \frac{1}{2}\mu\omega^2 r^2. \tag{3.285}$$

From Eq. 2.7, the force field is

$$\mathbf{F}(\mathbf{r}) = -\left(\frac{\partial V}{\partial r}\right)\mathbf{e}_r = -(kr)\mathbf{e}_r = -(\mu\omega^2 r)\mathbf{e}_r, \tag{3.286}$$

i.e., the particle is attracted radially toward the origin with a linear restoring force. This represents a three-dimensional isotropic harmonic oscillator having a classical frequency of $\omega = \sqrt{k/\mu}$.

Since this is a central potential problem, the stationary-state wavefunctions are given by Eq. 3.282. It is only necessary to solve the radial equation (Eq. 3.280) to find the energy spectrum and associated eigenfunctions. For a given value of l, the radial equation for the oscillator is

$$-\frac{\hbar^2}{2\mu}\frac{d^2u}{dr^2} + \left[\frac{1}{2}\mu\omega^2 r^2 + \frac{l(l+1)\hbar^2}{2\mu r^2}\right]u = E\,u, \tag{3.287}$$

where, for the time being, we drop the n, l subscripts of $u(r)$. Before proceeding with a solution, cast the equation into a dimensionless form by defining the variables

$$\rho = \beta^2 r^2 \qquad \text{and} \qquad \epsilon = \frac{E}{\hbar\omega}, \tag{3.288}$$

where $\beta = \sqrt{\mu\omega/\hbar}$. Equation 3.287 now takes the form

$$\frac{d^2u}{d\rho^2} + \frac{1}{2\rho}\frac{du}{d\rho} + \left[-\frac{1}{4} + \frac{\epsilon}{2\rho} - \frac{l(l+1)}{4\rho^2}\right]u = 0. \tag{3.289}$$

The $du/d\rho$-term can be eliminated by making the substitution

$$u(\rho) = \rho^{-1/4}w(\rho). \tag{3.290}$$

This leads to

$$\frac{d^2w}{d\rho^2} + \left[-\frac{1}{4} + \frac{\epsilon}{2\rho} - \left(\frac{l}{2} - \frac{1}{4}\right)\left(\frac{l}{2} + \frac{3}{4}\right)\frac{1}{\rho^2}\right]w = 0. \tag{3.291}$$

The next step is to examine the asymptotic behavior of $w(\rho)$, and hence $u(\rho)$, at both small and large values of ρ. For small ρ, one retains the last term in the brackets of Eq. 3.291; for large ρ, only the first term is retained. In these limits, the solutions are

$$\underset{(\text{small } \rho)}{w(\rho) \to \rho^{\frac{l}{2}+\frac{3}{4}},\ \rho^{-\frac{l}{2}+\frac{1}{4}}} \qquad \text{and} \qquad \underset{(\text{large } \rho)}{w(\rho) \to e^{\pm\rho/2},} \tag{3.292}$$

or, using Eq. 3.290,

$$\underset{(\text{small } \rho)}{u(\rho) \to \rho^{\frac{l}{2}+\frac{1}{2}},\ \rho^{-\frac{l}{2}}} \qquad \text{and} \qquad \underset{(\text{large } \rho)}{u(\rho) \to e^{\pm\rho/2}.} \tag{3.293}$$

At small ρ, the solution $\rho^{-l/2}$ is unacceptable since it does not vanish at $\rho = 0$, and hence at $r = 0$; at large ρ, the solution $e^{+\rho/2}$ is unacceptable because it blows up as $\rho \to \infty$. Thus, it is appropriate to write the general solution for all ρ as

$$u(\rho) = \rho^{(l+1)/2}\, e^{-\rho/2}\, v(\rho). \tag{3.294}$$

This form brings out the limiting behavior of $u(\rho)$ explicitly. The unknown function $v(\rho)$ must vary slowly compared to $e^{-\rho/2}$ at large ρ, and be finite at $\rho = 0$.

Substituting Eq. 3.294 into Eq. 3.289 yields a differential equation for $v(\rho)$:

$$\frac{d^2v}{d\rho^2} + \left(\frac{l+\frac{3}{2}}{\rho} - 1\right)\frac{dv}{d\rho} + \left[\frac{\epsilon - \left(l+\frac{3}{2}\right)}{2\rho}\right] v = 0. \tag{3.295}$$

This can be solved exactly by assuming a power-series solution

$$v(\rho) = \sum_{i=0}^{\infty} a_i \rho^i. \tag{3.296}$$

The result is the following recursion relation for the coefficients:

$$a_{i+1} = \frac{i - \frac{1}{2}\left[\epsilon - \left(l+\frac{3}{2}\right)\right]}{(i+1)\left(i+l+\frac{3}{2}\right)}\, a_i. \tag{3.297}$$

Now examine the ratio a_{i+1}/a_i for large i in order to determine the behavior of $v(\rho)$ for large ρ. We see that a_{i+1}/a_i approaches the value $1/i$, which is identical to the behavior exhibited by the power series for the function $e^{+\rho}$. Therefore, from Eq. 3.294, we have $u(\rho) \sim e^{+\rho/2}$ for large ρ, which was the behavior we previously found to be unacceptable! The only way to avoid this situation is to demand that the power series for $v(\rho)$ terminate and be a polynomial of some finite order, N. According to Eq. 3.297, this will occur if

$$N - \frac{1}{2}\left[\epsilon - \left(l + \frac{3}{2}\right)\right] = 0 \tag{3.298}$$

for $N = 0, 1, 2, ...$, etc. Solving for ϵ, we find that the sought after energy eigenvalues are quantized:

$$E = \epsilon\hbar\omega = \left(2N + l + \frac{3}{2}\right)\hbar\omega. \tag{3.299}$$

Finally, we define the *principal quantum number* n:

$$n = 2N + l. \tag{3.300}$$

Since N and l are restricted to the values 0, 1, 2, ..., then so is the value of n. The energy spectrum can then be written as

$$E_n = \left(n + \frac{3}{2}\right)\hbar\omega. \tag{3.301}$$

The levels are equally spaced by an amount $\hbar\omega$, just as in the case of the one-dimensional oscillator—only the zero-point energy of $3\hbar\omega/2$ is different. Note, however, that for each n there is, in general, more than one corresponding value of l. This means that the three-dimensional oscillator's energy levels are degenerate, i.e., for each energy E_n there are multiple eigenstates. From Eq. 3.300, the value of l is restricted to

$$l \leq n \qquad \text{and} \qquad \begin{cases} l \text{ is even, if } n \text{ is even} \\ l \text{ is odd, if } n \text{ is odd.} \end{cases} \tag{3.302}$$

From our previous developments, it then follows that the radial wavefunctions can be written in the form

$$R_{n,l}(r) \propto r^l\, e^{-\beta^2 r^2/2}\, L^{l+\frac{1}{2}}_{\frac{n-l}{2}}\left(\beta^2 r^2\right), \tag{3.303}$$

where

$$L^{l+\frac{1}{2}}_{\frac{n-l}{2}}\left(\beta^2 r^2\right) = \sum_{i=0}^{\frac{n-l}{2}} a_i \left(\beta^2 r^2\right)^i \qquad \text{with} \qquad a_{i+1} = \frac{i - \frac{n-l}{2}}{(i+1)\left(i + l + \frac{3}{2}\right)}\, a_i. \tag{3.304}$$

The $L^{l+\frac{1}{2}}_{\frac{n-l}{2}}(x)$ are the *associated Laguerre polynomials* of half-integer order. A few of the $R_{n,l}$'s and $r^2\left|R_{n,l}\right|^2$'s are sketched in Fig. 3.9. One interprets $r^2\left|R_{n,l}(r)\right|^2 dr$ as the probability that the particle is located within a spherical shell between r and $r + dr$. Thus, the radial wavefunctions are subject to the normalization condition

$$\int_0^\infty dr\, \left|R_{n,l}(r)\right|^2 r^2 = 1. \tag{3.305}$$

Let us now examine the degree of degeneracy of each energy level, i.e., find $D(n)$, the number of eigenstates corresponding to each principal quantum number n. Since each value of l has $(2l+1)$ values of m associated with it, we have

$$D(n) = \begin{cases} \sum\limits_{l=0,2,4,\ldots,n} (2l+1), & \text{for even } n \\ \sum\limits_{l=1,3,5,\ldots,n} (2l+1), & \text{for odd } n. \end{cases} \tag{3.306}$$

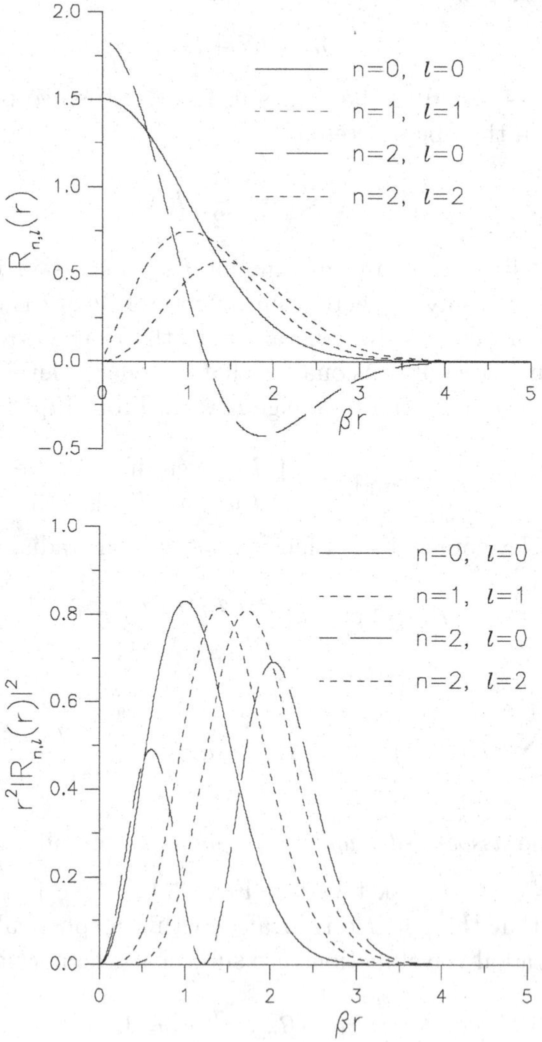

Figure 3.9 Radial wavefunctions and probability densities for the 3-D isotropic oscillator.

The number of states can be calculated as

$$D(n) = [\text{number of terms in sum}] \times [\text{mean value of the terms}] . \tag{3.307}$$

$$\begin{array}{ccc} & \text{n even} \swarrow \quad \searrow \text{n odd} & \\ \left[\frac{n-0}{2}+1\right]\left[2\left(\frac{n}{2}\right)+1\right] & & \left[\frac{n-1}{2}+1\right]\left[2\left(\frac{n-1}{2}\right)+1\right] \\ \searrow & & \swarrow \end{array}$$

$$D(n) = \frac{(n+1)(n+2)}{2}, \tag{3.308}$$

i.e., the result is the same for both even and odd n.

An interesting property of the three-dimensional oscillator is that the time-independent Schrödinger equation, Eq. 3.164, is not only separable in the spherical (r, θ, ϕ) coordinate system, but also in Cartesian (x, y, z) coordinates. For the latter case, it is easy to show that the energy eigenvalues take the form

$$E = \left(n_x + n_y + n_y + \frac{3}{2}\right)\hbar\omega \qquad (n_x, n_y, n_z = 0, 1, 2, ...) \tag{3.309}$$

with corresponding eigenfunctions

$$\varphi_{n_x,n_y,n_z}(x,y,z) \propto e^{-\beta^2(x^2+y^2+z^2)/2} H_{n_x}(\beta x)\, H_{n_y}(\beta y)\, H_{n_z}(\beta z). \tag{3.310}$$

As in the case of spherical coordinates, we define a principal quantum number, however, here the definition is

$$n = n_x + n_y + n_z. \tag{3.311}$$

However, since n_x, n_y, and n_z are all 0, 1, 2, ..., etc., so is n, and just as in the case of spherical coordinates, the spectrum is represented by Eq. 3.301. What about the degeneracy of the levels? Table 3.2 shows the degeneracy of the first few energy levels for both the spherical and Cartesian separations. In both cases, the degeneracy is the same, and is given by Eq. 3.308. The eigenfunctions representing the individual states are quite different, however, for the two solutions. The reason is that, for a given E_n, each of the two sets of wavefunctions $\varphi_{n_x,n_y,n_z}(x,y,z)$ and $\psi_{n,l,m}(r,\theta,\phi)$ represents a different set of basis states in the same subspace (i.e., they are connected by a unitary transformation). A wavefunction in one set is just a linear combination of wavefunctions in the other. For example, consider the normalized wavefunctions for the three-fold degenerate $n = 1$ level:

$$\begin{array}{ll} \underline{(x,y,z)\text{ - set}} & \underline{(r,\theta,\phi)\text{ - set}} \\ \varphi_{1,0,0} = \frac{\sqrt{2}}{\pi^{3/4}}\beta^{5/2}\, x\, e^{-\beta^2(x^2+y^2+z^2)/2} & \psi_{1,1,1} = \frac{\beta^{5/2}}{\pi^{3/4}}\sin\theta\; e^{i\phi}\, r\, e^{-\beta^2 r^2/2} \\ \varphi_{0,1,0} = \frac{\sqrt{2}}{\pi^{3/4}}\beta^{5/2}\, y\, e^{-\beta^2(x^2+y^2+z^2)/2} & \psi_{1,1,0} = \frac{\sqrt{2}\beta^{5/2}}{\pi^{3/4}}\cos\theta\; r\, e^{-\beta^2 r^2/2} \\ \varphi_{0,0,1} = \frac{\sqrt{2}}{\pi^{3/4}}\beta^{5/2}\, z\, e^{-\beta^2(x^2+y^2+z^2)/2} & \psi_{1,1,-1} = \frac{\beta^{5/2}}{\pi^{3/4}}\sin\theta\; e^{i\phi}\, r\, e^{-\beta^2 r^2/2}. \end{array} \tag{3.312}$$

Table 3.2 Determination of degeneracy D of first few energy levels of 3-D oscillator using the sets of quantum numbers (n_x, n_y, n_z) and (N, l, m) for the Cartesian and spherical coordinate separations, respectively.

n	n_x	n_y	n_z	N	l	m	$D(n)$
0	0	0	0	0	0	0	1
	1	0	0	0	1	-1	
1	0	1	0	0	1	0	3
	0	0	1	0	1	1	
	2	0	0	0	2	-2	
	0	2	0	0	2	-1	
2	0	0	2	0	2	0	6
	1	1	0	0	2	1	
	1	0	1	0	2	2	
	0	1	1	1	0	0	

Thus, we see that the $\psi_{n,l,m}(r, \theta, \phi)$'s are given by the following linear combinations of the $\varphi_{n_x,n_y,n_z}(x, y, z)$'s:

$$\begin{aligned} \psi_{1,1,1} &= \frac{1}{\sqrt{2}}\left(\varphi_{1,0,0} + i\varphi_{0,1,0}\right) \\ \psi_{1,1,0} &= \varphi_{0,0,1} \\ \psi_{1,1,-1} &= \frac{1}{\sqrt{2}}\left(\varphi_{1,0,0} - i\varphi_{0,1,0}\right). \end{aligned} \tag{3.313}$$

The three-dimensional harmonic oscillator provides a starting point for understanding the basic structure of atomic nuclei. For example, it is known that nuclei are especially stable when the number of protons (Z) or the number of neutrons (N) is equal to one of the following so-called *magic numbers*: 2, 8, 20, 28, 50, 82, or 126. In the *shell model* of the nucleus, it is assumed that the motion of any one nucleon is essentially unaffected by the motion of any other nucleon, i.e., one considers each proton and neutron to be moving independently within a common potential. By choosing the potential to be that of the isotropic oscillator, one observes that the first three magic numbers are a direct consequence of the degeneracy of the energy levels. For a given type of nucleon (i.e., proton or neutron), the *Pauli exclusion principle* allows one spin-up and one spin-down particle to occupy each quantum state. This fact, along with Eq. 3.308 for the degeneracy of the energy levels, predicts the occupancies of the $n = 0$, 1, and 2 levels to be 2, 6, and 12, respectively. As a result, the first three levels are successively filled when Z or N reaches the observed magic numbers of 2, 8, and 20. This, however, is where the agreement stops. The occupancy of the $n = 3$ level predicts that the next filling occurs at a Z or N of 40, which does not agree with the observed magic number of 28. In order to correctly predict the entire list of magic numbers, it becomes necessary to incorporate the important coupling between the intrinsic spin angular momentum and orbital angular momentum of each nucleon. This has the effect of splitting the energy levels, with the result of producing the exact values of the observed magic numbers.[††]

Suggested References

Two standard first-year graduate texts on quantum mechanics are

[a] E. Merzbacher, *Quantum Mechanics*, 3rd ed. (John Wiley and Sons, New York, 1997).

[b] L. I. Schiff, *Quantum Mechanics*, 3rd ed. (McGraw-Hill, New York, 1968).

Here are some excellent books written at the advanced undergraduate level:

[c] C. Cohen-Tannoudji, B. Diu, and F. Laloe, *Quantum Mechanics* (John Wiley and Sons, France, 1977).

[d] S. Gasiorowicz, *Quantum Physics*, 3rd ed. (John Wiley and Sons, New York, 2003).

[e] R. L. Liboff, *Introductory Quantum Mechanics*, 4th ed. (Addison-Wesley, Reading, Mass., 2002).

[f] D. J. Griffiths, *Introduction to Quantum Mechanics*, 2nd ed. (Pearson Prentice Hall, Upper Saddle River, NJ, 2005).

[g] J. S. Townsend, *A Modern Approach to Quantum Mechanics*, 2nd ed. (University Science Books, Sausalito, CA, 2000).

Finally, those seriously interested in quantum mechanics should take the time to read Dirac's book:

[h] P. A. M. Dirac, *The Principles of Quantum Mechanics*, 4th ed. (Oxford University Press, Oxford, 1982).

Problems

1. (a) Work out the commutator relation between the kinetic energy and the potential energy operators for the one-dimensional harmonic oscillator. Is there an uncertainty relation between these energies?

   (b) If the oscillator is in state $|n\rangle$, what is the expectation for each of these energies?

[††]For an energy-level diagram of the isotropic oscillator with spin-orbit coupling see, for example, Goswami [6].

2. The Hamiltonian for a one-dimensional harmonic oscillator driven by an external electric field E is given by

$$\hat{H} = \frac{1}{2m}\hat{p}^2 + \frac{1}{2}m\omega^2\hat{q}^2 - eE\hat{q}.$$

   (a) Use the equation of motion in the Heisenberg picture to solve for the time dependence of the position operator $\hat{q}(t)$. Show that

$$[\hat{q}(t_1), \hat{q}(t_2)] = \frac{i\hbar}{m\omega}\sin[\omega(t_2 - t_1)].$$

   This means that an operator that commutes at the same time need not commute at different times.

   (b) Similarly, derive the two-time commutator for the momentum.

3. (a) Starting from the fundamental commutation relation between $\hat{q}$ and $\hat{p}$, derive the operator relation

$$[\hat{p}, F(\hat{q})] = -i\hbar\frac{\partial F(\hat{q})}{\partial\hat{q}}.$$

   (b) Let $F(\hat{q}) = e^{i\xi\hat{q}/\hbar}$, where ξ is a constant. Then, from the above commutator relation, show that $e^{i\xi\hat{q}/\hbar}$ is a *translational operator* in momentum space in the sense that

$$e^{i\xi\hat{q}/\hbar}|p\rangle = |p + \xi\rangle.$$

   (c) With $|\phi\rangle$ being an arbitrary state vector, use the latter result to show that

$$\langle p \mid \hat{q} \mid \phi\rangle = i\hbar\frac{\partial}{\partial p}\langle p \mid \phi\rangle.$$

   (d) In particular, show that the momentum-space wavefunction $\phi_q(p) \equiv \langle p \mid q\rangle$ satisfies the differential equation

$$\frac{d\phi_q(p)}{dp} = -\frac{i}{\hbar}q\phi_q(p).$$

   (e) $\phi_q(p)$ is often normalized according to

$$\int_{-\infty}^{+\infty}\phi_{q'}^*(p)\,\phi_q(p)\,dp = \delta(q' - q).$$

   In this case, show that the normalized momentum-space wavefunction takes the form

$$\phi_q(p) \equiv \langle p \mid q\rangle = \frac{1}{\sqrt{2\pi\hbar}}e^{-iqp/\hbar}.$$

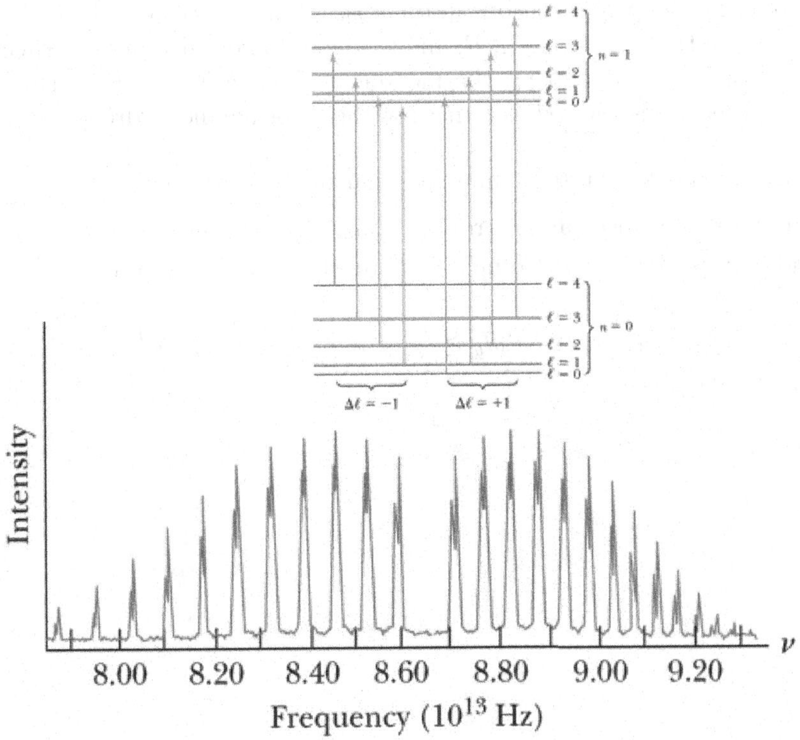

Figure 3.10 Transitions from the ground ($n = 0$) to the first excited ($n = 1$) vibrational state of the HCl molecule, along with the associated infrared absorption spectrum, for Problem 4. The peaks on the left side of the spectrum correspond to a change in the rotational angular momentum quantum number of $\Delta\ell = -1$; those on the right side correspond to $\Delta\ell = +1$. The fine splitting seen in each peak occurs because of the presence of the two chlorine isotopes: Cl^{35} (75.5%) and Cl^{37} (24.5%).

4. Transitions between molecular vibrational and rotational states of the HCl molecule and the associated IR absorption spectrum are shown in Fig. 3.10. The *selection rules* for allowed transitions (for example, see Problem 1 at the end of Chapter 8) demand that changes in the vibrational and rotational quantum numbers are restricted by $\Delta n = \pm 1$ and $\Delta l = \pm 1$.

   (a) For the case of $l \to l - 1$ transitions between the ground ($n = 0$) and the first excited ($n = 1$) vibrational state of the HCl molecule (i.e., the peaks on the left side of the absorption spectrum), what are the possible energies of the absorbed photon (express in terms of I, the moment of inertia of the molecule, the vibrational frequency, ω, and quantum numbers).

(b) From the peak positions in the absorption spectrum, calculate the equilibrium HCl bond length, R, and the bond force constant (i.e., effective spring constant). Note that the moment of inertia is determined by $I = mR^2$, where m is the reduced mass of the chlorine and hydrogen atoms.

5. If a system is known to be in an angular momentum eigenstate $|j, m\rangle$, calculate explicitly the uncertainty product $\triangle J_x \triangle J_y$ (i.e., use $\triangle J_x = \sqrt{\langle J_x^2 \rangle - \langle J_x \rangle^2}$, and likewise for $\triangle J_y$). Compare the result to the general inequality

$$\triangle J_x \triangle J_y \geq \frac{1}{2} \left| \langle j, m \mid \left[\hat{J}_x, \hat{J}_y \right] \mid j, m \rangle \right| .$$

Chapter 4
CLASSICAL TREATMENT OF ELECTROMAGNETIC FIELDS AND RADIATION

Maxwell's equations are the cornerstone for a complete description of the classical theory of electromagnetic fields and their interaction with matter. The field can either be described by electric and magnetic field vectors, or equivalently, in terms of electromagnetic potential functions. We demonstrate how a system composed of an electromagnetic field interacting with charged particles is cast into the general Hamiltonian framework developed in Chapter 2. Through the use of Maxwell's equations in a material medium, it will be learned that electromagnetic radiation is generated whenever a medium exhibits an oscillating polarization or dipole moment. This phenomenon, in turn, is closely related to the mechanism at work when light is scattered from a dielectric. The problem of light scattering from particles is then addressed in some detail.

4.1 Electromagnetic Field Equations and Conservation Laws

In general, to describe an electromagnetic (EM) field one needs to specify the electric field vector $\mathbf{E}$ and the magnetic field vector $\mathbf{B}$ as a function of position $\mathbf{r}$ and time t. The sources of the EM field are charges (e_i) and their associated currents. Charge density $\rho(\mathbf{r},t)$ and current density $\mathbf{j}(\mathbf{r},t)$ are defined as follows:

$$\rho(\mathbf{r},t) = \sum_i e_i \delta(\mathbf{r} - \mathbf{r}_i) \tag{4.1}$$

$$\mathbf{j}(\mathbf{r},t) = \sum_i e_i \mathbf{v}_i \delta(\mathbf{r} - \mathbf{r}_i). \tag{4.2}$$

Here, $\mathbf{r}_i$ and $\mathbf{v}_i$ represent the position and velocity, respectively, of charge e_i. For a given charge-density and current-density distribution, the resulting field is determined by a set of field equations, namely, *Maxwell's equations*. In vacuum they can be written as*

$$\nabla \cdot \mathbf{B} = 0 \tag{4.3}$$

*The equations given are in Gaussian units. For example, the electric field has units of "statvolts/cm" and the magnetic field has units of "gauss."

$$\nabla \times \mathbf{E} + \frac{1}{c}\frac{\partial \mathbf{B}}{\partial t} = 0 \tag{4.4}$$

$$\nabla \cdot \mathbf{E} = 4\pi\rho \tag{4.5}$$

$$\nabla \times \mathbf{B} - \frac{1}{c}\frac{\partial \mathbf{E}}{\partial t} = \frac{4\pi}{c}\mathbf{j}. \tag{4.6}$$

c is the speed of light in vacuum. The first two equations involve no charge or current sources. Equation 4.3 asserts the non-existence of free magnetic monopoles and Eq. 4.4 is *Faraday's law of induction*. Equations 4.5 and 4.6, which do involve sources, are *Gauss's law* and the extended version of *Ampere's law*, respectively. The term $\frac{1}{c}(\partial \mathbf{E}/\partial t)$ in Eq. 4.6 accounts for so-called *displacement currents*.

The motion of charged particles located in the EM field is determined by the *Lorentz force law*:

$$\mathbf{f} = \rho\left(\mathbf{E} + \frac{1}{c}\mathbf{v}\times\mathbf{B}\right). \tag{4.7}$$

$\mathbf{f}$ is the force density exerted by the field on the charged-particle distribution, i.e., it is the force per unit volume.

4.1.1 Conservation of Charge

Taking the divergence of Eq. 4.6, we have

$$\nabla \cdot (\nabla \times \mathbf{B}) - \frac{1}{c}\frac{\partial}{\partial t}\nabla \cdot \mathbf{E} = \frac{4\pi}{c}\nabla \cdot \mathbf{j}. \tag{4.8}$$

The first term vanishes since the divergence of any curl is zero. Making use of Gauss's law then produces

$$\frac{\partial \rho}{\partial t} + \nabla \cdot \mathbf{j} = 0. \tag{4.9}$$

This is the *continuity equation* for charge. Alternatively, by integrating this equation over a volume V enclosed by a surface S, and applying the *divergence theorem*, the continuity equation becomes

$$-\frac{\partial}{\partial t}\int_V \rho \, d^3r = \oint_S \mathbf{j} \cdot d\mathbf{a}. \tag{4.10}$$

Here, $d\mathbf{a} = \mathbf{n}\, da$, where da is an area element of the surface S and $\mathbf{n}$ is the unit outward normal to the surface element. Equation 4.10 is a mathematical statement of charge conservation, i.e., the rate at which charge decreases inside a volume V is equivalent to the rate at which charge flows out through the closed surface S bounding that volume.

4.1.2 Conservation of Energy

Take the scalar product of Eq. 4.4 and Eq. 4.6 with **B** and **E**, respectively. The result of subtracting the two equations is

$$\frac{1}{c}\left(\mathbf{B}\cdot\frac{\partial \mathbf{B}}{\partial t}+\mathbf{E}\cdot\frac{\partial \mathbf{E}}{\partial t}\right)=-\frac{4\pi}{c}\mathbf{j}\cdot\mathbf{E}-\left[\mathbf{B}\cdot(\nabla\times\mathbf{E})-\mathbf{E}\cdot(\nabla\times\mathbf{B})\right]. \tag{4.11}$$

Since

$$\nabla\cdot(\mathbf{E}\times\mathbf{B})=\mathbf{B}\cdot(\nabla\times\mathbf{E})-\mathbf{E}\cdot(\nabla\times\mathbf{B}), \tag{4.12}$$

we have

$$\frac{1}{8\pi}\frac{\partial}{\partial t}\left(E^2+B^2\right)=-\mathbf{j}\cdot\mathbf{E}-\frac{c}{4\pi}\nabla\cdot(\mathbf{E}\times\mathbf{B}). \tag{4.13}$$

Let us define the following two quantities:

$$\mathbf{S}=\frac{c}{4\pi}\mathbf{E}\times\mathbf{B} \tag{4.14}$$

$$\mathcal{E}=\frac{1}{8\pi}\left(E^2+B^2\right). \tag{4.15}$$

Then Eq. 4.13 becomes

$$\frac{\partial \mathcal{E}}{\partial t}=-\mathbf{j}\cdot\mathbf{E}-\nabla\cdot\mathbf{S}. \tag{4.16}$$

Now integrate this result over a volume V and use the divergence theorem to write

$$\frac{\partial}{\partial t}\int_V \mathcal{E}\, d^3r=-\int_V \mathbf{j}\cdot\mathbf{E}\, d^3r-\oint_S \mathbf{S}\cdot d\mathbf{a}. \tag{4.17}$$

If the volume contains particles with charges q_i moving with velocities $\mathbf{v}_i$ then, using Eq. 4.2, we have

$$\int_V \mathbf{j}\cdot\mathbf{E}\, d^3r=\sum_i e_i\mathbf{v}_i\cdot\mathbf{E}_i, \tag{4.18}$$

where $\mathbf{E}_i\equiv\mathbf{E}(\mathbf{r}_i)$. Furthermore, let us denote the kinetic energy of the ith particle by T_i. Then, using the Lorentz force ($\mathbf{F}_i$) on each particle, gives

$$\frac{\partial T_i}{\partial t}=\mathbf{F}_i\cdot\mathbf{v}_i=e_i\mathbf{v}_i\cdot\left(\mathbf{E}_i+\frac{1}{c}\mathbf{v}_i\times\mathbf{B}_i\right)=e_i\mathbf{v}_i\cdot\mathbf{E}_i. \tag{4.19}$$

Combining this result with Eq. 4.18, allows us to rewrite Eq. 4.17 in the form

$$\frac{\partial}{\partial t}\left(\int_V \mathcal{E}\, d^3r+\sum_i T_i\right)=-\oint_S \mathbf{S}\cdot d\mathbf{a}. \tag{4.20}$$

This equation can be interpreted as one of energy balance, i.e., the rate at which energy increases within the volume V is equivalent the rate at which energy flows inward through the enclosing boundary S. Thus, it is plausible to identify $\mathcal{E}$ as the energy density of the EM field, while **S**, known as the *Poynting vector*, represents energy flow (having Gaussian units of ergs/(cm^2·sec)).

4.1.3 Conservation of Momentum

Let $\mathbf{p}$ denote the total momentum carried by charges in a volume V. Then, the equation of motion takes the form

$$\frac{d\mathbf{p}}{dt} = \int_V \mathbf{f}\, d^3r, \tag{4.21}$$

where $\mathbf{f}$ can be written in the form

$$\mathbf{f} = \rho\mathbf{E} + \frac{1}{c}\mathbf{j}\times\mathbf{B}. \tag{4.22}$$

The results of solving for ρ and $\mathbf{j}$ in Eqs. 4.5 and 4.6, respectively, can be inserted to rewrite the force density as

$$\begin{aligned} \mathbf{f} &= \frac{1}{4\pi}\left[(\nabla\cdot\mathbf{E})\,\mathbf{E} + (\nabla\times\mathbf{B})\times\mathbf{B} - \frac{1}{c}\frac{\partial\mathbf{E}}{\partial t}\times\mathbf{B}\right] \\ &= \frac{1}{4\pi}\left[(\nabla\cdot\mathbf{E})\,\mathbf{E} - \mathbf{B}\times(\nabla\times\mathbf{B}) + \mathbf{E}\times\frac{1}{c}\frac{\partial\mathbf{B}}{\partial t} - \frac{\partial}{\partial t}\left(\frac{1}{c}\mathbf{E}\times\mathbf{B}\right)\right]. \end{aligned} \tag{4.23}$$

From Faraday's law (Eq. 4.4), the third term in the square brackets becomes $-\mathbf{E}\times(\nabla\times$ We may also add $(\nabla\cdot\mathbf{B})\,\mathbf{B}$ to the square brackets since $\nabla\cdot\mathbf{B} = 0$. The result is

$$\mathbf{f} = \frac{1}{4\pi}\left[(\nabla\cdot\mathbf{E})\,\mathbf{E} - \mathbf{E}\times(\nabla\times\mathbf{E}) + (\nabla\cdot\mathbf{B})\,\mathbf{B} - \mathbf{B}\times(\nabla\times\mathbf{B})\right] - \frac{\partial\mathbf{g}}{\partial t}, \tag{4.24}$$

where we have defined the vector

$$\mathbf{g} = \frac{1}{4\pi c}(\mathbf{E}\times\mathbf{B}) = \frac{1}{c^2}\mathbf{S}. \tag{4.25}$$

The equation of motion now becomes

$$\frac{d}{dt}(\mathbf{p}+\mathbf{G}) = \int_V \frac{1}{4\pi}\left[(\nabla\cdot\mathbf{E})\,\mathbf{E} - \mathbf{E}\times(\nabla\times\mathbf{E}) + (\nabla\cdot\mathbf{B})\,\mathbf{B} - \mathbf{B}\times(\nabla\times\mathbf{B})\right] d^3r, \tag{4.26}$$

with

$$\mathbf{G} = \int_V \mathbf{g}\, d^3r. \tag{4.27}$$

$\mathbf{G}$ represents the momentum of the EM field and $\mathbf{g}$ corresponds to the field's momentum density. The integrand in Eq. 4.26 proves to be the divergence of a second-rank tensor, i.e.,

$$\frac{d}{dt}(\mathbf{p}+\mathbf{G}) = \int_V \nabla\cdot\widetilde{\widetilde{\mathbf{T}}}\, d^3r. \tag{4.28}$$

$\tilde{\tilde{\mathbf{T}}}$ is the *Maxwell stress tensor.* Its elements are given by

$$T_{ij} = \frac{1}{4\pi}\left[E_iE_j + B_iB_j - \frac{1}{2}\left(E^2 + B^2\right)\delta_{ij}\right]. \tag{4.29}$$

Finally, we can recast Eq. 4.28 using the divergence theorem as

$$\frac{d}{dt}\left(\mathbf{p} + \mathbf{G}\right) = \oint_S \tilde{\tilde{\mathbf{T}}}\cdot d\mathbf{a}. \tag{4.30}$$

This is a statement of conservation of momentum, i.e., the rate at which the particles and the EM field enclosed within the surface S gain momentum is equal to the rate at which momentum flows inward across the boundary.†

Example 4.1 Radiation Pressure. Here we determine the pressure produced by a plane, linearly polarized EM wave normally incident onto a flat, perfectly absorbing surface. The assumed coordinate directions are as shown in Fig. 4.1. The force per unit area exerted on the plane surface is determined by calculating $-\tilde{\tilde{\mathbf{T}}}\cdot\mathbf{n}$. The Cartesian components of the unit surface normal $\mathbf{n}$ and the field vectors are

$$\mathbf{n} = (1,0,0) \qquad \mathbf{E} = (0,E,0) \qquad \mathbf{B} = (0,0,B). \tag{4.31}$$

If the wave is propagating in vacuum, then $B = E$, and from Eq. 4.29, the only non-vanishing element in the Maxwell stress tensor turns out to be $T_{11} = -E^2/4\pi$. Thus,

$$-\tilde{\tilde{\mathbf{T}}}\cdot\mathbf{n} = -T_{11}\mathbf{e}_1 = \frac{1}{4\pi}E^2\mathbf{e}_1. \tag{4.32}$$

Since $E = E_0\cos(\omega t + \varphi)$, we have the following simple result for the time-averaged radiation pressure:

$$\left\langle -\tilde{\tilde{\mathbf{T}}}\cdot\mathbf{n}\right\rangle = \frac{1}{8\pi}E_0^2\mathbf{e}_1. \tag{4.33}$$

4.2 Electromagnetic Potentials

Due to the underlying form of Maxwell's equations, a general EM field can be cast in terms of a scalar potential ϕ and a vector potential $\mathbf{A}$. In particular, the fact that $\nabla\cdot\mathbf{B} = 0$ implies that

$$\mathbf{B} = \nabla\times\mathbf{A}, \tag{4.34}$$

since the divergence of a curl vanishes. As a consequence, Faraday's law (Eq. 4.4) takes the form

$$\nabla\times\left(\mathbf{E} + \frac{1}{c}\frac{\partial\mathbf{A}}{\partial t}\right) = 0. \tag{4.35}$$

†The momentum flow is equivalent to the force exerted on the system boundary.

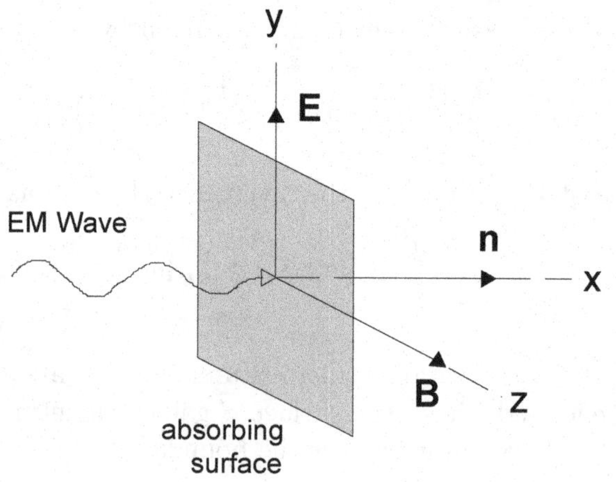

Figure 4.1 Linearly polarized electromagnetic wave striking an absorbing surface.

However, it immediately follows that the quantity in parentheses is derivable from a scalar potential ϕ, i.e., $\mathbf{E} + \frac{1}{c}\left(\partial \mathbf{A}/\partial t\right) = -\boldsymbol{\nabla}\phi$, since $\boldsymbol{\nabla}\times\left(\boldsymbol{\nabla}\phi\right) = 0$. Thus,

$$\mathbf{E} = -\boldsymbol{\nabla}\phi - \frac{1}{c}\frac{\partial \mathbf{A}}{\partial t}. \tag{4.36}$$

It is important to realize that, as things stand, the vector potential $\mathbf{A}$ is not unique, since any other potential of the form

$$\mathbf{A}' = \mathbf{A} + \boldsymbol{\nabla}\chi \tag{4.37}$$

also satisfies Eq. 4.34. Substituting into Eq. 4.36 gives

$$\mathbf{E} = -\boldsymbol{\nabla}\phi - \frac{1}{c}\frac{\partial}{\partial t}\left(\mathbf{A}' - \boldsymbol{\nabla}\chi\right) = -\boldsymbol{\nabla}\phi' - \frac{1}{c}\frac{\partial \mathbf{A}'}{\partial t}, \tag{4.38}$$

where

$$\phi' = \phi - \frac{1}{c}\frac{\partial \chi}{\partial t}. \tag{4.39}$$

The physically meaningful fields $\mathbf{E}$ and $\mathbf{B}$ are unaltered by the transformations of Eqs. 4.37 and 4.39, known as *gauge transformations.* One says that the fields exhibit *gauge invariance.* Up to this point, the function χ has been completely arbitrary. However, one can choose to restrict its value by imposing an additional condition on $\mathbf{A}$. This is called the choice of a *gauge.*

4.2.1 The Coulomb Gauge

A gauge that will prove to be particularly suitable for quantizing the electromagnetic field (next chapter) is the *Coulomb* or *transverse gauge.* In this case, one requires that

$$\nabla \cdot \mathbf{A} = \nabla \cdot \mathbf{A}' = 0. \tag{4.40}$$

Since the divergence of Eq. 4.37 leads to

$$\nabla \cdot \mathbf{A}' = \nabla \cdot \mathbf{A} + \nabla^2 \chi, \tag{4.41}$$

the Coulomb gauge demands that

$$\nabla^2 \chi = 0. \tag{4.42}$$

That is, the function χ must satisfy Laplace's equation.

In this gauge, by taking the divergence of Eq. 4.36, we have

$$\nabla \cdot \mathbf{E} = -\nabla^2 \phi - \frac{1}{c}\frac{\partial}{\partial t} \nabla \cdot \mathbf{A} = -\nabla^2 \phi. \tag{4.43}$$

Then, from Gauss's law (Eq. 4.5), we see that, at any given instant, the scalar potential satisfies Poisson's equation:

$$\nabla^2 \phi = -4\pi \rho. \tag{4.44}$$

The reason for the name "Coulomb gauge" is that Eq. 4.44 is identical to the equation for the Coulomb potential for a static charge distribution. In the present case, however, ϕ is determined by the *instantaneous* configuration of a time-varying charge distribution. Thus, the scalar potential at a point in space is calculated as though any variations in the charge distribution over time are sensed immediately at that point. At first glance, it might appear that this behavior is at odds with the theory of special relativity which demands that any disturbance in the EM field can only propagate at the finite speed of light, c. This apparent contradiction is resolved, however, by realizing that the potentials ϕ and $\mathbf{A}$ are not physically observable quantities. As long as changes in the physically meaningful fields, $\mathbf{E}$ and $\mathbf{B}$, are preceded by an appropriate time delay, there is no contradiction. For other gauges, in fact, the equation for the scalar potential involves the time explicitly (see, for example, the section on the Lorentz gauge which follows), and the latter perceived problem never surfaces. Attaching too much physical meaning to the electromagnetic potentials themselves can be misleading.

One can use Eqs. 4.34 and 4.36 to rewrite Eq. 4.6 for the fields in terms of the scalar and vector potentials:

$$\nabla \times (\nabla \times \mathbf{A}) - \frac{1}{c}\frac{\partial}{\partial t}\left(-\nabla \phi - \frac{1}{c}\frac{\partial \mathbf{A}}{\partial t}\right) = \frac{4\pi}{c}\mathbf{j}. \tag{4.45}$$

Using the basic identity

$$\nabla \times (\nabla \times \mathbf{A}) = \nabla(\nabla \cdot \mathbf{A}) - \nabla^2 \mathbf{A}, \tag{4.46}$$

in the Coulomb gauge Eq. 4.45 becomes

$$-\nabla^2\mathbf{A} + \frac{1}{c^2}\frac{\partial^2\mathbf{A}}{\partial t^2} = \frac{4\pi}{c}\mathbf{j} - \frac{1}{c}\boldsymbol{\nabla}\frac{\partial\phi}{\partial t}. \tag{4.47}$$

Let us now decompose the current density into a longitudinal component $\mathbf{j}_L$ and a transverse component $\mathbf{j}_T$, i.e.,

$$\mathbf{j} = \mathbf{j}_L + \mathbf{j}_T, \tag{4.48}$$

where

$$\boldsymbol{\nabla}\times\mathbf{j}_L = 0 \qquad \text{and} \qquad \boldsymbol{\nabla}\cdot\mathbf{j}_T = 0. \tag{4.49}$$

This follows from *Helmholtz' theorem* which states that any vector field can be written as the sum of an irrotational (longitudinal) component having zero curl and a transverse component having zero divergence. We now show that the last term in Eq. 4.47 exactly cancels the longitudinal component of the current density. First, differentiate Eq. 4.44 with respect to time and rearrange:

$$\boldsymbol{\nabla}\cdot\left(\frac{1}{4\pi}\boldsymbol{\nabla}\frac{\partial\phi}{\partial t}\right) + \frac{\partial\rho}{\partial t} = 0. \tag{4.50}$$

Now use the continuity equation (Eq. 4.9) and Eq. 4.49 to write

$$\boldsymbol{\nabla}\cdot\left(\frac{1}{4\pi}\boldsymbol{\nabla}\frac{\partial\phi}{\partial t}\right) - \boldsymbol{\nabla}\cdot\mathbf{j} = \boldsymbol{\nabla}\cdot\left(\frac{1}{4\pi}\boldsymbol{\nabla}\frac{\partial\phi}{\partial t} - \mathbf{j}_L\right) = 0. \tag{4.51}$$

Thus,

$$\boldsymbol{\nabla}\frac{\partial\phi}{\partial t} = 4\pi\,\mathbf{j}_L, \tag{4.52}$$

and Eq. 4.47 for the vector potential reduces to the wave equation

$$-\nabla^2\mathbf{A} + \frac{1}{c^2}\frac{\partial^2\mathbf{A}}{\partial t^2} = \frac{4\pi}{c}\mathbf{j}_T. \tag{4.53}$$

The Coulomb gauge is sometimes called the "transverse gauge" because the source term in the wave equation depends solely on the transverse component of the current density.

For the case of radiation present in vacuum, we have $\nabla^2\phi = 0$, making it convenient to take $\phi = 0$. To determine the radiation field, one only needs to find the vector potential by solving the source-free wave equation

$$\nabla^2\mathbf{A} - \frac{1}{c^2}\frac{\partial^2\mathbf{A}}{\partial t^2} = 0 \tag{4.54}$$

subject to the transverse gauge condition $\boldsymbol{\nabla}\cdot\mathbf{A} = 0$. One can then solve for $\mathbf{A}$ by assuming a travelling plane-wave solution of the form

$$\mathbf{A}(\mathbf{r}, t) = \mathbf{A}_0\, e^{i(\mathbf{k}\cdot\mathbf{r} - \omega t)}, \tag{4.55}$$

where $\mathbf{k}$ and ω represent the wavevector and angular frequency, respectively, of the harmonic wave. Substitution of this expression into the wave equation produces the dispersion relation $\omega = ck$. The gauge condition produces the constraint

$$\nabla \cdot \mathbf{A} = i\mathbf{k} \cdot \mathbf{A}_0 \, e^{i(\mathbf{k} \cdot \mathbf{r} - \omega t)} = 0. \tag{4.56}$$

Therefore $\mathbf{k} \cdot \mathbf{A}_0 = 0$, and the wave is transverse.

4.2.2 The Lorentz Gauge

One is free to choose a gauge that turns out to be invariant under the Lorentz transformation of special relativity. The gauge is specified by the so-called *Lorentz condition*

$$\nabla \cdot \mathbf{A} + \frac{1}{c}\frac{\partial \phi}{\partial t} = \nabla \cdot \mathbf{A}' + \frac{1}{c}\frac{\partial \phi'}{\partial t} = 0. \tag{4.57}$$

Then, from the gauge transformation (Eqs. 4.37 and 4.39), we have

$$\nabla \cdot \mathbf{A}' + \frac{1}{c}\frac{\partial \phi'}{\partial t} = \nabla \cdot \mathbf{A} + \nabla^2 \chi + \frac{1}{c}\frac{\partial \phi}{\partial t} - \frac{1}{c^2}\frac{\partial^2 \chi}{\partial t^2} = 0, \tag{4.58}$$

which demands that the function χ satisfy

$$\nabla^2 \chi - \frac{1}{c^2}\frac{\partial^2 \chi}{\partial t^2} = 0. \tag{4.59}$$

In the Lorentz gauge, by taking the divergence of Eq. 4.36, we have

$$\nabla \cdot \mathbf{E} = -\nabla^2 \phi - \frac{1}{c}\frac{\partial}{\partial t}\nabla \cdot \mathbf{A} = -\nabla^2 \phi - \frac{1}{c}\frac{\partial}{\partial t}\left(-\frac{1}{c}\frac{\partial \phi}{\partial t}\right). \tag{4.60}$$

Then, from Gauss's law, we see that the scalar potential obeys a wave equation with charge density acting as a source term:

$$\nabla^2 \phi - \frac{1}{c^2}\frac{\partial^2 \phi}{\partial t^2} = -4\pi\rho. \tag{4.61}$$

It is straightforward to show that Eqs. 4.34, 4.36, 4.46 and 4.58 can be used to transform Eq. 4.6 for the $\mathbf{E}$ and $\mathbf{B}$ fields into a wave equation for the vector potential alone, with the source term supplied by the current density:

$$\nabla^2 \mathbf{A} - \frac{1}{c^2}\frac{\partial^2 \mathbf{A}}{\partial t^2} = -\frac{4\pi}{c}\mathbf{j}. \tag{4.62}$$

Thus, in the Lorentz gauge, the equations for ϕ and $\mathbf{A}$ naturally decouple.

For electromagnetic radiation in vacuum, the last two equations reduce to

$$\nabla^2 \phi - \frac{1}{c^2}\frac{\partial^2 \phi}{\partial t^2} = 0 \qquad \text{and} \qquad \nabla^2 \mathbf{A} - \frac{1}{c^2}\frac{\partial^2 \mathbf{A}}{\partial t^2} = 0. \tag{4.63}$$

The equation for the vector potential is identical to Eq. 4.54 derived from the Coulomb gauge. In the present case, it is also convenient to demand that ϕ vanish everywhere—then the Lorentz gauge condition (Eq. 4.57) reduces to that of the Coulomb gauge, and the vector potential is again subject to the transversality condition $\nabla \cdot \mathbf{A} = 0$. As before, the basic solution to the wave equation is a travelling, transverse plane wave.

Example 4.2 Hamiltonian for a Charged Particle in an EM Field. The Lorentz force on a single charged particle (mass m, charge e) moving under the influence of an EM field is

$$\mathbf{F} = e\mathbf{E} + \frac{e}{c}\mathbf{v}\times\mathbf{B}. \tag{4.64}$$

Newton's second law produces the component equations of motion. For example, the x-component equation is

$$m\ddot{x} = eE_x + \frac{e}{c}\left(v_y B_z - v_z B_y\right). \tag{4.65}$$

The resulting equations of motion can be re-expressed in terms of the scalar potential $\phi(x, y, z)$ and the components of the vector potential $\mathbf{A}(x, y, z)$ by using Eqs. 4.34 and 4.36. For example, the x-component equation above can be written as

$$m\ddot{x} = e\left(-\frac{\partial\phi}{\partial x} - \frac{1}{c}\frac{\partial A_x}{\partial t}\right) + \frac{e}{c}v_y\left(\frac{\partial A_y}{\partial x} - \frac{\partial A_x}{\partial y}\right) - \frac{e}{c}v_z\left(\frac{\partial A_x}{\partial z} - \frac{\partial A_z}{\partial x}\right). \tag{4.66}$$

We now demonstrate that these equations of motion are derivable from Lagrange's equations (Eq. 2.14) if one assumes a particle Lagrangian (non-relativistic) of the form

$$\mathcal{L} = T - e\phi + \frac{e}{c}\mathbf{v}\cdot\mathbf{A}, \tag{4.67}$$

where the last term is a velocity-dependent contribution to the potential. First observe that

$$\frac{\partial\mathcal{L}}{\partial\dot{x}} = \frac{\partial T}{\partial\dot{x}} + \frac{e}{c}A_x = m\dot{x} + \frac{e}{c}A_x \tag{4.68}$$

and

$$\frac{d}{dt}\left(\frac{\partial\mathcal{L}}{\partial\dot{x}}\right) = m\ddot{x} + \frac{e}{c}\left(\frac{\partial A_x}{\partial t} + \frac{\partial A_x}{\partial x}\dot{x} + \frac{\partial A_x}{\partial y}\dot{y} + \frac{\partial A_x}{\partial z}\dot{z}\right). \tag{4.69}$$

Then notice that

$$-\frac{\partial\mathcal{L}}{\partial x} = e\frac{\partial\phi}{\partial x} - \frac{e}{c}\left(v_x\frac{\partial A_x}{\partial x} + v_y\frac{\partial A_y}{\partial x} + v_z\frac{\partial A_z}{\partial x}\right). \tag{4.70}$$

Inserting these results into Lagrange's equation produces

$$\frac{d}{dt}\left(\frac{\partial\mathcal{L}}{\partial\dot{x}}\right) - \frac{\partial\mathcal{L}}{\partial x} = m\ddot{x} + e\frac{\partial\phi}{\partial x} + \frac{e}{c}\frac{\partial A_x}{\partial t} + \frac{e}{c}v_y\left(\frac{\partial A_x}{\partial y} - \frac{\partial A_y}{\partial x}\right) + \frac{e}{c}v_z\left(\frac{\partial A_x}{\partial z} - \frac{\partial A_z}{\partial x}\right) = 0, \tag{4.71}$$

which, after rearranging, reduces to the equation of motion, Eq. 4.66 (and similarly for the other two component equations). The generalized momentum associated with the coordinate x is

$$p_x = \frac{\partial \mathcal{L}}{\partial \dot{x}} = m\dot{x} + \frac{e}{c}A_x. \tag{4.72}$$

Because of the similar form for the other components, we write the momentum canonically conjugate to the position vector $\mathbf{r}$ as

$$\mathbf{p} = m\mathbf{v} + \frac{e}{c}\mathbf{A}. \tag{4.73}$$

Note that $\mathbf{p} \neq m\mathbf{v}$! Finally, referring to Eq. 2.17, the Hamiltonian function for the particle is obtained as follows:

$$\begin{aligned} H &= \sum_k v_k \frac{\partial \mathcal{L}}{\partial v_k} - \mathcal{L} \\ &= \sum_k \left(m v_k^2 + \frac{e}{c} v_k A_k \right) - T + e\phi - \frac{e}{c} \sum_k v_k A_k \\ &= \frac{1}{2} m v^2 + e\phi \\ &= \frac{1}{2m} \left| \mathbf{p} - \frac{e}{c}\mathbf{A} \right|^2 + e\phi. \end{aligned} \tag{4.74}$$

Example 4.3 Hamiltonian Formulation for EM Field + Charged Particles. Here, we show how the electromagnetic field and its interaction with charge and current density can be cast into the Lagrangian/Hamiltonian formalism. At the outset, let us focus our attention on the Lagrangian for the EM field. Since it is defined at every point in space, the field is equivalent to a mechanical system possessing an infinite number of degrees of freedom. Such a situation is reminiscent of the one encountered in Section 2.5 for a one-dimensional system, where the concept of a Lagrangian density, $\mathbb{L}$, was introduced. Extending this idea to three-dimensions, the Lagrangian for the field is

$$\mathcal{L} = \iiint \mathbb{L}\, dx\, dy\, dz. \tag{4.75}$$

One specifies the field at each instant of time by attaching the independent field variables A_x, A_y, A_z, and ϕ to each point in x-y-z space; thus, these can be chosen as the generalized coordinates for the field. The Lagrangian density is then a function of these field variables, along with their respective space and time derivatives. It becomes convenient to introduce the *field four-vector* A_μ and the *space-time four-vector* x_μ:

$$A_\mu = (A_1, A_2, A_3, A_4) = (A_x, A_y, A_z, i\phi) \tag{4.76}$$

$$x_\mu = (x_1, x_2, x_3, x_4) = (x, y, z, ict)\,. \tag{4.77}$$

Then, the Lagrangian density takes on the following compact functional form:

$$\mathbb{L} = \mathbb{L}\left(A_\mu, \frac{\partial A_\mu}{\partial x_\nu}\right) \qquad (\mu, \nu = 1, 2, 3, 4). \tag{4.78}$$

A_μ and $\partial A_\mu / \partial x_\nu$ replace the variables q_k and $\dot{q}_k$ appearing in the Lagrangian of a point-particle system. By analogy with Eq. 2.107, Lagrange's equations for the field are then

$$\sum_{\nu=1}^{4} \frac{\partial}{\partial x_\nu}\left[\frac{\partial \mathbb{L}}{\partial\left(\partial A_\mu / \partial x_\nu\right)}\right] - \frac{\partial \mathbb{L}}{\partial A_\mu} = 0 \qquad (\mu = 1, 2, 3, 4). \tag{4.79}$$

We now claim that the Lagrangian density, which applies to the field and any interactions it has with existing charge and current densities, has the form

$$\mathbb{L} = \frac{1}{8\pi}\left(E^2 - B^2\right) + \frac{1}{c}\mathbf{j}\cdot\mathbf{A} - \rho\phi, \tag{4.80}$$

where $\mathbf{B}$ and $\mathbf{E}$ are understood to be related to the potentials by Eqs. 4.34 and 4.36, respectively. This assertion can be verified by showing that the two Maxwell's equations, Eqs. 4.5 and 4.6, are recovered upon substitution of the postulated Lagrangian density into Eqs. 4.79. Note that the other two Maxwell's equations are automatically satisfied whenever the EM field is written in terms of the scalar and vector potentials. So, first let us show that Eq. 4.5 follows from our Lagrangian density. Consider the Lagrange's equation for the case $A_\mu = A_4 = i\phi$, i.e.,

$$\sum_{\nu=1}^{4} \frac{\partial}{\partial x_\nu}\left[\frac{\partial \mathbb{L}}{\partial\left(\partial \phi / \partial x_\nu\right)}\right] = \frac{\partial \mathbb{L}}{\partial \phi}. \tag{4.81}$$

Inspection of the form of $\mathbb{L}$ leads to

$$\frac{1}{8\pi}\sum_{\nu=1}^{4} \frac{\partial}{\partial x_\nu}\left[\frac{\partial\left(E^2\right)}{\partial\left(\partial \phi / \partial x_\nu\right)}\right] = -\rho. \tag{4.82}$$

This, in turn, reduces to

$$-\frac{1}{4\pi}\sum_{k=1}^{3} \frac{\partial E_k}{\partial x_k} = -\rho, \tag{4.83}$$

which is Gauss's law, as required. Next, consider Lagrange's equation with $A_\mu = A_1 = A_x$, and note that

$$E^2 - B^2 = \left|-\boldsymbol{\nabla}\phi - \frac{1}{c}\frac{\partial \mathbf{A}}{\partial t}\right|^2 - \left(\frac{\partial A_z}{\partial y} - \frac{\partial A_y}{\partial z}\right)^2 - \left(\frac{\partial A_x}{\partial z} - \frac{\partial A_z}{\partial x}\right)^2 - \left(\frac{\partial A_y}{\partial x} - \frac{\partial A_x}{\partial y}\right)^2. \tag{4.84}$$

Then, the required derivatives of the Lagrangian density follow:

$$\frac{\partial}{\partial x_4}\left[\frac{\partial \mathbb{L}}{\partial\left(\partial A_1 / \partial x_4\right)}\right] = \frac{\partial}{\partial t}\left[\frac{\partial \mathbb{L}}{\partial\left(\partial A_x / \partial t\right)}\right] = -\frac{1}{4\pi c}\frac{\partial E_x}{\partial t} \tag{4.85}$$

$$\begin{aligned}\sum_{k=1}^{3}\frac{\partial}{\partial x_k}\left[\frac{\partial \mathbb{L}}{\partial\left(\partial A_1/\partial x_k\right)}\right] &= \frac{1}{8\pi}\left[-2\frac{\partial}{\partial z}\left(\frac{\partial A_x}{\partial z}-\frac{\partial A_z}{\partial x}\right)+2\frac{\partial}{\partial y}\left(\frac{\partial A_y}{\partial x}-\frac{\partial A_x}{\partial y}\right)\right] \\ &= \frac{1}{4\pi}\left(\frac{\partial B_z}{\partial y}-\frac{\partial B_y}{\partial z}\right)=\frac{1}{4\pi}\left(\boldsymbol{\nabla}\times\mathbf{B}\right)_x \end{aligned} \tag{4.86}$$

$$\frac{\partial \mathbb{L}}{\partial A_1}=\frac{\partial \mathbb{L}}{\partial A_x}=\frac{1}{c}\,j_x. \tag{4.87}$$

Thus, the $\mu = 1$ Lagrange's equation produces

$$\frac{1}{4\pi}\left(\boldsymbol{\nabla}\times\mathbf{B}\right)_x-\frac{1}{4\pi c}\frac{\partial E_x}{\partial t}=\frac{1}{c}\,j_x, \tag{4.88}$$

which is just the x-component of Eq. 4.6. Taken together, the $\mu = 1$, 2, and 3 Lagrange's equations produce this Maxwell's equation in its complete vector form. We have now demonstrated that Eq. 4.80 is the correct form for $\mathbb{L}$.

The expression $\frac{1}{c}\mathbf{j}\cdot\mathbf{A} - \rho\phi$ that appears in the Lagrangian density can be identified as the interaction portion of the Lagrangian found in the previous example for a charged particle moving under the influence of a field. To see this, use Eqs. 4.1 and 4.2 for the charge and current densities. Then, in the present case, the interaction part of the Lagrangian reduces as follows:

$$\begin{aligned}\mathcal{L}_{int} &= \int\left(\frac{1}{c}\mathbf{j}\cdot\mathbf{A}-\rho\phi\right)d^3r=\int\left[\sum_i\frac{e_i}{c}\mathbf{v}_i\cdot\mathbf{A}\,\delta\left(\mathbf{r}-\mathbf{r}_i\right)-\sum_i e_i\phi\,\delta\left(\mathbf{r}-\mathbf{r}_i\right)\right]d^3r \\ &= \sum_i\frac{e_i}{c}\mathbf{v}_i\cdot\mathbf{A}_i-\sum_i e_i\phi_i, \end{aligned} \tag{4.89}$$

($\mathbf{A}_i$ and ϕ_i are the potentials evaluated at the positions $\mathbf{r}_i$ corresponding to the charged particles residing in the field.) The interaction portion is the same, independent of whether one considers the Lagrangian for the particle or for the field. Thus, $\mathcal{L}_{int}$ represents the *mutual* interaction between the charged particles and the EM field. It is then reasonable to assimilate Eqs. 4.67 and 4.80 to form a single Lagrangian and a single Lagrangian density for the combined system of EM field plus charged particles, i.e.,

$$\mathcal{L}=\frac{1}{2}\sum_i m_i v_i^2+\frac{1}{8\pi}\int\left(E^2-B^2\right)d^3r+\sum_i e_i\left(\frac{1}{c}\mathbf{v}_i\cdot\mathbf{A}_i-\phi\right) \tag{4.90}$$

$$\mathbb{L}=\frac{1}{2}\sum_i m_i v_i^2\delta\left(\mathbf{r}-\mathbf{r}_i\right)+\frac{1}{8\pi}\left(E^2-B^2\right)+\left(\frac{1}{c}\mathbf{j}\cdot\mathbf{A}-\rho\phi\right). \tag{4.91}$$

The first term in each of the relations above corresponds to the Lagrangian for charged particles that are free, the second term applies to the EM field alone, while the last term represents the mutual charge-field interaction. The Lagrangian is a function of the $3N$ generalized coordinates needed to locate the N charged particles and of the four coordinates A_μ at each of the infinite number of locations in x-y-z space. When the totality of these coordinates are considered, Lagrange's equations reduce to the Lorentz force laws for the particles and Maxwell's equations for the field.

In our previous example, we found the Hamiltonian for a charged particle interacting with the field (Eq. 4.74). Let us now find the contribution to the Hamiltonian function due to the field alone. Start with the Lagrangian density for the field:

$$\mathbb{L}_f = \frac{1}{8\pi}\left(E^2 - B^2\right) = \frac{1}{8\pi}\left|-\nabla\phi - \frac{1}{c}\frac{\partial \mathbf{A}}{\partial t}\right|^2 - \frac{1}{8\pi}\left|\nabla\times\mathbf{A}\right|^2. \tag{4.92}$$

Then, following Eq. 2.17, the corresponding Hamiltonian density for the field is obtained:

$$\mathbb{H}_f = \sum_{\mu=1}^{4}\frac{\partial A_\mu}{\partial t}\frac{\partial \mathbb{L}_f}{\partial\left(\partial A_\mu/\partial t\right)} - \mathbb{L}_f = \sum_{k=1}^{3}\frac{\partial A_k}{\partial t}\frac{\partial \mathbb{L}_f}{\partial\left(\partial A_k/\partial t\right)} - \mathbb{L}_f. \tag{4.93}$$

Notice, however, that

$$\frac{\partial A_k}{\partial t}\frac{\partial \mathbb{L}_f}{\partial\left(\partial A_k/\partial t\right)} = \frac{1}{4\pi c}\frac{\partial A_k}{\partial t}\left(\frac{\partial \phi}{\partial x_k} + \frac{1}{c}\frac{\partial A_k}{\partial t}\right), \tag{4.94}$$

which leads to

$$\mathbb{H}_f = -\frac{1}{4\pi}\mathbf{E}\cdot\frac{1}{c}\frac{\partial \mathbf{A}}{\partial t} + \frac{1}{8\pi}B^2 - \frac{1}{8\pi}E^2. \tag{4.95}$$

Finally, for the first term use

$$\mathbf{E}\cdot\frac{1}{c}\frac{\partial \mathbf{A}}{\partial t} = -\mathbf{E}\cdot\nabla\phi + \mathbf{E}\cdot\left(\nabla\phi + \frac{1}{c}\frac{\partial \mathbf{A}}{\partial t}\right) = -\,\mathbf{E}\cdot\nabla\phi - \mathbf{E}\cdot\mathbf{E} = -\,\mathbf{E}\cdot\nabla\phi - E^2 \tag{4.96}$$

to write

$$\mathbb{H}_f = \frac{1}{4\pi}\mathbf{E}\cdot\nabla\phi + \frac{1}{8\pi}\left(E^2 + B^2\right). \tag{4.97}$$

Now, the total Hamiltonian for the field H_f is obtained by integrating this result over space:

$$H_f = \int \mathbb{H}_f\, d^3r = \frac{1}{4\pi}\int \mathbf{E}\cdot\nabla\phi\, d^3r + \frac{1}{8\pi}\int\left(E^2 + B^2\right) d^3r. \tag{4.98}$$

However, we also have the identity

$$\int \mathbf{E}\cdot\nabla\phi\, d^3r = \int \nabla\cdot(\phi\mathbf{E})\, d^3r - \int \phi\,(\nabla\cdot\mathbf{E})\; d^3r. \tag{4.99}$$

But note that the first integral on the right-hand side vanishes since it can be expressed as a surface integral, i.e, $\int \nabla\cdot(\phi\mathbf{E})\, d^3r = \oint \phi\mathbf{E}\cdot d\mathbf{a}$, and $\mathbf{E} \to 0$ everywhere on the surface of an arbitrarily large enclosure. Thus,

$$\int \mathbf{E}\cdot\nabla\phi\, d^3r = -\int \phi\,(\nabla\cdot\mathbf{E})\; d^3r = -\int \phi\,(4\pi\rho)\; d^3r = -4\pi\sum_i e_i\phi_i \tag{4.100}$$

and the field Hamiltonian becomes

$$H_f = -\sum_i e_i\phi_i + \frac{1}{8\pi}\int \left(E^2 + B^2\right) d^3r. \tag{4.101}$$

Finally, adding this field Hamiltonian to the sum of the particle Hamiltonians, i.e., Eq. 4.74, we obtain the following Hamiltonian for the total system of field plus charges:

$$H = \frac{1}{8\pi}\int \left(E^2 + B^2\right) d^3r + \sum_i \frac{1}{2m_i}\left|\mathbf{p}_i - \frac{e_i}{c}\mathbf{A}_i\right|^2. \tag{4.102}$$

This means that when charged particles interact with each other via the EM field, the conserved quantity is the particle kinetic energies plus the integral of the EM field energy density.

4.3 Field Due to a Changing Polarization

We now consider the production of electromagnetic radiation by time-varying dipole moments.‡ In the section to follow, it will be shown that this phenomenon is, in fact, intimately related to the problem of light scattering from small dielectric particles.

To begin, consider electromagnetic phenomena in some material medium. In addition to the fundamental fields **E** and **B**, the atoms of the material contribute an *electric polarization* **P** and a *magnetization* **M**. These quantities represent the densities of electric and magnetic dipole moments, respectively. Moments may result from the presence of permanent dipoles and/or induced dipoles in the material. It turns out to be advantageous to define two new fields by combining **E** and **B** with the dipole densities **P** and **M** as follows:

$$\mathbf{D} = \mathbf{E} + 4\pi\mathbf{P} \tag{4.103}$$

$$\mathbf{H} = \mathbf{B} - 4\pi\mathbf{M}. \tag{4.104}$$

D and **H** are called the *electric displacement* and the *magnetic intensity*, respectively. The *macroscopic* Maxwell's equations that result for a material medium are then

$$\boldsymbol{\nabla}\cdot\mathbf{B} = 0 \tag{4.105}$$

$$\boldsymbol{\nabla}\times\mathbf{E} + \frac{1}{c}\frac{\partial\mathbf{B}}{\partial t} = 0 \tag{4.106}$$

$$\boldsymbol{\nabla}\cdot\mathbf{D} = 4\pi\rho \tag{4.107}$$

$$\boldsymbol{\nabla}\times\mathbf{H} - \frac{1}{c}\frac{\partial\mathbf{D}}{\partial t} = \frac{4\pi}{c}\mathbf{j}. \tag{4.108}$$

‡In a later chapter, we will address the related problem of radiation produced by an accelerating charged particle.

The symbols ρ and $\mathbf{j}$ represent the free charge and current densities, i.e., they do not include the charges and currents bound to dipoles associated with the polarization and magnetization of the medium.

Consider the fields in a region where dipoles may exist, but no free charges or currents exist, i.e., $\rho = 0$ and $\mathbf{j} = 0$. In this case, using Eqs. 4.103 and 4.104, we can rewrite the last two Maxwell's equations as

$$\boldsymbol{\nabla}\cdot\mathbf{E} = -4\pi\boldsymbol{\nabla}\cdot\mathbf{P} \tag{4.109}$$

$$\boldsymbol{\nabla}\times\mathbf{B} - \frac{1}{c}\frac{\partial \mathbf{E}}{\partial t} = \frac{4\pi}{c}\frac{\partial \mathbf{P}}{\partial t} + 4\pi\boldsymbol{\nabla}\times\mathbf{M}. \tag{4.110}$$

These are of the same form as Maxwell's equations in vacuum if one makes the identifications

$$\rho \to -\boldsymbol{\nabla}\cdot\mathbf{P} \qquad \text{and} \qquad \mathbf{j} \to \frac{\partial \mathbf{P}}{\partial t} + c\boldsymbol{\nabla}\times\mathbf{M}. \tag{4.111}$$

Notice that the continuity equation, Eq. 4.9, is automatically satisfied by the above relations. Now, instead of working with the EM potentials $\mathbf{A}$ and ϕ which naturally depend on the sources $\mathbf{j}$ and ρ, we transform to a pair of so-called *Hertz vectors* $\boldsymbol{\Pi}_e$ and $\boldsymbol{\Pi}_m$ defined by

$$\phi = -\boldsymbol{\nabla}\cdot\boldsymbol{\Pi}_e \tag{4.112}$$

$$\mathbf{A} = \frac{1}{c}\frac{\partial \boldsymbol{\Pi}_e}{\partial t} + \boldsymbol{\nabla}\times\boldsymbol{\Pi}_m. \tag{4.113}$$

By substituting Eqs. 4.111–4.113 into the inhomogeneous wave equations for ϕ and $\mathbf{A}$ (Eqs. 4.61 and 4.62), it follows that the Hertz vectors satisfy

$$\nabla^2\boldsymbol{\Pi}_e - \frac{1}{c^2}\frac{\partial^2 \boldsymbol{\Pi}_e}{\partial t^2} = -4\pi\mathbf{P} \tag{4.114}$$

$$\nabla^2\boldsymbol{\Pi}_m - \frac{1}{c^2}\frac{\partial^2 \boldsymbol{\Pi}_m}{\partial t^2} = -4\pi\mathbf{M}. \tag{4.115}$$

In other words, the Hertz vectors also satisfy inhomogeneous wave equations, except $\mathbf{P}$ and $\mathbf{M}$ now act as the sources. The solutions are

$$\boldsymbol{\Pi}_e(\mathbf{r}, t) = \int_V d^3r' \, \frac{\mathbf{P}\left(\mathbf{r}', t - \frac{|\mathbf{r}-\mathbf{r}'|}{c}\right)}{|\mathbf{r}-\mathbf{r}'|} \tag{4.116}$$

$$\boldsymbol{\Pi}_m(\mathbf{r}, t) = \int_V d^3r' \, \frac{\mathbf{M}\left(\mathbf{r}', t - \frac{|\mathbf{r}-\mathbf{r}'|}{c}\right)}{|\mathbf{r}-\mathbf{r}'|}, \tag{4.117}$$

where $\mathbf{P}$ and $\mathbf{M}$ are to be evaluated at the *retarded time* $t - |\mathbf{r}-\mathbf{r}'|/c$. This accounts for the propagation delay experienced by the light in traversing the distance $|\mathbf{r}-\mathbf{r}'|$. The fields can then be obtained from

$$\mathbf{E} = -\boldsymbol{\nabla}\phi - \frac{1}{c}\frac{\partial \mathbf{A}}{\partial t} = \boldsymbol{\nabla}\times\left(\boldsymbol{\nabla}\times\boldsymbol{\Pi}_e - \frac{1}{c}\frac{\partial \boldsymbol{\Pi}_m}{\partial t}\right) - 4\pi\mathbf{P} \tag{4.118}$$

and

$$\mathbf{B} = \nabla\times\mathbf{A} = \nabla\times\left(\frac{1}{c}\frac{\partial \mathbf{\Pi}_e}{\partial t} + \nabla\times\mathbf{\Pi}_m\right). \tag{4.119}$$

For a dielectric medium, we have $\mathbf{M} = 0$ and $\mathbf{P} \neq 0$; thus $\mathbf{\Pi}_m = 0$ and $\mathbf{\Pi}_e \neq 0$. Furthermore, for points outside the material medium, the fields reduce to

$$\mathbf{E} = \nabla\times(\nabla\times\mathbf{\Pi}_e) = \nabla(\nabla\cdot\mathbf{\Pi}_e) - \frac{1}{c^2}\frac{\partial^2 \mathbf{\Pi}_e}{\partial t^2} \tag{4.120}$$

and

$$\mathbf{B} = \frac{1}{c}\frac{\partial}{\partial t}\nabla\times\mathbf{\Pi}_e. \tag{4.121}$$

The example that follows investigates the field produced when the polarization is simply that of a point electric dipole.

Example 4.4 Electric Dipole Field and Radiation. Consider a time-varying electric dipole $\mathbf{p}(t) = p(t)\mathbf{e}_z$ located at the origin. The polarization at the retarded time of $t' = t - r/c$ is then represented by

$$\mathbf{P}(\mathbf{r}', t') = \mathbf{p}(t')\delta(\mathbf{r}'). \tag{4.122}$$

The Hertz vector $\mathbf{\Pi}_e$ is then obtained from Eq. 4.116 to be

$$\mathbf{\Pi}_e = \frac{\mathbf{p}\left(t - \frac{r}{c}\right)}{r} = \frac{p\left(t - \frac{r}{c}\right)}{r}\mathbf{e}_z = \frac{[p]}{r}\mathbf{e}_z. \tag{4.123}$$

$[p]$ denotes the dipole-moment component evaluated at the retarded time t'. Using spherical coordinates (r, θ, ϕ), we have the components

$$\Pi_r = \Pi\cos\theta = \frac{[p]}{r}\cos\theta, \qquad \Pi_\theta = -\Pi\sin\theta = -\frac{[p]}{r}\sin\theta, \qquad \Pi_\phi = 0. \tag{4.124}$$

The field vectors can now be calculated from Eqs. 4.120 and 4.121. It is left to the reader to perform the necessary differential operations in spherical coordinates. The resulting field components are

$$\begin{aligned} E_r &= 2\left(\frac{[\dot{p}]}{cr^2} + \frac{[p]}{r^3}\right)\cos\theta & B_r &= 0 \\ E_\theta &= \left(\frac{[\ddot{p}]}{c^2 r} + \frac{[\dot{p}]}{cr^2} + \frac{[p]}{r^3}\right)\sin\theta & B_\theta &= 0 \\ E_\phi &= 0 & B_\phi &= \left(\frac{[\ddot{p}]}{c^2 r} + \frac{[\dot{p}]}{cr^2}\right)\sin\theta. \end{aligned} \tag{4.125}$$

Thus, the $\mathbf{E}$ and $\mathbf{B}$ fields are mutually perpendicular. Now consider the points far from the dipole, i.e., at large values of r. This corresponds to the so-called *far field*

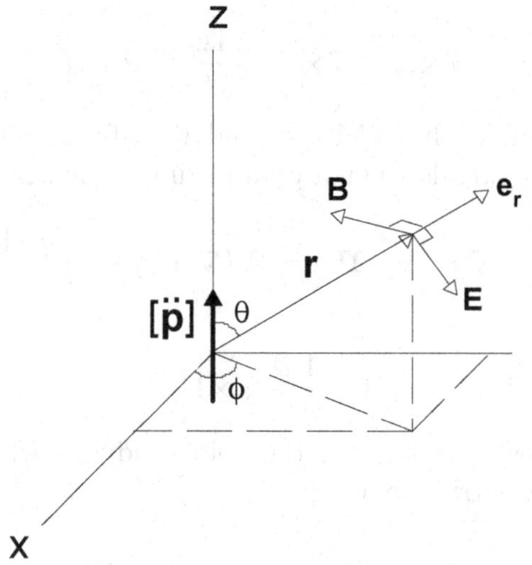

Figure 4.2 Field vectors in the radiation zone of a time-varying electric dipole.

or *radiation zone*. In this region, we only keep terms of order r^{-1}, and the radial component of the electric field vanishes. The only surviving field components are

$$E_\theta = B_\phi = \frac{[\ddot{p}]}{c^2 r}\sin\theta. \tag{4.126}$$

As shown in Fig. 4.2, **E** and **B** are transverse fields in the radiation zone; they can be expressed compactly as

$$\mathbf{B} = \frac{1}{c^2 r}\,[\ddot{\mathbf{p}}]\times\mathbf{e}_r \tag{4.127}$$

$$\mathbf{E} = \mathbf{B}\times\mathbf{e}_r. \tag{4.128}$$

These fields transport energy radially outward from the dipole at a rate per unit area determined by the Poynting vector, **S**:

$$\mathbf{S} = \frac{c}{4\pi}\mathbf{E}\times\mathbf{B} = \left(\frac{[\ddot{p}]^2}{4\pi c^3 r^2}\sin^2\theta\right)\mathbf{e}_r = \frac{c}{4\pi}E^2\mathbf{e}_r = \frac{c}{4\pi}B^2\mathbf{e}_r. \tag{4.129}$$

The power radiated by the dipole per unit solid angle $d\Omega$ is

$$\frac{dP}{d\Omega} = (\mathbf{S}\cdot\mathbf{e}_r)\,r^2 = \frac{[\ddot{p}]^2}{4\pi c^3}\sin^2\theta. \tag{4.130}$$

Figure 4.3 displays this $\sin^2\theta$ angular distribution for dipole radiation. It also follows

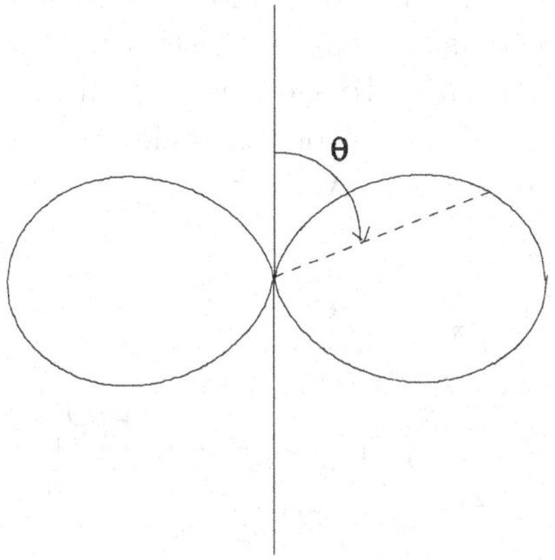

Figure 4.3 Radiation pattern from time-varying dipole.

that the total power emitted is

$$P = \int_{4\pi} \frac{dP}{d\Omega}\, d\Omega = \frac{[\ddot{p}]^2}{4\pi c^3} \int_{\theta=0}^{\pi} \int_{\phi=0}^{2\pi} (\sin^2\theta) \sin\theta \, d\theta \, d\phi = \frac{2\,[\ddot{p}]^2}{3c^3}. \tag{4.131}$$

In the case when the time-variation of the dipole moment is simple harmonic, i.e.,

$$[\mathbf{p}] = \mathbf{p}(t') = (p_0 \cos\omega t')\, \mathbf{e}_z, \tag{4.132}$$

it is particularly straightforward to compute the *average* power emitted, as long as an integral number of cycles is considered. Here, one need not be concerned about the evaluation of $\mathbf{p}$ at retarded time t' since all times are essentially equivalent. We then see that

$$\left\langle [\ddot{p}]^2 \right\rangle = p_0^2 \omega^4 \left\langle \cos^2\omega t' \right\rangle = \frac{1}{2} p_0^2 \omega^4, \tag{4.133}$$

and the average power radiated by the dipole is

$$\left\langle \frac{dP}{d\Omega} \right\rangle = \frac{p_0^2 \omega^4}{8\pi c^3} \sin^2\theta \tag{4.134}$$

or

$$\langle P \rangle = \frac{p_0^2 \omega^4}{3c^3}. \tag{4.135}$$

Since the oscillation frequency ω is inversely related to λ, the wavelength of the radiation, we see a characteristic λ^{-4} dependence for the power emitted by an oscillating electric dipole.

Example 4.5 Multipole Radiation. We can extend the analysis of radiation due to an oscillating dipole to that of a more general time-varying source distribution. In particular, one finds the characteristics of the far-field radiation to be a superposition of *multipolar fields.* To see this, return to the general expression that determines $\mathbf{\Pi}_e$ from the polarization of the dielectric source, i.e., Eq. 4.116. The temporal Fourier transform is then given by

$$\begin{aligned}
\mathbf{\Pi}_e(\mathbf{r},\omega) &= \frac{1}{2\pi}\int_{-\infty}^{+\infty} dt\, e^{i\omega t}\,\mathbf{\Pi}_e(\mathbf{r},t) \\
&= \frac{1}{2\pi}\int_{-\infty}^{+\infty} dt\, e^{i\omega t}\int_V d^3r'\, \frac{\mathbf{P}\left(\mathbf{r}', t-\frac{|\mathbf{r}-\mathbf{r}'|}{c}\right)}{|\mathbf{r}-\mathbf{r}'|} \\
&= \int_V d^3r'\, \frac{e^{i\omega|\mathbf{r}-\mathbf{r}'|/c}}{|\mathbf{r}-\mathbf{r}'|}\left[\frac{1}{2\pi}\int_{-\infty}^{+\infty} dt'\, e^{i\omega t'}\,\mathbf{P}(\mathbf{r}',t')\right] \\
&= \int_V d^3r'\, \frac{e^{ik|\mathbf{r}-\mathbf{r}'|}}{|\mathbf{r}-\mathbf{r}'|}\,\mathbf{P}(\mathbf{r}',\omega),
\end{aligned} \tag{4.136}$$

where $k = \omega/c$ and $\mathbf{P}(\mathbf{r}',\omega)$ is the Fourier transform of the polarization. In the radiation zone, i.e., where $r \gg r'$, we can make use of the following expansion for a spherical outgoing wave:

$$\frac{e^{ik|\mathbf{r}-\mathbf{r}'|}}{|\mathbf{r}-\mathbf{r}'|} = ik\sum_{\ell=0}^{\infty}(2\ell+1)\,P_\ell(\cos\theta)\,j_\ell(kr')\,h_\ell^{(1)}(kr). \tag{4.137}$$

Here, j_ℓ and $h_\ell^{(1)}$ are *spherical Bessel and Hankel functions*, respectively. The P_ℓ are *Legendre polynomials* with argument $\cos\theta \equiv \mathbf{e}_r\cdot\mathbf{e}_{r'}$. For the far field where $kr \gg 1$, one replaces $h_\ell^{(1)}$ with its asymptotic expansion

$$h_\ell^{(1)}(kr) \to (-i)^{\ell+1}\,\frac{e^{ikr}}{kr}, \tag{4.138}$$

and the Hertz vector becomes

$$\mathbf{\Pi}_e(\mathbf{r},\omega) = \frac{e^{ikr}}{r}\int_V d^3r'\sum_{\ell=0}^{\infty}(-i)^\ell\,(2\ell+1)\,P_\ell(\cos\theta)\,j_\ell(kr')\,\mathbf{P}(\mathbf{r}',\omega)\,. \tag{4.139}$$

Furthermore, when the wavelength of the emitted radiation is comparable to or larger than the size of the source, then $kr' < 1$, and one only need consider the limiting behavior of j_ℓ for small arguments:

$$j_\ell(kr') \to \frac{2^\ell\ell!}{(2\ell+1)!}\,(kr')^\ell\,. \tag{4.140}$$

Inserting this into Eq. 4.139 produces the *multipole expansion* of $\mathbf{\Pi}_e$, i.e.,

$$\mathbf{\Pi}_e(\mathbf{r},\omega) = \sum_{\ell=0}^{\infty} \mathbf{\Pi}_e^{(\ell)}(\mathbf{r},\omega) \tag{4.141}$$

where

$$\mathbf{\Pi}_e^{(\ell)}(\mathbf{r},\omega) = \frac{e^{ikr}}{r}\frac{2^\ell \ell!}{(2\ell)!}\int_V d^3r'\ (-ikr')^\ell\, P_\ell(\cos\theta)\,\mathbf{P}\,(\mathbf{r}',\omega)\,. \tag{4.142}$$

Consider the first few contributions to the expansion. $\ell = 0$ is the *electric dipole term*:

$$\mathbf{\Pi}_e^{(0)}(\mathbf{r},\omega) = \frac{e^{ikr}}{r}\int_V d^3r'\ \mathbf{P}\,(\mathbf{r}',\omega)\,. \tag{4.143}$$

The integral corresponds to the Fourier transform (in time) of the *net* electric dipole moment. Higher-order multipole terms depend on the specific distribution of electric dipoles contributing to the polarization of the source. For example, the $\ell = 1$ term is given by

$$\begin{aligned}\mathbf{\Pi}_e^{(1)}(\mathbf{r},\omega) &= -ik\frac{e^{ikr}}{r}\int_V d^3r'\ (r'\cos\theta)\ \mathbf{P}\,(\mathbf{r}',\omega)\\ &= -ik\frac{e^{ikr}}{r^2}\int_V d^3r'\ (\mathbf{r}\cdot\mathbf{r}')\,\mathbf{P}\,(\mathbf{r}',\omega)\,.\end{aligned} \tag{4.144}$$

This corresponds to *magnetic dipole* plus *electric quadrupole* radiation. By making use of Eqs. 4.120 and 4.121, one can transform the multipole expansion of the Hertz vector into a corresponding expansion of multipole radiation fields.

4.4 Light Scattering from Dielectric Particles

Consider a situation like that shown in Fig. 4.4 where a dielectric particle having index of refraction n is immersed in a homogeneous background medium of refractive index n_0. The geometry shown is representative of a typical light-scattering experiment. A beam of polarized light, having a wavelength λ_0 in vacuum, is incident on the particle. The propagation of the incoming beam is chosen to be in the $+z$-direction and it is characterized by the wavenumber $k = k_v n_0$, where $k_v = 2\pi/\lambda_0 = \omega/c$ is the wavenumber in vacuum. As a result, the incident light can be treated as a polarized plane wave of the form

$$\mathbf{E}_i(\mathbf{r},t) = \mathbf{E}_0\, e^{i(\mathbf{k}\cdot\mathbf{r}-\omega t)}\,. \tag{4.145}$$

In our set-up, the experimenter has chosen to work in what is known as the *VV-geometry.* Here, the incoming electric field is polarized along the x-direction ($\mathbf{E}_0 = E_0\mathbf{e}_x$) and one measures the vertically polarized field-component (i.e., x-component)

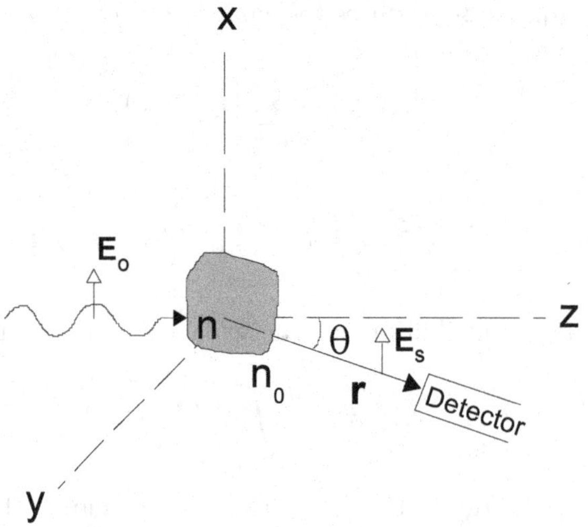

Figure 4.4 Basic VV-geometry for light scattering from a dielectric particle.

of radiation scattered into the y-z plane (also referred to as the scattering plane). This is the case in a majority of light-scattering experiments. Now, given the particle to be situated at the origin, the general problem at hand is to calculate the scattered wave at time t received by a detector at position $\mathbf{r}$ in the scattering plane that is far outside the scattering volume.

A solution to the scattering problem can be developed by applying the Maxwell's equations

$$\nabla \times \mathbf{E} + \frac{1}{c}\frac{\partial \mathbf{B}}{\partial t} = 0 \tag{4.146}$$

$$\nabla \times \mathbf{H} - \frac{1}{c}\frac{\partial \mathbf{D}}{\partial t} = \frac{4\pi}{c}\mathbf{j}. \tag{4.147}$$

to the regions both inside and outside the particle. The vectors $\mathbf{D}$ and $\mathbf{H}$ can be expressed phenomenologically by the pair of equations

$$\mathbf{D} = \varepsilon \mathbf{E} \tag{4.148}$$

$$\mathbf{H} = \frac{1}{\mu}\mathbf{B}. \tag{4.149}$$

ε is the *dielectric constant* and μ is the *magnetic permeability* of the medium. These relations are just an alternative to Eqs. 4.103 and 4.104 which previously defined $\mathbf{D}$ and $\mathbf{H}$ in terms of the medium's polarization $\mathbf{P}$ and magnetization $\mathbf{M}$. The use of Eqs. 4.148 and 4.149 is more appropriate here because the refractive index of a medium is simply related to ε and μ through

$$n = \sqrt{\mu\varepsilon}. \tag{4.150}$$

We will only consider scattering from media that are non-magnetic ($\mu = 1$) and void of free charge ($\rho = \mathbf{j} = 0$). In keeping with the form for the incident wave (Eq. 4.145), one assigns an $e^{-i\omega t}$ time-dependence to the total field and the Maxwell's equations become

$$\nabla\times\mathbf{E} - \frac{i\omega}{c}\mathbf{B} = 0 \tag{4.151}$$

$$\nabla\times\mathbf{B} + \frac{in^2\omega}{c}\mathbf{E} = 0. \tag{4.152}$$

Eliminating **B** from this pair of equations results in a single equation for the electric field at an arbitrary point in space:

$$\nabla\times(\nabla\times\mathbf{E}) - n^2k_v^2\mathbf{E} = 0. \tag{4.153}$$

4.4.1 Integral Formulation of the Scattered Field

An approach to solving Eq. 4.153 is to first rewrite it as

$$\nabla\times(\nabla\times\mathbf{E}) - n_0^2k_v^2\mathbf{E} = \left(n^2 - n_0^2\right)k_v^2\mathbf{E}. \tag{4.154}$$

By writing the equation in this way, it becomes apparent that the term on the right-hand side acts as the source term responsible for producing the scattered wave. Notice that if this term were not present, it would mean that n, the particle's refractive index, exactly matches n_0, the index of the surrounding medium, and the situation would be equivalent to replacing the particle with background material. From an optical standpoint, it would be as if there were no particle present at all. In that case, the incident plane wave propagates freely through the homogeneous medium, and there is no scattering. This statement is born out by the fact that the incident field, given by Eq. 4.145, satisfies Eq. 4.154 with no source term, i.e.,

$$\nabla\times(\nabla\times\mathbf{E}_i) = \nabla(\nabla\cdot\mathbf{E}_i) - \nabla^2\mathbf{E}_i = -\nabla^2\mathbf{E}_i = n_0^2k_v^2\mathbf{E}_i. \tag{4.155}$$

To find the field when the source term is present and the particle does scatter, we first solve the problem of finding the scattered wave reaching location $\mathbf{r}$ as the result of a point-scatterer being situated at $\mathbf{r}'$. This is accomplished by solving the Green's function equation

$$\nabla\times\left[\nabla\times\tilde{\tilde{\mathbf{G}}}(\mathbf{r}-\mathbf{r}')\right] - n_0^2k_v^2\tilde{\tilde{\mathbf{G}}}(\mathbf{r}-\mathbf{r}') = \tilde{\tilde{\mathbf{I}}}\delta(\mathbf{r}-\mathbf{r}') \tag{4.156}$$

subject to the boundary condition $\nabla\cdot\tilde{\tilde{\mathbf{G}}}(\mathbf{r}-\mathbf{r}') = 0$. The solution for the Green's tensor is [7]

$$\tilde{\tilde{\mathbf{G}}}(\mathbf{r}-\mathbf{r}') = \frac{e^{ik|\mathbf{r}-\mathbf{r}'|}}{4\pi|\mathbf{r}-\mathbf{r}'|}\left[\tilde{\tilde{\mathbf{I}}}f_1(|\mathbf{r}-\mathbf{r}'|) + \frac{(\mathbf{r}-\mathbf{r}')(\mathbf{r}-\mathbf{r}')}{|\mathbf{r}-\mathbf{r}'|^2}f_2(|\mathbf{r}-\mathbf{r}'|)\right], \tag{4.157}$$

where

$$f_1(|\mathbf{r}-\mathbf{r}'|) = \frac{-1 + ik|\mathbf{r}-\mathbf{r}'| + k^2|\mathbf{r}-\mathbf{r}'|^2}{k^2|\mathbf{r}-\mathbf{r}'|^2} \tag{4.158}$$

$$f_2(|\mathbf{r}-\mathbf{r}'|) = \frac{3 - ik|\mathbf{r}-\mathbf{r}'| - k^2|\mathbf{r}-\mathbf{r}'|^2}{k^2|\mathbf{r}-\mathbf{r}'|^2}. \tag{4.159}$$

With this known, the total field satisfying Eq. 4.154 is

$$\mathbf{E}(\mathbf{r},t) = \mathbf{E}_i(\mathbf{r},t) + \int_V d^3r'\, \tilde{\tilde{\mathbf{G}}}(\mathbf{r}-\mathbf{r}') \cdot k_v^2 \left(n^2 - n_0^2\right) \mathbf{E}(\mathbf{r}',t). \tag{4.160}$$

The integral represents the scattered field—it is evaluated over the entire volume of the scattering particle.

Since the detector is placed at a location $\mathbf{r}$ far outside the particle, we are interested in determining the far-field solution, where we also demand that $k|\mathbf{r}-\mathbf{r}'| \gg 1$. In this case, $f_1 \to 1$ and $f_2 \to -1$ above, and $\mathbf{r}-\mathbf{r}' \to \mathbf{r}$ everywhere in Eq. 4.157 except in the term $e^{ik|\mathbf{r}-\mathbf{r}'|}$. In this latter case, we make use of

$$|\mathbf{r}-\mathbf{r}'| = \sqrt{r^2 + r'^2 - 2\mathbf{r}\cdot\mathbf{r}'} \simeq r\sqrt{1 - 2\frac{\mathbf{r}\cdot\mathbf{r}'}{r^2}} \simeq r - \mathbf{r}'\cdot\mathbf{e}_r. \tag{4.161}$$

The Green's tensor then simplifies to

$$\tilde{\tilde{\mathbf{G}}} = \left(\tilde{\tilde{\mathbf{I}}} - \mathbf{e}_r\mathbf{e}_r\right) \frac{e^{ikr}}{4\pi r} e^{-ik\mathbf{r}'\cdot\mathbf{e}_r}. \tag{4.162}$$

By letting $\mathbf{k}' = k\mathbf{e}_r$ denote the scattered wavevector, the fundamental equation for the scattered field at the detector becomes

$$\mathbf{E}_s(\mathbf{r},t) = k_v^2\left(n^2 - n_0^2\right) \frac{e^{ikr}}{4\pi r} \left(\tilde{\tilde{\mathbf{I}}} - \mathbf{e}_r\mathbf{e}_r\right) \cdot \int_V d^3r'\, e^{-i\mathbf{k}'\cdot\mathbf{r}'} \mathbf{E}(\mathbf{r}',t). \tag{4.163}$$

The factor of e^{ikr}/r is the characteristic radial-dependence for the amplitude of a spherical outgoing wave. The angular-dependence of the scattered field can only be ascertained by performing the given integral over the volume of the particle. This, in turn, depends on the internal field of the particle.

Example 4.6 Rayleigh-Gans-Debye Scattering. Consider scattering from particles that are small relative to the wavelength of the incident light and having a refractive index close to that of the surrounding medium. More specifically, we are presupposing the conditions

$$|m-1| \ll 1 \qquad \text{and} \qquad kd\,|m-1| \ll 1, \tag{4.164}$$

where $m = n/n_0$ and d is the characteristic dimension of the particle. In cases like this, one is able to apply the so-called *Rayleigh-Gans-Debye (RGD) approximation* to Eq. 4.163, which replaces the internal field of the particle, $\mathbf{E}(\mathbf{r}', t)$, with the incident field, $\mathbf{E}_i(\mathbf{r}', t) = \mathbf{E}_0 e^{i(\mathbf{k}\cdot\mathbf{r}' - \omega t)}$. From a physical standpoint, the basic underlying assumption is that the scattering is weak enough such that only radiation undergoing a single scattering event within the particle contributes to the field at point $\mathbf{r}$. Under the RGD approximation, the scattered field is

$$\mathbf{E}_s(\mathbf{r}, t) = k^2 (m^2 - 1) \frac{e^{ikr}}{4\pi r} \left(\tilde{\tilde{\mathbf{I}}} - \mathbf{e}_r \mathbf{e}_r\right) \cdot \mathbf{E}_0 \int_V d^3r' \, e^{i(\mathbf{Q}\cdot\mathbf{r}' - \omega t)}. \tag{4.165}$$

$\mathbf{Q}$ is called the *scattering vector*; it corresponds to the difference between the incident and scattered wavevectors, i.e.,

$$\mathbf{Q} = \mathbf{k} - \mathbf{k}'. \tag{4.166}$$

If the experiment is conducted in the VV-geometry where the polarization of the incident electric field is normal to the scattering plane (i.e., the plane defined by $\mathbf{k}$ and $\mathbf{k}'$), then $\left(\tilde{\tilde{\mathbf{I}}} - \mathbf{e}_r \mathbf{e}_r\right) \cdot \mathbf{E}_0 = \mathbf{E}_0$, and we have

$$\mathbf{E}_s(\mathbf{r}, t) = k^2 (m^2 - 1) \frac{e^{ikr}}{4\pi r} \mathbf{E}_0 \int_V d^3r' \, e^{i(\mathbf{Q}\cdot\mathbf{r}' - \omega t)}. \tag{4.167}$$

This tells us that when the RGD approximation is valid, the polarization of the scattered wave is also normal to the scattering plane, i.e., no depolarized scattering component ever emerges.

At this point, it becomes convenient to rewrite the positions of points internal to the particle in terms of relative position vectors $\boldsymbol{\rho}$ with respect to the center-of-mass position $\mathbf{R}$, i.e.,

$$\mathbf{r}' = \mathbf{R} + \boldsymbol{\rho}. \tag{4.168}$$

Then Eq. 4.165 becomes

$$\mathbf{E}_s(\mathbf{r}, t) = F(\mathbf{Q}) \frac{e^{ikr}}{r} \mathbf{E}_0 e^{i(\mathbf{Q}\cdot\mathbf{R} - \omega t)}, \tag{4.169}$$

where $F(\mathbf{Q})$ is called the *form factor* for the particle, given by

$$F(\mathbf{Q}) = \frac{k^2(m^2 - 1)}{4\pi} \int_V d^3\rho \, e^{i\mathbf{Q}\cdot\boldsymbol{\rho}}. \tag{4.170}$$

The form factor is determined by (a) the size, shape, and composition of the particle and (b) the orientation of the particle relative to the scattering vector $\mathbf{Q}$.

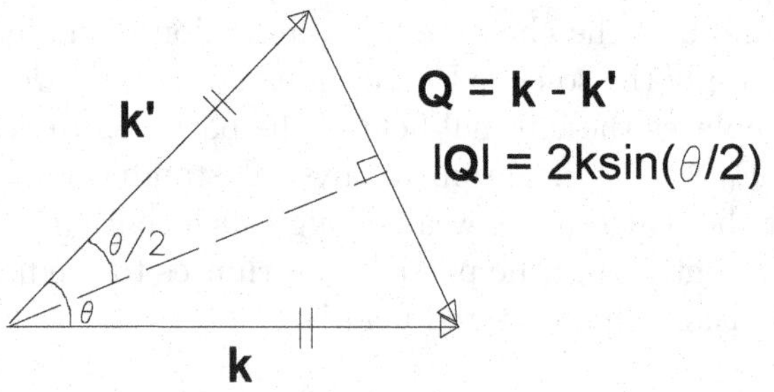

Figure 4.5 Relationship between the scattering vector **Q** and the incident and scattered wavevectors.

If the scatterer is a small, homogeneous sphere, then the form factor is only a function of the particle radius a and the magnitude of the scattering vector. As shown in Fig. 4.5, $Q \equiv |\mathbf{Q}|$ is determined by the scattering angle θ:

$$Q = 2k \sin\frac{\theta}{2}. \tag{4.171}$$

The integral can be evaluated by working in spherical coordinates (ρ, θ, ϕ) (here, V denotes the particle volume):

$$\frac{1}{V}\int_V d^3\rho\, e^{i\mathbf{Q}\cdot\boldsymbol{\rho}} = \frac{1}{V}\int_{\phi=0}^{2\pi} d\phi \int_{\rho=0}^{a} \rho d\rho \int_{\theta=0}^{\pi} e^{iQ\rho\cos\theta}\rho\sin\theta\, d\theta = \frac{3j_1(Qa)}{Qa}. \tag{4.172}$$

Here, $j_1(x)$ is the first-order spherical Bessel function; it is given by

$$j_1(x) = \frac{\sin x - x\cos x}{x^2}. \tag{4.173}$$

The scattered wave now has the form

$$\mathbf{E}_s(\mathbf{r}, t) = k^2\left(m^2 - 1\right)a^3\left[\frac{j_1(Qa)}{Qa}\right]\frac{e^{ikr}}{r}\mathbf{E}_0 e^{i(\mathbf{Q}\cdot\mathbf{R}-\omega t)}. \tag{4.174}$$

The detected intensity of the scattered light is proportional to $|\mathbf{E}_s|^2$, i.e.,

$$I \propto \frac{1}{r^2}k^4\left(m^2-1\right)^2 a^6\left[\frac{j_1(Qa)}{Qa}\right]^2 E_0^2. \tag{4.175}$$

A plot of this distribution is displayed in Fig. 4.6.

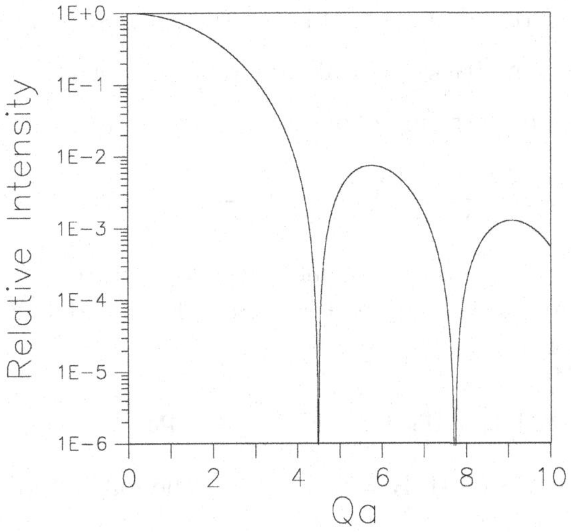

Figure 4.6 Scattered light intensity from a homogeneous sphere in the RGD approximation.

Example 4.7 Rayleigh Scattering Limit. *Rayleigh scattering* corresponds to the case where the particle size is very much smaller than the wavelength of the light. This is sometimes referred to as the long wavelength limit. For a spherical particle of radius a, this restriction means that $2ka \ll 1$. It then follows from Eq. 4.171 that $Qa \ll 1$. If, along with the condition for Rayleigh scattering, one was to impose the additional limitation $|m-1| \ll 1$ on the relative refractive index, then the RGD approximation would be valid. In that case, we see that

$$\frac{j_1(Qa)}{Qa} \simeq \frac{1}{3}\left(1 - \frac{1}{10}Q^2a^2 + \ldots\right) \to \frac{1}{3}, \tag{4.176}$$

and the particle's form factor would exhibit no angular dependence. From Eq. 4.174, the scattered field would then be

$$\mathbf{E}_s(\mathbf{r}, t) = k^2\left(\frac{m^2-1}{3}\right)a^3\,\frac{e^{ikr}}{r}\mathbf{E}_0 e^{i(\mathbf{Q}\cdot\mathbf{R}-\omega t)}. \tag{4.177}$$

This is not quite the correct result for Rayleigh scattering, however, because the first RGD condition, $|m-1| \ll 1$, is unnecessarily restrictive. In Example 4.9, the rigorous calculation for scattering from a sphere of arbitrary size and refractive index will be examined (the so-called *Mie scattering* problem). In the long wavelength limit [8] one finds that the factor $(m^2-1)/3$ in Eq. 4.177 should, in fact, be replaced by $(m^2-1)/(m^2+2)$. This alteration now makes the above expression for the scattered field correct for the case of the VV-geometry, where $\mathbf{k}' \perp \mathbf{E}_0$. If one also wants to consider the wave scattered above and below the previously defined scattering plane, then it is necessary to go back and make the replacement $\mathbf{E}_0 \rightarrow \left(\tilde{\tilde{\mathbf{I}}} - \mathbf{e}_r\mathbf{e}_r\right) \cdot \mathbf{E}_0$. The more general Rayleigh scattering expression is therefore

$$\mathbf{E}_s(\mathbf{r},t) = k^2 \left(\frac{m^2-1}{m^2+2}\right) a^3 \frac{e^{ikr}}{r} \left(\tilde{\tilde{\mathbf{I}}} - \mathbf{e}_r\mathbf{e}_r\right) \cdot \mathbf{E}_0 e^{i(\mathbf{Q}\cdot\mathbf{R}-\omega t)}. \tag{4.178}$$

Let us compare this result to the far-field radiation emitted by an oscillating electric dipole $\mathbf{p}_0 e^{-i\omega t}$ situated at the origin. Using Eqs. 4.127 and 4.128, the dipole field can be written in the form

$$\mathbf{E}_{\text{dipole}}(\mathbf{r},t) = -k^2 \frac{e^{i(kr-\omega t)}}{r} (\mathbf{p}_0 \times \mathbf{e}_r) \times \mathbf{e}_r. \tag{4.179}$$

Comparing to Eq. 4.178 (with $\mathbf{R} = 0$) identifies the following correspondence:

$$(\mathbf{p}_0 \times \mathbf{e}_r) \times \mathbf{e}_r \leftrightarrow -\left(\frac{m^2-1}{m^2+2}\right) a^3 \left(\tilde{\tilde{\mathbf{I}}} - \mathbf{e}_r\mathbf{e}_r\right) \cdot \mathbf{E}_0. \tag{4.180}$$

This shows that the incident radiation field, $\mathbf{E}_0 e^{-i\omega t}$, induces an oscillating dipole moment, $\mathbf{p}_0 e^{-i\omega t}$, in the scattering particle. The light scattered by the particle is, in actuality, the radiation emitted by the induced dipole. The resulting dipole moment is in the same direction as the polarization of the incident field, with magnitude

$$p_0 = \alpha E_0, \tag{4.181}$$

where $\alpha = a^3(m^2-1)/(m^2+2)$ is the so-called *polarizability* of the sphere.

For a particle at the origin, Eq. 4.178 can be rewritten in the form

$$\mathbf{E}_s(\mathbf{r},t) = \mathbf{f}(\theta) \frac{e^{i(kr-\omega t)}}{r} E_0, \tag{4.182}$$

where we have introduced the *scattering amplitude* $\mathbf{f}(\theta)$ given by

$$\mathbf{f}(\theta) = k^2 \alpha \left(\tilde{\tilde{\mathbf{I}}} - \mathbf{e}_r\mathbf{e}_r\right) \cdot \frac{\mathbf{E}_0}{E_0}. \tag{4.183}$$

One is often interested in calculating the angular *differential cross-section* $d\sigma/d\Omega$, where $(d\sigma/d\Omega)\, d\Omega$ is the power scattered into solid angle $d\Omega$ divided by the intensity of the incident beam, i.e.,

$$\frac{d\sigma}{d\Omega} = \frac{\langle dP/d\Omega \rangle}{cE_0^2/8\pi} = \frac{\langle (\mathbf{S}_s \cdot \mathbf{e}_r)\, r^2 \rangle}{cE_0^2/8\pi} = \frac{r^2 \langle |\mathbf{E}_s|^2 \rangle}{E_0^2} = \langle |\mathbf{f}(\theta)|^2 \rangle. \tag{4.184}$$

Here, $\langle\ \rangle$ denotes a time average. For Rayleigh scattering, it then follows that

$$\frac{d\sigma}{d\Omega} = k^4 \alpha^2 \left| \left(\tilde{\tilde{\mathbf{I}}} - \mathbf{e}_r\mathbf{e}_r\right) \cdot \frac{\mathbf{E}_0}{E_0} \right|^2. \tag{4.185}$$

For the VV-geometry of Fig. 4.4, we have $\mathbf{E}_0 \perp \mathbf{e}_r$, and $\left|\left(\tilde{\tilde{\mathbf{I}}} - \mathbf{e}_r\mathbf{e}_r\right)\cdot\frac{\mathbf{E}_0}{E_0}\right| = 1$. Thus, the total Rayleigh scattering cross-section is

$$\sigma = \int\limits_{4\pi} \frac{d\sigma}{d\Omega} d\Omega = 4\pi k^4 \alpha^2, \tag{4.186}$$

or

$$\sigma = \pi a^2 \left[2k^2a^2 \left(\frac{m^2 - 1}{m^2 + 2}\right)\right]^2. \tag{4.187}$$

Notice that the scattering cross-section is much smaller than the projected geometric area of the particle πa^2 since $ka \ll 1$ in the Rayleigh scattering limit. The characteristic k^4-dependence is also apparent.

Example 4.8 Extension to Scattering by Fluctuations. Return to Eq. 4.165, the general expression for the scattered field in the RGD approximation, and substitute $n_0 k_v$ for k. In order to further generalize the result, let us also place the refractive index factor $m^2 - 1$ back into the integral. Then,

$$\mathbf{E}_s(\mathbf{r}, t) = k_v^2 \frac{e^{ikr}}{4\pi r} \left(\tilde{\tilde{\mathbf{I}}} - \mathbf{e}_r\mathbf{e}_r\right) \cdot \mathbf{E}_0 e^{-i\omega t} \int\limits_V d^3r' \left(n^2 - n_0^2\right) e^{i\mathbf{Q}\cdot\mathbf{r}'}. \tag{4.188}$$

In this form, the expression is amenable to examining the problem of scattering from local fluctuations in a dielectric medium. To see this, use Eq. 4.150 to rewrite the refractive index term as the difference between the instantaneous value $\varepsilon(\mathbf{r}', t)$ of the dielectric constant at various points in the scattering volume, V, and the background dielectric constant ε_0:

$$n^2 - n_0^2 = \varepsilon(\mathbf{r}', t) - \varepsilon_0 \equiv \Delta\varepsilon(\mathbf{r}', t). \tag{4.189}$$

In the case of a fluid, one also has the relationship [9]

$$\Delta\varepsilon(\mathbf{r}', t) = \left(\frac{\partial\varepsilon}{\partial\rho}\right)_T \Delta\rho(\mathbf{r}', t), \tag{4.190}$$

where

$$\Delta\rho(\mathbf{r}', t) = \rho(\mathbf{r}', t) - \rho_0 \tag{4.191}$$

represents the local thermal density-fluctuations in the medium having average density ρ_0. Furthermore, if we denote the spatial Fourier transform of $\Delta\rho(\mathbf{r}', t)$ by

$$\Delta\rho(\mathbf{Q}, t) = \int\limits_V d^3r' \, \Delta\rho(\mathbf{r}', t) \, e^{i\mathbf{Q}\cdot\mathbf{r}'}, \tag{4.192}$$

then the scattered field becomes

$$\mathbf{E}_s(\mathbf{r},t) = k_v^2 \frac{e^{ikr}}{4\pi r}\left(\tilde{\tilde{\mathbf{I}}} - \mathbf{e}_r\mathbf{e}_r\right)\cdot \mathbf{E}_0 e^{-i\omega t}\left(\frac{\partial \varepsilon}{\partial \rho}\right)_T \Delta\rho(\mathbf{Q},t). \tag{4.193}$$

Equation 4.193 shows that the scattering of light is caused by local density fluctuations in the medium.

Referring back to Eq. 4.182, we can write the scattering amplitude as

$$\mathbf{f}(\theta) = \frac{1}{4\pi}k_v^2\left(\tilde{\tilde{\mathbf{I}}} - \mathbf{e}_r\mathbf{e}_r\right)\cdot\frac{\mathbf{E}_0}{E_0}\left(\frac{\partial \varepsilon}{\partial \rho}\right)_T \Delta\rho(\mathbf{Q},t), \tag{4.194}$$

so the differential cross-section for scattering in the VV-geometry becomes

$$\frac{d\sigma}{d\Omega} = \left\langle |\mathbf{f}(\theta)|^2 \right\rangle = \frac{1}{16\pi^2}k_v^4\left(\frac{\partial \varepsilon}{\partial \rho}\right)_T^2 \langle \Delta\rho(\mathbf{Q})\,\Delta\rho(-\mathbf{Q})\rangle. \tag{4.195}$$

The time average $\langle \Delta\rho(\mathbf{Q})\,\Delta\rho(-\mathbf{Q})\rangle$ is statistically equivalent to an ensemble average, and is usually written as the so-called *structure factor* for the medium [10]:

$$S(\mathbf{Q}) = \frac{1}{N}\langle \Delta\rho(\mathbf{Q})\,\Delta\rho(-\mathbf{Q})\rangle. \tag{4.196}$$

N is the number of molecules in the scattering volume V. For light scattering, Q is very small, therefore we can use the following result from thermodynamic fluctuation theory [11]:

$$\lim_{Q\to 0} S(\mathbf{Q}) = \rho_0 k_B T \chi_T. \tag{4.197}$$

Here, $\chi_T = \rho_0^{-1}(\partial\rho/\partial P)_{T,\,\rho=\rho_0}$ is the *isothermal compressibility* of the fluid. Thus, the cross-section is

$$\frac{d\sigma}{d\Omega} = |\mathbf{f}(\theta)|^2 = \frac{N}{16\pi^2}k_v^4\left(\frac{\partial \varepsilon}{\partial \rho}\right)_T^2 \rho_0 k_B T \chi_T, \tag{4.198}$$

a result first obtained by Einstein in 1910 [12]. Equation 4.198 illustrates that light scattering can be used to measure macroscopic thermodynamic properties associated with the equation of state of a substance.

4.4.2 Differential Formulation of the Scattered Field

There is an alternate approach to solving the problem of light scattering from a dielectric particle. It involves constructing the solution to Maxwell's equations both inside and outside the particle, subject to the boundary conditions for the field components at the surface of the particle. The method is rigorous, and leads to an exact solution of the scattering problem for particles of any size and refractive index. The usefulness of the technique is limited, however, by its inherent mathematical complexity, i.e., the problem has only been completely solved in the cases of scattering from homogeneous spheres [13], infinite cylinders [14] [15], and prolate/oblate spheroids [16].

For this formulation of the scattering problem, we return to the basic vector wave equation for the electric field, as given by Eq. 4.153. From Eqs. 4.151 and 4.152, it is easy to show that the magnetic field vector satisfies the same wave equation. Because of the conditions $\boldsymbol{\nabla}\cdot\mathbf{E} = 0$ and $\boldsymbol{\nabla}\cdot\mathbf{B} = 0$, we can then write the field equations as

$$\begin{aligned} \nabla^2\mathbf{E} + n^2k_v^2\mathbf{E} &= 0 \\ \nabla^2\mathbf{B} + n^2k_v^2\mathbf{B} &= 0. \end{aligned} \tag{4.199}$$

These equations are valid at points both inside and outside the particle. Here, be aware that n denotes the refractive index for the region in question.

It turns out that solutions to the vector wave equation can be constructed from solutions to the scalar wave equation

$$\nabla^2\psi + n^2k_v^2\psi = 0, \tag{4.200}$$

where ψ represents any rectangular component of $\mathbf{E}$ or $\mathbf{B}$. Specifically, given a solution ψ of the scalar equation, the following two solutions of the vector wave equation can be formed:[§]

$$\mathbf{M}_\psi = \boldsymbol{\nabla}\times(\mathbf{r}\psi) \qquad \text{and} \qquad \mathbf{N}_\psi = \frac{1}{nk_v}\boldsymbol{\nabla}\times\mathbf{M}_\psi. \tag{4.201}$$

In addition,

$$\mathbf{M}_\psi = \frac{1}{nk_v}\boldsymbol{\nabla}\times\mathbf{N}_\psi. \tag{4.202}$$

In order to determine the electric and magnetic fields satisfying the vector wave equation, it has been demonstrated that one must first determine two independent solutions, ψ_1 and ψ_2, to the scalar equation, and then find the corresponding pairs of solutions to the vector wave equation, i.e., $\mathbf{M}_{\psi_1}$, $\mathbf{N}_{\psi_1}$ and $\mathbf{M}_{\psi_2}$, $\mathbf{N}_{\psi_2}$. The field vectors are then given by

$$\mathbf{E} = \mathbf{M}_{\psi_2} + i\mathbf{N}_{\psi_1} \qquad \text{and} \qquad \mathbf{B} = n\left(-\mathbf{M}_{\psi_1} + i\mathbf{N}_{\psi_2}\right). \tag{4.203}$$

Example 4.9 Mie Scattering. The application of the latter formalism to the general problem of scattering by a homogeneous sphere is referred to as *Mie scattering.*[¶] Polarized light scatters from a sphere of radius a and relative refractive index m that is situated at the origin of an (r, θ, ϕ) spherical coordinate system. As before, we focus on the VV-geometry of Fig. 4.4. To solve the scattering problem exactly, one first determines the general form for the fields both inside and outside the particle. The specification of the fields is then completed by demanding that the tangential

[§]See, for example, Stratton [17].

[¶]The elements of Mie scattering presented here essentially follow the major points outlined in the treatment by van de Hulst [18].

components of the internal and external fields be continuous at the interface between the particle and the surrounding medium. Because the particle is spherical, these boundary conditions reduce to

$$\begin{aligned} \mathbf{e}_r \times \mathbf{E}_{\text{in}}(r=a) &= \mathbf{e}_r \times \mathbf{E}_{\text{out}}(r=a) \\ \mathbf{e}_r \times \mathbf{B}_{\text{in}}(r=a) &= \mathbf{e}_r \times \mathbf{B}_{\text{out}}(r=a), \end{aligned} \tag{4.204}$$

where the subscripts "in" and "out" denote the fields *inside* and *outside* the particle.

Outside the sphere, one has both the incident fields and the scattered fields, i.e.,

$$\mathbf{E}_{\text{out}} = \mathbf{E}_i + \mathbf{E}_s \qquad \text{and} \qquad \mathbf{B}_{\text{out}} = \mathbf{B}_i + \mathbf{B}_s. \tag{4.205}$$

One can show that the known form for the incident fields, $\mathbf{E}_i = E_0 e^{i(kz-\omega t)}\mathbf{e}_x$ and $\mathbf{B}_i = B_0 e^{i(kz-\omega t)}\mathbf{e}_y = n_0 E_0 e^{i(kz-\omega t)}\mathbf{e}_y$, can be derived from Eqs. 4.201–4.203 by choosing the following expansions for the two solutions to the scalar wave equation:

$$\begin{aligned} \psi_1 &= E_0 e^{-i\omega t}\cos\phi \sum_{\ell=1}^{\infty}(-i)^{\ell}\left[\frac{2\ell+1}{\ell(\ell+1)}\right]P_{\ell}^{(1)}(\cos\theta)\, j_{\ell}(kr) \\ \psi_2 &= E_0 e^{-i\omega t}\sin\phi \sum_{\ell=1}^{\infty}(-i)^{\ell}\left[\frac{2\ell+1}{\ell(\ell+1)}\right]P_{\ell}^{(1)}(\cos\theta)\, j_{\ell}(kr). \end{aligned} \tag{4.206}$$

For the scattered wave, the two solutions can also be expressed as a pair of infinite series:

$$\begin{aligned} \psi_1 &= -E_0 e^{-i\omega t}\cos\phi \sum_{\ell=1}^{\infty} a_{\ell}(-i)^{\ell}\left[\frac{2\ell+1}{\ell(\ell+1)}\right]P_{\ell}^{(1)}(\cos\theta)\, h_{\ell}^{(2)}(kr) \\ \psi_2 &= -E_0 e^{-i\omega t}\sin\phi \sum_{\ell=1}^{\infty} b_{\ell}(-i)^{\ell}\left[\frac{2\ell+1}{\ell(\ell+1)}\right]P_{\ell}^{(1)}(\cos\theta)\, h_{\ell}^{(2)}(kr), \end{aligned} \tag{4.207}$$

where the a_{ℓ}'s and b_{ℓ}'s are undetermined coefficients. The choice of spherical Hankel functions, $h_{\ell}^{(2)}(kr)$, hinges on the boundary conditions at the particle surface and at infinity.

The solution inside the sphere must be finite at $r = 0$; the form consistent with this condition is

$$\begin{aligned} \psi_1 &= mE_0 e^{-i\omega t}\cos\phi \sum_{\ell=1}^{\infty} c_{\ell}(-i)^{\ell}\left[\frac{2\ell+1}{\ell(\ell+1)}\right]P_{\ell}^{(1)}(\cos\theta)\, j_{\ell}(mkr) \\ \psi_2 &= mE_0 e^{-i\omega t}\sin\phi \sum_{\ell=1}^{\infty} d_{\ell}(-i)^{\ell}\left[\frac{2\ell+1}{\ell(\ell+1)}\right]P_{\ell}^{(1)}(\cos\theta)\, j_{\ell}(mkr), \end{aligned} \tag{4.208}$$

where the c_{ℓ}'s and d_{ℓ}'s are two more sets of undetermined coefficients.

Now that the solutions to the scalar wave equation are specified inside and outside the particle, the functional form for the internal and external fields can be

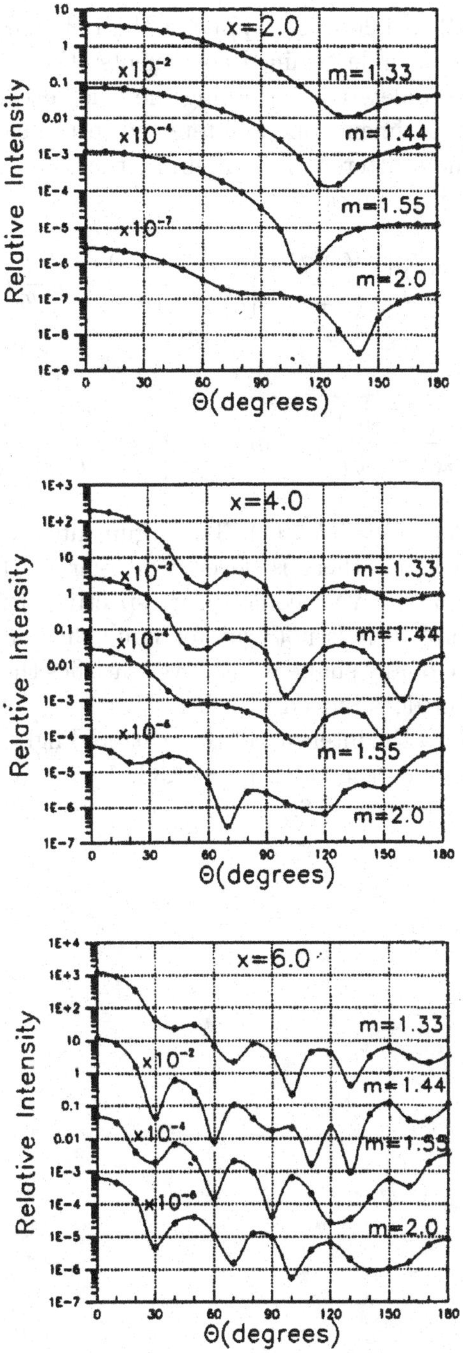

Figure 4.7 Mie calculation of scattered light intensity from a homogeneous sphere ($x = ka$). (Data is taken from Lowan [19].)

determined from Eqs. 4.201–4.203. By applying the boundary conditions given by Eqs. 4.204, the four sets of undetermined coefficients (the a_ℓ's, b_ℓ's, c_ℓ's, and d_ℓ's) can be derived, thereby completely determining the fields everywhere.

For a scattering experiment, one is interested in the electric field vector at a distance very far from the scatterer. In this limit, the Mie result for the scattered field in the VV-geometry is given below:

$$\mathbf{E}_s(\mathbf{r},t) = \mathbf{E}_0 \frac{e^{i(kr-\omega t)}}{-ikr} \sum_{\ell=1}^{\infty} \left[\frac{2\ell+1}{\ell(\ell+1)}\right] \left[a_\ell \frac{P_\ell^1(\cos\theta)}{\sin\theta} + b_\ell \frac{d}{d\theta} P_\ell^1(\cos\theta)\right]. \quad (4.209)$$

The coefficients are

$$a_\ell = \frac{\gamma_\ell(x)\,\gamma'_\ell(y) - m\,\gamma'_\ell(x)\,\gamma_\ell(y)}{\zeta_\ell(x)\,\gamma'_\ell(y) - m\,\zeta'_\ell(x)\,\gamma_\ell(y)} \quad \text{and} \quad b_\ell = \frac{\gamma'_\ell(x)\,\gamma_\ell(y) - m\,\gamma_\ell(x)\,\gamma'_\ell(y)}{m\,\zeta_\ell(x)\,\gamma'_\ell(y) - \zeta'_\ell(x)\,\gamma_\ell(y)}, \quad (4.210)$$

where $\gamma_\ell(z) = z\,j_\ell(z)$ and $\zeta_\ell(z) = z\,h_\ell^{(2)}(z)$. The arguments are $x = ka$ and $y = mka$. It is important to observe that there is never a depolarized field component when scattering from a sphere in the VV-geometry, irrespective of the particle's size and refractive index. In general, this is not true for non-spherical scatterers. However, recall that when a particle of any shape whatsoever satisfies the RGD approximation, the depolarized component does indeed vanishes.

Figure 4.7 shows how the intensity of the scattered light, which is proportional to $|\mathbf{E}_s|^2$, varies with angle for the cases $m = 1.33$, 1.44, 1.55, and 2, evaluated at different values of ka between 2 and 6.

Suggested References

These two texts provide excellent overviews of general electromagnetic theory. The first book is primarily aimed at the graduate level, while the others are written more for advanced undergraduates:

[a] J. D. Jackson, *Classical Electrodynamics*, 3rd ed. (John Wiley and Sons, New York, 1998).

[b] J. R. Reitz, F. J. Milford, and R. W. Christy, *Foundations of Electromagnetic Theory*, 4th ed. (Addison-Wesley, Reading, MA, 1992).

[c] J. B. Marion, *Classical Electromagnetic Radiation* (Academic Press, New York, 1965).

The following are general references on light scattering from fluctuations and particles:

[d] B. Chu, *Laser Light Scattering: Basic Principles and Practice*, 2nd ed. (Academic Press, San Diego, 1991).

[e] B. Crosignani, P. Di Porto, and M. Bertolotti, *Statistical Properties of Scattered Light* (Academic Press, New York, 1975).

[f] C. F. Bohren and D. R. Huffman, Absorption and Scattering of Light by Small Particles (John Wiley and Sons, New York, 1983).

[g] M. Kerker, *The Scattering of Light and Other Electromagnetic Radiation* (Academic Press, New York, 1969).

[h] H. C. van de Hulst, *Light Scattering by Small Particles* (John Wiley and Sons, New York, 1957).

[i] W. Brown, ed., *Light Scattering: Principles and Development* (Oxford University Press, Oxford, 1996).

Problems

1. The Lagrangian of a relativistic point particle of rest energy mc^2 and charge e moving with velocity $\mathbf{v}$ in an electromagnetic field (vector potential $\mathbf{A}$, scalar potential ϕ) is

$$\mathcal{L} = -\frac{mc^2}{\gamma} + \frac{e}{c}\mathbf{v}\cdot\mathbf{A} - e\phi,$$

where $\gamma = \left(1 - \beta^2\right)^{-1/2}$ and $\beta = |\mathbf{v}|/c$.

(a) Show that this Lagrangian produces the correct equation of motion for the charged particle in the EM field.

(b) Using the standard procedure, construct the Hamiltonian for the particle.

(c) Determine the non-relativistic limit of the Hamiltonian.

2. The beam from a helium-neon laser ($\lambda = 632.8$ nm) is scattered from a dielectric sphere of diameter $d = 0.25$ μm in air.

(a) In order to apply the RGD approximation to the computation of the scattered field, what restriction must be placed on the refractive index of the sphere?

(b) Using the RGD approximation, plot the relative scattering intensity $I(\theta)/I(0$ vs. the scattering angle θ for incident radiation polarized normal to the scattering plane.

(c) Repeat the last calculation for incident radiation with its polarization lying in the scattering plane.

3. In the RGD approximation, the form factor for scattering from a homogeneous dielectric sphere of radius a and relative refractive index $m = n/n_0$ (see Eqs. 4.170 and 4.172) can be written as

$$F(Q) = \frac{k^2(m^2-1)V}{4\pi}A(Q),$$

where $V = 4\pi a^3/3$ is the volume of the sphere and

$$A(Q) = \frac{3j_1(Qa)}{Qa}.$$

Furthermore, by applying the condition $|m-1| \ll 1$, as required under the RGD approximation, one may write

$$F(Q) \simeq \frac{2}{3}k^2(m-1)a^3 A(Q).$$

(a) Instead of a homogeneous sphere, consider scattering from a coated sphere, as shown in Fig. 4.8. The index of the particle core relative to the background medium is $m_2 = n_2/n_0$, and the index of the outer shell relative to the background is $m_1 = n_1/n_0$. Let f represent the ratio of the inner radius to the outer radius of the particle, i.e., $f = 1 - t/a$, where t is the thickness of the shell. Show that the form factor for a coated sphere is given by

$$F(Q) \simeq \frac{2}{3}k^2(m_1-1)a^3\left[\frac{3j_1(Qa)}{Qa} + f^3\left(\frac{m_2-m_1}{m_1-1}\right)\frac{3j_1(fQa)}{fQa}\right].$$

Hint: *The problem is very straightforward if one considers scattering from the coated sphere to be a superposition of scattering from a certain combination of homogeneous spheres.*

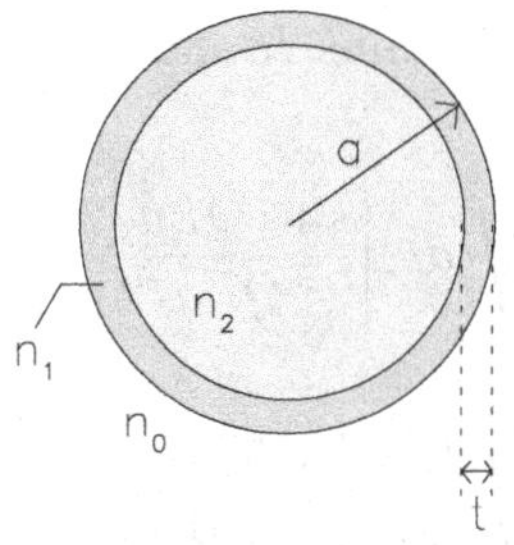

Figure 4.8 Coated sphere for Problem 3.

(b) For RGD scattering in the VV-geometry, plot the relative scattered light intensity $\log\left[I\left(Q\right)/I\left(0\right)\right]$ vs. the scattering vector Q for the following set of parameters: $n_1 = 1.420$, $n_2 = 1.342$, $n_0 = 1.333$, $a = 700$ nm, and $t = 21$ nm. On the same plot show the scattering from the corresponding homogeneous sphere (i.e., same parameters as above, except with $t \to 0$). Compare the plots for the coated and the homogeneous sphere and note that even the presence of just a very thin coating produces a substantial change in the scattered light intensity. The parameters listed above are typical for the case of a bacterium in water$^{\|}$ where the latter effect of the outer coating on the scattering intensity has been clearly observed [20].

4. If the conditions for the RGD approximation are met then, similar to the last problem, the intensity of light scattered from a small dielectric particle of arbitrary shape is proportional to a normalized *particle structure factor* given by

$$P\left(Q\right) = \left\langle \left|A\left(\mathbf{Q}\right)\right|^2 \right\rangle$$

and the bracket indicates an average over the (assumed) random orientation of the particle. For a given particle orientation (relative to the **Q**-vector), $A\left(\mathbf{Q}\right)$ is determined by

$$A\left(\mathbf{Q}\right) = \frac{1}{V} \int_V d^3\rho \, e^{i\mathbf{Q}\cdot\boldsymbol{\rho}}$$

where V is the volume of the particle and $\boldsymbol{\rho}$ is the position vector of an arbitrary point in the interior of the particle.

(a) Show that the particle structure factor for a randomly-oriented uniform

$^{\|}$In actuality, the bacteria referred to are best modelled as prolate spheroids having an axial ratio of about 2 to 1. However, it has been observed that the scattered intensity is much more sensitive to variations in the thickness and refractive index of the outer coating than to the details of the particle shape.

cylindrical particle of radius R and length L is given by

$$P(Q) = \frac{1}{2} \int_{-1}^{1} d\mu \left[\frac{\sin(QL\mu/2)}{QL\mu/2}\right]^2 \left[\frac{2J_1\left(QR\sqrt{1-\mu^2}\right)}{QR\sqrt{1-\mu^2}}\right]^2$$

where $J_1(x)$ is the first-order Bessel function.

(b) Show that for a long, thin rod, $P(Q)$ reduces asymptotically as follows:

$$P(Q) \underset{QL>2\pi}{\longrightarrow} \frac{\pi}{QL}\left[\frac{2J_1(QR)}{QR}\right]^2 \underset{QR<1}{\longrightarrow} \frac{\pi}{QL} e^{-Q^2R^2/4}.$$

Indicate how the radius of the rod can be extracted from a plot of $\ln[QI(Q)]$ vs. Q^2, where $I(Q)$ is the intensity as a function of the scattering vector.

(c) Show that for a thin disk, the asymptotic form becomes

$$P(Q) \underset{QR>1}{\longrightarrow} \frac{2}{Q^2R^2}\left[\frac{\sin(QL/2)}{QL/2}\right]^2 \underset{QL<1}{\longrightarrow} \frac{2}{Q^2R^2} e^{-Q^2L^2/12}.$$

Indicate how the thickness of the disk can be extracted from a plot of $\ln[Q^2I(Q)]$ vs. Q^2.

(d) In the polymer light scattering literature, a formula often used is that corresponding to the extreme limit where $R \to 0$. Show that in this so-called "stiff thin rod limit" one obtains

$$P(Q) \underset{R\to 0}{\longrightarrow} \int_{0}^{1} d\mu \left[\frac{\sin(QL\mu/2)}{QL\mu/2}\right]^2 = \frac{1}{x}\int_{0}^{2x} \frac{\sin u}{u} du - \left(\frac{\sin x}{x}\right)^2$$

where $x = QL/2$. Explore how one should most effectively plot the intensity in order to extract the length of the rod.

Chapter 5
QUANTUM PROPERTIES OF THE FIELD

A detailed treatment of the interaction of radiation with atomic systems requires that the continuous field variables discussed in the preceding chapter undergo quantization. This is accomplished by decomposing the vector potential, and ultimately the classical field Hamiltonian, into a superposition of field modes. We show that each mode is equivalent to a simple harmonic oscillator, which can be quantized using methods paralleling those outlined in Chapter 3. The concept of the *photon* is introduced to represent a quantum of excitation that can promote or de-excite the energy, or equivalently the number state, of an existing mode. An analogous treatment is also used to formalize the lattice displacement field in an atomic solid. In this case, one speaks of vibrational excitation quanta, or *phonons*. These exhibit rather interesting dispersion characteristics. Finally, we discuss how individual photon states can be superimposed to produce quasi-classical states featuring wavelike properties—these are minimum uncertainty states known as *coherent states*. Other minimum uncertainty states, namely, the so-called *squeezed states*, are also introduced.

5.1 Canonical Formulation of a Pure Radiation Field

Making the transition from a classical electromagnetic field to a quantum one is based on replacing the vector potential $\mathbf{A}$ by a corresponding quantum-mechanical operator $\hat{\mathbf{A}}$. Before doing this, however, we first cast the classical field into a canonical form that is readily amenable to the quantization procedure which will be outlined in the next section. The manipulations, at present, are purely classical.

In the preceding chapter, we saw that the free electromagnetic field can be described by a homogeneous wave equation for the vector potential (i.e., Eq. 4.54) subject to the Coulomb gauge condition $\boldsymbol{\nabla}\cdot\mathbf{A} = 0$. The most general solution is a superposition of travelling plane waves of the form $\mathcal{A}_{\mathbf{k}\lambda} \exp\left[i\left(\mathbf{k}\cdot\mathbf{r} - \omega_k t\right)\right] \boldsymbol{\epsilon}_{\mathbf{k}\lambda}$. Each is a mode characterized by its own wavevector $\mathbf{k}$, with associated amplitude $\mathcal{A}_{\mathbf{k}\lambda}$ and frequency $\omega_k = ck$. The polarization, or direction, of the vector potential for each $\mathbf{k}$ is specified by the unit vector $\boldsymbol{\epsilon}_{\mathbf{k}\lambda}$. Recall that the gauge condition demands that the field is transverse (i.e., $\mathbf{k}\cdot\boldsymbol{\epsilon}_{\mathbf{k}\lambda} = 0$), thus the polarization vector is perpendicular to $\mathbf{k}$. Since there exist two linearly-independent choices for the polarization, each wavevector $\mathbf{k}$ has a pair of travelling-wave modes associated with it. The subscript λ takes on the value 1 or 2, depending on the polarization of the mode.

For a general radiation field in free space not subject to any boundary conditions there are an uncountably infinite number of available modes, each representing a degree of freedom of the system. However, to better facilitate the ensuing field quantization process, it will prove advantageous to work in a cubic cavity of side L, its edges aligned along the three Cartesian coordinate axes. The field is then subject to periodic boundary conditions

$$\mathbf{A}(x, y, z, t) = \mathbf{A}(x + L, y, z, t) = \mathbf{A}(x, y + L, z, t) = \mathbf{A}(x, y, z + L, t), \tag{5.1}$$

leading to

$$e^{ik_x L} = e^{ik_y L} = e^{ik_z L} = 1. \tag{5.2}$$

As a result, the components of the wavevector are restricted to the following discrete values:

$$k_x = \frac{2\pi n_x}{L}, \qquad k_y = \frac{2\pi n_y}{L}, \qquad k_z = \frac{2\pi n_z}{L} \qquad (n_x,\ n_y,\ n_z = 0,\ \pm 1,\ \pm 2,\ \ldots). \tag{5.3}$$

Thus, the degrees of freedom, although still infinite in number, have now become enumerable. This makes it natural to decompose the vector potential into a Fourier series constructed from the permitted cavity modes, i.e.,

$$\mathbf{A}(\mathbf{r}, t) = \sum_{\mathbf{k},\lambda} \left[\mathcal{A}_{\mathbf{k}\lambda}\, e^{i(\mathbf{k}\cdot\mathbf{r} - \omega_k t)} + \mathcal{A}_{-\mathbf{k}\lambda}\, e^{-i(\mathbf{k}\cdot\mathbf{r} - \omega_k t)}\right] \boldsymbol{\epsilon}_{\mathbf{k}\lambda}. \tag{5.4}$$

The two terms represent waves travelling in opposite directions. In addition, we have $\mathcal{A}_{-\mathbf{k}\lambda} = \mathcal{A}^*_{\mathbf{k}\lambda}$, and the vector potential is real..

It is helpful to represent the allowable $\mathbf{k}$-vectors as points on a cubic lattice in a k_x-k_y-k_z space, as indicated in Fig. 5.1. Each point corresponds to a pair of cavity modes because of the two available polarizations. Later on it will be important to know the *mode density* or *density of states* which relates the number of modes to a given wavenumber (or frequency) interval. This is equivalent to determining the number of lattice points between two spherical shells of radii k and $k+dk$, as shown in the figure. To find this, observe that there is one wavevector (and hence two modes) for each unit cell of size $(2\pi/L)^3$ in $\mathbf{k}$-space. The number of modes dN within a volume $dk_x dk_y dk_z$ is therefore

$$dN = 2\left(\frac{L}{2\pi}\right)^3 dk_x dk_y dk_z. \tag{5.5}$$

By transforming to spherical coordinates in $\mathbf{k}$-space, the number of modes in the interval between k and $k+dk$ that correspond to a wave-propagation direction within solid angle $d\Omega$ about $\mathbf{k}$ is seen to be

$$dN = 2\left(\frac{L}{2\pi}\right)^3 k^2 dk\, d\Omega. \tag{5.6}$$

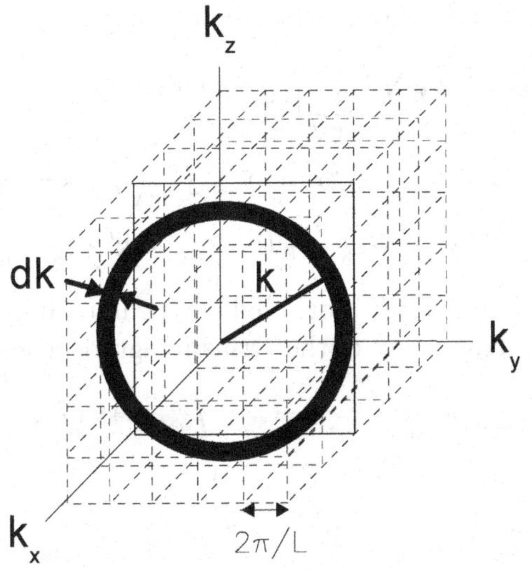

Figure 5.1 Points in **k**-space representing the modes of a cubical cavity.

However, since $k^2 dk = (\omega_k^2/c^3)\, d\omega_k$, one can also write down the number of modes in the frequency interval ω_k to $\omega_k + d\omega_k$ that propagate within $d\Omega$. With $V = L^3$, the result is

$$g(\omega_k)\, d\omega_k = \frac{2V}{(2\pi c)^3}\, \omega_k^2\, d\omega_k\, d\Omega. \tag{5.7}$$

$g(\omega_k)$ is the desired mode density.

From the expression for the vector potential (Eq. 5.4), the electric and magnetic field vectors immediately follow:

$$\mathbf{E}(\mathbf{r},t) = -\frac{1}{c}\frac{\partial \mathbf{A}}{\partial t} = i\sum_{\mathbf{k},\lambda} k\left[\mathcal{A}_{\mathbf{k}\lambda}\, e^{i(\mathbf{k}\cdot\mathbf{r}-\omega_k t)} - \mathcal{A}_{-\mathbf{k}\lambda}\, e^{-i(\mathbf{k}\cdot\mathbf{r}-\omega_k t)}\right]\boldsymbol{\epsilon}_{\mathbf{k}\lambda} \tag{5.8}$$

$$\mathbf{B}(\mathbf{r},t) = \boldsymbol{\nabla}\times\mathbf{A} = i\sum_{\mathbf{k},\lambda}\left[\mathcal{A}_{\mathbf{k}\lambda}\, e^{i(\mathbf{k}\cdot\mathbf{r}-\omega_k t)} - \mathcal{A}_{-\mathbf{k}\lambda}\, e^{-i(\mathbf{k}\cdot\mathbf{r}-\omega_k t)}\right](\mathbf{k}\times\boldsymbol{\epsilon}_{\mathbf{k}\lambda})\,. \tag{5.9}$$

In obtaining the expression for **B**, we made use of

$$\boldsymbol{\nabla}\times\left(e^{\pm i\mathbf{k}\cdot\mathbf{r}}\boldsymbol{\epsilon}_{\mathbf{k}\lambda}\right) = \left(\boldsymbol{\nabla}e^{\pm i\mathbf{k}\cdot\mathbf{r}}\right)\times\boldsymbol{\epsilon}_{\mathbf{k}\lambda} = \pm i e^{\pm i\mathbf{k}\cdot\mathbf{r}}\,(\mathbf{k}\times\boldsymbol{\epsilon}_{\mathbf{k}\lambda})\,. \tag{5.10}$$

We are now in the position of being able to find the classical Hamiltonian of the EM field in terms of the various cavity modes. From Chapter 4, recall that

$$H = \frac{1}{8\pi}\int_V \left(E^2 + B^2\right) d^3r. \tag{5.11}$$

The result is an integral over a cumbersome double-sum:

$$\begin{aligned} H &= -\frac{1}{8\pi}\int_V d^3r \sum_{\mathbf{k},\lambda}\sum_{\mathbf{k}',\lambda'} \left[\mathcal{A}_{\mathbf{k}\lambda}\, e^{i(\mathbf{k}\cdot\mathbf{r}-\omega_k t)} - \mathcal{A}_{-\mathbf{k}\lambda}\, e^{-i(\mathbf{k}\cdot\mathbf{r}-\omega_k t)}\right] \\ &\quad \times \left[\mathcal{A}_{\mathbf{k}'\lambda'}\, e^{i(\mathbf{k}'\cdot\mathbf{r}-\omega_{k'} t)} - \mathcal{A}_{-\mathbf{k}'\lambda'}\, e^{-i(\mathbf{k}'\cdot\mathbf{r}-\omega_{k'} t)}\right] \\ &\quad \times \left[kk'\boldsymbol{\epsilon}_{\mathbf{k}\lambda}\cdot\boldsymbol{\epsilon}_{\mathbf{k}'\lambda'} + (\mathbf{k}\times\boldsymbol{\epsilon}_{\mathbf{k}\lambda})\cdot(\mathbf{k}'\times\boldsymbol{\epsilon}_{\mathbf{k}'\lambda'})\right]. \end{aligned} \tag{5.12}$$

At first, evaluation of this expression might appear to be a rather daunting task. However, note that a great many of the terms integrate to zero because

$$\int_V e^{\pm i(\mathbf{k}-\mathbf{k}')\cdot\mathbf{r}}\, d^3r = V\,\delta_{\mathbf{k}\mathbf{k}'} \qquad \text{and} \qquad \int_V e^{\pm i(\mathbf{k}+\mathbf{k}')\cdot\mathbf{r}}\, d^3r = V\,\delta_{-\mathbf{k},\mathbf{k}'}. \tag{5.13}$$

The expression that survives is

$$\begin{aligned} H &= \frac{V}{8\pi}\sum_{\mathbf{k},\lambda}\sum_{\lambda'} \left[k^2\boldsymbol{\epsilon}_{\mathbf{k}\lambda}\cdot\boldsymbol{\epsilon}_{\mathbf{k}\lambda'} + (\mathbf{k}\times\boldsymbol{\epsilon}_{\mathbf{k}\lambda})\cdot(\mathbf{k}\times\boldsymbol{\epsilon}_{\mathbf{k}\lambda'})\right](\mathcal{A}_{\mathbf{k}\lambda}\mathcal{A}_{-\mathbf{k}\lambda'} + \mathcal{A}_{-\mathbf{k}\lambda}\mathcal{A}_{\mathbf{k}\lambda'}) \\ &\quad - \frac{V}{8\pi}\sum_{\mathbf{k},\lambda}\sum_{\lambda'} \left[k^2\boldsymbol{\epsilon}_{\mathbf{k}\lambda}\cdot\boldsymbol{\epsilon}_{-\mathbf{k}\lambda'} + (\mathbf{k}\times\boldsymbol{\epsilon}_{\mathbf{k}\lambda})\cdot(-\mathbf{k}\times\boldsymbol{\epsilon}_{-\mathbf{k}\lambda'})\right] \\ &\quad \times \left(\mathcal{A}_{\mathbf{k}\lambda}\mathcal{A}_{-\mathbf{k}\lambda'}\, e^{-2i\omega_k t} + \mathcal{A}_{-\mathbf{k}\lambda}\mathcal{A}_{\mathbf{k}\lambda'}\, e^{+2i\omega_k t}\right), \end{aligned} \tag{5.14}$$

where we have used the fact that $\omega_{-k} = \omega_k$. In the first double-sum, notice that both dot-products reduce to the same simple form, namely $k^2\delta_{\lambda\lambda'}$. In addition, the reader should verify that $(\mathbf{k}\times\boldsymbol{\epsilon}_{\mathbf{k}\lambda})\cdot(-\mathbf{k}\times\boldsymbol{\epsilon}_{-\mathbf{k}\lambda'}) = -k^2\boldsymbol{\epsilon}_{\mathbf{k}\lambda}\cdot\boldsymbol{\epsilon}_{-\mathbf{k}\lambda'}$, thus all terms in the second double-sum vanish and, as expected, the Hamiltonian becomes time-independent. The resulting expression has the following particularly simple form:

$$H = \frac{V}{2\pi c^2}\sum_{\mathbf{k},\lambda} \omega_k^2 \mathcal{A}_{-\mathbf{k}\lambda}\mathcal{A}_{\mathbf{k}\lambda}. \tag{5.15}$$

From Eqs. 4.25 and 4.27, classically, the total momentum of the field is given by

$$\mathbf{G} = \frac{1}{4\pi c}\int_V (\mathbf{E}\times\mathbf{B})\, d^3r. \tag{5.16}$$

Following arguments similar to the ones above for the Hamiltonian,* the momentum is also expressible as a sum over modes. Since we are dealing with an isolated system, the result is again independent of time:

$$\mathbf{G} = \frac{V}{2\pi c^2}\sum_{\mathbf{k},\lambda} \omega_k \mathcal{A}_{-\mathbf{k}\lambda}\mathcal{A}_{\mathbf{k}\lambda}\mathbf{k}. \tag{5.17}$$

*The detailed steps are given by Louisell [21].

At this point, let us redefine the mode amplitudes to incorporate the characteristic time-dependence, i.e.,

$$A_{\mathbf{k}\lambda}(t) = \mathcal{A}_{\mathbf{k}\lambda}e^{-i\omega_k t} \qquad \text{and} \qquad A_{-\mathbf{k}\lambda}(t) = \mathcal{A}_{-\mathbf{k}\lambda}e^{+i\omega_k t}. \tag{5.18}$$

Note that this does not change the basic form of the energy and momentum of the field:

$$H = \frac{V}{2\pi c^2}\sum_{\mathbf{k},\lambda}\omega_k^2 A_{-\mathbf{k}\lambda}A_{\mathbf{k}\lambda} \tag{5.19}$$

$$\mathbf{G} = \frac{V}{2\pi c^2}\sum_{\mathbf{k},\lambda}\omega_k A_{-\mathbf{k}\lambda}A_{\mathbf{k}\lambda}\mathbf{k}. \tag{5.20}$$

The canonical form of the classical radiation field is now obtained by introducing the following transformation of variables:

$$Q_{\mathbf{k}\lambda}(t) = \sqrt{\frac{V}{4\pi c^2}}\left(A_{\mathbf{k}\lambda} + A_{-\mathbf{k}\lambda}\right) \tag{5.21}$$

$$P_{\mathbf{k}\lambda}(t) = \dot{Q}_{\mathbf{k}\lambda} = -i\omega_k\sqrt{\frac{V}{4\pi c^2}}\left(A_{\mathbf{k}\lambda} - A_{-\mathbf{k}\lambda}\right). \tag{5.22}$$

Alternatively, we have

$$A_{\mathbf{k}\lambda}(t) = \sqrt{\frac{\pi c^2}{V}}\left(Q_{\mathbf{k}\lambda} + \frac{iP_{\mathbf{k}\lambda}}{\omega_k}\right) \qquad \text{and} \qquad A_{-\mathbf{k}\lambda}(t) = \sqrt{\frac{\pi c^2}{V}}\left(Q_{\mathbf{k}\lambda} - \frac{iP_{\mathbf{k}\lambda}}{\omega_k}\right). \tag{5.23}$$

Substituting into Eq. 5.19 for the field Hamiltonian and Eq. 5.20 for the field momentum gives

$$H = \sum_{\mathbf{k},\lambda}\frac{1}{2}\left(P_{\mathbf{k}\lambda}^2 + \omega_k^2 Q_{\mathbf{k}\lambda}^2\right) \tag{5.24}$$

and

$$\mathbf{G} = \sum_{\mathbf{k},\lambda}\frac{1}{2\omega_k}\left(P_{\mathbf{k}\lambda}^2 + \omega_k^2 Q_{\mathbf{k}\lambda}^2\right)\mathbf{k}. \tag{5.25}$$

Observe that each field-mode contributes a term to the Hamiltonian that is identical in form to that of a harmonic oscillator. In fact, the equivalence between the radiation field and a collection of oscillators is not just fortuitous—rather, it is a completely formal one. This is evident once one notices that

$$\frac{\partial H}{\partial Q_{\mathbf{k}\lambda}} = \omega_k^2 Q_{\mathbf{k}\lambda} = -\dot{P}_{\mathbf{k}\lambda} \qquad \text{and} \qquad \frac{\partial H}{\partial P_{\mathbf{k}\lambda}} = P_{\mathbf{k}\lambda} = \dot{Q}_{\mathbf{k}\lambda}, \tag{5.26}$$

which is exactly the form of Hamilton's equations of motion for a system specified by generalized coordinates $Q_{\mathbf{k}\lambda}$ with canonically conjugate generalized momenta $P_{\mathbf{k}\lambda}$.

5.2 Quantization of a Pure Radiation Field

Historically, the measured spectral distribution of thermal radiation from a blackbody radiator was unexplained until the year 1900, at which time Planck first suggested the notion of quantization [22]. A classical calculation of the blackbody spectrum led to the so-called *ultraviolet catastrophe*, i.e., the theory predicted, to a severe extent, an excessive amount of radiation at the high frequencies. Planck was able to successfully resolve this dilemma by postulating that the walls of a blackbody cavity are composed of oscillators possessing quantum characteristics. In particular, for an oscillator of angular frequency ω, the energy could only take on integral multiples of $\hbar\omega$. A few years later, Einstein was able to explain the photoelectric effect by similarly quantizing the electromagnetic field itself [23]. Eventually, the name *photon* was coined to denote a quantum of the radiation field.†

The fact that the energy carried by a photon is proportional to the frequency of the radiation can be seen from an application of the position-momentum uncertainty principle. Imagine one measures the position of an electron; this can be accomplished by illuminating it with light of wavelength λ and collecting a quantum of the scattered light into the optics of a microscope (see Fig. 5.2). The uncertainty associated with the electron's position is given by the resolving power of the microscope's objective lens, i.e., $\triangle x \simeq \lambda/\sin\alpha$, where 2α is the angle subtended by the lens aperture. Now, because the collected photon could have entered the lens anywhere within this angle, the x-component of the scattered photon's momentum is uncertain by an amount $2p_\lambda \sin\alpha$, where we have assumed that the momentum of the photon before the scattering is known to be exactly p_λ. But momentum conservation tells us that the uncertainty in the x-component of the recoil momentum transferred to the electron is the same, so $\triangle p_x \simeq 2p_\lambda \sin\alpha$. Thus, $\triangle x \triangle p_x \simeq 2\lambda p_\lambda$. But the Heisenberg uncertainty principle requires that $\triangle x \triangle p_x \sim \hbar/2$. We then conclude that the momentum of the scattered photon is $p_\lambda \sim h/\lambda$, and its energy is $E_\lambda \sim cp_\lambda \sim h(c/\lambda) \sim \hbar\omega_\lambda$.

Let us now return to the problem at hand, i.e., that of formally quantizing a general EM field. As mentioned before, this is accomplished by replacing the classical vector potential $\mathbf{A}$ by a quantum-mechanical operator $\hat{\mathbf{A}}$. Toward this end, we begin by introducing the non-Hermitian operator $\hat{A}_{\mathbf{k}\lambda}$ and its adjoint $\hat{A}^\dagger_{\mathbf{k}\lambda}$ as replacements for the classical mode amplitudes, $A_{\mathbf{k}\lambda}$ and $A_{-\mathbf{k}\lambda}$, respectively:

$$\hat{A}_{\mathbf{k}\lambda}(t) = \sqrt{\frac{\pi c^2}{V}}\left(\hat{Q}_{\mathbf{k}\lambda} + \frac{i\hat{P}_{\mathbf{k}\lambda}}{\omega_k}\right) \quad \text{and} \quad \hat{A}^\dagger_{\mathbf{k}\lambda}(t) = \sqrt{\frac{\pi c^2}{V}}\left(\hat{Q}_{\mathbf{k}\lambda} - \frac{i\hat{P}_{\mathbf{k}\lambda}}{\omega_k}\right). \tag{5.27}$$

The canonically conjugate position and momentum for each mode are now also operators. They obey the commutator relations

$$\left[\hat{Q}_{\mathbf{k}\lambda}, \hat{P}_{\mathbf{k}\lambda}\right] = i\hbar\delta_{\mathbf{k}\mathbf{k}'}\delta_{\lambda\lambda'} \tag{5.28}$$

†The word "photon" first appeared in an article by Lewis [24].

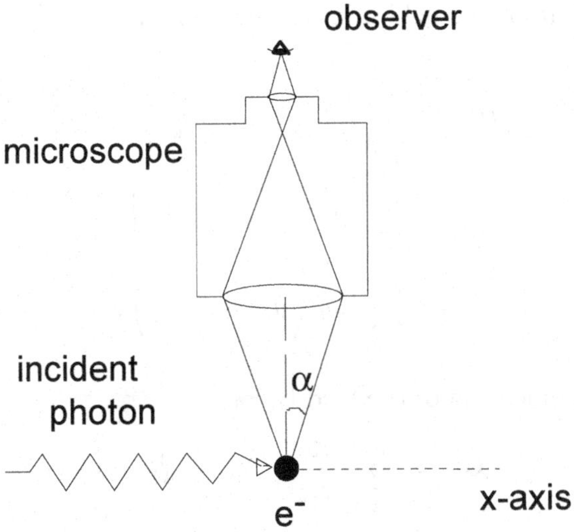

Figure 5.2 Measuring the position of an electron by detecting a scattered photon with a light microscope.

and

$$\left[\hat{Q}_{\mathbf{k}\lambda}, \hat{Q}_{\mathbf{k}'\lambda'}\right] = \left[\hat{P}_{\mathbf{k}\lambda}, \hat{P}_{\mathbf{k}'\lambda'}\right] = 0. \tag{5.29}$$

We then see that $\hat{A}_{\mathbf{k}\lambda}$ and $\hat{A}^\dagger_{\mathbf{k}\lambda}$ do not commute, i.e.,

$$\left[\hat{A}_{\mathbf{k}\lambda}, \hat{A}^\dagger_{\mathbf{k}'\lambda'}\right] = \frac{2\pi\hbar c^2}{V\omega_k}\delta_{\mathbf{k}\mathbf{k}'}\delta_{\lambda\lambda'}. \tag{5.30}$$

For the sake of convenience, let us replace $\hat{A}_{\mathbf{k}\lambda}$ and $\hat{A}^\dagger_{\mathbf{k}\lambda}$ by a set of dimensionless operators, $\hat{a}_{\mathbf{k}\lambda}$ and $\hat{a}^\dagger_{\mathbf{k}\lambda}$:

$$\hat{A}_{\mathbf{k}\lambda} = \sqrt{\frac{2\pi\hbar c^2}{V\omega_k}}\hat{a}_{\mathbf{k}\lambda} \qquad \text{and} \qquad \hat{A}^\dagger_{\mathbf{k}\lambda} = \sqrt{\frac{2\pi\hbar c^2}{V\omega_k}}\hat{a}^\dagger_{\mathbf{k}\lambda}. \tag{5.31}$$

The commutators that follow are

$$\left[\hat{a}_{\mathbf{k}\lambda}, \hat{a}^\dagger_{\mathbf{k}'\lambda'}\right] = \delta_{\mathbf{k}\mathbf{k}'}\delta_{\lambda\lambda'} \tag{5.32}$$

and

$$\left[\hat{a}_{\mathbf{k}\lambda}, \hat{a}_{\mathbf{k}'\lambda'}\right] = \left[\hat{a}^\dagger_{\mathbf{k}\lambda}, \hat{a}^\dagger_{\mathbf{k}'\lambda'}\right] = 0. \tag{5.33}$$

The generalized coordinates and momenta now become

$$\hat{Q}_{\mathbf{k}\lambda} = \sqrt{\frac{\hbar}{2\omega_k}}\left(\hat{a}_{\mathbf{k}\lambda} + \hat{a}^\dagger_{\mathbf{k}\lambda}\right) \qquad \text{and} \qquad \hat{P}_{\mathbf{k}\lambda} = -i\sqrt{\frac{\hbar\omega_k}{2}}\left(\hat{a}_{\mathbf{k}\lambda} - \hat{a}^\dagger_{\mathbf{k}\lambda}\right) \tag{5.34}$$

and the Hamiltonian for the field immediately follows:

$$\begin{aligned}\hat{H} &= \sum_{\mathbf{k},\lambda}\frac{1}{2}\left(\hat{P}^2_{\mathbf{k}\lambda}+\omega_k^2\hat{Q}^2_{\mathbf{k}\lambda}\right) \\ &= \sum_{\mathbf{k},\lambda}\frac{1}{2}\hbar\omega_k\left(\hat{a}_{\mathbf{k}\lambda}\hat{a}^\dagger_{\mathbf{k}\lambda}+\hat{a}^\dagger_{\mathbf{k}\lambda}\hat{a}_{\mathbf{k}\lambda}\right) \\ &= \sum_{\mathbf{k},\lambda}\hbar\omega_k\left(\hat{a}^\dagger_{\mathbf{k}\lambda}\hat{a}_{\mathbf{k}\lambda}+\frac{1}{2}\right).\end{aligned} \tag{5.35}$$

The total momentum of the field becomes

$$\begin{aligned}\hat{\mathbf{G}} &= \sum_{\mathbf{k},\lambda}\frac{1}{2\omega_k}\left(\hat{P}^2_{\mathbf{k}\lambda}+\omega_k^2\hat{Q}^2_{\mathbf{k}\lambda}\right)\mathbf{k} \\ &= \sum_{\mathbf{k},\lambda}\hbar\mathbf{k}\left(\hat{a}^\dagger_{\mathbf{k}\lambda}\hat{a}_{\mathbf{k}\lambda}+\frac{1}{2}\right) \\ &= \sum_{\mathbf{k},\lambda}\hbar\mathbf{k}\hat{a}^\dagger_{\mathbf{k}\lambda}\hat{a}_{\mathbf{k}\lambda}.\end{aligned} \tag{5.36}$$

In the last step we have used the fact that $\sum_{\mathbf{k},\lambda}\hbar\mathbf{k}=0$, since for each vector $\mathbf{k}$ in the sum there is an oppositely directed vector $-\mathbf{k}$. Finally, the classical fields are replaced by the following quantum-mechanical operators:

$$\hat{\mathbf{A}}(\mathbf{r},t)=\sum_{\mathbf{k},\lambda}\sqrt{\frac{2\pi\hbar c^2}{V\omega_k}}\left[\hat{a}_{\mathbf{k}\lambda}\,e^{i\mathbf{k}\cdot\mathbf{r}}+\hat{a}^\dagger_{\mathbf{k}\lambda}e^{-i\mathbf{k}\cdot\mathbf{r}}\right]\boldsymbol{\epsilon}_{\mathbf{k}\lambda} \tag{5.37}$$

$$\hat{\mathbf{E}}(\mathbf{r},t)=\sum_{\mathbf{k},\lambda}i\sqrt{\frac{2\pi\hbar\omega_k}{V}}\left[\hat{a}_{\mathbf{k}\lambda}\,e^{i\mathbf{k}\cdot\mathbf{r}}-\hat{a}^\dagger_{\mathbf{k}\lambda}\,e^{-i\mathbf{k}\cdot\mathbf{r}}\right]\boldsymbol{\epsilon}_{\mathbf{k}\lambda} \tag{5.38}$$

$$\hat{\mathbf{B}}(\mathbf{r},t)=\sum_{\mathbf{k},\lambda}i\sqrt{\frac{2\pi\hbar c^2}{V\omega_k}}\left[\hat{a}_{\mathbf{k}\lambda}\,e^{i\mathbf{k}\cdot\mathbf{r}}-\hat{a}^\dagger_{\mathbf{k}\lambda}e^{-i\mathbf{k}\cdot\mathbf{r}}\right](\mathbf{k}\times\boldsymbol{\epsilon}_{\mathbf{k}\lambda})\,. \tag{5.39}$$

The time-dependence is carried by the operators $\hat{a}_{\mathbf{k}\lambda}$ and $\hat{a}^\dagger_{\mathbf{k}\lambda}$ in that

$$\hat{a}_{\mathbf{k}\lambda}(t)=\hat{a}_{\mathbf{k}\lambda}(0)e^{-i\omega_k t}\qquad\text{and}\qquad\hat{a}^\dagger_{\mathbf{k}\lambda}(t)=\hat{a}^\dagger_{\mathbf{k}\lambda}(0)e^{+i\omega_k t}. \tag{5.40}$$

The physical interpretation of the results are as follows: As was the case for the simple one-dimensional oscillator (refer back to Section 3.2), the operators $\hat{a}_{\mathbf{k}\lambda}$ and $\hat{a}^\dagger_{\mathbf{k}\lambda}$ act as lowering and raising operators, respectively. Note, however, that each $(\mathbf{k},\lambda)$-mode of the field behaves as an independent oscillator, hence there is a lowering and raising operator for each mode. Like before, the operator $\hat{N}_{\mathbf{k}\lambda}\equiv\hat{a}^\dagger_{\mathbf{k}\lambda}\hat{a}_{\mathbf{k}\lambda}$

is a number operator; this time its eigenvalues $n_{\mathbf{k}\lambda}$ represent the *number of photons* (i.e., 0, 1, 2, ..., etc.) with wavevector $\mathbf{k}$ and polarization $\boldsymbol{\epsilon}_{\mathbf{k}\lambda}$ present in the field. A photon in mode ($\mathbf{k}$,λ) has energy $\hbar\omega_k$ and momentum $\hbar\mathbf{k}$. An eigenstate of the photon-number operator is denoted by $|n_{\mathbf{k}_1\lambda_1},\, n_{\mathbf{k}_2\lambda_2},\, \ldots\rangle$; this may also be written as a product of the eigenstates belonging to the independent modes, i.e., $|n_{\mathbf{k}_1\lambda_1},\, n_{\mathbf{k}_2\lambda_2},\, \ldots\rangle = |n_{\mathbf{k}_1\lambda_1}\rangle\, |n_{\mathbf{k}_2\lambda_2}\rangle \cdots$. The actions of the raising and lowering operators on a state $|n_{\mathbf{k}_1\lambda_1},\, n_{\mathbf{k}_2\lambda_2},\, \ldots\rangle$ are to add or remove a single photon from a mode of the field, as shown below:

$$\hat{a}^{\dagger}_{\mathbf{k}\lambda}\,|\ldots,\, n_{\mathbf{k}\lambda},\, \ldots\rangle = \sqrt{n_{\mathbf{k}\lambda}+1}\,|\ldots,\, n_{\mathbf{k}\lambda}+1,\, \ldots\rangle \tag{5.41}$$

$$\hat{a}_{\mathbf{k}\lambda}\,|\ldots,\, n_{\mathbf{k}\lambda},\, \ldots\rangle = \sqrt{n_{\mathbf{k}\lambda}}\,|\ldots,\, n_{\mathbf{k}\lambda}-1,\, \ldots\rangle\,. \tag{5.42}$$

Given the form of Eqs. 5.37–5.39, the field is itself a quantum-mechanical operator that creates and annihilates photons.

Example 5.1 Quantization of a Lattice Displacement Field. Here, we consider the longitudinal vibrations of a one-dimensional lattice, like that shown in Fig. 5.3. The Lagrangian of this system was previously given by Eq. 2.98; it is rewritten here, except with a change of notation:

$$\mathcal{L} = \sum_i \frac{1}{2}m\dot{q}_i^2 - \sum_i \frac{1}{2}m\omega_0^2\left(q_{i+1}-q_i\right)^2. \tag{5.43}$$

Situated at each lattice point is a mass m that interacts with its neighbor via an effective spring constant $m\omega_0^2$. The displacement of a given mass from its equilibrium position x_i is now denoted by the coordinate q_i; the equilibrium separation is $x_{i+1} - x_i = a$. The Hamiltonian is obtained from the Lagrangian and the canonical momentum $p_i = \partial L/\partial\dot{q}_i = m\dot{q}_i$ to be

$$\begin{aligned} H &= \sum_i \dot{q}_i p_i - \mathcal{L} = \sum_i m\dot{q}_i^2 - \sum_i \frac{1}{2}m\dot{q}_i^2 + \frac{1}{2}m\omega_0^2\sum_i\left(q_{i+1}-q_i\right)^2 \\ &= \frac{1}{2m}\sum_i p_i^2 + \frac{1}{2}m\omega_0^2\sum_i\left(q_{i+1}-q_i\right)^2. \end{aligned} \tag{5.44}$$

A canonical formulation of the problem is developed by transforming to a superposition of lattice waves

$$q_i = \frac{1}{2\sqrt{Nm}}\sum_k\left(Q_k\, e^{ikx_i} + Q_{-k}\, e^{-ikx_i}\right), \tag{5.45}$$

where N is the total number of lattice points being considered. The individual terms in the sum represent modes of the lattice displacement field, each with a propagation

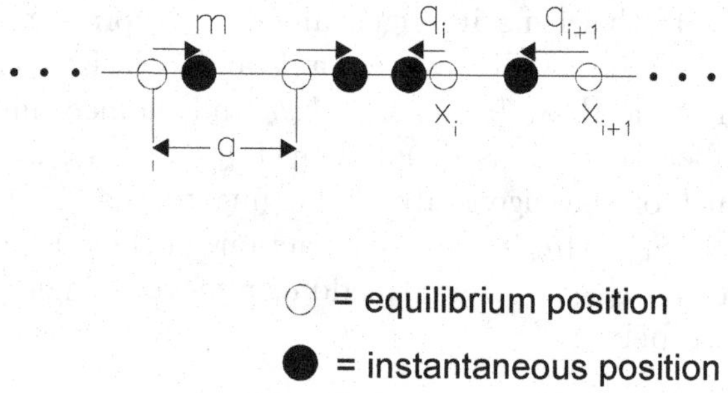

Figure 5.3 Longitudinal vibrations of a one-dimensional lattice.

constant k and amplitude Q_k. The two types of terms represent longitudinal waves travelling in opposite directions. To assure that the displacement q_i is real, we require that $Q_{-k} = Q_k^*$. The inverted form of Eq. 5.45 is

$$Q_k = \sqrt{\frac{m}{N}} \sum_i q_i \, e^{-ikx_i}, \tag{5.46}$$

which follows from the identity $\sum_k e^{\pm ik(x_i - x_j)} = N\delta_{ij}$. The possible values of k are restricted by the cyclic boundary condition $q_{i+N} = q_i$. From Eq. 5.45, this requires that $e^{ikx_i} = e^{ikx_{i+N}}$, thus $e^{ikNa} = 1$. As a result, the values of k are limited to

$$k = \frac{2\pi}{a}\frac{n}{N} \qquad (n = 0,\ \pm 1,\ \pm 2,\ \ldots). \tag{5.47}$$

With the aid of

$$\sum_i e^{\pm i(k-k')x_i} = N\delta_{kk'} \qquad \text{and} \qquad \sum_i e^{\pm i(k+k')x_i} = N\delta_{-kk'} \tag{5.48}$$

it becomes rather straightforward to show that the Lagrangian of the system can be re-expressed as the following summation over the various modes of the lattice displacement field:

$$\mathcal{L} = \sum_k \left[\frac{1}{2} \dot{Q}_k \dot{Q}_{-k} - \omega_0^2 Q_k Q_{-k} \left(1 - \cos ka\right) \right]. \tag{5.49}$$

The Hamiltonian is obtained by treating the Q_k's and Q_{-k}'s as generalized coordinates with canonically conjugate momenta

$$P_k = \frac{\partial \mathcal{L}}{\partial \dot{Q}_k} = \dot{Q}_{-k} \qquad \text{and} \qquad P_{-k} = \dot{Q}_k. \tag{5.50}$$

The result is

$$\begin{aligned} H &= \sum_k P_k \dot{Q}_k - \mathcal{L} \\ &= \sum_k \frac{1}{2}\left(P_k P_{-k} + \omega_k^2 Q_k Q_{-k}\right), \end{aligned} \tag{5.51}$$

where

$$\omega_k^2 = 2\omega_0^2 (1 - \cos ka) \tag{5.52}$$

and $\omega_{-k} = \omega_k$.[‡] Also, note that the generalized momenta are

$$P_k = \dot{Q}_{-k} = \sqrt{\frac{m}{N}} \sum_i \dot{q}_i e^{ikx_i} = \frac{1}{\sqrt{Nm}} \sum_i p_i e^{ikx_i}. \tag{5.53}$$

To quantize the field, we replace the classical observables q_i and p_i with corresponding quantum-mechanical Hermitian operators $\hat{q}_i$ and $\hat{p}_i$ that satisfy the commutators

$$[\hat{q}_i, \hat{p}_j] = i\hbar\delta_{ij} \tag{5.54}$$

$$[\hat{q}_i, \hat{q}_j] = [\hat{p}_i, \hat{p}_j] = 0. \tag{5.55}$$

The wave coordinates Q_k and momenta P_k are also replaced by operators. From Eqs. 5.46 and 5.53, we have

$$\hat{Q}_k = \sqrt{\frac{m}{N}} \sum_i \hat{q}_i\, e^{-ikx_i} \qquad \text{and} \qquad \hat{P}_k = \frac{1}{\sqrt{Nm}} \sum_i \hat{p}_i e^{ikx_i}. \tag{5.56}$$

From the Hermitian property of the $\hat{q}_i$'s and $\hat{p}_i$'s, we see that the adjoint operators are

$$\hat{Q}_k^\dagger = \sqrt{\frac{m}{N}} \sum_i \hat{q}_i\, e^{+ikx_i} = \hat{Q}_{-k} \qquad \text{and} \qquad \hat{P}_k^\dagger = \frac{1}{\sqrt{Nm}} \sum_i \hat{p}_i e^{-ikx_i} = \hat{P}_{-k}. \tag{5.57}$$

Thus, $\hat{Q}_k$ and $\hat{P}_k$ are non-Hermitian. The commutators involving these operators become

$$\begin{aligned} \left[\hat{Q}_k, \hat{P}_{k'}\right] &= \frac{1}{N}\left[\sum_i \hat{q}_i\, e^{-ikx_i}, \sum_j \hat{p}_j e^{ik'x_j}\right] \\ &= \frac{1}{N}\sum_{i,j} [\hat{q}_i, \hat{p}_j]\, e^{-ikx_i} e^{+ik'x_j} = \frac{i\hbar}{N}\sum_{i,j} \delta_{ij} e^{-ikx_i} e^{+ik'x_j} \\ &= i\hbar\delta_{kk'} \end{aligned} \tag{5.58}$$

[‡]This dispersion relation is identical to the one derived in Problem 2 at the end of Chapter 2 for the longitudinal vibrations of a classical 1D-lattice.

and

$$\left[\hat{Q}_k, \hat{Q}_{k'}\right] = \left[\hat{P}_k, \hat{P}_{k'}\right] = 0. \tag{5.59}$$

The displacement field can now be transformed to a set of uncoupled quantum oscillators by transforming to a new set of non-Hermitian operators, $\hat{a}_k$ and $\hat{a}_k^\dagger$, as follows:

$$\hat{a}_k = \frac{1}{\sqrt{2\hbar\omega_k}}\left(\omega_k\hat{Q}_k + i\hat{P}_{-k}\right) \qquad \text{and} \qquad \hat{a}_k^\dagger = \frac{1}{\sqrt{2\hbar\omega_k}}\left(\omega_k\hat{Q}_{-k} - i\hat{P}_k\right). \tag{5.60}$$

Then

$$\hbar\omega_k\hat{a}_k^\dagger\hat{a}_k = \frac{1}{2}\left(\omega_k^2\hat{Q}_{-k}\hat{Q}_k + \hat{P}_k\hat{P}_{-k} + i\omega_k\hat{Q}_{-k}\hat{P}_{-k} - i\omega_k\hat{P}_k\hat{Q}_k\right) \tag{5.61}$$

and

$$\begin{aligned}\sum_k \hbar\omega_k\hat{a}_k^\dagger\hat{a}_k &= \sum_k \frac{1}{2}\left(\hat{P}_k\hat{P}_{-k} + \omega_k^2\hat{Q}_k\hat{Q}_{-k}\right) + \sum_k \frac{i\omega_k}{2}\left[\hat{Q}_k, \hat{P}_k\right] \\ &= \hat{H} - \sum_k \frac{1}{2}\hbar\omega_k. \end{aligned} \tag{5.62}$$

The final Hamiltonian then has a quantized form that mirrors that of the EM field:

$$\hat{H} = \sum_k \hbar\omega_k\left(\hat{a}_k^\dagger\hat{a}_k + \frac{1}{2}\right). \tag{5.63}$$

In addition, Eqs. 5.60 can be inverted so that

$$\hat{Q}_k = \sqrt{\frac{\hbar}{2\omega_k}}\left(\hat{a}_k + \hat{a}_{-k}^\dagger\right) \qquad \text{and} \qquad \hat{P}_k = -i\sqrt{\frac{\hbar\omega_k}{2}}\left(\hat{a}_{-k} - \hat{a}_k^\dagger\right) \tag{5.64}$$

and the displacement operator $\hat{q}_i$ can be obtained from Eq. 5.45 as follows:

$$\begin{aligned}\hat{q}_i &= \frac{1}{2\sqrt{Nm}}\sum_k \sqrt{\frac{\hbar}{2\omega_k}}\left[\left(\hat{a}_k + \hat{a}_{-k}^\dagger\right)e^{+ikx_i} + \left(\hat{a}_{-k} + \hat{a}_k^\dagger\right)e^{-ikx_i}\right] \\ &= \frac{1}{2}\sum_k \sqrt{\frac{\hbar}{2Nm\omega_k}}\left(\hat{a}_k e^{+ikx_i} + \hat{a}_k^\dagger e^{-ikx_i}\right) \\ &\quad + \frac{1}{2}\sum_k \sqrt{\frac{\hbar}{2Nm\omega_k}}\left(\hat{a}_{-k} e^{-ikx_i} + \hat{a}_{-k}^\dagger e^{+ikx_i}\right). \end{aligned} \tag{5.65}$$

However, since we are summing over both positive and negative values of k, the two summations in the latter expression are identical, and the result is more simply

$$\hat{q}_i = \sum_k \sqrt{\frac{\hbar}{2Nm\omega_k}}\left(\hat{a}_k e^{+ikx_i} + \hat{a}_k^\dagger e^{-ikx_i}\right). \tag{5.66}$$

This equation should remind the reader of the quantized form for the vector potential of the radiation field given by Eq. 5.37.

As in the case of the EM field, $\hat{a}_k$ and $\hat{a}_k^\dagger$ are again lowering and raising operators satisfying the commutator relations

$$\begin{aligned}\left[\hat{a}_k, \hat{a}_k^\dagger\right] &= \frac{1}{2\hbar\omega_k}\left\{\omega_k^2\left[\hat{Q}_k, \hat{Q}_{-k}\right] - \left[\hat{P}_k, \hat{P}_{-k}\right] - i\omega_k\left[\hat{Q}_{-k}, \hat{P}_{-k}\right] - i\omega_k\left[\hat{Q}_k, \hat{P}_k\right]\right\} \\ &= \delta_{kk'} \end{aligned} \tag{5.67}$$

and

$$[\hat{a}_k, \hat{a}_{k'}] = \left[\hat{a}_k^\dagger, \hat{a}_{k'}^\dagger\right] = 0. \tag{5.68}$$

The operator $\hat{a}_k^\dagger$ serves to create a single *phonon*, or lattice vibration quantum, in mode k of the lattice displacement field. Similarly, $\hat{a}_k$ destroys one phonon in the same mode. The vibrational modes of the one-dimensional lattice are equivalent to a set of uncoupled harmonic oscillators, just as was the case for modes of a radiation field. A major difference, however, lies in the dispersion relation for the two cases. Whereas photons are characterized simply by $\omega_k = ck$, phonons of the 1-D lattice satisfy Eq. 5.52. Figure 5.4 displays the ω-k diagram for the lattice waves. The vibration frequency is a repeating function of k with period $2\pi/a$, and the maximum frequency attainable by the points of the lattice is $2\omega_0$. The interval $-\pi/a < k < +\pi/a$ is called the *first Brillouin zone.* For lattice waves having values of k near the center of this zone, the phase velocity (ω_k/k) and the group velocity $(d\omega_k/dk)$ both have a constant magnitude close to $a\omega_0$. As k shifts outward toward an edge of the Brillouin zone, the phase velocity becomes somewhat reduced, while the group velocity approaches zero.

Example 5.2 Phonons in Three-Dimensional Solids. We now extend the formalism of our previous example to the three-dimensional lattice of an atomic solid. In this case, the classical Hamiltonian is of the form

$$H = \sum_{\alpha,s,\nu} \frac{1}{2m_\nu} p_{\alpha s\nu}^2 + \sum_{\alpha,s,\nu} \sum_{\alpha',s',\nu'} \frac{1}{2} \mathcal{G}_{\alpha' s' \nu'}^{\alpha\, s\, \nu}\, q_{\alpha s\nu} q_{\alpha' s' \nu'}. \tag{5.69}$$

Here, an atom is labelled by a pair of indices s and ν; s indicates the unit cell and ν denotes the site within that unit cell where the atom is situated. $q_{\alpha s\nu}$ is the atom's displacement-component along the αth Cartesian coordinate-direction relative to the equilibrium position $\mathbf{r}_{s\nu} = \mathbf{r}_s + \mathbf{r}_\nu$ (as shown in Fig 5.5, $\mathbf{r}_s$ is some point in the sth unit cell and $\mathbf{r}_\nu$ is the equilibrium position of the νth atom relative to that point); $p_{\alpha s\nu}$ is the atom's canonical momentum conjugate to $q_{\alpha s\nu}$. The particular form of Eq. 5.69 results from expanding the potential energy function V of the lattice in a Taylor series about the equilibrium configuration, then keeping terms up to second order.

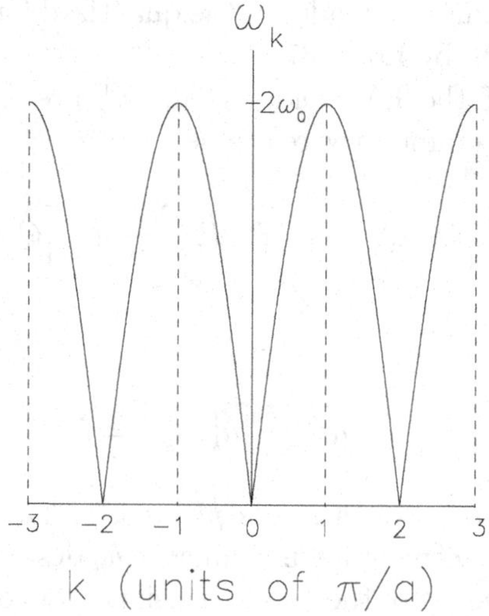

Figure 5.4 Phonon dispersion curve for one-dimensional lattice.

As discussed earlier in Section 2.6, the coefficients $\mathcal{G}_{\alpha' s' \nu'}^{\alpha s \nu}$ are then given by

$$\mathcal{G}_{\alpha' s' \nu'}^{\alpha s \nu} = \frac{\partial^2 V}{\partial q_{\alpha s \nu} \partial q_{\alpha' s' \nu'}}. \tag{5.70}$$

Hamilton's equations, i.e.,

$$\frac{\partial H}{\partial p_{\alpha s \nu}} = \dot{q}_{\alpha s \nu} \qquad \text{and} \qquad \frac{\partial H}{\partial q_{\alpha s \nu}} = -\dot{p}_{\alpha s \nu}, \tag{5.71}$$

lead to the equation of motion

$$m_\nu \ddot{q}_{\alpha s \nu} = - \sum_{\alpha', s', \nu'} \mathcal{G}_{\alpha' s' \nu'}^{\alpha s \nu} q_{\alpha' s' \nu'}. \tag{5.72}$$

In keeping with the procedure for previous cases, we transform to a set of coordinates $\mathsf{q}_{\alpha \mathbf{k} \nu}$ by

$$q_{\alpha s \nu} = \frac{1}{\sqrt{m_\nu}} \sum_{\mathbf{k}} \mathsf{q}_{\alpha \mathbf{k} \nu} e^{i(\mathbf{k} \cdot \mathbf{r}_s - \omega_{\mathbf{k}} t)}. \tag{5.73}$$

The exponential involves the coordinate $\mathbf{r}_s$ rather than $\mathbf{r}_{s\nu}$ because the factor $e^{i\mathbf{k} \cdot \mathbf{r}_\nu}$ is included in $\mathsf{q}_{\alpha \mathbf{k} \nu}$. As before, the possible $\mathbf{k}$-vectors are restricted; they are determined by imposing cyclic boundary conditions on a region containing N unit cells. Letting $\mathbf{b}_1$, $\mathbf{b}_2$, and $\mathbf{b}_3$ denote the fundamental reciprocal lattice vectors, the permissible wavevectors are

$$\mathbf{k} = \frac{1}{N} (n_1 \mathbf{b}_1 + n_2 \mathbf{b}_2 + n_3 \mathbf{b}_3) \qquad (n_1,\, n_2,\, n_3 = 0,\, \pm 1,\, \pm 2,\, \ldots). \tag{5.74}$$

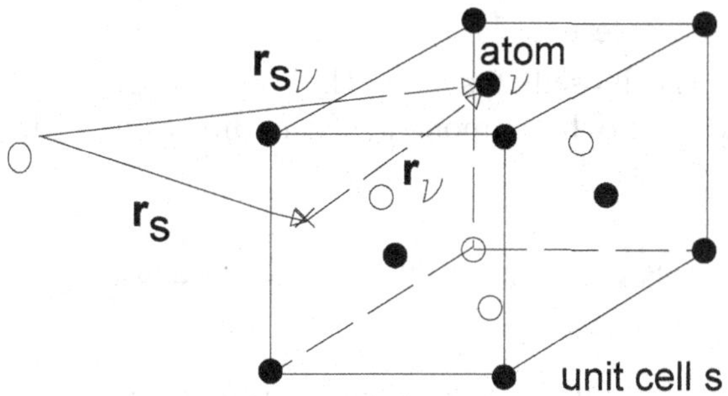

Figure 5.5 Coordinate system used for a three-dimensional lattice.

Substituting Eq. 5.73 into Eq. 5.72 produces

$$\omega_{\mathbf{k}}^2 \mathsf{q}_{\alpha\mathbf{k}\nu} = \sum_{\alpha',s',\nu'} \frac{1}{\sqrt{m_\nu m_{\nu'}}} \mathcal{G}_{\substack{\alpha\, s\, \nu\\ \alpha' s' \nu'}} e^{-i\mathbf{k}\cdot(\mathbf{r}_s - \mathbf{r}_{s'})} \mathsf{q}_{\alpha'\mathbf{k}\nu'}. \tag{5.75}$$

The $\mathcal{G}_{\substack{\alpha\, s\, \nu\\ \alpha' s' \nu'}}$'s, however, only depend on the difference $\mathbf{R}_\ell = \mathbf{r}_s - \mathbf{r}_{s'}$, therefore $\mathcal{G}_{\substack{\alpha\, s\, \nu\\ \alpha' s' \nu'}} = \mathcal{G}_{\substack{\alpha\, \nu\\ \alpha' \nu'}}(\mathbf{R}_\ell)$. Letting

$$\mathcal{D}_{\substack{\alpha\, \nu\\ \alpha' \nu'}}(\mathbf{k}) \equiv \frac{1}{\sqrt{m_\nu m_{\nu'}}} \sum_\ell \mathcal{G}_{\substack{\alpha\, \nu\\ \alpha' \nu'}}(\mathbf{R}_\ell) e^{i\mathbf{k}\cdot\mathbf{R}_\ell}, \tag{5.76}$$

Eq. 5.75 becomes the following eigenvalue problem:

$$\sum_{\alpha',\nu'} \mathcal{D}_{\substack{\alpha\, \nu\\ \alpha' \nu'}}(\mathbf{k}) \mathsf{u}^{(j)}_{\alpha'\mathbf{k}\nu'} = \left[\omega^{(j)}_{\mathbf{k}}\right]^2 \mathsf{u}^{(j)}_{\alpha\mathbf{k}\nu}. \tag{5.77}$$

The solution is a set of lattice displacement modes, each having a particular wavevector $\mathbf{k}$ and belonging to a so-called *dispersion branch* j. A nice qualitative discussion of the different observed dispersion branches can be found in the book by Marcuse [25]; the details will not be related here. Rather, let it suffice to say that the ω-$\mathbf{k}$ diagram for a 3-D lattice is rather complex; the Brillouin zones are now three-dimensional and they contain different branches representing either *acoustic* or *optical* phonon modes.[§] There are three acoustic branches; they correspond to the longitudinal vibrations and two possible transverse vibrations of the crystal lattice. For any mode in an acoustic branch, all the atoms in a given unit cell oscillate approximately in phase. In contrast, modes of an optical branch are distinguished by the fact that some atoms oscillate close to 180° out of phase with other atoms in the same unit cell. High vibrational frequencies are associated with these branches and, as a consequence, optical modes can often be excited by the absorption of light. Associated with each acoustic or

[§] A basic treatment of acoustic and optical phonon modes in a classical 1D diatomic lattice is presented in Problem 3 at the end of Chapter 2.

optical mode is a characteristic lattice vibration frequency, $\omega_{\mathbf{k}}^{(j)}$. In this mode, the νth atom of each unit cell oscillates along the direction given by the eigenvector $\mathsf{u}_{\alpha\mathbf{k}\nu}^{(j)}$ (specifically, it gives the αth component of the direction). The eigenvectors satisfy the relations

$$\sum_{\alpha,\nu} \mathsf{u}_{\alpha\mathbf{k}\nu}^{(j)*}\mathsf{u}_{\alpha\mathbf{k}\nu}^{(j')} = \delta_{jj'} \qquad \text{and} \qquad \sum_{j} \mathsf{u}_{\alpha\mathbf{k}\nu}^{(j)*}\mathsf{u}_{\alpha\mathbf{k}\nu'}^{(j)} = \delta_{\alpha\alpha'}\delta_{\nu\nu'}. \tag{5.78}$$

We now make a transformation to phonon coordinates, i.e.,

$$q_{\alpha\ell\nu} = \frac{1}{\sqrt{Nm_\nu}} \sum_{\mathbf{k},j} Q_{\mathbf{k}j}\mathsf{u}_{\alpha\mathbf{k}\nu}^{(j)} e^{i\mathbf{k}\cdot\mathbf{R}_\ell} \tag{5.79}$$

and

$$p_{\alpha\ell\nu} = \sqrt{\frac{m_\nu}{N}} \sum_{\mathbf{k},j} P_{\mathbf{k}j}\mathsf{u}_{\alpha\mathbf{k}\nu}^{(j)} e^{i\mathbf{k}\cdot\mathbf{R}_\ell}, \tag{5.80}$$

with $Q_{\mathbf{k}j}^* = Q_{-\mathbf{k}j}$, $P_{\mathbf{k}j}^* = P_{-\mathbf{k}j}$, and $\mathsf{u}_{\alpha\mathbf{k}\nu}^{(j)} = \mathsf{u}_{\alpha-\mathbf{k}\nu}^{(j)}$. The mode momenta are given by $P_{\mathbf{k}j} = \dot{Q}_{\mathbf{k}j}$. After substituting Eqs. 5.79 and 5.80 into the Hamiltonian of Eq. 5.69, we find that

$$H = \sum_{\mathbf{k},j} \frac{1}{2}\left\{P_{\mathbf{k}j}P_{-\mathbf{k}j} + \left[\omega_{\mathbf{k}}^{(j)}\right]^2 Q_{\mathbf{k}j}Q_{-\mathbf{k}j}\right\}. \tag{5.81}$$

The field is quantized by following a procedure that essentially mimics the one performed for the 1-D lattice of the previous example (i.e., refer to Eqs. 5.54–5.68 and the associated discussion). Again, the result is that the lattice vibrational modes form a collection of uncoupled harmonic oscillators. The final field operators have the forms given below:

$$\hat{Q}_{\mathbf{k}j} = \sqrt{\frac{\hbar}{2\omega_{\mathbf{k}}^{(j)}}}\left(\hat{a}_{\mathbf{k}j} + \hat{a}_{-\mathbf{k}j}^\dagger\right) \qquad \text{and} \qquad \hat{P}_{\mathbf{k}j} = -i\sqrt{\frac{\hbar\omega_{\mathbf{k}}^{(j)}}{2}}\left(\hat{a}_{-\mathbf{k}j} - \hat{a}_{\mathbf{k}j}^\dagger\right) \tag{5.82}$$

$$\hat{H} = \sum_{\mathbf{k},j} \hbar\omega_{\mathbf{k}}^{(j)}\left(\hat{a}_{\mathbf{k}j}^\dagger\hat{a}_{\mathbf{k}j} + \frac{1}{2}\right) \tag{5.83}$$

$$\hat{q}_{\alpha\ell\nu} = \sum_{\mathbf{k}j} \sqrt{\frac{\hbar}{2Nm_\nu\omega_{\mathbf{k}}^{(j)}}}\left(\hat{a}_{\mathbf{k}j}\mathsf{u}_{\alpha\mathbf{k}\nu}^{(j)}e^{+i\mathbf{k}\cdot\mathbf{R}_\ell} + \hat{a}_{\mathbf{k}j}^\dagger\mathsf{u}_{\alpha\mathbf{k}\nu}^{(j)*}e^{-i\mathbf{k}\cdot\mathbf{R}_\ell}\right). \tag{5.84}$$

The operators, $\hat{a}_{\mathbf{k}j}$ and $\hat{a}_{\mathbf{k}j}^\dagger$, annihilate or create a single phonon of wavevector $\mathbf{k}$ belonging to branch j.

5.3 Coherent States of the Radiation Field

We have shown that a charge-free radiation field can be transformed into a set of harmonic oscillators, one for each mode specified by wavevector $\mathbf{k}$ and polarization $\boldsymbol{\epsilon}_{\mathbf{k}\lambda}$. The eigenstates of the number operator $\hat{N}_{\mathbf{k}\lambda}$ are the number states $|n_{\mathbf{k}\lambda}\rangle$. These can be identified as *photon states*, i.e., in the state $|n_{\mathbf{k}\lambda}\rangle$ we know definitely that there are n photons in mode $(\mathbf{k}, \lambda)$. We now pose the following question: Does the state $|n_{\mathbf{k}\lambda}\rangle$ contain information about the wave properties of the field? To answer this, consider the expectation values for $Q(t)$ and $P(t)$, the oscillator's "position" and "momentum" when the field is in one of the number states.¶ Recalling Eqs. 5.34 for the corresponding operators,

$$\hat{Q}(t) = \sqrt{\frac{\hbar}{2\omega}}\left[\hat{a}(t) + \hat{a}^\dagger(t)\right] \qquad \text{and} \qquad \hat{P}(t) = -i\sqrt{\frac{\hbar\omega}{2}}\left[\hat{a}(t) - \hat{a}^\dagger(t)\right], \tag{5.85}$$

we immediately see that

$$\langle n \mid \hat{Q}(t) \mid n\rangle = \langle n \mid \hat{P}(t) \mid n\rangle = 0. \tag{5.86}$$

Thus, an oscillatory time-dependence, like that expected classically, fails to appear. However, this result should not be surprising. As previously discussed in Section 3.3, the expectation values of physical observables will only exhibit an oscillatory time-dependence when the system is in a superposition of energy eigenstates which, for the case of the harmonic oscillator, are the number states. By specifying the exact value of n, the number of photons in the field, the "position," or equivalently the *phase*, of the oscillator is completely lost! To emphasize this point, consider the operator for a single mode of the electric field $\mathbf{E}$ (see Eq. 5.38):

$$\hat{\mathbf{E}}(\mathbf{r}, t) = i\sqrt{\frac{2\pi\hbar\omega}{V}}\left[\hat{a}(t)\, e^{i\mathbf{k}\cdot\mathbf{r}} - \hat{a}^\dagger(t)\, e^{-i\mathbf{k}\cdot\mathbf{r}}\right]\boldsymbol{\epsilon}. \tag{5.87}$$

In a number state, the expectation value of $\mathbf{E}$ also vanishes, i.e.,

$$\langle n \mid \hat{\mathbf{E}} \mid n\rangle = 0. \tag{5.88}$$

In the n-representation, attention is focused on the particle-like aspects of the field embodied in the photon. This conceptualization of the field is an extreme quantum one. In many situations, however, it is more fitting to talk about the field as a wave having a well-defined phase and amplitude. To achieve this picture while retaining the quantum character of the field, it will prove fruitful to consider a representation based on the normalized eigenstates $|\alpha\rangle$ of the annihilation operator $\hat{a}$, i.e., we investigate the eigenvalue problem

$$\hat{a}\,|\alpha\rangle = \alpha\,|\alpha\rangle. \tag{5.89}$$

¶At this point, and for the remainder of this section, we simplify our notation by dropping the $\mathbf{k}\lambda$-subscripts from all operators and number states. The presentation can be applied to any particular mode of the radiation field.

The states $|\alpha\rangle$ are called *coherent states.* To determine the meaning of the eigenvalue α, first form the Hermitian conjugate of Eq. 5.89

$$\langle\alpha|\,\hat{a}^\dagger = \langle\alpha|\,\alpha^*. \tag{5.90}$$

(α is in general a complex number.) Then, recalling the explicit time-dependence of the lowering and raising operators to be

$$\hat{a}(t) = \hat{a}e^{-i\omega t} \qquad \text{and} \qquad \hat{a}^\dagger(t) = \hat{a}^\dagger e^{+i\omega t}, \tag{5.91}$$

we observe that the expectation values of $Q(t)$ and $P(t)$ in the state $|\alpha\rangle$ reduce to

$$\langle Q\rangle = \langle\alpha \mid \hat{Q}(t) \mid \alpha\rangle = Q_0\left(\alpha e^{-i\omega t} + \alpha^* e^{+i\omega t}\right) \tag{5.92}$$

and

$$\langle P\rangle = \langle\alpha \mid \hat{P}(t) \mid \alpha\rangle = -i\omega Q_0\left(\alpha e^{-i\omega t} - \alpha^* e^{+i\omega t}\right), \tag{5.93}$$

where $Q_0 = \sqrt{\hbar/2\omega}$. Note that $\langle P\rangle = d\langle Q\rangle/dt$. Now express the complex number α as the product of a magnitude and a phase factor, i.e., $\alpha = |\alpha|\,e^{i\phi}$. Then

$$\langle Q\rangle = 2Q_0\,|\alpha|\cos(\omega t - \phi) \qquad \text{and} \qquad \langle P\rangle = -2\omega Q_0\,|\alpha|\sin(\omega t - \phi). \tag{5.94}$$

Similarly, in a coherent state, the expectation value of $\mathbf{E}$ becomes

$$\begin{aligned}\langle\mathbf{E}\rangle &= \langle\alpha \mid \hat{\mathbf{E}} \mid \alpha\rangle = i\sqrt{\frac{2\pi\hbar\omega}{V}}\left[\alpha e^{+i(\mathbf{k}\cdot\mathbf{r}-\omega t)} - \alpha^* e^{-i(\mathbf{k}\cdot\mathbf{r}-\omega t)}\right]\boldsymbol{\epsilon} \\ &= -2\sqrt{\frac{2\pi\hbar\omega}{V}}\,|\alpha|\sin(\mathbf{k}\cdot\mathbf{r} - \omega t + \phi)\,\boldsymbol{\epsilon}.\end{aligned} \tag{5.95}$$

The results show that in a state $|\alpha\rangle$, the variables Q and P oscillate in a classical fashion, while the electric field propagates as a travelling wave. In each case, the phase and amplitude of the motion correspond to the phase and magnitude of the state's eigenvalue α.

In a coherent state, the number of photons n is not specified exactly. To see this, note that the mean number of photons in state $|\alpha\rangle$ is

$$\langle n\rangle = \langle\alpha \mid \hat{N} \mid \alpha\rangle = \langle\alpha \mid \hat{a}^\dagger\hat{a} \mid \alpha\rangle = |\alpha|^2. \tag{5.96}$$

Also,

$$\begin{aligned}\langle n^2\rangle &= \langle\alpha \mid \hat{N}^2 \mid \alpha\rangle = \langle\alpha \mid \hat{a}^\dagger\hat{a}\hat{a}^\dagger\hat{a} \mid \alpha\rangle \\ &= |\alpha|^2\langle\alpha \mid \hat{a}\hat{a}^\dagger \mid \alpha\rangle = |\alpha|^2\langle\alpha \mid \left(1 + \hat{a}^\dagger\hat{a}\right) \mid \alpha\rangle \\ &= |\alpha|^2 + |\alpha|^4.\end{aligned} \tag{5.97}$$

Therefore, the uncertainty in the number of photons is represented by the variance

$$(\Delta n)^2 = \langle n^2\rangle - \langle n\rangle^2 = |\alpha|^2, \tag{5.98}$$

or, from Eq. 5.96,

$$(\triangle n)^2 = \langle n \rangle\,. \tag{5.99}$$

This is precisely the result one would expect if the probability of observing n photons in state $|\alpha\rangle$ constitutes a *Poisson distribution* about the mean $\langle n\rangle$. To see that this is in fact the case, let us expand a state $|\alpha\rangle$ in terms of the n-representation, i.e.,

$$|\alpha\rangle = \sum_{n=0}^{\infty} |n\rangle \langle n \mid \alpha\rangle = \sum_{n=0}^{\infty} c_n(\alpha)\,|n\rangle\,. \tag{5.100}$$

The squared coefficient $|c_n(\alpha)|^2 = |\langle n \mid \alpha\rangle|^2$ is the probability of finding the oscillator in the number state $|n\rangle$ when a measurement is performed on state $|\alpha\rangle$. Now, by operating on this equation with the lowering operator, we see that

$$\hat{a}\,|\alpha\rangle = \sum_{n=0}^{\infty} c_n(\alpha)\hat{a}\,|n\rangle = \sum_{n=1}^{\infty} c_n(\alpha)\sqrt{n}\,|n-1\rangle = \sum_{n=0}^{\infty} c_{n+1}(\alpha)\sqrt{n+1}\,|n\rangle\,. \tag{5.101}$$

But applying Eq. 5.100 to the eigenvalue equation for $\hat{a}$ gives

$$\hat{a}\,|\alpha\rangle = \alpha\,|\alpha\rangle = \sum_{n=0}^{\infty} \alpha c_n(\alpha)\,|n\rangle\,. \tag{5.102}$$

Now compare terms in the last two equations. The result is that the $c_n(\alpha)$'s are related as follows:

$$c_{n+1}(\alpha) = \frac{\alpha}{\sqrt{n+1}}c_n(\alpha). \tag{5.103}$$

Observe the pattern that emerges:

$$\begin{aligned}
c_1(\alpha) &= \frac{\alpha}{\sqrt{1}}c_0(\alpha) \\
c_2(\alpha) &= \frac{\alpha}{\sqrt{2}}c_1(\alpha) = \frac{\alpha}{\sqrt{1}}\frac{\alpha}{\sqrt{2}}c_0(\alpha) \\
c_3(\alpha) &= \frac{\alpha}{\sqrt{3}}c_2(\alpha) = \frac{\alpha}{\sqrt{1}}\frac{\alpha}{\sqrt{2}}\frac{\alpha}{\sqrt{3}}c_0(\alpha) \\
&\vdots \\
c_n(\alpha) &= \frac{\alpha^n}{\sqrt{n!}}c_0(\alpha).
\end{aligned} \tag{5.104}$$

$c_0(\alpha)$ is obtained by the normalization

$$\langle \alpha \mid \alpha\rangle = |c_0(\alpha)|^2 \sum_{n,m=0}^{\infty} \frac{\alpha^{*m}\alpha^n}{\sqrt{m!\,n!}}\delta_{mn} = |c_0(\alpha)|^2 \sum_{n=0}^{\infty} \frac{|\alpha|^{2n}}{n!} = |c_0(\alpha)|^2 e^{|\alpha|^2} = 1, \tag{5.105}$$

so

$$c_0(\alpha) = e^{-|\alpha|^2/2}. \tag{5.106}$$

Thus, Eq. 5.100 for the expansion of $|\alpha\rangle$ in terms of n-states becomes

$$|\alpha\rangle = e^{-\frac{1}{2}|\alpha|^2} \sum_{n=0}^{\infty} \frac{\alpha^n}{\sqrt{n!}} |n\rangle . \tag{5.107}$$

The probability for obtaining n photons in a measurement follows the anticipated Poisson distribution:

$$P(n) = |c_n(\alpha)|^2 = \frac{|\alpha|^{2n}}{n!} e^{-|\alpha|^2} = \frac{\langle n\rangle^n}{n!} e^{-\langle n\rangle}. \tag{5.108}$$

The last step follows from Eq. 5.96. The coherent light produced by a laser is approximately in an $|\alpha\rangle$ state. If a laser beam is detected by a photon counter, the probability of detecting n photons in a certain time interval follows Eq. 5.108.

We can also derive an uncertainty relation between photon-number and oscillator phase. This is accomplished by first expressing $\hat{a}$ and $\hat{a}^\dagger$ in terms of the number operator $\hat{N}$ and phase operators $\hat{e}^{+i\phi}$ and $\hat{e}^{-i\phi}$:

$$\hat{a} = \hat{e}^{+i\phi} \hat{N}^{\frac{1}{2}} \qquad \text{and} \qquad \hat{a}^\dagger = \hat{N}^{\frac{1}{2}} \hat{e}^{-i\phi}. \tag{5.109}$$

The reader should take note that the operator symbol is attached to the phase operator as a whole, not to the symbol ϕ. This is done to emphasize that the phase operators are not exponential functions of some operator $\hat{\phi}$. In fact, no such operator exists. To make this clear, observe from the latter relations that

$$\hat{e}^{+i\phi} = \hat{a}\hat{N}^{-\frac{1}{2}} \qquad \text{and} \qquad \hat{e}^{-i\phi} = \hat{N}^{-\frac{1}{2}} \hat{a}^\dagger. \tag{5.110}$$

Then clearly

$$\hat{e}^{-i\phi}\hat{e}^{+i\phi} = \hat{N}^{-\frac{1}{2}}\hat{a}^\dagger\hat{a}\hat{N}^{-\frac{1}{2}} = \hat{N}^{-\frac{1}{2}}\hat{N}\hat{N}^{-\frac{1}{2}} = 1; \tag{5.111}$$

however, one might be surprised to learn

$$\hat{e}^{+i\phi}\hat{e}^{-i\phi} = \hat{a}\hat{N}^{-\frac{1}{2}}\hat{N}^{-\frac{1}{2}}\hat{a}^\dagger = \hat{a}\hat{N}^{-1}\hat{a}^\dagger \neq 1. \tag{5.112}$$

In light of these facts, notice that $\hat{a}^\dagger\hat{a} = \hat{N}$, as required. Realize though, that as things are presently defined, we will not be able to determine a phase-number uncertainty principle. This is because the operators corresponding to the observables must be Hermitian, and the phase operators are not. To resolve this problem, we construct the following new pair of phase operators that are Hermitian:

$$\hat{cos}\,\phi = \frac{1}{2}\left(\hat{e}^{+i\phi} + \hat{e}^{-i\phi}\right) \qquad \text{and} \qquad \hat{sin}\,\phi = \frac{1}{2i}\left(\hat{e}^{+i\phi} - \hat{e}^{-i\phi}\right). \tag{5.113}$$

We can then, for example, form the commutator

$$\left[\hat{cos}\,\phi,\hat{N}\right]=\frac{1}{2}\left[\hat{e}^{+i\phi},\hat{N}\right]+\frac{1}{2}\left[\hat{e}^{-i\phi},\hat{N}\right]. \tag{5.114}$$

But from $\left[\hat{a},\hat{a}^{\dagger}\right]=1$ and Eqs. 5.109, we write

$$\hat{e}^{+i\phi}\hat{N}\hat{e}^{-i\phi}-\hat{N}=1 \tag{5.115}$$

or

$$\hat{e}^{+i\phi}\hat{N}-\hat{N}\hat{e}^{+i\phi}=\hat{e}^{+i\phi}, \tag{5.116}$$

so

$$\left[\hat{e}^{+i\phi},\hat{N}\right]=\hat{e}^{+i\phi}. \tag{5.117}$$

Similarly,

$$\left[\hat{e}^{-i\phi},\hat{N}\right]=-\hat{e}^{-i\phi}. \tag{5.118}$$

As a result, the commutator of Eq. 5.114 becomes

$$\left[\hat{cos}\,\phi,\hat{N}\right]=\frac{1}{2}\left(\hat{e}^{+i\phi}-\hat{e}^{-i\phi}\right)=i\,\hat{sin}\,\phi. \tag{5.119}$$

It is left to the reader to show that

$$\left[\hat{sin}\,\phi,\hat{N}\right]=-i\,\hat{cos}\,\phi. \tag{5.120}$$

Finally, referring to Eq. 3.77, we find the following two phase-number uncertainty relations:

$$\triangle n\,\triangle\cos\phi\geq\frac{1}{2}\left|\langle\sin\phi\rangle\right|\quad\text{and}\quad\triangle n\,\triangle\sin\phi\geq\frac{1}{2}\left|\langle\cos\phi\rangle\right|. \tag{5.121}$$

In other words, it is impossible to precisely specify both the oscillator phase and the number of photons contained in any state of the radiation field.

Presented below are some important features of the coherent states:

- **The coherent states are complete, but not orthogonal.**
  First let us prove that the coherent states form a complete set of basis states by using the n-representation of Eq. 5.107 (since $\alpha=|\alpha|\,e^{i\phi}$, the ensuing integration is over the entire complex plane):

$$\begin{aligned}\frac{1}{\pi}\int|\alpha\rangle\langle\alpha|\,d^2\alpha &= \sum_{n,m=0}^{\infty}\frac{|n\rangle\langle m|}{\pi\sqrt{n!m!}}\int e^{-|\alpha|^2}\alpha^{*m}\alpha^n d^2\alpha\\ &= \sum_{n,m=0}^{\infty}\frac{|n\rangle\langle m|}{\pi\sqrt{n!m!}}\int_{|\alpha|=0}^{\infty}e^{-|\alpha|^2}|\alpha|^{n+m}|\alpha|\,d|\alpha|\\ &\qquad\times\int_{\phi=0}^{2\pi}e^{i(n-m)\phi}d\phi.\end{aligned} \tag{5.122}$$

Using $\int_{\phi=0}^{2\pi} e^{i(n-m)\phi}d\phi = 2\pi\delta_{nm}$, and letting $x = |\alpha|^2$, simplifies the result to

$$\frac{1}{\pi}\int |\alpha\rangle\langle\alpha|\, d^2\alpha = \sum_{n=0}^{\infty}\frac{|n\rangle\langle n|}{n!}\int_0^{\infty} e^{-x}x^n dx. \tag{5.123}$$

But $\int_{x=0}^{\infty} e^{-x}x^n dx = n!$, so

$$\frac{1}{\pi}\int |\alpha\rangle\langle\alpha|\, d^2\alpha = \sum_{n=0}^{\infty}|n\rangle\langle n| = \hat{I}, \tag{5.124}$$

and the completeness condition is satisfied. Now check the coherent states for orthogonality:

$$\begin{aligned}\langle\alpha \mid \beta\rangle &= e^{-\frac{1}{2}\left(|\beta|^2+|\alpha|^2\right)}\sum_{n,m=0}^{\infty}\frac{\alpha^{*n}\beta^m}{\sqrt{n!m!}}\langle n \mid m\rangle \\ &= e^{-\frac{1}{2}\left(|\beta|^2+|\alpha|^2\right)}\sum_{n=0}^{\infty}\frac{(\beta\alpha^*)^n}{n!} = e^{-\frac{1}{2}\left(|\beta|^2+|\alpha|^2\right)}e^{\beta\alpha^*}.\end{aligned} \tag{5.125}$$

So we find that $\langle\alpha \mid \beta\rangle$ does not vanish for $\alpha \neq \beta$, which means that the coherent states are not orthogonal. Note, however, that since

$$|\langle\alpha \mid \beta\rangle|^2 = e^{-|\beta-\alpha|^2}, \tag{5.126}$$

the states approach orthogonality as $|\beta - \alpha|$ becomes large.

- **The coherent states can be generated by using a "displacement operator."**
  By combining Eqs. 3.97 and 5.107, we produce

$$|\alpha\rangle = e^{-\frac{1}{2}|\alpha|^2}\sum_{n=0}^{\infty}\frac{\alpha^n}{\sqrt{n!}}\frac{\left(\hat{a}^\dagger\right)^n}{\sqrt{n!}}|0\rangle = e^{-\frac{1}{2}|\alpha|^2}e^{\alpha\hat{a}^\dagger}|0\rangle. \tag{5.127}$$

Now use the *Baker-Hausdorff theorem*‖

$$e^{\hat{A}}e^{\hat{B}} = e^{\hat{A}+\hat{B}+\frac{1}{2}[\hat{A},\hat{B}]} \tag{5.128}$$

‖See, for example, Louisell [21, p. 137]

and identify $\hat{A} \equiv \alpha\hat{a}^\dagger$ and $\hat{B} \equiv -\alpha^*\hat{a}$. Then, $\left[\hat{A}, \hat{B}\right] = |\alpha|^2 \left[\hat{a}, \hat{a}^\dagger\right] = |\alpha|^2$, so that $e^{\alpha\hat{a}^\dagger} e^{-\alpha^*\hat{a}} = e^{\alpha\hat{a}^\dagger - \alpha^*\hat{a} + \frac{1}{2}|\alpha|^2}$, and we are able to write

$$e^{-\frac{1}{2}|\alpha|^2} e^{\alpha\hat{a}^\dagger} e^{-\alpha^*\hat{a}} |0\rangle = e^{\alpha\hat{a}^\dagger - \alpha^*\hat{a}} |0\rangle . \tag{5.129}$$

But using the fact $e^{-\alpha^*\hat{a}} |0\rangle = (1 - \alpha^*\hat{a} + \ldots) |0\rangle = |0\rangle$, Eqs. 5.127 and 5.129 combine to give

$$|\alpha\rangle = e^{\alpha\hat{a}^\dagger - \alpha^*\hat{a}} |0\rangle . \tag{5.130}$$

This shows that a coherent state $|\alpha\rangle$ can be generated from the vacuum state $|0\rangle$ by applying the unitary operator

$$\hat{D}(\alpha) \equiv e^{\alpha\hat{a}^\dagger - \alpha^*\hat{a}}, \tag{5.131}$$

i.e.,

$$|\alpha\rangle = \hat{D}(\alpha) |0\rangle . \tag{5.132}$$

We call $\hat{D}(\alpha)$ a *displacement operator.* Its action on the state $|0\rangle$ is parallel to that of the operator $\left(\hat{a}^\dagger\right)^n / \sqrt{n!}$ (see Eq. 3.97) which generates a number state, rather than a coherent state, from the vacuum state. In addition to Eq. 5.132, the reader should also verify the following properties of the displacement operator:

$$\hat{D}^\dagger(\alpha) = \hat{D}(-\alpha) \tag{5.133}$$

$$\hat{D}^\dagger(\alpha)\hat{a}\hat{D}(\alpha) = \hat{a} + \alpha. \tag{5.134}$$

- **Coherent states are minimum uncertainty states.**
  The uncertainty relation between Q and P for any coherent state is $\Delta Q \, \Delta P = \hbar/2$, i.e., the product has the smallest value allowed by the uncertainty principle. To demonstrate this fact observe that

$$\begin{aligned}
(\Delta Q)^2 &= \langle Q^2 \rangle - \langle Q \rangle^2 = \langle \alpha \mid \hat{Q}^2 \mid \alpha \rangle - \langle \alpha \mid \hat{Q} \mid \alpha \rangle^2 \\
&= \langle \alpha \mid \frac{\hbar}{2\omega} \left(\hat{a} + \hat{a}^\dagger\right)^2 \mid \alpha \rangle - \langle \alpha \mid \sqrt{\frac{\hbar}{2\omega}} \left(\hat{a} + \hat{a}^\dagger\right) \mid \alpha \rangle^2 \\
&= \frac{\hbar}{2\omega} \langle \alpha \mid \left(\hat{a}^2 + \hat{a}^{\dagger 2} + \hat{a}\hat{a}^\dagger + \hat{a}^\dagger\hat{a}\right) \mid \alpha \rangle - \frac{\hbar}{2\omega} \langle \alpha \mid \left(\hat{a} + \hat{a}^\dagger\right) \mid \alpha \rangle^2 \\
&= \frac{\hbar}{2\omega} \left(\alpha^2 + \alpha^{*2} + 2\alpha\alpha^* + 1\right) - \frac{\hbar}{2\omega} \left(\alpha^2 + \alpha^{*2} + 2\alpha\alpha^*\right) \\
&= \frac{\hbar}{2\omega}
\end{aligned} \tag{5.135}$$

and similarly

$$(\Delta P)^2 = \langle \alpha \mid \hat{P}^2 \mid \alpha \rangle - \langle \alpha \mid \hat{P} \mid \alpha \rangle^2$$

$$
\begin{aligned}
&= \langle\alpha \,|\, -\frac{\hbar\omega}{2}\left(\hat{a}-\hat{a}^{\dagger}\right)^{2} \,|\, \alpha\rangle - \langle\alpha \,|\, -i\sqrt{\frac{\hbar\omega}{2}}\left(\hat{a}-\hat{a}^{\dagger}\right) \,|\, \alpha\rangle^{2} \\
&= -\frac{\hbar\omega}{2}\left(\alpha^{2}+\alpha^{*2}-2\alpha\alpha^{*}-1\right)+\frac{\hbar\omega}{2}\left(\alpha^{2}+\alpha^{*2}-2\alpha\alpha^{*}\right) \\
&= \frac{\hbar\omega}{2}.
\end{aligned}
\tag{5.136}
$$

Thus,

$$
\Delta Q\,\Delta P = \sqrt{\frac{\hbar}{2\omega}}\sqrt{\frac{\hbar\omega}{2}} = \frac{\hbar}{2}. \tag{5.137}
$$

This means that a coherent state $|\alpha\rangle$ comes as close to a classical state as one can possibly approach for the radiation field. By comparison, a number state $|n\rangle$ is at the other extreme—it is essentially the most quantum-like state.

Example 5.3 Coherent State Wavefunctions. From Eqs. 5.85, the annihilation operator is

$$
\hat{a}(t) = \frac{1}{\sqrt{2\hbar\omega}}\left(\omega\hat{Q}+i\hat{P}\right), \tag{5.138}
$$

so the position-space representation of $\hat{a}(t)\,|\alpha\rangle = \hat{a}e^{-i\omega t}\,|\alpha\rangle = \alpha e^{-i\omega t}\,|\alpha\rangle$ becomes

$$
\frac{1}{\sqrt{2\hbar\omega}}\langle Q \,|\, \left(\omega\hat{Q}+i\hat{P}\right) \,|\, \alpha\rangle = \alpha e^{-i\omega t}\langle Q\,|\alpha\rangle. \tag{5.139}
$$

$\langle Q\,|\alpha\rangle$ is the coherent state wavefunction, $\psi_\alpha(Q,t)$. Recalling that $\langle Q \mid \hat{Q} \mid \alpha\rangle = Q\psi_\alpha(Q,t)$ and $\langle Q \mid \hat{P} \mid \alpha\rangle = -i\hbar\frac{\partial}{\partial Q}\psi_\alpha(Q,t)$ (see Section 3.4), rearranging Eq. 5.139 produces the following differential equation for the wavefunction:

$$
\frac{\partial}{\partial Q}\psi_\alpha(Q,t) = \left(\sqrt{\frac{2\omega}{\hbar}}\alpha e^{-i\omega t} - \frac{\omega}{\hbar}Q\right)\psi_\alpha(Q,t). \tag{5.140}
$$

Integration of Eq. 5.140 produces the solution

$$
\psi_\alpha(Q,t) = C(\alpha,t)\exp\left(-\frac{\omega}{2\hbar}Q^2 + \sqrt{\frac{2\omega}{\hbar}}\alpha e^{-i\omega t}Q\right), \tag{5.141}
$$

where $C(\alpha,t)$ is determined by normalization to be

$$
C(\alpha,t) = \left(\frac{\omega}{\pi\hbar}\right)^{1/4}\exp\left[-\left|\alpha\right|^2\cos^2\left(\omega t-\phi\right)\right] \tag{5.142}
$$

(within an arbitrary phase factor). Notice that ψ_0 is just the ground-state wavefunction of Eq. 3.172. The position-space probability density is given by

$$
\left|\psi_\alpha(Q,t)\right|^2 = \sqrt{\frac{\omega}{\pi\hbar}}\exp\left\{-\frac{\omega}{\hbar}\left[Q-\sqrt{\frac{2\hbar}{\omega}}\left|\alpha\right|\cos\left(\omega t-\phi\right)\right]^2\right\}. \tag{5.143}
$$

Observe that this is a Gaussian of constant width $\Delta Q = \sqrt{\hbar/2\omega}$, as expected from Eq. 5.135, and the mean of the distribution exhibits the oscillatory behavior $\langle Q \rangle = \sqrt{2\hbar/\omega}\,|\alpha| \cos(\omega t - \phi)$, precisely as predicted by Eq. 5.94! The coherent states are unique in that they are the only quantum states in nature that can oscillate in this manner while maintaining their shape.

5.4 Squeezed States

In the last section, we learned that the coherent states are states of minimum uncertainty. Said another way, they exhibit the least possible amount of *quantum noise* consistent with the uncertainty principle. A natural question one might ask is whether the coherent states are unique in this regard, or do other states of the radiation field also exist that qualify as minimum uncertainty states. The fact is that other such states do exist, namely, a subset of the so called *squeezed states.* To understand how the squeezed and coherent states differ requires that we take a close look at how quantum noise manifests itself in each case.

A coherent state for a given mode of the radiation field is characterized by a complex amplitude $\alpha = |\alpha|\, e^{i\phi}$, or equivalently,

$$\alpha = \alpha_1 + i\alpha_2, \tag{5.144}$$

where $\alpha_1 = |\alpha| \cos\phi$ and $\alpha_2 = |\alpha| \sin\phi$. The quantum noise carried by the corresponding coherent state $|\alpha\rangle$ is evident if one tries to measure both real and imaginary parts of α. To see this, it becomes convenient to define the Hermitian operators $\hat{X}_C$ and $\hat{X}_S$, known as *quadrature amplitudes,* which represent the real and imaginary components of the complex field amplitude:**

$$\begin{aligned} \hat{X}_C &= \frac{1}{2}\left(\hat{a} + \hat{a}^\dagger\right) = \sqrt{\frac{\omega}{2\hbar}}\hat{Q} \\ \hat{X}_S &= \frac{1}{2i}\left(\hat{a} - \hat{a}^\dagger\right) = \frac{1}{\sqrt{2\hbar\omega}}\hat{P}. \end{aligned} \tag{5.145}$$

Also, let $|x_c\rangle$ and $|x_s\rangle$ be the eigenstates of $\hat{X}_C$ and $\hat{X}_S$, respectively, with corresponding eigenvalues x_c and x_s. Then, notice that in a coherent state, the mean and uncertainty associated with measuring the real and imaginary parts of the field amplitude are

$$\begin{aligned} \langle x_c \rangle &= \langle \alpha \mid \hat{X}_C \mid \alpha \rangle = \frac{1}{2}\left(\langle \alpha \mid \hat{a} \mid \alpha \rangle + \langle \alpha \mid \hat{a}^\dagger \mid \alpha \rangle\right) = \alpha_1 \\ \langle x_s \rangle &= \langle \alpha \mid \hat{X}_S \mid \alpha \rangle = \frac{1}{2i}\left(\langle \alpha \mid \hat{a} \mid \alpha \rangle - \langle \alpha \mid \hat{a}^\dagger \mid \alpha \rangle\right) = \alpha_2 \end{aligned} \tag{5.146}$$

**The subscripts "*C*" and "*S*" for the two quadrature amplitudes are chosen to label the *C*osine and *S*ine components of the complex field amplitude.

and[††]

$$\Delta x_c = \sqrt{\frac{\omega}{2\hbar}}\Delta Q = \sqrt{\frac{\omega}{2\hbar}}\sqrt{\frac{\hbar}{2\omega}} = \frac{1}{2}$$
$$\Delta x_s = \frac{1}{\sqrt{2\hbar\omega}}\Delta P = \frac{1}{\sqrt{2\hbar\omega}}\sqrt{\frac{\hbar\omega}{2}} = \frac{1}{2}. \tag{5.147}$$

The uncertainty product is therefore $\Delta x_c \Delta x_s = 1/4$. We see that this result is consistent with a minimum uncertainty state since the commutator between the quadrature amplitudes is

$$\left[\hat{X}_C, \hat{X}_S\right] = \frac{1}{2\hbar}\left[\hat{Q}, \hat{P}\right] = \frac{i}{2}, \tag{5.148}$$

and, from Eq. 3.77, the general uncertainty relation is of the form

$$\Delta x_c \Delta x_s \geq 1/4. \tag{5.149}$$

For the coherent states, the fluctuations in x_c and x_s are always equal—this is conveniently visualized as a circle of minimum uncertainty in x_c-x_s phase space, as shown in Fig. 5.6. In the case of the squeezed states, however, the relative fluctuations exhibited by the two quadrature amplitudes can be controlled so that, in general, they turn out to be unequal. Consequently, a squeezed state is represented by an uncertainty ellipse in phase space (see below). Because of the uncertainty relation given by Eq. 5.149, it is impossible to simultaneously reduce both Δx_c and Δx_s below the value of 1/2, however, one of these noise components can be squeezed below this value at the expense of the other. Although squeezed states are not in general minimum uncertainty states, they do fall into this category when the squeezing is performed in certain ways, as we will describe shortly.

Squeezed states are classified as being of one of two types—*scaling squeezed states* or *coherent squeezed states* [26]. In each case, a different "squeezing process" is involved, as we explain below:

- **Scaling squeezed states**
  These states are denoted by $|\alpha_{\rm ss}; \mu, \nu\rangle$. They are generated by starting with some coherent state $|\alpha_{\rm in}\rangle$ as input, and performing the following transformation:

$$|\alpha_{\rm ss}; \mu, \nu\rangle = \hat{S}(\mu, \nu)\,|\alpha_{\rm in}\rangle\,. \tag{5.150}$$

  $\hat{S}(\mu, \nu)$ is the unitary *squeezing operator* defined as

$$\hat{S}(\mu, \nu) = e^{\frac{1}{2}\left(\varsigma^* \hat{a}^2 - \varsigma a^{\dagger 2}\right)}, \tag{5.151}$$

  where

$$\varsigma = s e^{i\theta}. \tag{5.152}$$

[††]In Eqs. 5.147, we have made use of Eqs. 5.145, and then Eqs. 5.135 and 5.136.

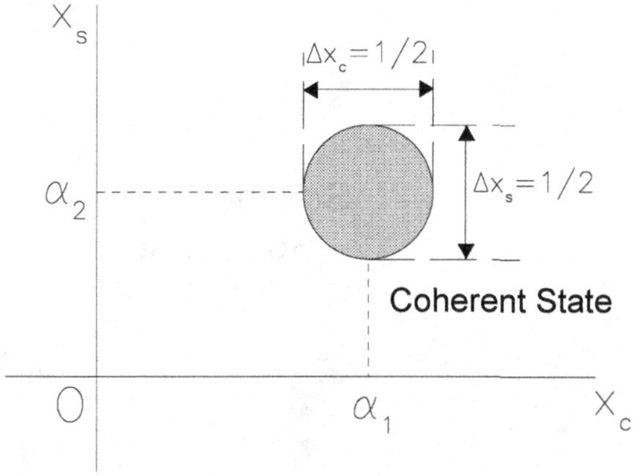

Figure 5.6 Phase space for coherent state.

s and θ are called the *degree of squeezing* and the *squeezing angle*, respectively. The *squeezing parameters* μ and ν are given by

$$\mu = \cosh s \qquad \text{and} \qquad \nu = e^{i\theta} \sinh s. \tag{5.153}$$

The action of the squeezing operator to form a scaling squeezed state is best visualized in the context of phase space. As shown in Fig. 5.7, the uncertainty circle of the coherent state $|\alpha_{\text{in}}\rangle$ is squeezed into the shape of an ellipse and, at the same time, the center of the uncertainty region is shifted from the complex amplitude $\alpha_{\text{in}} = |\alpha_{\text{in}}| e^{i\phi_{\text{in}}}$ to the value

$$\alpha_{\text{ss}} = |\alpha_{\text{ss}}| e^{i\phi_{\text{ss}}} = \mu\alpha_{\text{in}} + \nu e^{i\theta}\alpha_{\text{in}}^{*}. \tag{5.154}$$

The major and minor axes of the ellipse take on the lengths e^{s} and e^{-s}, respectively, and they are oriented at an angle $\theta/2$ relative to the x_c- and x_s-axes. As conveyed in the figure, the uncertainties in x_c and x_s that one actually observes are given by the projections of the ellipse onto the corresponding coordinate axes. We see that the redistribution of quantum noise for the two quadrature amplitudes (i.e., shape and orientation of the ellipse) and the shift of the center-point in phase space are not independent for the scaling squeezed states. This type of squeezing is characteristic of the *optical parametric process* used for certain optical amplification devices [27].

- **Coherent squeezed states**
  The coherent squeezed states, represented by $|\alpha_{\text{cs}}; \mu, \nu\rangle$, are generated by the process

$$|\alpha_{\text{cs}}; \mu, \nu\rangle = \hat{D}(\alpha_{\text{cs}})\,\hat{S}(\mu, \nu)\,|0\rangle = \hat{D}(\alpha_{\text{cs}})\,|0; \mu, \nu\rangle\,. \tag{5.155}$$

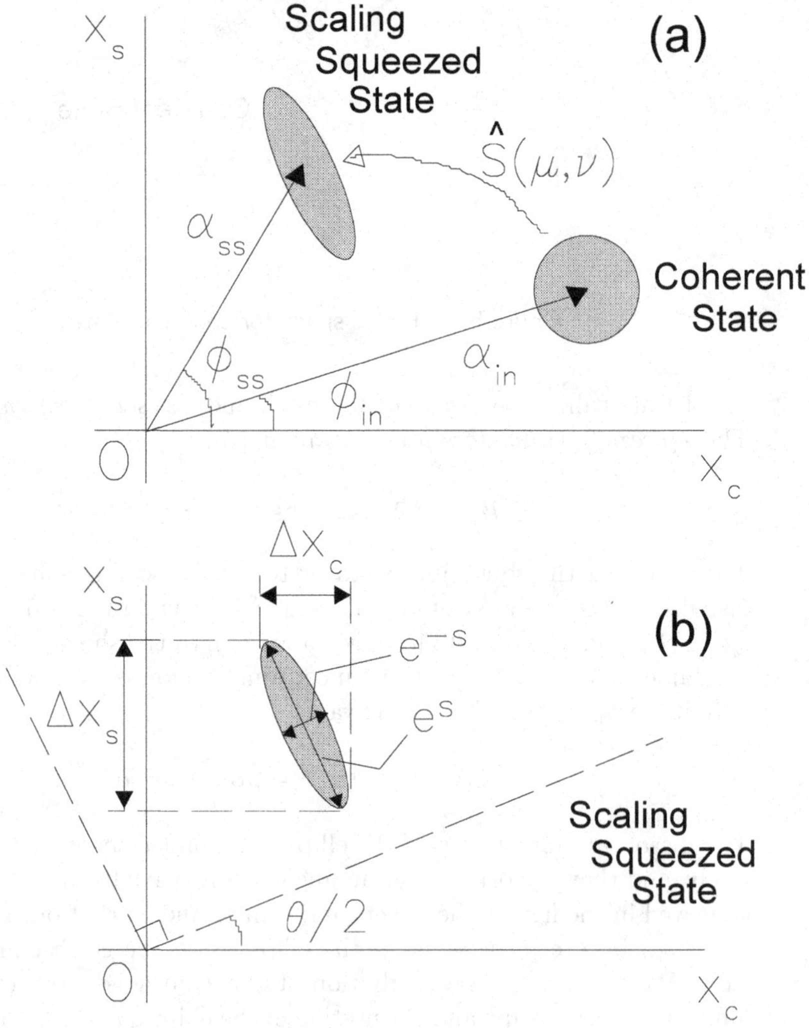

Figure 5.7 (a) Generation of a scaling squeezed state illustrated in phase space. (b) Degree of squeezing (s) and squeezing angle (θ) as depicted in phase space.

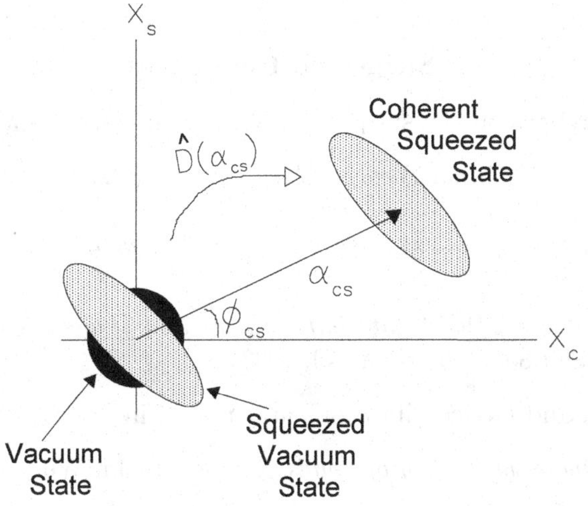

Figure 5.8 Formation of a coherent squeezed state as depicted in phase space.

$\hat{D}$ is the unitary displacement operator previously defined by Eq. 5.131. As illustrated in Fig. 5.8, the formation of a coherent squeezed state consists of two steps: First, the operator $\hat{S}(\mu, \nu)$ squeezes the uncertainty circle of the vacuum state $|0\rangle$ into an ellipse; this results in a squeezed vacuum state $|0; \mu, \nu\rangle$, with center still maintained at the origin of phase space. In the second step, the operator $\hat{D}(\alpha_{\text{cs}})$ displaces the center of the ellipse to the amplitude $\alpha_{\text{cs}} = |\alpha_{\text{cs}}| e^{i\phi_{\text{cs}}}$. We see, therefore, that the squeezing and shifting processes are controlled independently for the coherent squeezed states. Squeezing of this type occurs in the so-called *optical bistability process* [28].

When the squeezing angle θ takes on the value of either 0 or π, the two axes of the uncertainty ellipse will lie parallel to the coordinate axes. The noise components, Δx_c and Δx_s, are then simply given by the lengths of the semimajor and semiminor axes, $\frac{1}{2}e^{-s}$ and $\frac{1}{2}e^{s}$. As a result, one has $\Delta x_c \Delta x_s = 1/4$, which means that squeezed states of this type are minimum uncertainty states.[‡‡] For θ not equal to 0 or π, however, the squeezed states do not fall into this category.

This has certainly been a very brief introduction to the concept of squeezed states. The topic will come up again when photon counting statistics are discussed in Chapter 11. At that time, it will be shown that squeezed light leads to a counting distribution that is very different from that produced by a coherent state—in other words, one observes non-Poisson counting statistics. The reader interested in a more detailed treatment of squeezed light should consult the reference suggested at the end of this chapter.

[‡‡]These squeezed states of minimum uncertainty are sometimes referred to as *ideal squeezed states.*

Suggested References

Good general references on the subjects of field quantization and coherent states are

[a] R. Loudon, *The Quantum Theory of Light*, 3rd ed. (Oxford University Press, Oxford, 2000).

[b] W. H. Louisell, *Quantum Statistical Properties of Radiation* (John Wiley and Sons, New York, 1973).

[c] J. Weiner and P.-T. Ho, *Light-Matter Interaction: Fundamentals and Applications* (John Wiley and Sons, Hoboken, NJ, 2003).

Field quantization and lattice vibrations are treated in

[d] C. Kittel, *Quantum Theory of Solids*, 2nd ed. (John Wiley and Sons, New York, 1987).

[e] D. Marcuse, *Engineering Quantum Electrodynamics* (Harcourt, Brace and World, New York, 1970).

A recent general reference on squeezed light is

[f] O. Hirota, ed., *Squeezed Light* (Elsevier Science Publishers, Amsterdam, 1992).

Problems

1. In this chapter, it was shown how to quantize a radiation field when the field is enclosed in an imaginary cubic box with sides of length L and a periodic boundary condition is applied. The aim of this problem is to work out the quantization procedure when the radiation field is enclosed in a cubic cavity having conducting walls. In this case, one expands the field as

$$\mathbf{A}(\mathbf{r},t) = \sum_{\ell} A_\ell(t)\,\mathbf{U}_\ell(\mathbf{r})$$

where $\mathbf{U}_\ell(\mathbf{r})$ is the eigenfunction corresponding to a cavity mode denoted by index ℓ. The $\mathbf{U}_\ell(\mathbf{r})$'s satisfy a cavity equation, namely,

$$\nabla^2\mathbf{U}_\ell(\mathbf{r}) + k_\ell^2\mathbf{U}_\ell(\mathbf{r}) = 0$$

subject to the transversality condition

$$\boldsymbol{\nabla}\cdot\mathbf{U}_\ell(\mathbf{r}) = 0.$$

The boundary conditions to be used for $\mathbf{U}_\ell(\mathbf{r})$ are derived from the fact that the tangential component of the electric field, $\mathbf{E}$, and the normal component of the magnetic field, $\mathbf{B}$, must vanish at the conducting walls.

(a) Show that the above conditions on **E** and **B** lead to the following boundary conditions on $\mathbf{U}_\ell(\mathbf{r})$:

- At a wall, the tangential component of $\mathbf{U}_\ell(\mathbf{r})$ vanishes.
- At a wall, the normal component of $\boldsymbol{\nabla}\times\mathbf{U}_\ell(\mathbf{r})$ vanishes.

Also show that the eigenfunctions satisfy the orthonormality condition

$$\int_V \mathbf{U}_\ell(\mathbf{r})\cdot\mathbf{U}_m(\mathbf{r})\,d^3r = \delta_{\ell m}.$$

(b) Derive the equation satisfied by the amplitudes, $A_\ell(t)$.

(c) Show that the Hamiltonian of the field can be written as

$$H = \frac{1}{8\pi}\int_V \left(E^2 + B^2\right) d^3r = \frac{V}{4\pi c^2}\sum_\ell \frac{1}{2}\left(\dot{A}_\ell^2 + \omega_\ell^2 A_\ell^2\right).$$

(d) Make a transformation to a pair of canonical variables, P_ℓ and Q_ℓ, such that the Hamiltonian becomes

$$H = \sum_\ell \frac{1}{2}\left(P_\ell^2 + \omega_\ell^2 Q_\ell^2\right)$$

with

$$P_\ell = \frac{\partial H_\ell}{\partial \dot{Q}_\ell} = \dot{Q}_\ell.$$

2. Consider the longitudinal vibrations of an elastic rod having linear mass-density μ and Young's modulus Y. The Lagrangian density was previously given (see Eq. 2.103) as

$$\mathbb{L} = \frac{1}{2}\left[\mu\dot{\eta}^2 - Y\left(\frac{\partial\eta}{\partial x}\right)^2\right],$$

where $\eta(x,t)$ is the longitudinal displacement at position x and time t. Go through the procedure of quantizing the displacement field of the rod.

3. Given the Hamiltonian for a unit-mass harmonic oscillator,

$$\hat{H} = \frac{1}{2}\hat{P}^2 + \frac{1}{2}\omega^2\hat{Q}^2,$$

one can construct an operator $\hat{U}(p,q)$ such that

$$\hat{U}(p,q) = e^{i\left(p\hat{Q} - q\hat{P}\right)/\hbar},$$

where p and q are complex numbers. Starting from the ground state, $|0\rangle$, of the oscillator, one is able to generate a state $|p,q\rangle$ by

$$|p,q\rangle = \hat{U}(p,q)\,|0\rangle\,.$$

The objective of this problem is to establish that the state $|p,q\rangle$ has all the properties of the coherent state $|\alpha\rangle$.

(a) First prove the operator identity

$$e^{\hat{A}}\hat{B}e^{-\hat{A}} = \hat{B} + \left[\hat{A},\hat{B}\right] + \frac{1}{2!}\left[\hat{A},\left[\hat{A},\hat{B}\right]\right] + \cdots.$$

(b) Use the latter identity to prove that

$$\hat{U}^{-1}(p,q)\left[\alpha\hat{P} + \beta\hat{Q}\right]\hat{U}(p,q) = \alpha\left(\hat{P}+p\right) + \beta\left(\hat{Q}+q\right),$$

showing that $\hat{U}(p,q)$ is a translational operator.

(c) Now consider the modified Hamiltonian

$$\hat{H}' = \frac{1}{2}\hat{P}^2 + \frac{1}{2}\omega^2\hat{Q}^2 - \frac{1}{2}\hbar\omega,$$

and show that

$$\hat{H}'\,|0\rangle = 0.$$

That is, the ground state $|0\rangle$ is an eigenstate of $\hat{H}'$ with an eigenvalue of zero.

(d) Prove that

$$\begin{aligned} 0 &= \hat{U}(p,q)\,\hat{H}'\,|0\rangle = \hat{U}\hat{H}'\hat{U}^{-1}\hat{U}\,|0\rangle = \hat{U}\hat{H}'\hat{U}^{-1}\,|p,q\rangle \\ &= \left[\frac{1}{2}\left(\hat{P}-p\right)^2 + \frac{1}{2}\omega^2\left(\hat{Q}-q\right)^2 - \frac{1}{2}\hbar\omega\right]|p,q\rangle\,. \end{aligned}$$

In other words, $|p,q\rangle$ is an eigenstate of the harmonic oscillator Hamiltonian in which the coordinate is displaced by q and the momentum is displaced by p. Also prove that the expectation values of the position and momentum in state $|p,q\rangle$ are q and p, respectively, i.e.,

$$\langle p,q \mid \hat{Q} \mid p,q\rangle = q \qquad \text{and} \qquad \langle p,q \mid \hat{P} \mid p,q\rangle = p.$$

(e) Finally, show that $|p,q\rangle$ is the coherent state $|\alpha\rangle$, with α given by

$$\alpha = \sqrt{\frac{\omega}{2\hbar}}\left(q + \frac{ip}{\omega}\right).$$

Hint: *See Eq. 5.130.*

4. We have seen that the coherent states, $|\alpha\rangle$, are the eigenstates of the destruction operator, $\hat{a}$. On the other hand, the creation operator, $\hat{a}^\dagger$, has no normalizable eigenstates. Prove this last statement.

Chapter 6
TIME-DEPENDENT PERTURBATION THEORY, TRANSITION PROBABILITIES, AND SCATTERING

The central tool for calculating transition rates in quantum mechanics is known as *Fermi's golden rule*; it is a result that comes directly from *time-dependent perturbation theory*. To develop the relevant formalism, it is convenient to introduce the *interaction picture* of quantum mechanics, an alternative to the Schrödinger and Heisenberg pictures first presented in Chapter 3. A major application of the golden rule is to the calculation of scattering cross-sections. As an example, in this chapter we will derive the so-called *double-differential cross-section* for thermal neutron scattering. It will be seen that the probability that a neutron transfers a particular momentum and energy to the scattering medium depends on a knowledge of *dynamic structure factors*; these are functions that appear in light and x-ray scattering calculations as well.

6.1 The Interaction Picture in Quantum Mechanics

In the Schrödinger picture of quantum mechanics, the time-evolution of a system's state vector $|\psi_s(t)\rangle$ is dictated by the Schrödinger equation, i.e., Eq. 3.120. In Section 3.3 it was shown that given the state vector at $t = 0$, i.e., $|\psi_s(0)\rangle$, the solution for the state vector at time t can be expressed as $|\psi_s(t)\rangle = \hat{U}(t)\,|\psi_s(0)\rangle$, where $\hat{U}(t)$ is a unitary time-evolution operator satisfying the differential equation

$$i\hbar\frac{d}{dt}\hat{U}(t) = \hat{H}\hat{U}(t). \tag{6.1}$$

In cases when the Hamiltonian $\hat{H}$ is independent of time, Eq. 6.1 can be integrated and the solution is simply $\hat{U}(t) = \exp\left(-i\hat{H}t/\hbar\right)$. However, certain situations, e.g., many problems encountered in quantum electrodynamics, share the common feature that the Hamiltonian can be split into two pieces, namely, a time-independent Hamiltonian $\hat{H}_0$ and a time-dependent potential $\hat{V}(t)$, i.e.,

$$\hat{H} = \hat{H}_0 + \hat{V}(t). \tag{6.2}$$

Think of $\hat{H}_0$ as the Hamiltonian for the system when no disturbance, or perturbation, is present; we assume that the eigenstates of $\hat{H}_0$ are known. The term $\hat{V}(t)$ represents some small perturbing influence acting on the system. In cases like this where the Hamiltonian of the system is time-dependent, be aware that

$$\hat{U}(t) \neq \exp\left[-\frac{i}{\hbar}\int_0^t \hat{H}(t')\,dt'\right], \tag{6.3}$$

since $\hat{H}(t')$ does not commute with $\hat{H}(t)$. To correctly deduce the time evolution of the system, we transform from the Schrödinger picture to the so-called *interaction picture* of quantum mechanics.

In going to the interaction picture, $\hat{H}_0$ is eliminated from the Hamiltonian. This is accomplished by defining a new unitary operator $\hat{U}_0(t)$ that connects state vectors in the Schrödinger and interaction pictures:

$$|\psi_s(t)\rangle = \hat{U}_0(t)\,|\psi_I(t)\rangle\,. \tag{6.4}$$

We postulate that $\hat{U}_0(t)$ satisfies the following operator equation

$$i\hbar\frac{d}{dt}\hat{U}_0(t) = \hat{H}_0\hat{U}_0(t) \tag{6.5}$$

so that

$$\hat{U}_0(t) = e^{-i\hat{H}_0 t/\hbar}. \tag{6.6}$$

Equations 6.4 and 6.2 can now be inserted into the Schrödinger equation. Then, with the aid of Eq. 6.5, one gets the equation of motion for the state vector in the interaction picture, i.e.,

$$i\hbar\frac{d}{dt}|\psi_I(t)\rangle = \hat{V}_I(t)\,|\psi_I(t)\rangle\,, \tag{6.7}$$

where

$$\hat{V}_I(t) = \hat{U}_0^\dagger(t)\hat{V}(t)\hat{U}_0(t). \tag{6.8}$$

In fact, since the expectation value of any quantity A is independent of the picture used for the state vectors, we have

$$\langle\psi_I(t)\mid\hat{A}_I(t)\mid\psi_I(t)\rangle = \langle\psi_s(t)\mid\hat{A}\mid\psi_s(t)\rangle = \langle\psi_I(t)\mid\hat{U}_0^\dagger(t)\hat{A}\hat{U}_0(t)\mid\psi_I(t)\rangle, \tag{6.9}$$

making use of Eq. 6.4. It then follows that the general transformation between operators in the Schrödinger and interaction pictures is

$$\hat{A}_I(t) = \hat{U}_0^\dagger(t)\hat{A}\hat{U}_0(t). \tag{6.10}$$

Now introduce $\hat{U}_I(t)$, a time-evolution operator in the interaction picture, by

$$|\psi_I(t)\rangle = \hat{U}_I(t)\,|\psi_I(0)\rangle\,. \tag{6.11}$$

From Eq. 6.4 notice that $|\psi_s(0)\rangle = |\psi_I(0)\rangle \equiv |\psi(0)\rangle$. Substituting the expression above into Eq. 6.7 gives the following important equation of motion for the evolution operator:

$$i\hbar\frac{d}{dt}\hat{U}_I(t) = \hat{V}_I(t)\hat{U}_I(t). \tag{6.12}$$

The task at hand is to solve the above equation for $\hat{U}_I(t)$—this determines the time evolution of the perturbed system. More specifically, recall that the eigenstates of the Hamiltonian with the perturbation turned off are known, e.g.,

$$\hat{H}_0 |n\rangle = E_n^{(0)} |n\rangle . \tag{6.13}$$

We may then expand the state vector in the interaction picture in terms of the complete set $\{|n\rangle\}$, i.e.,

$$|\psi_I(t)\rangle = \sum_n c_n(t) |n\rangle , \tag{6.14}$$

where

$$c_n(t) = \langle n \mid \psi_I(t)\rangle. \tag{6.15}$$

From Eqs. 6.4 and 6.6, we now see that

$$|\psi_s(t)\rangle = e^{-i\hat{H}_0 t/\hbar} |\psi_I(t)\rangle = \sum_n c_n(t)e^{-iE_n^{(0)}t/\hbar} |n\rangle , \tag{6.16}$$

so

$$\langle n \mid \psi_s(t)\rangle = c_n(t)e^{-iE_n^{(0)}t/\hbar}. \tag{6.17}$$

Comparing Eqs. 6.15 and 6.17 shows that

$$|c_n(t)|^2 = |\langle n \mid \psi_s(t)\rangle|^2 = |\langle n \mid \psi_I(t)\rangle|^2 . \tag{6.18}$$

Therefore,

$$|c_n(t)|^2 = \left|\langle n \mid \hat{U}_I(t) \mid \psi(0)\rangle\right|^2 . \tag{6.19}$$

This result can be interpreted as follows: If, at $t = 0$, a measurement is made finding the system in state $|\psi(0)\rangle$, then, after a time t, the action of the switched-on perturbation is to induce transitions into the various energy eigenstates $\{|n\rangle\}$ with probabilities given by the $|c_n(t)|^2$'s. In a case where the system starts out in some initial eigenstate $|m\rangle$, the probability of making a transition into a final state $|n\rangle$ after time t is

$$|c_{nm}(t)|^2 = \left|\langle n \mid \hat{U}_I(t) \mid m\rangle\right|^2 . \tag{6.20}$$

6.2 Perturbation Expansion of the Time-Evolution Operator

With the aim of determining $\hat{U}_I(t)$, we integrate both sides of Eq. 6.12 from $t = 0$ to time t. Then,

$$\hat{U}_I(t) = 1 + \frac{1}{i\hbar}\int_0^t dt'\hat{V}_I(t')\hat{U}_I(t'). \tag{6.21}$$

Assume that $\hat{V}_I(t')$ is a small quantity. Then we expect that

$$\left|\frac{1}{i\hbar}\int_0^t dt'\hat{V}_I(t')\hat{U}_I(t')\right| \ll 1, \tag{6.22}$$

and we can solve for $\hat{U}_I(t)$ iteratively:

$$\begin{aligned}\hat{U}_I(t) &= 1 + \frac{1}{i\hbar}\int_0^t dt_1\hat{V}_I(t_1) + \left(\frac{1}{i\hbar}\right)^2\int_0^t dt_1\int_0^{t_1} dt_2\hat{V}_I(t_1)\hat{V}_I(t_2) + \ldots \\ &+ \left(\frac{1}{i\hbar}\right)^m\int_0^t dt_1\int_0^{t_1} dt_2\int_0^{t_2} dt_3\cdots\int_0^{t_{m-1}} dt_m\hat{V}_I(t_1)\cdots\hat{V}_I(t_m) + \cdots\end{aligned} \tag{6.23}$$

This is commonly referred to as the *time-dependent perturbation expansion.*

Our job now is to find the transition probability $|c_{nm}(t)|^2$ for the different order approximations to $\hat{U}_I(t)$. Observe that to zeroth-order the result is uninteresting, i.e., $\hat{U}_I^{(0)}(t) = 1$, and from Eq. 6.20, $\left|c_{nm}^{(0)}(t)\right|^2 = |\langle n \mid m\rangle|^2 = \delta_{nm}$, which means that no transition takes place. The effects of the perturbation $\hat{V}(t)$ show up in the higher-order terms. Specifically, the first non-trivial contribution comes from

$$\hat{U}_I^{(1)}(t) = \frac{1}{i\hbar}\int_0^t dt_1\hat{V}_I(t_1). \tag{6.24}$$

It gives rise to the first-order transition amplitude from state $|m\rangle$ to state $|n\rangle$, i.e.,

$$\begin{aligned}c_{nm}^{(1)}(t) &= \frac{1}{i\hbar}\int_0^t dt_1\langle n \mid \hat{V}_I(t_1) \mid m\rangle = \frac{1}{i\hbar}\int_0^t dt_1\langle n \mid e^{+i\hat{H}_0t_1/\hbar}\hat{V}(t_1)e^{-i\hat{H}_0t_1/\hbar} \mid m\rangle \\ &= \frac{1}{i\hbar}\int_0^t dt_1\langle n \mid \hat{V}(t_1) \mid m\rangle e^{i\omega_{nm}t_1},\end{aligned} \tag{6.25}$$

where

$$\omega_{nm} = \frac{E_n^{(0)} - E_m^{(0)}}{\hbar}. \tag{6.26}$$

Even if the matrix element $\langle n \mid \hat{V}(t) \mid m\rangle$ vanishes, transitions from initial state $|m\rangle$ to final state $|n\rangle$ may still be possible. For this to occur, however, the transition must be higher than first order. It is usually only necessary to look as far as the second order; in this case, the system makes the transition $|m\rangle \to |n\rangle$ by way of various intermediate states, say $|l\rangle$, with matrix elements $\langle l \mid \hat{V}(t) \mid m\rangle$ and $\langle n \mid \hat{V}(t) \mid l\rangle$ that are non-zero. Consider the second-order contribution to $\hat{U}_I(t)$, namely,

$$\hat{U}_I^{(2)}(t) = \left(\frac{1}{i\hbar}\right)^2 \int_0^t dt_1 \int_0^{t_1} dt_2 \hat{V}_I(t_1)\hat{V}_I(t_2). \tag{6.27}$$

Then,

$$c_{nm}^{(2)}(t) = \langle n \mid \hat{U}_I^{(2)}(t) \mid m\rangle = \left(\frac{1}{i\hbar}\right)^2 \int_0^t dt_1 \int_0^{t_1} dt_2 \langle n \mid \hat{V}_I(t_1)\hat{V}_I(t_2) \mid m\rangle. \tag{6.28}$$

Now introduce the intermediate states by inserting a completeness relation $\sum_l |l\rangle \langle l|$. Then,

$$\begin{aligned} c_{nm}^{(2)}(t) &= \left(\frac{1}{i\hbar}\right)^2 \int_0^t dt_1 \int_0^{t_1} dt_2 \sum_l \langle n \mid \hat{V}_I(t_1) \,|l\rangle \,\langle l|\, \hat{V}_I(t_2) \mid m\rangle \\ &= \left(\frac{1}{i\hbar}\right)^2 \int_0^t dt_1 \int_0^{t_1} dt_2 \sum_l \langle n \mid e^{+i\hat{H}_0 t_1/\hbar}\hat{V}(t_1)e^{-i\hat{H}_0 t_1/\hbar} \,|l\rangle \\ &\quad \times \left\langle l \mid e^{+i\hat{H}_0 t_2/\hbar}\hat{V}(t_2)e^{-i\hat{H}_0 t_2/\hbar} \mid m \right\rangle \\ &= \left(\frac{1}{i\hbar}\right)^2 \int_0^t dt_1 \int_0^{t_1} dt_2 \sum_l \langle n \mid \hat{V}(t_1) \,|l\rangle \,\langle l|\, \hat{V}(t_2) \mid m\rangle e^{i\omega_{nl}t_1} e^{i\omega_{lm}t_2}. \end{aligned} \tag{6.29}$$

The higher-order terms can be handled in a similar fashion.

6.3 Fermi's Golden Rule

6.3.1 First-Order Transitions

In many situations of interest, we will find that $\hat{V}$ does not explicitly depend on time. In this case, the integral given by Eq. 6.25 for $c_{nm}^{(1)}(t)$ reduces to

$$c_{nm}^{(1)}(t) = \frac{1 - e^{i\omega_{nm}t}}{\hbar\omega_{nm}} \langle n \mid \hat{V} \mid m\rangle \tag{6.30}$$

and the first-order transition probability becomes

$$\left|c_{nm}^{(1)}(t)\right|^2 = \frac{4}{\hbar^2} \left|\langle n \mid \hat{V} \mid m\rangle\right|^2 \left[\frac{\sin\left(\omega_{nm}t/2\right)}{\omega_{nm}}\right]^2. \tag{6.31}$$

Figure 6.1 shows $\left|c_{nm}^{(1)}(t)\right|^2$ plotted as a function of ω_{nm}. Notice that the transition probability is highly-peaked around $\omega_{nm} = 0$, which means that the energy difference between the initial and final state of the system tends to be very small. The width of the central peak is proportional to t^{-1}; thus, in accord with the time-energy uncertainty principle ($\triangle E \triangle t \geq \hbar$ or $\triangle\omega_{nm} \triangle t \geq 1$), the violation of energy conservation is a distinct possibility at short times. As t progresses, however, the function becomes more and more sharply peaked, with a central height that grows as $\sim t^2$. In the limit $t \to \infty$, the transition probability becomes a delta-function at $\omega_{nm} = 0$, and energy is conserved.

Equation 6.31 is valid as long as one is considering transitions from some well-defined initial state $|m\rangle$ to an equally well-defined final state $|n\rangle$. In a great deal of cases, however, the final state is not perfectly sharp, rather it consists of a range of closely-spaced energy eigenstates. It is therefore appropriate to calculate $\left|c_{\bar{n}m}^{(1)}(t)\right|^2 = \sum\limits_n \left|c_{nm}^{(1)}(t)\right|^2$, the probability of making a transition to some range of final states. Treating the final states as a continuum of energy levels, let $g\left(E_n^{(0)}\right)$ be the density of final states. Then the number of states in the energy range $E_n^{(0)} \to E_n^{(0)} + dE_n^{(0)}$ is given by $g\left(E_n^{(0)}\right) dE_n^{(0)}$, and $\left|c_{\bar{n}m}^{(1)}(t)\right|^2$ is obtained by replacing the summation by an integral, i.e.,

$$\begin{aligned} \left|c_{\bar{n}m}^{(1)}(t)\right|^2 &= \frac{4}{\hbar^2}\int \left|\langle n \mid \hat{V} \mid m\rangle\right|^2 \left[\frac{\sin\left(\omega_{nm}t/2\right)}{\omega_{nm}}\right]^2 g\left(E_n^{(0)}\right) dE_n^{(0)} \\ &= \frac{4}{\hbar}\int \left|\langle n \mid \hat{V} \mid m\rangle\right|^2 \left[\frac{\sin\left(\omega_{nm}t/2\right)}{\omega_{nm}}\right]^2 g\left(E_m^{(0)} + \hbar\omega_{nm}\right) d\omega_{nm}. \end{aligned} \tag{6.32}$$

If t is sufficiently long, then the only significant transitions are those where energy is conserved (i.e., $E_n^{(0)} \simeq E_m^{(0)}$). As we have seen, in this case the function $\left[\sin^2\left(\omega_{nm}t/2\right)\right]/\omega_{nm}^2$ exhibits a very narrow peak at $\omega_{nm} = 0$, so one is justified in pulling the much more slowly varying terms $\left|\langle n \mid \hat{V} \mid m\rangle\right|^2$ and $g\left(E_m^{(0)} + \hbar\omega_{nm}\right) = g\left(E_n^{(0)}\right)$ out of the integral. The remaining integral reduces to the form

$$\int_{-\infty}^{+\infty} \frac{\sin^2 \alpha x}{x^2} dx = \pi\alpha, \tag{6.33}$$

producing the result

$$\left|c_{\bar{n}m}^{(1)}(t)\right|^2 = \frac{2\pi t}{\hbar}\left|\langle n \mid \hat{V} \mid m\rangle\right|^2 g\left(E_n^{(0)}\right). \tag{6.34}$$

As a final step, we can write the expression for $W_{\bar{n}m}^{(1)}$, which is the transition probability

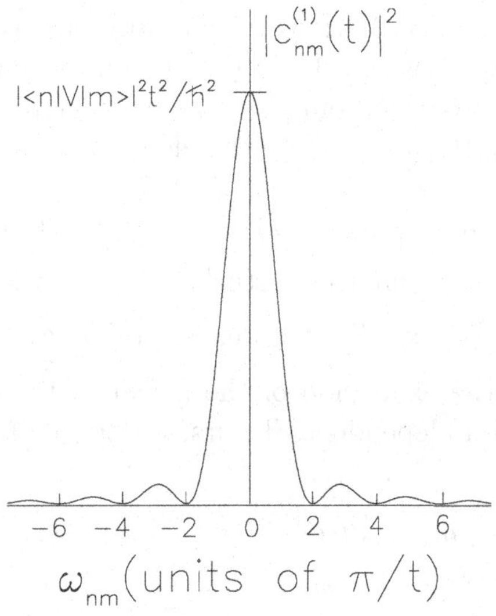

Figure 6.1 Probability of making the transition from state $|m\rangle$ to state $|n\rangle$ after a time t as a function of $\omega_{nm} = \left(E_n^{(0)} - E_m^{(0)}\right)/\hbar$.

per unit time (to first order). It is given by the important expression

$$W_{\bar{n}m}^{(1)} = \frac{2\pi}{\hbar}\left|\langle n \mid \hat{V} \mid m\rangle\right|^2 g\left(E_n^{(0)}\right) \tag{6.35}$$

known as *Fermi's golden rule.**

6.3.2 *Extension to Scattering Problems*

Here, we develop a more general form of Fermi's golden rule applicable to scattering calculations. In a typical scattering situation one considers a stream of quanta (i.e., neutrons, photons, or charged particles) entering, and then leaving, the field set up by some material medium. The quantum system is composed of the *scattered "particles" + scattering medium.* Both long before and long after the scattering interaction takes place, the system is in an energy eigenstate of $\hat{H}_0$, the Hamiltonian for particles and scatterer far apart, where there is no mutual interaction between the two. While quanta approach and recede from the scatterer, an interaction potential $\hat{V}$ exists that induces transitions into various scattered states. The perturbation is turned on gradually starting at $t_0 = -\infty$, and then turned off gradually until $t = +\infty$. This slow turn-on and turn-off of the interaction potential is sometimes referred to as *adiabatic switching*. The probability that scattering takes place between some initial system state $|m\rangle$ to some final state $|n\rangle$ is given by Eq. 6.20 in conjunction with the

*More specifically, this result was coined "golden rule no. 2" by Fermi [29].

perturbation expansion for $\hat{U}_I(t)$. However, because our starting time is at $t_0 = -\infty$ rather than the usual $t_0 = 0$, we need to adjust the lower limits of all the perturbation integrals to reflect this. At this point, it is convenient to specify the amplitude for scattering between state $|m\rangle$ and $|n\rangle$ by the elements of a *scattering matrix* $\hat{\mathsf{S}}$ given by

$$\langle n \mid \hat{\mathsf{S}} \mid m\rangle \equiv \langle n \mid \hat{U}_I(-\infty, +\infty) \mid m\rangle, \tag{6.36}$$

where $\hat{U}_I(-\infty, +\infty)$ means that the perturbation integrals from time 0 to t should be evaluated from $-\infty$ to $+\infty$. The scattering probability is then $\left|\langle n \mid \hat{\mathsf{S}} \mid m\rangle\right|^2$.

Let us examine the expansion of the $\hat{\mathsf{S}}$-matrix to various order with the assumption that $\hat{V}$ is time independent. To first order (see Eq. 6.25)

$$\langle n \mid \hat{\mathsf{S}}^{(1)} \mid m\rangle = \frac{1}{i\hbar}\langle n \mid \hat{V} \mid m\rangle \int_{-\infty}^{+\infty} dt_1 e^{i\left(E_n^{(0)} - E_m^{(0)}\right)t_1/\hbar} = -2\pi i \langle n \mid \hat{V} \mid m\rangle \delta\left(E_n^{(0)} - E_m^{(0)}\right). \tag{6.37}$$

The second-order contribution is

$$\begin{aligned}\left\langle n \mid \hat{\mathsf{S}}^{(2)} \mid m\right\rangle &= \left(\frac{1}{i\hbar}\right)^2 \sum_l \langle n \mid \hat{V} \,|l\rangle\, \langle l|\, \hat{V} \mid m\rangle \int_{-\infty}^{+\infty} dt_1\, e^{i\left(E_n^{(0)} - E_l^{(0)}\right)t_1/\hbar} \\ &\times \int_{-\infty}^{t_1} dt_2\, e^{i\left(E_l^{(0)} - E_m^{(0)}\right)t_2/\hbar}.\end{aligned} \tag{6.38}$$

To handle the last integral, we make the replacement

$$\int_{-\infty}^{t_1} dt_2\, e^{i\left(E_l^{(0)} - E_m^{(0)}\right)t_2/\hbar} \to \int_{-\infty}^{t_1} dt_2\, e^{i\left(E_l^{(0)} - E_m^{(0)} - i\eta\right)t_2/\hbar} = -i\hbar \frac{e^{i\left(E_l^{(0)} - E_m^{(0)} - i\eta\right)t_1/\hbar}}{E_l^{(0)} - E_m^{(0)} - i\eta} \tag{6.39}$$

with the understanding that $\eta \to 0^+$. Then Eq. 6.38 reduces to

$$\langle n \mid \hat{\mathsf{S}}^{(2)} \mid m\rangle = -2\pi i\, \delta\left(E_n^{(0)} - E_m^{(0)}\right) \sum_l \frac{\langle n \mid \hat{V} \,|l\rangle\, \langle l|\, \hat{V} \mid m\rangle}{E_m^{(0)} - E_l^{(0)} + i\eta}. \tag{6.40}$$

The higher-order expansion terms can be similarly treated to give

$$\langle n \mid \hat{\mathsf{S}} \mid m\rangle = -2\pi i\, \delta\left(E_n^{(0)} - E_m^{(0)}\right) \langle n \mid \hat{\mathsf{T}} \mid m\rangle, \tag{6.41}$$

with

$$\begin{aligned}\left\langle n \,|\, \hat{\mathsf{T}} \,|\, m\right\rangle &= \langle n \,|\, \hat{V} \,|\, m\rangle + \sum_l \frac{\langle n \mid \hat{V} \,|l\rangle\, \langle l|\, \hat{V} \mid m\rangle}{E_m^{(0)} - E_l^{(0)} + i\eta} \\ &+ \sum_{l,l'} \frac{\langle n \mid \hat{V} \,|l\rangle\, \langle l|\, \hat{V} \mid l'\rangle\, \langle l'|\, \hat{V} \mid m\rangle}{\left(E_m^{(0)} - E_{l'}^{(0)} + i\eta\right)\left(E_m^{(0)} - E_l^{(0)} + i\eta\right)} + \cdots.\end{aligned} \tag{6.42}$$

$\langle n \mid \hat{\mathsf{T}} \mid m \rangle$ is called the *transition matrix* or T-*matrix*. As it stands, Eq. 6.41 is essentially exact. The meaning of the δ-function is that the initial and final state of the system must have the same total energy. However, because the intermediate states have only a fleeting existence, the time-energy uncertainty principle (Eq. 3.119) permits them to violate energy conservation.

The scattering (transition) probability during the interval $-\infty < t < +\infty$ is now given as

$$\left|\langle n \mid \hat{\mathsf{S}} \mid m \rangle\right|^2 = 4\pi^2 \left|\langle n \mid \hat{\mathsf{T}} \mid m \rangle\right|^2 \delta^2\left(E_n^{(0)} - E_m^{(0)}\right). \tag{6.43}$$

However, if one of the δ-functions is replaced by its integral representation, i.e.,

$$\delta^2\left(E_n^{(0)} - E_m^{(0)}\right) = \left[\frac{1}{2\pi\hbar}\int_{-\infty}^{+\infty} e^{i\left(E_n^{(0)} - E_m^{(0)}\right)t/\hbar} dt\right] \delta\left(E_n^{(0)} - E_m^{(0)}\right), \tag{6.44}$$

we can make use of the general property $f(x)\delta(x-a) = f(a)\delta(x-a)$ to see that

$$\delta^2\left(E_n^{(0)} - E_m^{(0)}\right) = \left[\frac{1}{2\pi\hbar}\int_{-\infty}^{+\infty} dt\right] \delta\left(E_n^{(0)} - E_m^{(0)}\right) = \lim_{\mathcal{T}\to\infty}\left(\frac{2\mathcal{T}}{2\pi\hbar}\right) \delta\left(E_n^{(0)} - E_m^{(0)}\right). \tag{6.45}$$

Therefore,

$$\left|\langle n \mid \hat{\mathsf{S}} \mid m \rangle\right|^2 = \lim_{\mathcal{T}\to\infty}\left(\frac{4\pi\mathcal{T}}{\hbar}\right) \left|\langle n \mid \hat{\mathsf{T}} \mid m \rangle\right|^2 \delta\left(E_n^{(0)} - E_m^{(0)}\right), \tag{6.46}$$

and the expression for the scattering probability per unit time becomes

$$W_{nm} = \frac{\left|\langle n \mid \hat{\mathsf{S}} \mid m \rangle\right|^2}{\lim\limits_{\mathcal{T}\to\infty} 2\mathcal{T}} = \frac{2\pi}{\hbar}\left|\langle n \mid \hat{\mathsf{T}} \mid m \rangle\right|^2 \delta\left(E_n^{(0)} - E_m^{(0)}\right). \tag{6.47}$$

In scattering problems, the final states form a continuum. Thus we need to integrate the latter expression over the final states of density $g\left(E_n^{(0)}\right)$, i.e.,

$$W_{\bar{n}m} = \frac{2\pi}{\hbar}\int \left|\langle n \mid \hat{\mathsf{T}} \mid m \rangle\right|^2 \delta\left(E_n^{(0)} - E_m^{(0)}\right) g\left(E_n^{(0)}\right) dE_n^{(0)}. \tag{6.48}$$

Our final result is the following generalization of Fermi's golden rule that is, in principle, exact for scattering problems (recalling that $\hat{V}$ is assumed to be constant):†

$$W_{\bar{n}m} = \frac{2\pi}{\hbar}\left|\langle n \mid \hat{\mathsf{T}} \mid m \rangle\right|^2 g\left(E_n^{(0)}\right). \tag{6.49}$$

The first-order contribution is identical to Eq. 6.35.

†The reader may be interested in noting that "Fermi's golden rule no. 1" is the name reserved for the second-order contribution to $W_{\bar{n}m}$.

6.4 Double-Differential Scattering Cross-Sections

We now show how Fermi's golden rule can be used to develop the quantity $\frac{d^2\sigma}{d\Omega\, dE'}$, the *double-differential cross-section* for a scattering experiment. $\left(\frac{d^2\sigma}{d\Omega\, dE'}\right) d\Omega\, dE'$ is the number of quanta, or particles, scattered per unit time into a differential solid angle $d\Omega$ with energy between E' and $E' + dE'$, divided by the flux of the incident beam (refer to Fig. 6.2). The cone defined by $d\Omega$ is centered on the scattered wavevector $\mathbf{k}'$.

To apply the results of the previous section, we take the system's "unperturbed" Hamiltonian $\hat{H}_0$ to be

$$\hat{H}_0 = \hat{H}_S + \hat{H}_R, \tag{6.50}$$

where $\hat{H}_S$ and $\hat{H}_R$ are the Hamiltonians of the scattering medium and radiation, respectively, assuming no interaction between the two. The combined system is in an eigenstate of $\hat{H}_0$ both before the interaction $\hat{V}$ is switched on and after it is turned off. We let $|E_i\rangle$ and $|E_f\rangle$ denote the initial and final energy eigenstates of the scatterer, i.e.,

$$\hat{H}_S\, |E_i\rangle = E_i\, |E_i\rangle \qquad \text{and} \qquad \hat{H}_S\, |E_f\rangle = E_f\, |E_f\rangle\,. \tag{6.51}$$

A quantum of the incident radiation is appropriately represented by an eigenstate $|\mathbf{k}\lambda\rangle$, with definite momentum $\hbar\mathbf{k}$ and polarization (or spin) λ. The scattered quantum falls into state $|\mathbf{k}'\lambda'\rangle$. These are also energy eigenstates, so

$$\hat{H}_R\, |\mathbf{k}\lambda\rangle = E\, |\mathbf{k}\lambda\rangle \qquad \text{and} \qquad \hat{H}_R\, |\mathbf{k}'\lambda'\rangle = E'\, |\mathbf{k}'\lambda'\rangle\,, \tag{6.52}$$

where E and E' are the energies of the incident and scattered radiation. For the composite system, the initial state $|m\rangle$ and final state $|n\rangle$ are given by the products

$$|m\rangle = |\mathbf{k}\lambda\rangle\, |E_i\rangle \qquad \text{and} \qquad |n\rangle = |\mathbf{k}'\lambda'\rangle\, |E_f\rangle\,. \tag{6.53}$$

These are eigenstates of $\hat{H}_0$, so

$$\hat{H}_0\, |m\rangle = E_m\, |m\rangle \qquad \text{and} \qquad \hat{H}_0\, |n\rangle = E_n\, |n\rangle\,, \tag{6.54}$$

where the initial and final energies of the system are

$$E_m = E_i + E \qquad \text{and} \qquad E_n = E_f + E', \tag{6.55}$$

respectively.

The probability per unit time of scattering between two specified eigenstates of the system, $|m\rangle$ and $|n\rangle$, can be obtained by returning to Eq. 6.47. If we let

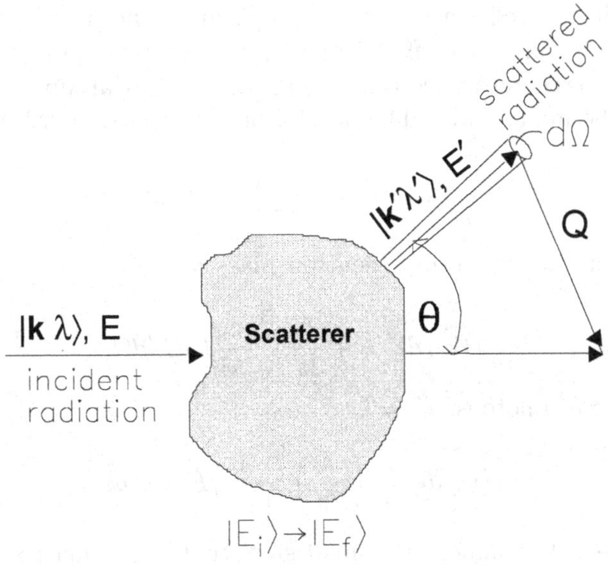

Figure 6.2 Double-differential scattering set-up.

$\hat{\mathsf{T}}_{\mathbf{k}'\lambda',\mathbf{k}\lambda} \equiv \langle \mathbf{k}'\lambda' \mid \hat{\mathsf{T}} \mid \mathbf{k}\lambda\rangle$, then

$$\begin{aligned} W_{nm} &= \frac{2\pi}{\hbar}\langle m \mid \hat{\mathsf{T}}^{\dagger} \mid n\rangle\langle n \mid \hat{\mathsf{T}} \mid m\rangle\delta\left(E_n - E_m\right) \\ &= \frac{2\pi}{\hbar}\langle E_i \mid \hat{\mathsf{T}}^{\dagger}_{\mathbf{k}\lambda,\mathbf{k}'\lambda'} \mid E_f\rangle\langle E_f \mid \hat{\mathsf{T}}_{\mathbf{k}'\lambda',\mathbf{k}\lambda} \mid E_i\rangle \\ &\quad \times \frac{1}{2\pi\hbar}\int_{-\infty}^{+\infty} e^{i\left(E_f - E_i\right)t/\hbar} e^{i(E' - E)t/\hbar} dt. \end{aligned} \tag{6.56}$$

Also, we have

$$\begin{aligned} \left\langle E_f \mid \hat{\mathsf{T}}_{\mathbf{k}'\lambda',\mathbf{k}\lambda} \mid E_i\right\rangle e^{i\left(E_f - E_i\right)t/\hbar} &= \langle E_f \mid e^{i\hat{H}_S t/\hbar}\hat{\mathsf{T}}_{\mathbf{k}'\lambda',\mathbf{k}\lambda} e^{-i\hat{H}_S t/\hbar} \mid E_i\rangle \\ &= \langle E_f \mid \hat{\mathsf{T}}_{\mathbf{k}'\lambda',\mathbf{k}\lambda}(t) \mid E_i\rangle, \end{aligned} \tag{6.57}$$

so that

$$W_{nm} = \frac{1}{\hbar^2}\int_{-\infty}^{+\infty} dt\, e^{-i\omega t}\langle E_i \mid \hat{\mathsf{T}}^{\dagger}_{\mathbf{k}\lambda,\mathbf{k}'\lambda'}(0) \mid E_f\rangle\langle E_f \mid \hat{\mathsf{T}}_{\mathbf{k}'\lambda',\mathbf{k}\lambda}(t) \mid E_i\rangle. \tag{6.58}$$

In this last equation, $\hbar\omega$ is the energy transferred to the material medium by a quantum of the radiation, i.e.,

$$\hbar\omega \equiv E - E'. \tag{6.59}$$

To find the desired scattering cross-section, we need to know the probability of scattering into a solid angle $d\Omega$ with energy in the interval dE'. This means that the beam is scattered into a continuum of final states around the wavevector $\mathbf{k}'$, represented by the interval $d\mathbf{k}'$. The density of final states, $g(\mathbf{k}')$, is[‡] (see Eq. 5.6)

$$g(\mathbf{k}')\, d\mathbf{k}' = \left(\frac{L}{2\pi}\right)^3 k'^2 dk' d\Omega. \tag{6.60}$$

In the case of neutron scattering (neutron mass m_n), we have $E' = \hbar^2 k'^2/2m_n$, so that

$$g(\mathbf{k}')\, d\mathbf{k}' = \left(\frac{L}{2\pi}\right)^3 \frac{m_n}{\hbar^2} k' dE' d\Omega. \tag{6.61}$$

For the scattering of photons, $E' = \hbar c k'$ and

$$g(\mathbf{k}')\, d\mathbf{k}' = \left(\frac{L}{2\pi}\right)^3 \frac{1}{(\hbar c)^3} E'^2 dE' d\Omega. \tag{6.62}$$

If the solid angle $d\Omega$ is sufficiently small such that the T-matrix element does not change significantly within $d\mathbf{k}'$, then the total probability (per unit time) of scattering into $d\mathbf{k}'$ is obtained by simply multiplying W_{nm} by $g(\mathbf{k}')\, d\mathbf{k}'$. This result, however, is incomplete. In a real scattering experiment direct measurements are only made on the scattered quanta—thus, it is also necessary to sum over all the final states, $|E_f\rangle$, of the scattering medium consistent with the energy conservation requirement $E_n = E_m$. Furthermore, the initial state of the scatterer is normally one of thermal equilibrium at some temperature T.[§] Therefore, it is also appropriate to sum over initial states, $|E_i\rangle$, weighted according to the Maxwell-Boltzmann probability distribution

$$P_i = \frac{e^{-E_i/k_B T}}{\sum_i e^{-E_i/k_B T}}. \tag{6.63}$$

Combining all these considerations produces a complete expression for $P(\mathbf{k}\lambda \to \mathbf{k}'\lambda')\, d\mathbf{k}'$ the probability per unit time that the quantum scatters into $d\mathbf{k}'$:

$$P(\mathbf{k}\lambda \to \mathbf{k}'\lambda')\, d\mathbf{k}' = g(\mathbf{k}')\, d\mathbf{k}' \sum_i P_i \sum_f W_{nm}. \tag{6.64}$$

The double summation in this expression can be reduced to

$$\sum_i P_i \sum_f W_{nm} = \frac{1}{\hbar^2} \int_{-\infty}^{+\infty} dt\, e^{-i\omega t} \sum_i P_i \sum_f \langle E_i \mid \hat{\mathsf{T}}^{\dagger}_{\mathbf{k}\lambda,\mathbf{k}'\lambda'}(0) \mid E_f\rangle \langle E_f \mid \hat{\mathsf{T}}_{\mathbf{k}'\lambda',\mathbf{k}\lambda}(t) \mid E_i\rangle$$

[‡]Recall that the density of states depends on L^3, the volume of the "box" enclosing the system. This is completely arbitrary and presents no problem here since, as will be shown shortly, the final calculated cross-sections do not depend on L. Also, the expression given is for a single polarization (or spin) of the scattered wave.

[§]As will be discussed in the next chapter, this is an example of a "mixed" quantum state.

$$= \frac{1}{\hbar^2} \int_{-\infty}^{+\infty} dt\, e^{-i\omega t} \sum_i P_i \langle E_i \mid \hat{\mathsf{T}}^\dagger_{\mathbf{k}\lambda,\mathbf{k}'\lambda'}(0)\, \hat{\mathsf{T}}_{\mathbf{k}'\lambda',\mathbf{k}\lambda}(t) \mid E_i \rangle$$

$$= \frac{1}{\hbar^2} \int_{-\infty}^{+\infty} dt\, e^{-i\omega t} \left\langle \mathsf{T}^\dagger_{\mathbf{k}\lambda,\mathbf{k}'\lambda'}(0)\, \mathsf{T}_{\mathbf{k}'\lambda',\mathbf{k}\lambda}(t) \right\rangle, \tag{6.65}$$

where the brackets $\langle \quad \rangle$ indicate an ensemble average consistent with the temperature T of the scattering medium.¶ The expression for the double-differential scattering cross-section is now obtained by

$$\left(\frac{d^2\sigma}{d\Omega\, dE'}\right) d\Omega\, dE' = \frac{P\,(\mathbf{k}\lambda \to \mathbf{k}'\lambda')\, d\mathbf{k}'}{\Phi_{\text{inc}}}, \tag{6.66}$$

where Φ_{inc} is the flux (number of particles per unit area per unit time) presented by a single quantum of the incident radiation. For neutron scattering, the flux is

$$\Phi_{\text{inc}} = \hbar k / m_n L^3, \tag{6.67}$$

and for photons, it is given by

$$\Phi_{\text{inc}} = c / L^3. \tag{6.68}$$

Example 6.1 Thermal Neutron Scattering. For the scattering of neutrons, Eqs. 6.61 and 6.67 can be applied to Eqs. 6.64–6.66 so that the double-differential cross-section can be written in the form

$$\frac{d^2\sigma}{d\Omega\, d\omega} = \frac{1}{2\pi} \left(\frac{m_n L^3}{2\pi\hbar^2}\right)^2 \left(\frac{k'}{k}\right) \int_{-\infty}^{+\infty} dt\, e^{-i\omega t} \left\langle \mathsf{T}^\dagger_{\mathbf{k}\lambda,\mathbf{k}'\lambda'}(0)\, \mathsf{T}_{\mathbf{k}'\lambda',\mathbf{k}\lambda}(t) \right\rangle, \tag{6.69}$$

where $\dfrac{d^2\sigma}{d\Omega\, d\omega} = \hbar \dfrac{d^2\sigma}{d\Omega\, dE'}$. We now express the elements of the T-matrix in their position-space representation. Since $\mathsf{T}_{\mathbf{k}'\lambda',\mathbf{k}\lambda}(t) \equiv \langle \mathbf{k}'\lambda' \mid \hat{\mathsf{T}}(t) \mid \mathbf{k}\lambda \rangle$, we can make use of Eq. 3.158 in three-dimensions to write

$$\mathsf{T}_{\mathbf{k}'\lambda',\mathbf{k}\lambda}(t) = \int_{L^3} d^3r\, \psi^*_{\mathbf{k}'}(\mathbf{r})\, \langle \lambda' \mid \hat{\mathsf{T}}(\mathbf{r}, t) \mid \lambda \rangle \psi_{\mathbf{k}}(\mathbf{r})\,. \tag{6.70}$$

¶The quantity $\left\langle \mathsf{T}^\dagger_{\mathbf{k}\lambda,\mathbf{k}'\lambda'}(0)\, \mathsf{T}_{\mathbf{k}'\lambda',\mathbf{k}\lambda}(t) \right\rangle_T$ can be expressed as the following trace of a matrix: $\text{Tr}\left[\hat{\rho}_S \hat{\mathsf{T}}^\dagger_{\mathbf{k}\lambda,\mathbf{k}'\lambda'}(0)\, \hat{\mathsf{T}}_{\mathbf{k}'\lambda',\mathbf{k}\lambda}(t)\right]$. In the next chapter, it will be shown that $\hat{\rho}_S$ is the so-called "density operator" of the scatterer, given by $\hat{\rho}_S = \sum_i P_i \left|E_i\right\rangle \left\langle E_i\right|$.

$\psi_{\mathbf{k}}(\mathbf{r})$ is the wavefunction for the momentum eigenstate $|\mathbf{k}\rangle$, i.e., $\psi_{\mathbf{k}}(\mathbf{r}) = L^{-3/2}e^{i\mathbf{k}\cdot\mathbf{r}}$. It then follows that $\mathsf{T}_{\mathbf{k}'\lambda',\mathbf{k}\lambda}(t)$ and $\langle\lambda' \mid \hat{\mathsf{T}}(\mathbf{r},t) \mid \lambda\rangle$ are related by the spatial Fourier transform

$$\mathsf{T}_{\mathbf{k}'\lambda',\mathbf{k}\lambda}(t) = \frac{1}{L^3}\int_{L^3} d^3r\, e^{i\mathbf{Q}\cdot\mathbf{r}}\langle\lambda' \mid \hat{\mathsf{T}}(\mathbf{r},t) \mid \lambda\rangle, \tag{6.71}$$

where $\mathbf{Q}$ is the wavevector transfer

$$\mathbf{Q} = \mathbf{k} - \mathbf{k}'. \tag{6.72}$$

Likewise,

$$\mathsf{T}^{\dagger}_{\mathbf{k}\lambda,\mathbf{k}'\lambda'}(0) = \frac{1}{L^3}\int_{L^3} d^3r\, e^{-i\mathbf{Q}\cdot\mathbf{r}}\langle\lambda \mid \hat{\mathsf{T}}^{\dagger}(\mathbf{r},t) \mid \lambda'\rangle. \tag{6.73}$$

$\hbar\mathbf{Q}$ is the momentum transferred by the neutron to the material medium.

A general expression for the scattering cross-section can now be determined once the neutron-target interaction potential $\hat{V}(\mathbf{r},t)$, and hence $\hat{\mathsf{T}}(\mathbf{r},t)$, is known. Toward this end, we digress, and first take a look at the more basic problem of the scattering of neutrons by a single, "free" atomic nucleus,[‖] i.e., a nucleus not bound within any molecule or crystal lattice—one that recoils freely. In particular, consider the scattering of *thermal neutrons*, i.e., those with energies in the neighborhood of k_BT (typically ~25 meV) or less. In this case, the neutron-nucleus interaction can be handled by making the so-called *Fermi approximation* [30]. Its underlying premise is based on the arguments briefly summarized here:[**] The neutron-nucleus interaction is characterized by an extremely strong force acting over a very short distance. Typically, the interaction potential can be approximated by a square-well of depth $V_0 \sim 36$ MeV and range $r_0 \sim 2$ F. From a calculational standpoint, the huge strength of the interaction seems to present a major difficulty in that it prohibits one from applying the standard perturbation techniques, as these are only valid for weak coupling. However, the fact that the interaction is of such a short range permits us to use a little bit of a "trick." Rather than dealing with the actual potential of depth V_0 and range r_0, we replace it with a much shallower, but more smeared-out, *Fermi pseudo-potential* with depth and range parameters $\tilde{V}_0$ and $\tilde{r}_0$, respectively (see Fig. 6.3). This is permitted because low-energy neutron scattering, as defined by the condition $kr_0 \ll 1$ (specifically, $kr_0 \sim 10^{-5}$–10^{-4} for the parameters previously stated), is insensitive to the precise shape of the interaction potential. Instead, the cross-section is determined only by a single parameter a, known as the *scattering length* of the free target nucleus,[††] which is proportional to the product $V_0r_0^3$. So, for example, we can still satisfy the low-energy scattering condition by choosing a range $\tilde{r}_0 \sim 100r_0$ for

[‖]Unless the target is a magnetic material, the primary interaction is between the neutron and the target nuclei, so the effects of the atomic electrons can be neglected.

[**]Complete details are given by Foderaro [31] and Sachs [32].

[††]Almost any standard text on nuclear physics will elaborate on the concept of "scattering length."

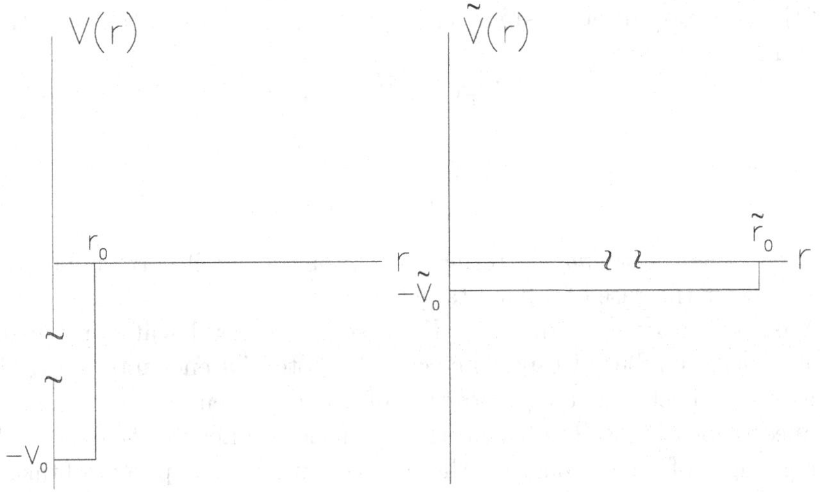

Figure 6.3 Comparison between the forms of the actual neutron-nucleus interaction potential $V(r)$ and the Fermi pseudopotential $\tilde{V}(r)$.

the pseudo-potential, since $k\tilde{r}_0 \sim 10^{-3}$–$10^{-2} \ll 1$. In addition, to maintain the original scattering cross-section, we must satisfy $\tilde{V}_0\tilde{r}_0^3 = V_0 r_0^3$. As a result, the required depth of the pseudo-potential is $\tilde{V}_0 = 10^{-6}V_0$ or $\tilde{V}_0 \sim 36$ eV, which is very much less than the depth of the actual potential. This makes a perturbation approach viable. Specifically, the fact that $\mu\tilde{V}_0\tilde{r}_0^2/\hbar^2 \ll 1$ (where μ is the neutron-nucleus reduced mass) allows one to apply the *(first) Born approximation* [33]; this basically assumes that the incident neutrons are scattered so weakly that the complete wavefunction inside the potential region is almost identical to the incident wavefunction. We thus see that the Fermi approximation consists of two parts, namely (1) replacing the real interaction potential with the Fermi pseudo-potential and (2) the application of the Born approximation. In order for the Fermi approximation to produce the correct scattering results [32], the proper form for the pseudo-potential turns out to be

$$\tilde{V}(\mathbf{r}) = \frac{2\pi\hbar^2}{\mu} a\,\delta(\mathbf{r}), \tag{6.74}$$

where $\mathbf{r}$ represents the position of the neutron relative to the nucleus. This choice is acceptable as long as the range $\tilde{r}_0$ is chosen to be much smaller than the zero-point vibration amplitude resulting from the nucleus being bound to a molecule or crystal. Typically, however, the vibration amplitude is in the neighborhood of ~ 0.1 angstroms. This is on the order of $\sim 10^2$ larger than $\tilde{r}_0$, so no problem is encountered in this regard.

The scattering length a is measured in an environment where the nucleus is free to recoil; thus a is sometimes called the *free* scattering length of the nucleus.

However, notice that for scattering from a nucleus of mass number A, the reduced mass of the neutron-nucleus system is $\mu = [A/(A+1)]\, m_n$, and the potential can also be written in the form

$$\tilde{V}(\mathbf{r}) = \frac{2\pi\hbar^2}{m_n} b\,\delta(\mathbf{r}), \tag{6.75}$$

where

$$b = \left(\frac{A+1}{A}\right) a. \tag{6.76}$$

b is referred to as the *bound* scattering length; physically, it corresponds to the scattering length for the case of a fixed target nucleus.

We now turn our attention to the case of a general scattering medium composed of a large number of target nuclei. The potential encountered by the incoming neutrons is effectively a superposition of pseudo-potentials, each centered at the instantaneous location $\mathbf{r}_\ell(t)$ of a particular nucleus. Letting b_ℓ denote the bound scattering length of the ℓth nucleus, the complete interaction potential takes the form

$$\hat{V}(\mathbf{r}, t) = \frac{2\pi\hbar^2}{m_n} \sum_\ell b_\ell \delta\left[\mathbf{r} - \mathbf{r}_\ell(t)\right]. \tag{6.77}$$

For a first-order process the T-matrix is identical to $\hat{V}$, so likewise

$$\hat{\mathsf{T}}(\mathbf{r}, t) = \frac{2\pi\hbar^2}{m_n} \sum_\ell b_\ell \delta\left[\mathbf{r} - \mathbf{r}_\ell(t)\right]. \tag{6.78}$$

Using Eqs. 6.71 and 6.73, we now have

$$\mathsf{T}_{\mathbf{k}'\lambda',\mathbf{k}\lambda}(t) = \frac{2\pi\hbar^2}{m_n L^3} \sum_\ell \langle \lambda' \mid b_\ell e^{i\mathbf{Q}\cdot\mathbf{r}_\ell(t)} \mid \lambda \rangle \tag{6.79}$$

and

$$\mathsf{T}^\dagger_{\mathbf{k}\lambda,\mathbf{k}'\lambda'}(0) = \frac{2\pi\hbar^2}{m_n L^3} \sum_\ell \langle \lambda \mid b_\ell e^{-i\mathbf{Q}\cdot\mathbf{r}_\ell(0)} \mid \lambda' \rangle. \tag{6.80}$$

As a result, the scattering cross-section becomes

$$\frac{d^2\sigma}{d\Omega\, d\omega} = \frac{1}{2\pi}\left(\frac{k'}{k}\right) \int_{-\infty}^{+\infty} dt\, e^{-i\omega t} \left\langle \sum_{\ell\ell'} \sum_{\lambda'} \langle \lambda \mid b_\ell e^{-i\mathbf{Q}\cdot\mathbf{r}_\ell(0)} \mid \lambda' \rangle \langle \lambda' \mid b_{\ell'} e^{i\mathbf{Q}\cdot\mathbf{r}_{\ell'}(t)} \mid \lambda \rangle \right\rangle. \tag{6.81}$$

We have inserted a summation over the final polarization λ' because most experiments do not analyze the spin of the scattered neutrons. Now from the closure relation $\sum_{\lambda'} |\lambda'\rangle\langle\lambda'| = \hat{I}$, it immediately follows that

$$\frac{d^2\sigma}{d\Omega\, d\omega} = \frac{1}{2\pi}\left(\frac{k'}{k}\right) \int_{-\infty}^{+\infty} dt\, e^{-i\omega t} \left\langle \sum_{\ell\ell'} \langle \lambda \mid b_\ell e^{-i\mathbf{Q}\cdot\mathbf{r}_\ell(0)} b_{\ell'} e^{i\mathbf{Q}\cdot\mathbf{r}_{\ell'}(t)} \mid \lambda \rangle \right\rangle. \tag{6.82}$$

Furthermore, if the incident beam is unpolarized, the result should be averaged over λ; this is denoted by the overbar in the expression below:

$$\frac{d^2\sigma}{d\Omega\, d\omega} = \frac{1}{2\pi}\left(\frac{k'}{k}\right)\int_{-\infty}^{+\infty} dt\; e^{-i\omega t}\left\langle \sum_{\ell\ell'} \overline{b_\ell b_{\ell'} e^{-i\mathbf{Q}\cdot\mathbf{r}_\ell(0)} e^{i\mathbf{Q}\cdot\mathbf{r}_{\ell'}(t)}} \right\rangle . \tag{6.83}$$

To proceed any further requires that we separate the latter expression into contributions from *coherent* and *incoherent* scattering, i.e., let us write

$$\frac{d^2\sigma}{d\Omega\, d\omega} = \left(\frac{d^2\sigma}{d\Omega\, d\omega}\right)_{\text{inc}} + \left(\frac{d^2\sigma}{d\Omega\, d\omega}\right)_{\text{coh}} . \tag{6.84}$$

Incoherence is an effect that arises because of two considerations. First of all, the neutron-nucleus interaction is spin-dependent, so the scattering length b_ℓ depends on the orientation of the neutron spin relative to the nuclear spin. Secondly, even if the scattering medium is chemically homogeneous, it will in general contain different isotopes of the same element, each with its own value of b_ℓ. As a consequence, it is necessary that the overbar in Eq. 6.83 represent an average over all possible nuclear spin orientations and isotope types. Since, in most cases, the spin and isotopic states of the various nuclei are completely uncorrelated with their positions, one can rewrite the cross-section as

$$\frac{d^2\sigma}{d\Omega\, d\omega} = \frac{1}{2\pi}\left(\frac{k'}{k}\right)\int_{-\infty}^{+\infty} dt\; e^{-i\omega t}\left\langle \sum_{\ell\ell'} \overline{b_\ell b_{\ell'}}\, e^{-i\mathbf{Q}\cdot\mathbf{r}_\ell(0)} e^{i\mathbf{Q}\cdot\mathbf{r}_{\ell'}(t)} \right\rangle . \tag{6.85}$$

For the average $\overline{b_\ell b_{\ell'}}$, notice that

$$\overline{b_\ell b_{\ell'}} = \overline{b_\ell^2} = \overline{b^2} \qquad \text{when } \ell = \ell' \tag{6.86}$$

and

$$\overline{b_\ell b_{\ell'}} = \left(\overline{b_\ell}\right)^2 = \overline{b}^2 \qquad \text{when } \ell \neq \ell' . \tag{6.87}$$

These two cases can be combined as

$$\overline{b_\ell b_{\ell'}} = \left(\overline{b^2} - \overline{b}^2\right)\delta_{\ell\ell'} + \overline{b}^2 . \tag{6.88}$$

Let us now identify a coherent and incoherent scattering length:

$$b_{\text{coh}} = \overline{b} \tag{6.89}$$

$$b_{\text{inc}} = \left(\overline{b^2} - \overline{b}^2\right)^{1/2} . \tag{6.90}$$

In other words, Eq. 6.88 becomes

$$\overline{b_\ell b_{\ell'}} = b_{\text{inc}}^2 \delta_{\ell\ell'} + b_{\text{coh}}^2 . \tag{6.91}$$

Equations 6.89 and 6.90 show that while $b_{\rm coh}$ is simply the scattering length averaged over spin and isotopic states, $b_{\rm inc}$ is the root-mean-square spread of the scattering lengths about this average. By substituting the last expression above into Eq. 6.85, we can now identify the incoherent and coherent contributions to the scattering cross-section to be

$$\left(\frac{d^2\sigma}{d\Omega\, d\omega}\right)_{\rm inc} = Nb_{\rm inc}^2 \left(\frac{k'}{k}\right) S_s\left(\mathbf{Q},\omega\right) \tag{6.92}$$

and

$$\left(\frac{d^2\sigma}{d\Omega\, d\omega}\right)_{\rm coh} = Nb_{\rm coh}^2 \left(\frac{k'}{k}\right) S\left(\mathbf{Q},\omega\right), \tag{6.93}$$

where

$$S_s\left(\mathbf{Q},\omega\right) = \frac{1}{2\pi}\int_{-\infty}^{+\infty} dt\; e^{-i\omega t}\left\langle \frac{1}{N}\sum_{\ell} e^{-i\mathbf{Q}\cdot\mathbf{r}_\ell(0)} e^{i\mathbf{Q}\cdot\mathbf{r}_\ell(t)}\right\rangle \tag{6.94}$$

and

$$S\left(\mathbf{Q},\omega\right) = \frac{1}{2\pi}\int_{-\infty}^{+\infty} dt\; e^{-i\omega t}\left\langle \frac{1}{N}\sum_{\ell\ell'} e^{-i\mathbf{Q}\cdot\mathbf{r}_\ell(0)} e^{i\mathbf{Q}\cdot\mathbf{r}_{\ell'}(t)}\right\rangle . \tag{6.95}$$

N is the number of nuclei in the target. $S\left(\mathbf{Q},\omega\right)$ is known as the *full dynamic structure factor*. Its form is basically determined by the spatial and temporal correlations between the atoms of the scattering medium. $S_s\left(\mathbf{Q},\omega\right)$ is called the *self dynamic structure factor.* Notice that if neutrons are being scattered from a homogeneous medium, the various nuclei are statistically equivalent and $S_s\left(\mathbf{Q},\omega\right)$ becomes simpler:

$$S_s\left(\mathbf{Q},\omega\right) = \frac{1}{2\pi}\int_{-\infty}^{+\infty} dt\; e^{-i\omega t}\left\langle e^{-i\mathbf{Q}\cdot\mathbf{r}(0)} e^{i\mathbf{Q}\cdot\mathbf{r}(t)}\right\rangle . \tag{6.96}$$

This structure factor is governed by the typical motion of a single atom within the scattering medium. Later on we will see that the dynamic structure factors also play a central role in the scattering of light and x-rays. Because of their general importance, they will be discussed in much greater detail in Chapter 12.

We conclude our discussion by working out the effects of spin incoherence on two different isotopes of hydrogen, specifically, light hydrogen (H^1) and heavy hydrogen, or deuterium, (D^2). For a given nuclear spin I, the nucleus-neutron system is in one of two possible spin states, each with its own characteristic scattering length. If the spins of the nucleus and neutron are aligned in a parallel sense, then the combined spin is $s_+ = I + \frac{1}{2}$ and the scattering length is represented by b_+; if the spins are oppositely oriented, the combined spin is $s_- = I - \frac{1}{2}$ and the scattering length is b_-. Recall, however, that for a given spin quantum number s, there are $2s+1$ values of the associated magnetic quantum number. It is therefore necessary to weight the two scattering lengths by the number of contributing states in each

Table 6.1 Scattering lengths and total cross-sections for light and heavy hydrogen. Note: 1 barn = 10^{-24} cm^2. (Data taken from Koester *et al.* [33])

	$b_+(10^{-12}\text{cm})$	$b_-(10^{-12}\text{cm})$	$\bar{b}(10^{-12}\text{cm})$	$\sigma_{\text{coh}}(\text{barns})$	$\sigma_{\text{inc}}(\text{barns})$
H^1	1.08	−4.74	−0.38	1.8	79.8
D^2	0.95	0.10	0.67	5.6	2.0

case, as follows: For b_+, there are $2\left(I+\frac{1}{2}\right)+1=2I+2$ states and for b_- there are $2\left(I-\frac{1}{2}\right)+1=2I$ states. Thus,

$$\bar{b}=\left(\frac{2I+2}{4I+2}\right)b_+ + \left(\frac{2I}{4I+2}\right)b_- \tag{6.97}$$

and

$$\overline{b^2}=\left(\frac{2I+2}{4I+2}\right)b_+^2 + \left(\frac{2I}{4I+2}\right)b_-^2. \tag{6.98}$$

The nuclear spins I for light and heavy hydrogen in the ground state are known to be $\frac{1}{2}$ and 1, respectively, producing the results in Table 6.1.

Here, $\sigma_{\text{coh}}=4\pi b_{\text{coh}}^2$ and $\sigma_{\text{inc}}=4\pi b_{\text{inc}}^2$ are the total (integrated) coherent and incoherent scattering cross-sections, respectively. Notice that $\sigma_{\text{coh}} \ll \sigma_{\text{inc}}$ in the case of H^1 because b_- is negative,‡‡ while b_+ is positive. Also, since the sign of the average (or coherent) scattering length $\bar{b}$ is different for H^1 and D^2, it is possible to mix light and heavy water, H_2O and D_2O, in various proportions so that the average scattering length of the mixture varies continuously from a negative to a positive value. Since $\bar{b}$ for O^{16} is 0.5804×10^{-12} cm, we have $\bar{b}_{H_2O}=-0.168\times10^{-12}$ cm and $\bar{b}_{D_2O}=-1.915\times10^{-12}$ cm, so b_{coh} of the mixture can be adjusted to take on any value between these two extremes. This is the basis of the so-called *contrast variation technique* employed in many small-angle neutron scattering experiments. Finally, to obtain an appreciation for how coherent scattering lengths and total cross-sections vary from one element to the next, one should study Figs. 6.4 and 6.5.

‡‡Physically, the negative value of b_- for H^1 means that a neutron and proton can not bind together when their spins are oppositely oriented, i.e., when $s=0$ (known as the "singlet state"). The positive value of b_+ indicates that binding can occur when the spins are aligned, and $s=1$ (referred to as the "triplet state"). In fact, this alignment explains why $I=1$ for the ground-state nuclear spin of D^2.

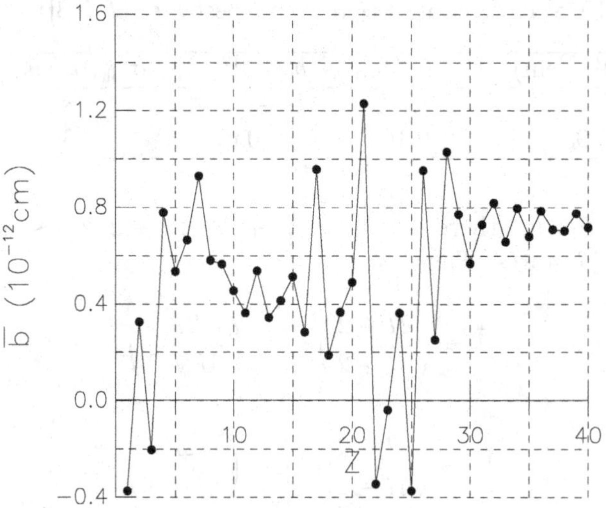

Figure 6.4 Coherent scattering length $\bar{b}$ as a function of atomic number Z, assuming a natural isotopic mixture. (Data taken from Koester et al. [33].)

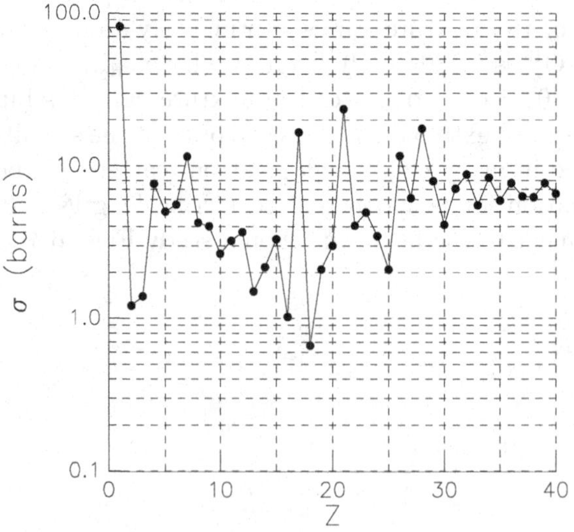

Figure 6.5 Total scattering cross-section σ as a function of atomic number Z, based on a natural isotopic mixture. (Data taken from Koester et al. [33].)

Suggested References

General treatments of time-dependent perturbation theory and the golden rule are given by

[a] E. Fermi, *Nuclear Physics* (University of Chicago Press, Chicago, 1950).

[b] L. I. Schiff, *Quantum Mechanics*, 3rd ed. (McGraw-Hill, New York, 1968).

More details on thermal neutron scattering can be found in

[c] A. Foderero, *The Elements of Neutron Interaction Theory* (M.I.T. Press, Cambridge, Mass., 1971).

[d] S. W. Lovesey, *Theory of Neutron Scattering from Condensed Matter*, Vol. 1 (Clarendon Press, Oxford, 1984).

[e] G. L. Squires, *Introduction to the Theory of Thermal Neutron Scattering* (Cambridge University Press, London, 1978).

Problems

1. Consider a two-level system with Hamiltonian $\hat{H}$ having only stationary states $|1\rangle$ and $|2\rangle$, with corresponding energies $\hbar\omega_1 < \hbar\omega_2$. At time $t = 0$, the system is in its ground state $|1\rangle$. A perturbation $\hat{W}$, not dependent on time, is switched on at $t = 0^+$. Calculate the probabilities for finding the system in each state after time t.
   Hint: *Solve*
$$i\hbar\frac{\partial}{\partial t}|\psi_s(t)\rangle = \left(\hat{H} + \hat{W}\right)|\psi_s(t)\rangle$$
   by letting
$$|\psi_s(t)\rangle = c_1(t)\, e^{-i\omega_1 t}|1\rangle + c_2(t)\, e^{-i\omega_2 t}|2\rangle$$
   with the initial conditions $c_1(0) = 1$ *and* $c_2(0) = 0$. *You are asked to calculate* $|c_1(t)|^2$ *and* $|c_1(t)|^2$.

2. Consider again the two-level system of the previous problem with the system in the ground state at $t = 0$. At time $t = 0^+$, a periodic perturbation $\hat{W}\cos\omega t$ is switched on with a frequency ω almost coinciding with the energy difference $\hbar\omega_0 = \hbar(\omega_2 - \omega_1)$ of the two levels. Work out the following steps:

(a) For $t > 0$, the system evolves according to the Schrödinger equation

$$i\hbar\frac{\partial}{\partial t}\left|\psi_s(t)\right\rangle = \left(\hat{H} + \hat{W}\cos\omega t\right)\left|\psi_s(t)\right\rangle .$$

Write the state vector in terms of the stationary states as

$$\left|\psi_s(t)\right\rangle = c_1(t)\, e^{-i\omega_1 t}\left|1\right\rangle + c_2(t)\, e^{-i\omega_2 t}\left|2\right\rangle$$

and derive the differential equations for the time-dependent amplitudes $c_1(t)$ and $c_2(t)$.

(b) Show that under the near-resonance condition ($\omega \approx \omega_0$) the equations simplify to

$$\begin{aligned} i\dot{c}_1(t) &= \frac{1}{2\hbar}\langle 1 \mid \hat{W} \mid 2\rangle e^{i\Delta\omega t} c_2(t) \\ i\dot{c}_2(t) &= \frac{1}{2\hbar}\langle 2 \mid \hat{W} \mid 1\rangle e^{-i\Delta\omega t} c_1(t) \end{aligned}$$

where $\Delta\omega = \omega - \omega_0$.

(c) Using the initial conditions $c_1(0) = 1$ and $c_2(0) = 0$, solve the above equations to obtain $c_1(t)$ and $c_2(t)$. Express the solutions in terms of the parameters

$$\Omega = \frac{1}{\hbar}\left|\langle 2 \mid \hat{W} \mid 1\rangle\right|$$

and

$$R = \left[\Omega^2 + (\Delta\omega)^2\right]^{1/2} .$$

(d) Determine the probability of finding the system in the excited state $|2\rangle$ at time t.

3. The usual perturbation expansion of the time-evolution operator, $\hat{U}_I(t)$, in the interaction picture is given by Eq. 6.23. It suffers from a shortcoming in that the truncated series is, in general, not unitary, whereas $\hat{U}_I(t)$, by definition, must be unitary. For some problems this might present a serious limitation. In these cases, one can expand $\hat{U}_I(t)$ in the so-called Magnus series such that

$$\begin{aligned} \hat{U}_I(t) &= e^{\hat{A}(t)} \\ \hat{A}^\dagger(t) &= -\hat{A}(t) \\ \hat{A}(t) &= \hat{A}_1(t) + \hat{A}_2(t) + \cdots . \end{aligned}$$

(a) Show that for the two lowest orders

$$\hat{A}_1(t) = \frac{1}{i\hbar}\int_0^t dt_1 \hat{V}_I(t_1)$$

$$\hat{A}_2(t) = \left(\frac{1}{i\hbar}\right)^2 \int_0^t dt_1 \int_0^{t_1} dt_2 \frac{1}{2}\left[\hat{V}_I(t_1), \hat{V}_I(t_2)\right].$$

(Recall that an operator that commutes at the same time need not commute at different times—see Problem 2 in Chapter 3.) The third-order term can be shown to be

$$\hat{A}_3(t) = \left(\frac{1}{i\hbar}\right)^3 \int_0^t dt_1 \int_0^{t_1} dt_2 \int_0^{t_2} \frac{1}{6}\left\{\left[\hat{V}_I(t_1), \left[\hat{V}_I(t_2), \hat{V}_I(t_3)\right]\right] + \left[\left[\hat{V}_I(t_1), \hat{V}_I(t_2)\right], \hat{V}_I(t_3)\right]\right\}$$

and, in general, the nth-order term contains commutators of the nth order. This is a useful expansion since if $\left[\hat{V}_I(t_1), \hat{V}_I(t_2)\right]$ is a constant, then all orders beyond the second vanish and the expansion reduces to only two terms.

(b) As an example, consider the problem of a driven harmonic oscillator given by the Hamiltonian

$$\hat{H} = \frac{1}{2m}\left(\hat{p}^2 + m^2\omega^2\hat{q}^2\right) + \sqrt{\frac{2m\omega}{\hbar}} f(t)\hat{q}.$$

By introducing the usual creation and annihilation operators, $\hat{a}^\dagger$ and $\hat{a}$, show that

$$\hat{H} = \hbar\omega\left(\hat{a}^\dagger\hat{a} + \frac{1}{2}\right) + f(t)\left(\hat{a}^\dagger + \hat{a}\right)$$

and

$$\hat{V}_I(t) = f(t)\left(e^{-i\omega t}\hat{a} + e^{i\omega t}\hat{a}^\dagger\right).$$

Furthermore, show that

$$\left[\hat{V}_I(t_1), \hat{V}_I(t_2)\right] = 2if(t_1)f(t_2)\sin\left[\omega(t_2 - t_1)\right],$$

which means that higher-order commutators all vanish. Hence, from Part (a),

$$\hat{U}_I(t) = e^{i\beta}e^{\left(\alpha\hat{a}^\dagger - \alpha^*\hat{a}\right)},$$

where

$$\alpha(t) = \frac{1}{i\hbar}\int_0^t dt_1 e^{i\omega t_1} f(t_1)$$

$$\beta(t) = -\frac{1}{\hbar^2}\int_0^t dt_1 \int_0^{t_1} dt_2 f(t_1) f(t_2) \sin[\omega(t_2 - t_1)].$$

(c) One can now evaluate $|c_{nm}(t)|^2 = \left|\langle n \mid \hat{U}_I(t) \mid m\rangle\right|^2$, i.e., the transition probability from state $|m\rangle$ to state $|n\rangle$. For example, the probability of making a transition from the ground state is

$$\begin{aligned} |c_{n0}(t)|^2 &= \left|\langle n \mid \hat{U}_I(t) \mid 0\rangle\right|^2 \\ &= \left|\langle n \mid e^{(\alpha\hat{a}^\dagger - \alpha^*\hat{a})} \mid 0\rangle\right|^2 \\ &= |\langle n \mid \alpha(t)\rangle|^2 \\ &= \frac{1}{n!}e^{-|\alpha(t)|^2}|\alpha(t)|^{2n}. \end{aligned}$$

The steps above follow by recalling that $e^{(\alpha\hat{a}^\dagger - \alpha^*\hat{a})}$ is in fact the displacement operator (introduced in Chapter 5) that takes the harmonic-oscillator ground state to a coherent state, i.e., $e^{(\alpha\hat{a}^\dagger - \alpha^*\hat{a})}|0\rangle = |\alpha\rangle$ (see Eq. 5.130). Find the general expression for $|c_{nm}(t)|^2$, the transition probability between any two harmonic-oscillator eigenstates.

4. Consider isotope incoherence in thermal neutron scattering. Suppose there are a variety of isotopic species of the same element, each with its own bound scattering length b_ℓ. Denote the fractional abundance of the ℓth species by c_ℓ. Then we have

$$\bar{b} = \sum_\ell c_\ell b_\ell \quad \text{and} \quad \overline{b^2} = \sum_\ell c_\ell b_\ell^2.$$

(a) Show that for the simplest case of only two isotopes, denoting the abundance of the first isotope by c, the incoherent scattering length is given by

$$b_{\text{inc}}^2 = c(1-c)(b_1 - b_2)^2.$$

(b) Take the element argon as an example. Argon has two principal isotopes, Ar^{36} and Ar^{40}, with $b_{36} = 2.43 \times 10^{-12}$ cm and $b_{40} = 0.193 \times 10^{-12}$ cm. The fractional abundances of these isotopes in natural argon are $c_{36} = 0.00337$ and $c_{40} = 0.996$. Calculate the ratio of coherent to incoherent scattering cross-sections.

(c) If the individual isotopes of argon are available commercially, one can make a sample having the maximum possible incoherent cross-section (and hence the minimum coherent cross-section). In order to do this, what should be the mixing ratio of the two isotopes and what are the resulting cross-sections?

5. The objective of this problem is to calculate the total cross-section for the scattering of cold neutrons from molecular hydrogen in the gas phase under the assumption that intermolecular correlations can be neglected. Follow the steps below:

(a) Define the *scattering-length operator* to be

$$\hat{b} = A + B\hat{\mathbf{s}}\cdot\hat{\imath},$$

where $\hat{\mathbf{s}}$ and $\hat{\imath}$ denote spin operators for the neutron and a target nucleus, respectively. A and B are constants to be determined. Now introduce the total neutron-nucleus spin operator $\hat{\mathbf{t}} = \hat{\mathbf{s}} + \hat{\imath}$, and note the operator relation

$$\hat{\mathbf{s}}\cdot\hat{\imath} = \frac{1}{2}\left(\hat{t}^2 - \hat{\imath}^2 - \hat{s}^2\right).$$

For a given spin eigenstate $|i, m_i\rangle\,|s, m_s\rangle$ of the system, we have the eigenvalue relation

$$\hat{\mathbf{s}}\cdot\hat{\imath}\,|i, m_i\rangle\,|s, m_s\rangle = \frac{1}{2}\left[t\,(t+1) - i\,(i+1) - s\,(s+1)\right]|i, m_i\rangle\,|s, m_s\rangle$$

where $s = 1/2$. By demanding that operator $\hat{b}$ has eigenvalue b_+ when $t = i + 1/2$ and eigenvalue b_- when $t = i - 1/2$, obtain the following values for the two constants:

$$A = \frac{i+1}{2i+1}b_+ + \frac{i}{2i+1}b_- \qquad \text{and} \qquad B = \frac{2}{2i+1}\left(b_+ - b_-\right).$$

(b) The differential cross-section for neutron scattering from molecular hydrogen, between an initial spin-state λ and a final spin-state λ' of the molecule-neutron system, is

$$\left(\frac{d\sigma}{d\Omega}\right)_{\lambda,\lambda'} = \frac{k'}{k}\left|\langle\lambda'|\sum_{\xi}\hat{b}_{\xi}e^{i\mathbf{Q}\cdot\mathbf{r}_{\xi}}\,|\lambda\rangle\right|^2,$$

where the index $\xi = 1, 2$ runs over the two protons of the molecule. However, if one is interested in the scattering of cold neutrons (energy ~3 meV), the wavelength (~5 Å) is large compared to the inter-proton separation of 0.74 Å in the hydrogen molecule, and one is justified in neglecting spatial interference effects due to the separation of the two protons. This means that the exponential factor appearing in the matrix element can simply be set to unity. Then, for a given initial state λ, the measured cross-section is obtained by summing over the final states λ', and we can write

$$\left(\frac{d\sigma}{d\Omega}\right)_\lambda = \frac{k'}{k}\langle\lambda|\left(\hat{b}_1+\hat{b}_2\right)^\dagger\left(\hat{b}_1+\hat{b}_2\right)|\lambda\rangle .$$

The relevant initial state of the molecule-neutron system is $|\lambda\rangle = |I, m_I\rangle\,|s, m_s$ where the first ket represents the total spin-state of the hydrogen molecule. Given that the total spin operator, $\hat{\mathbf{I}}$, for the molecule is the sum of the spin operators for the two protons, i.e., $\hat{\mathbf{I}} = \hat{\imath}_1 + \hat{\imath}_2$, show that

$$\left(\hat{b}_1+\hat{b}_2\right)^\dagger\left(\hat{b}_1+\hat{b}_2\right) = \left(2A + B\hat{\mathbf{s}}\cdot\hat{\mathbf{I}}\right)^2 = 4A^2 + \left(4AB - \frac{1}{2}B^2\right)\hat{\mathbf{s}}\cdot\hat{\mathbf{I}} + \frac{1}{4}B^2\hat{I}^2 .$$

To obtain this result, you must first show the operator relation

$$\left(\hat{\mathbf{I}}\cdot\hat{\mathbf{s}}\right)^2 = \frac{1}{4}\hat{I}^2 - \frac{1}{2}\hat{\mathbf{s}}\cdot\hat{\mathbf{I}} .$$

(c) For an unpolarized incident neutron beam, show that

$$\left(\frac{d\sigma}{d\Omega}\right)_{\text{unpol}} = \frac{k'}{k}\left[4A^2 + \frac{1}{4}B^2 I\,(I+1)\right] .$$

(d) There are two types of molecular hydrogen: When the two proton spins are parallel ($I = 1$), one has *orthohydrogen*; when the spins are anti-parallel ($I = 0$), one has *parahydrogen*. Show that

$$\left(\frac{d\sigma}{d\Omega}\right)_{\text{ortho}} = \frac{k'}{k}\left[\frac{1}{4}\left(3b_+ + b_-\right)^2 + \frac{1}{2}\left(b_+ - b_-\right)^2\right]$$

$$\left(\frac{d\sigma}{d\Omega}\right)_{\text{para}} = \frac{k'}{k}\left[\frac{1}{4}\left(3b_+ + b_-\right)^2\right] .$$

(e) Show that the total scattering cross-sections for the two types of molecular hydrogen are given by

$$\sigma_{\text{ortho}} = \frac{4\pi}{9}\left[\left(3b_+ + b_-\right)^2 + 2\left(b_+ - b_-\right)^2\right]$$

$$\sigma_{\text{para}} = \frac{4\pi}{9}\left(3b_+ + b_-\right)^2$$

and calculate the numerical values.

Chapter 7
THE DENSITY OPERATOR AND ITS ROLE IN QUANTUM STATISTICS

Up to this point, we have considered quantum systems where it is assumed that complete knowledge of the state vector exists at some instant of time. Quite often this is not the case; instead one may only have information related to certain macroscopic or thermodynamic properties such as temperature or average energy. These situations can be handled by introducing a so-called *density operator* into the framework of quantum mechanics. The action of this operator is to assign appropriate probabilities to the occupancy of available quantum states. We illustrate how to determine the density operator of a system by means of the *principle of maximum entropy*. Aspects of the technique are applied to a few examples such as radiation in a laser cavity both below and above threshold. We conclude this chapter by performing a perturbation expansion of the density operator. Transitions caused by a random perturbation are investigated.

7.1 Mixed States and the Density Operator

If a system is in a *pure quantum state* $|\psi\rangle$ and one measures a physical observable represented by a Hermitian operator $\hat{A}$, then the probability of obtaining eigenvalue a for the result is $P(a) = |\langle a \mid \psi\rangle|^2$. A large number of measurements of the variable A performed on an ensemble of systems, each identically prepared in the state $|\psi\rangle$, produces the average result $\langle A\rangle = \langle\psi \mid \hat{A} \mid \psi\rangle$. However, now consider a situation where one's a priori knowledge of the system's quantum state is less than perfect. Suppose, instead, that only the likelihood of being in certain pure states is known. In this case, one says that the system is in a *mixed state*—a member of the ensemble has a certain probability $P(\psi)$ of being in pure state $|\psi\rangle$.

For a mixed state, the expectation value of A for the ensemble is obtained by calculating

$$\langle A\rangle = \sum_{\psi} P(\psi)\langle\psi \mid \hat{A} \mid \psi\rangle. \tag{7.1}$$

This can be written in an alternate fashion by introducing the system's *density oper-*

ator, which is defined as

$$\hat{\rho} = \sum_{\psi} P(\psi) |\psi\rangle \langle\psi| . \tag{7.2}$$

Then, by inserting the identity operator $\sum_n |n\rangle \langle n|$ after the operator $\hat{A}$ in Eq. 7.1 (with $\{|n\rangle\}$ being any complete set of states), we find that

$$\begin{aligned} \langle A \rangle &= \sum_n \sum_{\psi} P(\psi)\langle\psi \mid \hat{A} \mid n\rangle\langle n \mid \psi\rangle = \sum_n \sum_{\psi} P(\psi)\langle n \mid \psi\rangle\langle\psi \mid \hat{A} \mid n\rangle \\ &= \sum_n \langle n \mid \hat{\rho}\hat{A} \mid n\rangle. \end{aligned} \tag{7.3}$$

The last expression is called the *trace* of the operator $\hat{\rho}\hat{A}$. It is the sum of the operator's diagonal matrix elements using any complete set of basis states. Letting $\mathrm{Tr}\left(\hat{\rho}\hat{A}\right)$ denote the trace of $\hat{\rho}\hat{A}$, the expectation value of A is expressed compactly as

$$\langle A \rangle = \mathrm{Tr}\left(\hat{\rho}\hat{A}\right) . \tag{7.4}$$

Some useful properties of the density operator are listed below:

- $\mathrm{Tr}\left(\hat{\rho}\hat{A}\right) = \mathrm{Tr}\left(\hat{A}\hat{\rho}\right)$ (This is a general property of any trace.)
- $\hat{\rho}^{\dagger} = \hat{\rho}$ (i.e., $\hat{\rho}$ is a Hermitian operator)
- $\mathrm{Tr}(\hat{\rho}) = 1$ (This is just the special case of $\hat{A} = 1$ in Eq. 7.4.)
- For a system in a pure state $|\psi\rangle$, the density operator is simply $\hat{\rho} = |\psi\rangle \langle\psi|$. Thus, $\hat{\rho}^2 = |\psi\rangle \langle\psi \mid \psi\rangle \langle\psi| = |\psi\rangle \langle\psi|$, or $\hat{\rho}^2 = \hat{\rho}$. As a result, for a pure state $\mathrm{Tr}(\hat{\rho}^2) = 1$. On the other hand, for a mixed state $\mathrm{Tr}(\hat{\rho}^2) < 1$.

Using a particular set of basis states, say $\{|n\rangle\}$, the density operator can be represented as a matrix having elements $\rho_{nm} = \langle n \mid \hat{\rho} \mid m\rangle$. The diagonal elements ρ_{nn} have a physical interpretation. Observe that

$$\rho_{nn} = \langle n \mid \hat{\rho} \mid n\rangle = \sum_{\psi} P(\psi)\langle n \mid \psi\rangle\langle\psi \mid n\rangle = \sum_{\psi} P(\psi) \left|\langle n \mid \psi\rangle\right|^2 . \tag{7.5}$$

Here, $\left|\langle n \mid \psi\rangle\right|^2$ is a conditional probability—it is the probability of finding the state $|n\rangle$ if it is known beforehand that the system is described by the state vector $|\psi\rangle$. In a mixed state, however, there is only statistical knowledge of an ensemble member's state vector $|\psi\rangle$, as given by the probability $P(\psi)$. These facts taken together mean that the matrix element ρ_{nn} is the total probability of finding the system in state

$|n\rangle$. Our previous statement above that $\mathrm{Tr}(\hat{\rho}) = \sum_n \rho_{nn} = 1$ is consistent with this physical interpretation. We can also now state that $0 \leq \rho_{nn} \leq 1$.

The density operator is unique in that, unlike other operators, it is time-dependent in the Schrödinger picture. To see this, recall from Eq. 3.110 that a state at time t in the Schrödinger picture, $|\psi_s(t)\rangle$, is obtained by projecting the initial state $|\psi(0)\rangle$ forward in time by using the time-evolution operator $\hat{U}(t)$, i.e., $|\psi_s(t)\rangle = \hat{U}(t)|\psi(0)\rangle$. Therefore, in this picture, Eq. 7.2 for the density operator becomes

$$\begin{aligned}\hat{\rho}(t) &= \sum_\psi P(\psi)\,|\psi_s(t)\rangle\,\langle\psi_s(t)| = \sum_\psi P(\psi)\hat{U}(t)\,|\psi(0)\rangle\,\langle\psi(0)|\,\hat{U}^\dagger(t) \\ &= \hat{U}(t)\hat{\rho}(0)\hat{U}^\dagger(t). \end{aligned} \tag{7.6}$$

The equation of motion for $\hat{\rho}(t)$ is obtained by making the substitution $\hat{U}(t) = e^{-i\hat{H}t/\hbar}$ (see Eq. 3.121), and then taking a time-derivative of Eq. 7.6. The result is

$$i\hbar\frac{\partial\hat{\rho}}{\partial t} = -\left[\hat{\rho}, \hat{H}\right]. \tag{7.7}$$

Compare this equation to Eq. 3.64; except for a minus sign, the density operator in the Schrödinger picture obeys an equation of motion like the one satisfied by operators in the Heisenberg picture. In addition, comparison with Eq. 2.142 shows that the above equation of motion is the quantum-mechanical analog of Liouville's theorem. The operator $\hat{\rho}$ in quantum mechanics corresponds to the density of points in classical phase space.

7.2 Entropy and Information Content—Determining the Density Operator of a System

A central variable in statistical mechanics is that of the *entropy* of a system, S. Classically, it can be defined as

$$S = -k_B \sum_i P_i \ln P_i, \tag{7.8}$$

where P_i is the probability of finding the system in state i (k_B is Boltzmann's constant). Entropy can be thought of as a measure of the lack of knowledge about the system. This interpretation is born out by examining the two extreme cases: When knowledge about the system is complete and one is sure, for example, that the system is in some state j, then $P_i = \delta_{ij}$, and the entropy given by Eq. 7.8 vanishes. On the other hand, when nothing is known about the state of the system, the entropy is at a maximum. This is so because δS, the variation in the entropy, is equal to zero when the entropy is maximized, i.e.,

$$\delta S = -k_B \sum_i (1 + \ln P_i)\,\delta P_i = 0 \tag{7.9}$$

subject to the constraint $\sum_i P_i = 1$, or

$$\sum_i \delta P_i = 0. \tag{7.10}$$

Now use the method of Lagrange multipliers; Eq. 7.10 is multiplied by an arbitrary constant λ, and the result is added to Eq. 7.9 giving

$$\sum_i (1 + \lambda + \ln P_i)\, \delta P_i = 0. \tag{7.11}$$

Because the variations δP_i are independent, the above equation is only satisfied if $P_i = e^{-(1+\lambda)} =$ constant. Thus, the entropy is maximized when it is equally likely for the system to be in any of its possible states, which is the same as saying that there is no information about the state of the system. If N is the number of possible states available to the system, then $\sum_{i=1}^{N} P_i = NP_i = 1$, and $P_i = 1/N$. As a result, from Eq. 7.8, the maximum value of the entropy is given by $S_{\max} = -k_B \sum_{i=1}^{N} \left(\frac{1}{N}\right) \ln\left(\frac{1}{N}\right) = k_B \ln N$.

In analogy with Eq. 7.8, the entropy in quantum statistical calculations is defined in terms of the density operator to be

$$S = -k_B \operatorname{Tr}(\hat{\rho} \ln \hat{\rho}). \tag{7.12}$$

To see the compatibility of Eq. 7.12 with Eq. 7.8, first evaluate the above trace using some complete set of basis states $\{|n\rangle\}$ so that

$$S = -k_B \sum_{n,m} \langle n \mid \hat{\rho} \mid m\rangle\langle m \mid \ln\hat{\rho} \mid n\rangle. \tag{7.13}$$

Then transform to another basis set $\{|i\rangle\}$ such that $\hat{\rho}\,|i\rangle = i\,|i\rangle$; in this representation the matrix for $\hat{\rho}$ is diagonal. The entropy then becomes

$$\begin{aligned}
S &= -k_B \sum_{n,m} \sum_{\substack{i,i' \\ i'',i'''}} \langle n \mid i\rangle\langle i \mid \hat{\rho} \mid i'\rangle\langle i' \mid m\rangle\langle m \mid i''\rangle\langle i'' \mid \ln\hat{\rho} \mid i'''\rangle\langle i''' \mid n\rangle \\
&= -k_B \sum_{n,m} \sum_{i,i''} \langle n \mid i\rangle \rho_{ii} \langle i \mid m\rangle\langle m \mid i''\rangle \ln\rho_{i''i''} \langle i'' \mid n\rangle \\
&= -k_B \sum_{i,i''} \sum_{n,m} \rho_{ii} \ln\rho_{i''i''} \langle i'' \mid n\rangle\langle n \mid i\rangle\langle i \mid m\rangle\langle m \mid i''\rangle \\
&= -k_B \sum_{i,i''} \rho_{ii} \ln\rho_{i''i''} \langle i'' \mid i\rangle\langle i \mid i''\rangle \\
&= -k_B \sum_i \rho_{ii} \ln\rho_{ii}.
\end{aligned} \tag{7.14}$$

Since ρ_{ii} is identical to P_i, or the probability of finding the system in state $|i\rangle$, the entropy reduces to $S = -k_B \sum_i P_i \ln P_i$, in agreement with the classical definition, Eq. 7.8.

To find the density operator for a system in equilibrium, we use the fact that the system's entropy is maximized, subject to the constraint $\mathrm{Tr}(\hat{\rho}) = 1$. If, in addition, there are measurements that provide knowledge about ensemble averages of any of the state variables, then Eq. 7.4 can be used to state additional constraints. For example, returning to the case where nothing is known about the system, we have $\mathrm{Tr}(\hat{\rho}) = 1$ as the sole constraint—it is the quantum analogue of the classical constraint $\sum_i P_i = 1$. To determine the density operator, we maximize S by setting $\delta S = 0$, i.e., using Eq. 7.12,

$$\mathrm{Tr}\left[(1 + \ln \hat{\rho})\, \delta\hat{\rho}\right] = 0. \tag{7.15}$$

Taking the variation of the constraint $\mathrm{Tr}(\hat{\rho}) = 1$ gives

$$\mathrm{Tr}\,(\delta\hat{\rho}) = 0. \tag{7.16}$$

The last two equations are the quantum versions of Eqs. 7.9 and 7.10. In a manner similar to before, by applying the method of Lagrange multipliers it is found that $\hat{\rho} =$ constant and the matrix elements in some representation, say $|n\rangle$, are then simply given by $\rho_{nm} =$ constant $\times \delta_{nm}$.

Now consider a system (having a fixed number of particles) for which the average energy of the ensemble is measured to be $\langle E\rangle$. Equation 7.4 then tells us that

$$\langle E\rangle = \mathrm{Tr}\left(\hat{H}\hat{\rho}\right). \tag{7.17}$$

This acts as an additional constraint in the determination of the density operator. Consequently, we need to supplement Eqs. 7.15 and 7.16 with

$$\mathrm{Tr}\left(\hat{H}\,\delta\hat{\rho}\right) = 0. \tag{7.18}$$

Again, using the method of Lagrange multipliers, we have

$$\mathrm{Tr}\left[\left(1 + \lambda + \ln\hat{\rho} + \beta\hat{H}\right)\delta\hat{\rho}\right] = 0, \tag{7.19}$$

where both λ and β are undetermined constants. For arbitrary $\delta\hat{\rho}$, we must have $1 + \lambda + \ln\hat{\rho} + \beta\hat{H} = 0$, or

$$\hat{\rho} = e^{-(1+\lambda)} e^{-\beta\hat{H}}. \tag{7.20}$$

The constant $e^{-(1+\lambda)}$ is determined by taking the trace of both sides and using $\mathrm{Tr}(\hat{\rho}) = 1$. We find that $e^{-(1+\lambda)} = 1/Z$, where $Z = \mathrm{Tr}\left(e^{-\beta\hat{H}}\right)$ is known as the *partition function* of the system. Z can be evaluated by using the system's energy eigenstates $|E_n\rangle$ for the trace calculation. We then have

$$Z = \sum_n \langle E_n \mid e^{-\beta\hat{H}} \mid E_n\rangle = \sum_n e^{-\beta E_n}, \tag{7.21}$$

and the density operator is given by

$$\hat{\rho} = \frac{e^{-\beta \hat{H}}}{Z}. \tag{7.22}$$

In the energy representation, the density matrix is diagonal, i.e.,

$$\rho_{nm} = \frac{\langle E_n \mid e^{-\beta \hat{H}} \mid E_m \rangle}{Z} = \frac{e^{-\beta E_n}}{\sum\limits_n e^{-\beta E_n}} \delta_{nm}. \tag{7.23}$$

The probability of finding the system in state $|E_n\rangle$ is therefore

$$P(E_n) = \rho_{nn} = \frac{e^{-\beta E_n}}{\sum\limits_n e^{-\beta E_n}}, \tag{7.24}$$

which is known as the *canonical* or *Maxwell-Boltzmann distribution*. $e^{-\beta E_n}$ is called the *Boltzmann factor*. The constant β is a function of the average energy $\langle E \rangle$; since $\langle E \rangle = \sum_n E_n P(E_n)$, there exists the following relationship between β and $\langle E \rangle$:

$$\langle E \rangle = \frac{\sum\limits_n E_n e^{-\beta E_n}}{\sum\limits_n e^{-\beta E_n}}. \tag{7.25}$$

This relation can also be seen from Eqs. 7.17 and 7.22. Observe that the average energy is completely determined by the partition function Z, since the above expression can be generated by

$$\langle E \rangle = -\frac{\partial}{\partial \beta} (\ln Z). \tag{7.26}$$

In fact, all the thermodynamic properties of the system can be found from the partition function. As another example of this, consider the maximized entropy of the system. From Eqs. 7.12 and 7.22, it is

$$\begin{aligned} S &= -k_B \mathrm{Tr} \left[\hat{\rho} \ln \left(e^{-\beta \hat{H}} / Z \right) \right] = k_B \mathrm{Tr} \left(\beta \hat{\rho} \hat{H} + \hat{\rho} \ln Z \right) \\ &= \beta k_B \mathrm{Tr} \left(\hat{\rho} \hat{H} \right) + k_B \ln Z \left[\mathrm{Tr} \left(\hat{\rho} \right) \right] \\ &= \beta k_B \langle E \rangle + k_B \ln Z. \end{aligned} \tag{7.27}$$

Example 7.1 EM Radiation in Thermal Equilibrium with a Cavity. Consider a case where the average energy of the system is not known; rather, the ensemble members exchange energy with a thermal reservoir at a known temperature T. This is commonly referred to as a *canonical ensemble*. The procedure for determining the density operator is identical to the case where $\langle E \rangle$ is known, except here the value of the average energy is completely determined by the temperature of the reservoir. Equations 7.21–7.25 still hold, but now the parameter β will be purely a function of T.

The particular system we will examine is that of a single mode of electromagnetic radiation in thermal equilibrium with the walls of a cavity at temperature T. This situation is encountered in a laser cavity below the oscillation threshold. In order to find how β depends on the temperature, recall that a mode of the EM field is identical to a harmonic oscillator—it has possible states $|E_n\rangle$ of energy $E_n = \hbar\omega\left(n + \frac{1}{2}\right)$. If we neglect the zero-point energy, the average energy of the mode, as determined from Eq. 7.25, is

$$\langle E\rangle = \frac{\hbar\omega \sum\limits_{n=0}^{\infty} n\left(e^{-\beta\hbar\omega}\right)^n}{\sum\limits_{n=0}^{\infty}\left(e^{-\beta\hbar\omega}\right)^n}. \tag{7.28}$$

Conveniently, the two infinite series converge and we have

$$\langle E\rangle = \frac{\hbar\omega}{e^{\beta\hbar\omega} - 1}. \tag{7.29}$$

β is obtained by examining the classical limit of this expression. This is done by treating $\hbar$ as very small, i.e., by setting $e^{\beta\hbar\omega} \to 1 + \beta\hbar\omega$; the result is $\langle E\rangle \to 1/\beta$. Note, however, that the equipartition theorem requires that, classically, each degree of freedom should contribute $\frac{1}{2}k_B T$ to the average energy of the system. Because the electric and magnetic fields each represent one degree of freedom, the average energy of the EM mode should be $k_B T$. We thus find that

$$\beta = \frac{1}{k_B T}. \tag{7.30}$$

For the canonical ensemble, the entropy, given by Eq. 7.27, then becomes

$$S = \frac{\langle E\rangle}{T} + k_B \ln Z. \tag{7.31}$$

Now consider the following very practical question: What is the probability $P(n)$ of counting n photons as a result of measuring the radiation field at temperature T with a photon detector? The answer is certainly ρ_{nn}, as given by Eq. 7.24. In this case, it is

$$P(n) = \rho_{nn} = \frac{\left(e^{-\beta\hbar\omega}\right)^n}{\sum\limits_{n=0}^{\infty}\left(e^{-\beta\hbar\omega}\right)^n} = x^n(1-x), \tag{7.32}$$

where $x = e^{-\beta\hbar\omega}$. However, notice that

$$\langle n\rangle = \sum_{n=0}^{\infty} nP(n) = (1-x)\sum_{n=0}^{\infty} nx^n = \frac{x}{1-x}, \tag{7.33}$$

or

$$x = \frac{\langle n \rangle}{1 + \langle n \rangle}. \tag{7.34}$$

Therefore, the photon-count probability takes on the following form, known as the *Bose-Einstein distribution*:

$$P(n) = \frac{\langle n \rangle^n}{(1 + \langle n \rangle)^{n+1}}. \tag{7.35}$$

Suppose instead of restricting the radiation to a single mode, we consider all possible modes that can exist within the confines of the cavity. Then, the enclosure is filled with a so-called *blackbody radiation spectrum*. Let us find the mean *energy density* $\langle u(\omega, T)\rangle \, d\omega$ (i.e., energy per unit volume) contributed to the field by photons having frequencies in the range $\omega \rightarrow \omega + d\omega$. For a cavity of volume V at temperature T, we write

$$\langle u(\omega, T)\rangle \, d\omega = \frac{1}{V} \langle E(\omega, T)\rangle \, g(\omega) d\omega, \tag{7.36}$$

where $g(\omega)$ is the density of available modes, as given by Eq. 5.7, and $\langle E(\omega, T)\rangle$ is the mean energy contributed to the radiation field by each mode, as given by Eq. 7.29. Substituting in the appropriate expressions and integrating over all directions of photon travel results in

$$\langle u(\omega, T)\rangle \, d\omega = \frac{\hbar}{\pi^2 c^3} \frac{\omega^3 d\omega}{e^{\beta\hbar\omega} - 1} = \frac{\hbar}{\pi^2 c^3} \left(\frac{k_B T}{\hbar}\right)^4 \frac{\eta^3 d\eta}{e^{\eta} - 1}. \tag{7.37}$$

In the final expression, we have introduced the dimensionless parameter $\eta \equiv \beta\hbar\omega = \hbar\omega / k_B T$. Equation 7.37 is Planck's famous spectral distribution for blackbody radiation, which is illustrated in Fig. 7.1. The peak position of the blackbody curve scales so that $\eta_{\max} = \omega_{\max}/T \approx 3$ is independent of the temperature. This statement is known as *Wien's displacement law*. In addition, $\langle u(T)\rangle \equiv \int_0^{\infty} \langle u(\omega, T)\rangle \, d\omega$, which is the mean energy density summed over all frequencies, is

$$\langle u(T)\rangle = \frac{\hbar}{\pi^2 c^3} \left(\frac{k_B T}{\hbar}\right)^4 \int_0^{\infty} \frac{\eta^3 d\eta}{e^{\eta} - 1}. \tag{7.38}$$

This shows that the result is proportional to T^4, a fact known as the *Stefan-Boltzmann law*. The integral, it turns out, is exactly $\pi^4/15$, so the integrated mean energy density has the final form

$$\langle u(T)\rangle = aT^4, \tag{7.39}$$

where $a = \pi^2 k_B^4 / 15 c^3 \hbar^3 = 7.5636 \times 10^{-15} \text{erg}\cdot\text{cm}^{-3}\cdot\text{K}^{-4}$.

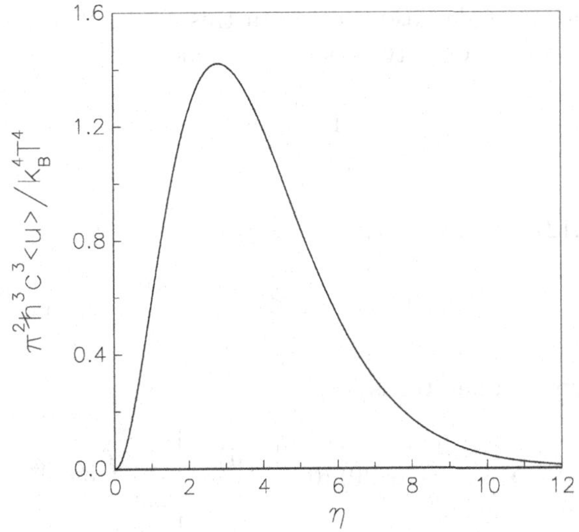

Figure 7.1 Mean energy density $\langle u \rangle$ (per unit dimensionless frequency range $d\eta$) as a function of $\eta = \hbar\omega/k_B T$ for a blackbody radiator.

Example 7.2 The Density Operator Applied to Coherent States. When a laser operates well above the oscillation threshold, the radiation field is in a coherent state $|\alpha\rangle$. Quantum mechanically, this corresponds to a pure state, and the density operator is simply

$$\hat{\rho} = |\alpha\rangle \langle\alpha| \,. \tag{7.40}$$

It is not surprising to find that the expectation value corresponding to the oscillator coordinate $\hat{Q}(t)$, as defined by Eq. 5.85, oscillates with a definite phase, i.e.,

$$\begin{aligned} \langle Q \rangle &= Q_0 \left[\mathrm{Tr}\left(\hat{a}(t) |\alpha\rangle \langle\alpha|\right) + \mathrm{Tr}\left(|\alpha\rangle \langle\alpha| \hat{a}^\dagger(t)\right)\right] \\ &= Q_0 \left(\alpha e^{-i\omega t} + \alpha^* e^{+i\omega t}\right) \mathrm{Tr}\left(\hat{\rho}\right) \\ &= 2Q_0 |\alpha| \cos\left(\omega t - \phi\right), \end{aligned} \tag{7.41}$$

since this matches our previous result, namely Eq. 5.94. In addition, the expected number of photons in the field is

$$\langle n \rangle = \mathrm{Tr}\left(\hat{a}^\dagger \hat{a} |\alpha\rangle \langle\alpha|\right) = |\alpha|^2 , \tag{7.42}$$

in agreement with Eq. 5.96, while the probability for detecting n photons certainly follows the Poisson distribution of Eq. 5.108, as shown below:

$$P(n) = \rho_{nn} = \langle n \mid \alpha\rangle\langle\alpha \mid n\rangle = |\langle n \mid \alpha\rangle|^2 = |c_n(\alpha)|^2 = \frac{|\alpha|^{2n}}{n!} e^{-|\alpha|^2} . \tag{7.43}$$

The more interesting case, however, is when an intensity or photon-counting measurement supplies a precise determination of $\langle n \rangle$ or $|\alpha|^2$, but no information is available on the phase of the laser beam. In this case, all phases between 0 and 2π are equally likely, and the density operator becomes

$$\hat{\rho} = \frac{1}{2\pi}\int_0^{2\pi} d\phi \, |\alpha\rangle\langle\alpha| \,. \tag{7.44}$$

Then, from Eq. 5.127, the integrand becomes

$$|\alpha\rangle\langle\alpha| = e^{-|\alpha|^2} e^{\alpha\hat{a}^\dagger} |0\rangle\langle 0| e^{\alpha^*\hat{a}} = e^{-|\alpha|^2} \sum_{n,m} \frac{\left(\alpha\hat{a}^\dagger\right)^n}{n!} |0\rangle\langle 0| \frac{\left(\alpha^*\hat{a}\right)^m}{m!}. \tag{7.45}$$

As a result, the density operator is

$$\begin{aligned}
\hat{\rho} &= e^{-|\alpha|^2} \sum_{n,m} \frac{\left(|\alpha|\,\hat{a}^\dagger\right)^n}{n!} |0\rangle\langle 0| \frac{\left(|\alpha|\,\hat{a}\right)^m}{m!} \left[\frac{1}{2\pi}\int_0^{2\pi} d\phi \, e^{i(n-m)\phi}\right] \\
&= e^{-|\alpha|^2} \sum_n \frac{|\alpha|^{2n}}{(n!)^2} \hat{a}^{\dagger n} |0\rangle\langle 0| \hat{a}^n = e^{-|\alpha|^2} \sum_n \frac{|\alpha|^{2n}}{(n!)^2} \sqrt{n!}\,|n\rangle\langle n|\sqrt{n!} \\
&= e^{-|\alpha|^2} \sum_n \frac{|\alpha|^{2n}}{n!} |n\rangle\langle n|
\end{aligned} \tag{7.46}$$

and the photon counting probability again reduces to the Poisson distribution since

$$P(n) = \langle n \mid \hat{\rho} \mid n\rangle = e^{-|\alpha|^2} \sum_m \frac{|\alpha|^{2m}}{m!} \langle n \mid m\rangle\langle m \mid n\rangle = \frac{|\alpha|^{2n}}{n!} e^{-|\alpha|^2}. \tag{7.47}$$

The relation between $\langle n\rangle$ and $|\alpha|^2$ is as it was before, i.e.,

$$\langle n\rangle = \mathrm{Tr}\left(\hat{a}^\dagger\hat{a}\hat{\rho}\right) = \frac{1}{2\pi}\int_0^{2\pi} d\phi \, \mathrm{Tr}\left(\hat{a}^\dagger\hat{a}\,|\alpha\rangle\langle\alpha|\right) = \frac{1}{2\pi}\int_0^{2\pi} d\phi \, |\alpha|^2 = |\alpha|^2 , \tag{7.48}$$

but in this case observe that

$$\begin{aligned}
\langle Q\rangle &= \mathrm{Tr}\left[\hat{Q}(t)\hat{\rho}\right] = \frac{Q_0}{2\pi}\int_0^{2\pi} d\phi \left[\mathrm{Tr}\left(\hat{a}(t)\,|\alpha\rangle\langle\alpha|\right) + \mathrm{Tr}\left(|\alpha\rangle\langle\alpha|\,\hat{a}^\dagger(t)\right)\right] \\
&= \frac{Q_0\,|\alpha|}{\pi} \mathrm{Tr}\left(|\alpha\rangle\langle\alpha|\right) \int_0^{2\pi} d\phi \cos\left(\omega t - \phi\right).
\end{aligned} \tag{7.49}$$

The integral over ϕ, however, vanishes, so in this case

$$\langle Q\rangle = 0, \tag{7.50}$$

and all phase information is lost.

7.3 Perturbation Expansion of the Density Operator

In the last chapter, we considered the time evolution of a system subject to a small perturbation $\hat{V}(t)$ by assuming a known, i.e., pure, initial quantum state. The basic procedure was to expand $\hat{U}_I(t)$, the time-evolution operator in the interaction picture, using an iterative procedure. Here we derive the analogous time-dependent perturbation expansion for the density operator. The result is more general because it handles systems that are initially in a mixed state.

Start by rewriting Eq. 7.7, the equation of motion for the density operator in the Schrödinger picture:

$$\frac{\partial \hat{\rho}(t)}{\partial t} = \frac{i}{\hbar}\left[\hat{\rho}(t),\, \hat{H}_0 + \hat{V}(t)\right]. \tag{7.51}$$

As before, $\hat{H}_0$ is the time-independent Hamiltonian of the unperturbed system. We can eliminate $\hat{H}_0$ by transforming to the interaction picture. According to Eqs. 6.10 and 6.6, the density operator in the interaction picture is given by

$$\hat{\rho}_I(t) = e^{+i\hat{H}_0 t/\hbar}\hat{\rho}(t)e^{-i\hat{H}_0 t/\hbar}, \tag{7.52}$$

and it is straightforward to show that

$$\frac{\partial \hat{\rho}_I(t)}{\partial t} = \frac{i}{\hbar}\left[\hat{H}_0, \hat{\rho}_I(t)\right] + e^{+i\hat{H}_0 t/\hbar}\frac{\partial \hat{\rho}(t)}{\partial t}e^{-i\hat{H}_0 t/\hbar}. \tag{7.53}$$

By substituting in the expression for $\partial\hat{\rho}(t)/\partial t$ given by Eq. 7.51, the equation of motion reduces to

$$\begin{aligned} \frac{\partial \hat{\rho}_I(t)}{\partial t} &= \frac{i}{\hbar}\left[\hat{H}_0, \hat{\rho}_I(t)\right] + \frac{i}{\hbar}e^{+i\hat{H}_0 t/\hbar}\left[\hat{\rho}(t),\, \hat{H}_0 + \hat{V}(t)\right]e^{-i\hat{H}_0 t/\hbar} \\ &= \frac{i}{\hbar}\left[\hat{H}_0, \hat{\rho}_I(t)\right] + \frac{i}{\hbar}\left[\hat{\rho}_I(t),\, \hat{H}_0 + \hat{V}_I(t)\right] \end{aligned} \tag{7.54}$$

or

$$\frac{\partial \hat{\rho}_I(t)}{\partial t} = \frac{i}{\hbar}\left[\hat{\rho}_I(t), \hat{V}_I(t)\right]. \tag{7.55}$$

We can now integrate both sides of this equation over the time interval 0 to t to get

$$\hat{\rho}_I(t) = \hat{\rho}_I(0) + \frac{i}{\hbar}\int_0^t dt'\left[\hat{\rho}_I(t'), \hat{V}_I(t')\right]. \tag{7.56}$$

The second term on the right-hand side is small. This allows one to expand $\hat{\rho}_I(t)$ by iteration. Using the fact that the density operators in the interaction and the Schrödinger picture are identical at $t = 0$, i.e., $\hat{\rho}_I(0) = \hat{\rho}(0)$, we obtain

$$\hat{\rho}_I(t) = \hat{\rho}(0) + \frac{i}{\hbar}\int_0^t dt_1\left[\hat{\rho}(0), \hat{V}_I(t_1)\right] + \left(\frac{i}{\hbar}\right)^2\int_0^t dt_1\int_0^{t_1} dt_2\left[\left[\hat{\rho}(0), \hat{V}_I(t_2)\right], \hat{V}_I(t_1)\right] + \ldots. \tag{7.57}$$

The primary utility of this result is in calculating matrix elements of the density operator at time t assuming the matrix elements at $t = 0$ are known. In the energy representation, the matrix elements in the two pictures are simply related by

$$\langle n \mid \hat{\rho}_I(t) \mid m\rangle = \langle n \mid e^{+i\hat{H}_0 t/\hbar}\hat{\rho}(t)e^{-i\hat{H}_0 t/\hbar} \mid m\rangle = e^{i(E_n - E_m)t/\hbar}\langle n \mid \hat{\rho}(t) \mid m\rangle. \quad (7.58)$$

Of particular interest are the diagonal matrix elements $\langle n \mid \hat{\rho}(t) \mid n\rangle = \langle n \mid \hat{\rho}_I(t) \mid n\rangle$, as they represent the probability of finding the system in state $|n\rangle$ at time t. For an initial state that is mixed, the $\langle n \mid \hat{\rho}(t) \mid n\rangle$'s serve the same function as the $|c_{nm}(t)|^2$'s, which are the transition probabilities appropriate for dealing with an initial state that is pure.

To see that Eq. 7.57 is in fact a generalization of Eq. 6.23 for the time-dependent perturbation expansion of $\hat{U}_I(t)$, consider a system in an initially pure energy eigenstate $|m\rangle$. In this case, all the matrix elements of $\hat{\rho}(0)$ vanish except for $\langle m \mid \hat{\rho}(0) \mid m\rangle = 1$, i.e., $\hat{\rho}(0) = |m\rangle\langle m|$. Taking diagonal elements of both sides of Eq. 7.57 produces the following expression, to second order, for $n \neq m$:

$$\begin{aligned}\langle n \mid \hat{\rho}(t) \mid n\rangle &= \langle n \mid m\rangle\langle m \mid n\rangle + \frac{i}{\hbar}\int_0^t dt_1\,\langle n \mid \left[\hat{\rho}(0), \hat{V}_I(t_1)\right] \mid n\rangle \\ &+ \left(\frac{i}{\hbar}\right)^2 \int_0^t dt_1 \int_0^{t_1} dt_2\,\left\langle n \mid \left[\left[\hat{\rho}(0), \hat{V}_I(t_2)\right], \hat{V}_I(t_1)\right] \mid n\right\rangle. \quad (7.59)\end{aligned}$$

The zeroth-order term vanishes since the states $|n\rangle$ and $|m\rangle$ are orthogonal. Recall that the zeroth-order contribution to $|c_{nm}(t)|^2$ also vanished when we examined the perturbation expansion of $\hat{U}_I(t)$. However, observe here that there is also no contribution from the first-order term since

$$\langle n \mid \left[\hat{\rho}(0), \hat{V}_I(t_1)\right] \mid n\rangle = \langle n \mid m\rangle\langle m \mid \hat{V}_I(t_1) \mid n\rangle - \langle n \mid \hat{V}_I(t_1) \mid m\rangle\langle m \mid n\rangle = 0. \quad (7.60)$$

The first non-vanishing contribution comes from the second-order term, so Eq. 7.59 reduces as follows:

$$\begin{aligned}\langle n \mid \hat{\rho}(t) \mid n\rangle &= \left(\frac{i}{\hbar}\right)^2 \int_0^t dt_1 \int_0^{t_1} dt_2\,\left\langle n \mid \left[\left[\hat{\rho}(0), \hat{V}_I(t_2)\right], \hat{V}_I(t_1)\right] \mid n\right\rangle \\ &= \frac{1}{\hbar^2}\int_0^t dt_1 \int_0^{t_1} dt_2\,\left\{\left\langle n \mid \hat{V}_I(t_2) \mid m\right\rangle\left\langle m \mid \hat{V}_I(t_1) \mid n\right\rangle\right. \\ &\left. + \left\langle n \mid \hat{V}_I(t_1) \mid m\right\rangle\left\langle m \mid \hat{V}_I(t_2) \mid n\right\rangle\right\} \\ &= \frac{1}{\hbar^2}\int_0^t dt_1 \int_0^{t_1} dt_2\end{aligned}$$

$$\times\left\{\left\langle n \mid e^{+i\hat{H}_0t_2/\hbar}\hat{V}(t_2)e^{-i\hat{H}_0t_2/\hbar} \mid m\right\rangle\left\langle m \mid e^{+i\hat{H}_0t_1/\hbar}\hat{V}(t_1)e^{-i\hat{H}_0t_1/\hbar} \mid n\right\rangle\right.$$
$$\left.+\left\langle n \mid e^{+i\hat{H}_0t_1/\hbar}\hat{V}(t_1)e^{-i\hat{H}_0t_1/\hbar} \mid m\right\rangle\left\langle m \mid e^{+i\hat{H}_0t_2/\hbar}\hat{V}(t_2)e^{-i\hat{H}_0t_2/\hbar} \mid n\right\rangle\right\}$$
$$= \frac{1}{\hbar^2}\int_0^t dt_1\int_0^{t_1} dt_2\left\{e^{-i\omega_{nm}(t_1-t_2)}\left\langle n \mid \hat{V}(t_2) \mid m\right\rangle\left\langle m \mid \hat{V}(t_1) \mid n\right\rangle\right.$$
$$\left.+e^{+i\omega_{nm}(t_1-t_2)}\left\langle n \mid \hat{V}(t_1) \mid m\right\rangle\left\langle m \mid \hat{V}(t_2) \mid n\right\rangle\right\}. \tag{7.61}$$

Let us now compare this expression with the first-order contribution to $|c_{nm}(t)|^2$, namely, $\left|c_{nm}^{(1)}(t)\right|^2$. Since we are considering the most general case of an explicitly time-dependent perturbation, we must use the square of Eq. 6.25, i.e.,

$$\left|c_{nm}^{(1)}(t)\right|^2 = \frac{1}{\hbar^2}\left|\int_0^t dt_1\langle n \mid \hat{V}(t_1) \mid m\rangle e^{i\omega_{nm}t_1}\right|^2, \tag{7.62}$$

which can also be rewritten as

$$\left|c_{nm}^{(1)}(t)\right|^2 = \frac{1}{\hbar^2}\int_0^t dt_1\int_0^t dt_2\langle n \mid \hat{V}(t_1) \mid m\rangle\langle m \mid \hat{V}(t_2) \mid n\rangle e^{i\omega_{nm}(t_1-t_2)}. \tag{7.63}$$

The integration should be performed over the square region in Fig. 7.2. If one breaks up the integration region into two triangles as shown, the result is the following sum of two integrals:

$$\left|c_{nm}^{(1)}(t)\right|^2 = \frac{1}{\hbar^2}\int_0^t dt_1\int_0^{t_1} dt_2\left\langle n \mid \hat{V}(t_1) \mid m\right\rangle\left\langle m \mid \hat{V}(t_2) \mid n\right\rangle e^{i\omega_{nm}(t_1-t_2)}$$
$$+\frac{1}{\hbar^2}\int_0^t dt_2\int_0^{t_2} dt_1\left\langle m \mid \hat{V}(t_2) \mid n\right\rangle$$
$$\times\left\langle n \mid \hat{V}(t_1) \mid m\right\rangle e^{i\omega_{nm}(t_1-t_2)}. \tag{7.64}$$

Finally, since t_1 and t_2 act as dummy variables, they can be interchanged in the second integral. The result is

$$\left|c_{nm}^{(1)}(t)\right|^2 = \frac{1}{\hbar^2}\int_0^t dt_1\int_0^{t_1} dt_2\left\langle n \mid \hat{V}(t_1) \mid m\right\rangle\left\langle m \mid \hat{V}(t_2) \mid n\right\rangle e^{+i\omega_{nm}(t_1-t_2)}$$
$$+\frac{1}{\hbar^2}\int_0^t dt_1\int_0^{t_1} dt_2\left\langle n \mid \hat{V}(t_2) \mid m\right\rangle$$
$$\times\left\langle m \mid \hat{V}(t_1) \mid n\right\rangle e^{-i\omega_{nm}(t_1-t_2)}, \tag{7.65}$$

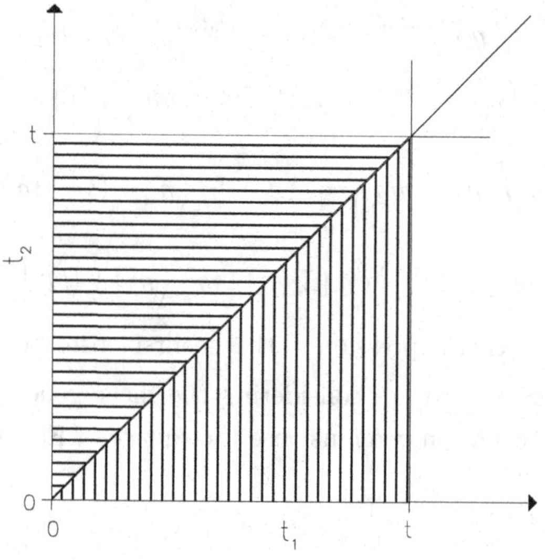

Figure 7.2 Region of integration in Eq. 7.63.

which is identical to the expression for $\langle n \mid \hat{\rho}(t) \mid n\rangle$ given by Eq. 7.61. So, in this case, we see that the second-order perturbation expansion of the density operator gives the same result as the first-order expansion of $\hat{U}_I(t)$.

Example 7.3 Transitions Due to a Randomly Fluctuating Disturbance. A situation that often arises is one where $\hat{V}(t)$ represents a random perturbation. For example, later on when we discuss nuclear magnetic resonance in Chapters 10 and 13, the influence of thermally-induced random magnetic-field fluctuations on a nuclear-spin system will be considered.

Consider a system in pure eigenstate $|m\rangle$ at $t = 0$. For a system under the influence of some random perturbation, we shall try to find W_{nm}, the probability per unit time for making a transition to state $|n\rangle$. To do this, it proves useful to begin with Eq. 7.61 (or equivalently Eq. 7.65) for the transition probability, and then take the time derivative to obtain

$$\begin{aligned} W_{nm} &= \frac{d}{dt}\langle n \mid \hat{\rho}(t) \mid n\rangle \\ &= \frac{1}{\hbar^2}\int_0^t dt' \left\{ e^{i\omega_{nm}(t'-t)} \left\langle n \mid \hat{V}(t') \mid m\right\rangle \left\langle m \mid \hat{V}(t) \mid n\right\rangle \right. \\ &\qquad \left. + e^{i\omega_{nm}(t-t')} \left\langle n \mid \hat{V}(t) \mid m\right\rangle \left\langle m \mid \hat{V}(t') \mid n\right\rangle \right\}. \end{aligned} \tag{7.66}$$

The stochastic nature of the disturbance can be handled by performing an ensemble average (denoted here by a bar over the matrix elements):

$$\begin{aligned} W_{nm} &= \frac{1}{\hbar^2}\int_0^t dt' \left\{ e^{i\omega_{nm}(t'-t)}\overline{\langle n \mid \hat{V}(t') \mid m\rangle\langle m \mid \hat{V}(t) \mid n\rangle} \right. \\ &\qquad \left. + e^{i\omega_{nm}(t-t')}\overline{\langle n \mid \hat{V}(t) \mid m\rangle\langle m \mid \hat{V}(t') \mid n\rangle} \right\}. \end{aligned} \tag{7.67}$$

Now introduce the change of variables $\tau = t - t'$, so that

$$\begin{aligned} W_{nm} &= \frac{1}{\hbar^2}\int_0^t d\tau \left\{ e^{-i\omega_{nm}\tau}\overline{\langle n \mid \hat{V}(t-\tau) \mid m\rangle\langle m \mid \hat{V}(t) \mid n\rangle} \right. \\ &\qquad \left. + e^{+i\omega_{nm}\tau}\overline{\langle m \mid \hat{V}(t-\tau) \mid n\rangle\langle n \mid \hat{V}(t) \mid m\rangle} \right\}. \end{aligned} \tag{7.68}$$

The term $\overline{\langle n \mid \hat{V}(t-\tau) \mid m\rangle\langle m \mid \hat{V}(t) \mid n\rangle}$ is an example of a *time-correlation function.* We will assert that the random fluctuations represented by $\hat{V}(t)$ are *stationary* in time. This essentially means that the underlying mechanism responsible for the fluctuations is independent of time and, as a consequence, the correlation function is independent of shifts in t. Only the variable τ comes into play, so the time-correlation function can be denoted as

$$G_{nm}(\tau) = \overline{\langle n \mid \hat{V}(t-\tau) \mid m\rangle\langle m \mid \hat{V}(t) \mid n\rangle}. \tag{7.69}$$

This allows us to rewrite Eq. 7.68 more compactly:

$$W_{nm} = \frac{1}{\hbar^2}\int_0^t \left[G_{nm}(\tau)e^{-i\omega_{nm}\tau} + G_{mn}(\tau)e^{+i\omega_{nm}\tau} \right] d\tau. \tag{7.70}$$

But note that

$$\begin{aligned} G_{mn}(\tau) &= \overline{\langle m \mid \hat{V}(t-\tau) \mid n\rangle\langle n \mid \hat{V}(t) \mid m\rangle} \\ &= \overline{\langle m \mid \hat{V}(t) \mid n\rangle\langle n \mid \hat{V}(t+\tau) \mid m\rangle} \\ &= \overline{\langle n \mid \hat{V}(t+\tau) \mid m\rangle\langle m \mid \hat{V}(t) \mid n\rangle} \\ &= G_{nm}(-\tau). \end{aligned} \tag{7.71}$$

Therefore, W_{nm} becomes

$$\begin{aligned} W_{nm} &= \frac{1}{\hbar^2}\int_0^t \left[G_{nm}(\tau)e^{-i\omega_{nm}\tau} + G_{nm}(-\tau)e^{+i\omega_{nm}\tau} \right] d\tau \\ &= \frac{1}{\hbar^2}\int_{-t}^{+t} G_{nm}(\tau)e^{-i\omega_{nm}\tau} d\tau. \end{aligned} \tag{7.72}$$

A non-periodic $\hat{V}(t)$ will give rise to a correlation function that decays with some finite *correlation time* τ_c, e.g., $G_{nm}(\tau) \sim e^{-|\tau|/\tau_c}$. Then for $t \gg \tau_c$, one may effectively take

$$W_{nm} = \frac{1}{\hbar^2} J_{nm}(\omega_{nm}), \tag{7.73}$$

where

$$J_{nm}(\omega) = \int_{-\infty}^{+\infty} G_{nm}(\tau) e^{-i\omega\tau} d\tau. \tag{7.74}$$

$J_{nm}(\omega)$, known as the *power spectral density* of the fluctuations, is the Fourier transform of $G_{nm}(\tau)$. The inverse transform is

$$G_{nm}(\tau) = \frac{1}{2\pi} \int_{-\infty}^{+\infty} J_{nm}(\omega) e^{+i\omega\tau} d\omega. \tag{7.75}$$

The fact that the correlation function and the power spectrum of the fluctuations are a Fourier transform pair is known as the *Wiener-Khintchine theorem*. The typical behaviors of the functions $V(t)$, $G_{nm}(\tau)$, and $J_{nm}(\omega)$ are sketched in Fig. 7.3. If, in fact, $G_{nm}(\tau)$ decays exponentially, then $J_{nm}(\omega)$ has a Lorentzian form, i.e.,

$$J_{nm}(\omega) \sim \frac{1}{1 + \tau_c^2 \omega^2}. \tag{7.76}$$

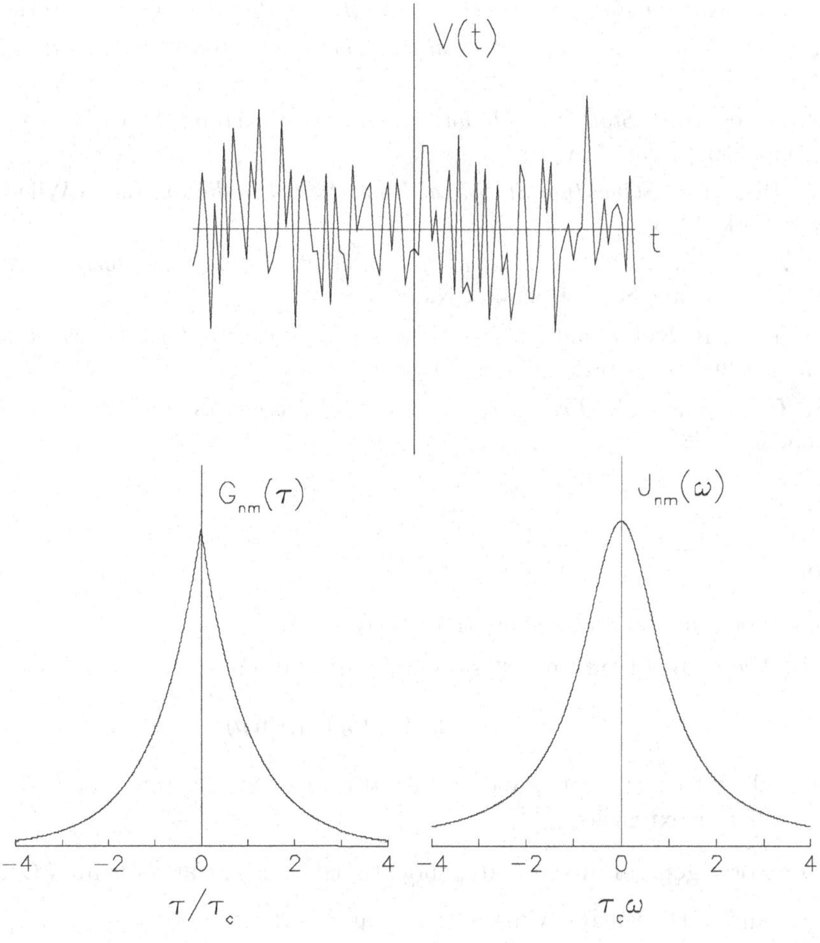

Figure 7.3 Randomly fluctuating $V(t)$ along with a representative (exponential) time-correlation function $G_{nm}(\tau)$ and (Lorentzian) power spectral density $J_{nm}(\omega)$.

Suggested References

More on quantum statistical physics and the density-operator formalism can be found in

[a] C. Kittel, *Elementary Statistical Physics* (Dover Publications, New York, 2004).

[b] F. Reif, *Fundamentals of Statistical and Thermal Physics* (McGraw-Hill, New York, 1965).

[c] R. P. Feynman, *Statistical Mechanics: A Set of Lectures*, 2nd ed. (Perseus Books Group, Cambridge, MA, 1998).

[d] W. H. Louisell, *Quantum Statistical Properties of Radiation* (John Wiley and Sons, New York, 1973).

[e] J. Weiner and P.-T. Ho, *Light-Matter Interaction: Fundamentals and Applications* (John Wiley and Sons, Hoboken, NJ, 2003).

[f] M. Toda, R. Kubo, and N. Saito, *Statistical Physics I: Equilibrium Statistical Mechanics*, 2nd ed. (Springer-Verlag, Berlin, 1992).

[g] R. C. Tolman, *The Principles of Statistical Mechanics* (Oxford University Press, London, 1942).

Problems

1. (a) For a mixed state, show that $\mathrm{Tr}\left(\hat{\rho}^2\right) < 1$.

   (b) From the quantum statistical definition of entropy,

   $$S = -k_B \mathrm{Tr}\left(\hat{\rho} \ln \hat{\rho}\right),$$

   show that the entropy of a pure state vanishes. (Hint: $\ln \hat{\rho}$ is defined by a Taylor expansion.)

   (c) For a general mixed state, show that the expectation value $\left\langle \hat{O}^\dagger \hat{O} \right\rangle$ is real and non-negative, where $\hat{O}$ is an arbitrary operator.

2. Consider a collection of atoms, each having only two levels, $|1\rangle$ and $|2\rangle$, with corresponding energies $\hbar\omega_1 < \hbar\omega_2$. 30% of the atoms are described by the state

   $$|\psi_A(t)\rangle = \frac{1}{\sqrt{2}}\left(e^{-i\omega_1 t}|1\rangle + e^{-i\omega_2 t}|2\rangle\right),$$

   50% are described by

   $$|\psi_B(t)\rangle = \frac{1}{\sqrt{10}}\left(e^{-i\omega_1 t}|1\rangle - 3e^{-i\omega_2 t}|2\rangle\right),$$

and 20% are described by

$$|\psi_C(t)\rangle = e^{-i\omega_2 t}|2\rangle .$$

(a) Determine the density matrix for this system using $|1\rangle$ and $|2\rangle$ as basis states.

(b) Show that $\hat{\rho}^2 \neq \hat{\rho}$.

(c) Find the probability that this system is in state $|\psi_A\rangle$.

3. The Helmholtz free energy $F(T, N, V)$ of a thermodynamic system is derivable from its partition function $Z(T, N, V)$ by

$$F = -k_B T \ln Z.$$

(a) For a single mode of the radiation field in thermal equilibrium at temperature T, show that

$$F = k_B T \ln\left(1 - e^{-\beta\hbar\omega}\right).$$

(In deriving this expression, the zero-point energy is ignored.) In the case of the multimode blackbody EM radiation field, the total free energy is obtained by simply integrating this expression over the various radiation modes, i.e.,

$$F = k_B T \int_0^\infty d\omega\, g(\omega) \ln\left(1 - e^{-\beta\hbar\omega}\right),$$

where

$$g(\omega)\, d\omega = \frac{8\pi V}{(2\pi c)^2}\omega^2 d\omega$$

is the mode density (see Eq. 5.7).

(b) Given the free energy, one can compute the pressure of a system by using the following basic relation from classical thermodynamics:

$$P = -\left(\frac{\partial F}{\partial V}\right)_{N,T}.$$

Show that the "radiation pressure" of the multimode "photon gas" is given by

$$P = \left(\frac{\pi^2 k_B^4}{45 c^3 \hbar^3}\right) T^4.$$

Notice that this is exactly one-third of the energy density of the gas (see Eq. 7.39).

4. (a) For a single mode of EM radiation in thermal equilibrium at temperature T, we know that the photon-count probability, $P(n)$, is given by the Bose-Einstein distribution of Eq. 7.35. From the definition of the density operator

$$\hat{\rho} = \sum_{n=0}^{\infty} P(n) |n\rangle \langle n|,$$

show that

$$\langle \alpha \mid \hat{\rho} \mid \alpha \rangle = \frac{1}{1+\langle n \rangle} e^{-|\alpha|^2/(1+\langle n \rangle)},$$

where $|\alpha\rangle$ is a coherent state.

(b) Define an antinormal ordering quasi-probability density function by

$$P_A(\alpha, \alpha^*) \equiv \frac{1}{\pi} \langle \alpha \mid \hat{\rho} \mid \alpha \rangle.$$

Show that

$$1 = \mathrm{Tr}(\hat{\rho}) = \frac{1}{\pi} \int \langle \alpha \mid \hat{\rho} \mid \alpha \rangle d^2\alpha = \int P_A(\alpha, \alpha^*) \, d^2\alpha,$$

where the integrals are over the complex plane.

(c) Introduce an associated characteristic function $C_A(\lambda, \lambda^*)$ defined by

$$C_A(\lambda, \lambda^*) = \mathrm{Tr}\left(\hat{\rho} e^{-\lambda^* \hat{a}} e^{\lambda \hat{a}^\dagger}\right).$$

Show that it is related to $P_A(\alpha, \alpha^*)$ by

$$C_A(\lambda, \lambda^*) = \int P_A(\alpha, \alpha^*) \, e^{\lambda \alpha^* - \lambda^* \alpha} d^2\alpha.$$

(d) Show that

$$C_A(\lambda, \lambda^*) = e^{-(1+\langle n \rangle)|\lambda|^2}.$$

(Hint: Evaluate the integral over the complex plane by letting $\alpha = \alpha_1 + i\alpha_2$ and $\lambda = \frac{1}{2}(v - iu)$, and note that the resulting integrals over α_1 and α_2 are Fourier transforms.)

(e) We next introduce the normal ordering characteristic function

$$C_N(\lambda, \lambda^*) = \mathrm{Tr}\left(\hat{\rho} e^{\lambda \hat{a}^\dagger} e^{-\lambda^* \hat{a}}\right).$$

By making use of the Baker-Hausdorff theorem (see Eq. 5.128), show that

$$C_N(\lambda, \lambda^*) = C_A(\lambda, \lambda^*) \, e^{|\lambda|^2} = e^{-\langle n \rangle |\lambda|^2}.$$

(f) Show that

$$C_N(\lambda, \lambda^*) = \int P_N(\alpha, \alpha^*)\, e^{\lambda\alpha^* - \lambda^*\alpha} d^2\alpha,$$

where $P_N(\alpha, \alpha^*)$ is called the *P-representation* of the density operator, defined as

$$\hat{\rho} = \int d^2\alpha\, P_N(\alpha, \alpha^*) |\alpha\rangle \langle\alpha| .$$

The function $P_N(\alpha, \alpha^*)$ associated with the coherent states is analogous to the function $P(n)$ associated with the occupation number states.

(g) Show that the relation between $C_N(\lambda, \lambda^*)$ and $P_N(\alpha, \alpha^*)$ can be inverted to give

$$P_N(\alpha, \alpha^*) = \frac{1}{\pi^2} \int C_N(\lambda, \lambda^*)\, e^{\lambda^*\alpha - \lambda\alpha^*} d^2\lambda.$$

(h) Finally, show that

$$P_N(\alpha, \alpha^*) = \frac{1}{\pi \langle n\rangle} e^{-|\alpha|^2/\langle n\rangle}.$$

This result shows that the P-representation of the density operator of a single-mode radiation field in thermal equilibrium is a Gaussian centered about the origin, with a width proportional to $\sqrt{\langle n\rangle}$.

5. Let us revisit the HCl infrared absorption spectrum previously shown in Fig. 3.10 (Problem 4, end of Chapter 3). Starting at the center of the spectrum and moving outward, notice that the intensity of the absorption peaks increases at first, goes through a maximum, and then drops off as the rotational angular momentum quantum number l of the lower energy state increases in value.

(a) Show that, compared to the lower vibrational state, the population of the upper vibrational state is negligibly low at room temperature. This means that the condition for a maximum peak intensity is that the number of molecules occupying the lower state in the transition is a maximum.

(b) Since, for a given value of l, there is a $(2l + 1)$-fold degeneracy, the higher-l states are statistically enhanced. However, at the same time, the population of higher-energy states is reduced by the Boltzmann factor. Given these two competing effects, show that, at temperature T, the value of l associated with the highest population is given by

$$l_{\max} = \frac{1}{2}\left(\frac{\sqrt{4Ik_BT}}{\hbar} - 1\right),$$

where I is the moment of inertia of the molecule. Calculate this value and comment on the result in the context of the observed HCl absorption spectrum.

Chapter 8
FIRST-ORDER RADIATION PROCESSES

We now begin our study of specific types of photon-atom interactions. The probability for the occurrence of a given type of process is governed by the interaction between the radiation field and the individual electrons bound within the atom, as represented by the Hamiltonian (see Eq. 4.102)

$$\hat{V} = \sum_i \left(-\frac{e}{mc} \hat{\mathbf{A}}_i \cdot \hat{\mathbf{p}}_i + \frac{e^2}{2mc^2} \hat{A}_i^2 \right), \tag{8.1}$$

where m is the electronic mass. The subscript i denotes a particular atomic electron having a momentum denoted by the operator $\hat{\mathbf{p}}_i$, and $\hat{\mathbf{A}}_i$ indicates the vector potential of the field evaluated at the location of electron i. The focus of this chapter is solely on processes arising from the interaction term $(-e/mc)\, \hat{\mathbf{A}}_i \cdot \hat{\mathbf{p}}_i$ that also produce a non-vanishing result for the first-order time-dependent perturbation calculation (as previously outlined in Chapter 6). We say that processes of this type are both *first-order in the interaction* and *first-order in the perturbation*. In these cases, the matrix element for the interaction $\langle n \mid \hat{V} \mid m \rangle$ is proportional to e, so the transition probability is proportional to e^2. In other words, the transition rate is proportional to the *fine structure constant* $\alpha = e^2/\hbar c = 1/137$. On the other hand, second-order processes, i.e., those that are second-order in the interaction (arising from the term $(e^2/2mc^2)\, \hat{A}_i^2$) and/or second-order in the perturbation (where one must consider matrix elements involving intermediate states), are characterized by transition rates proportional to e^4, or α^2—these will be discussed in the next chapter.

Processes that are first-order include the following:

1. Creation of a photon by an electron (or positron), e.g.,

   (a) emission of a photon by a bound electron in an atom

   (b) *bremsstrahlung*—the scattering of an electron by an external field, accompanied by the emission of a photon

2. Annihilation of a photon by an electron (or positron), e.g.,

   (a) absorption of a photon by a bound electron in an atom

(b) photoelectric effect

(c) *inverse bremsstrahlung*

3. Annihilation or creation of an electron-positron pair

In this chapter, we examine some of these processes, namely the emission and absorption of photons by atomic electrons and the photoelectric effect. Also, since it is a topic so closely tied to the physics of emission and absorption, we have included a section on radiation linewidth. The origins of electron-positron annihilation and pair production, as well as the subject of bremsstrahlung, are not presented in this book, but the interested reader is referred to Heitler's book [35].

8.1 Emission and Absorption of Photons by Atoms and Molecules

8.1.1 Emission

An atom or molecule in an excited state can lower its energy by emitting a photon. As illustrated in Fig. 8.1(a), the de-excitation of the atom from a state $|\mathcal{A}\rangle$ to a state $|\mathcal{B}\rangle$ is accompanied by the excitation of one mode of the radiation field from state $|n_{\mathbf{k}\lambda}\rangle$ to $|n_{\mathbf{k}\lambda}+1\rangle$. The emission process can be represented very simply by means of a *Feynman diagram*, as depicted in Fig. 8.1(b). Here, the straight lines correspond to the initial and final atomic states and the wiggly line indicates the emitted photon. It is understood that moving upward on the diagram is equivalent to proceeding later in time. A *vertex*, or junction of lines, appears for each interaction matrix element that contributes to the transition probability for the process. For example, for a process that is first-order in the perturbation, as in the present case, just one vertex is present since only the single matrix element $\langle n \mid \hat{V} \mid m\rangle$ between the initial and final states is involved. On the other hand, for processes that are second-order in the perturbation, such as the photon-scattering events to be described in the next chapter, there are transitions to an intermediate state $|l\rangle$. Consequently, the transition rate depends on the matrix elements $\langle n \mid \hat{V} \mid l\rangle$ and $\langle l \mid \hat{V} \mid m\rangle$, and the result is that two vertices appear on the Feynman diagram.

For the photon emission process of Fig. 8.1, the initial and final states of the combined atom-field system are $|m\rangle = |\mathcal{A}\rangle\,|n_{\mathbf{k}\lambda}\rangle$ and $|n\rangle = |\mathcal{B}\rangle\,|n_{\mathbf{k}\lambda}+1\rangle$, respectively. The transition is brought about by the interaction Hamiltonian as given by Eq. 8.1, with the vector potential of the field at the location of electron i given by Eq. 5.37, i.e.

$$\hat{\mathbf{A}}_i = \sqrt{\frac{2\pi\hbar c^2}{L^3\omega_k}}\left(\hat{a}_{\mathbf{k}\lambda}e^{i\mathbf{k}\cdot\mathbf{r}_i} + \hat{a}^\dagger_{\mathbf{k}\lambda}e^{-i\mathbf{k}\cdot\mathbf{r}_i}\right)\boldsymbol{\epsilon}_{\mathbf{k}\lambda}. \tag{8.2}$$

For the initial stages of the calculation, it is necessary to imagine that both the atom and radiation field are enclosed in a box of volume L^3, however we will shortly see that the box size does not appear in the final result for the photon emission rate. The probability per unit time that the atom emits a single photon of wavevector $\mathbf{k}$

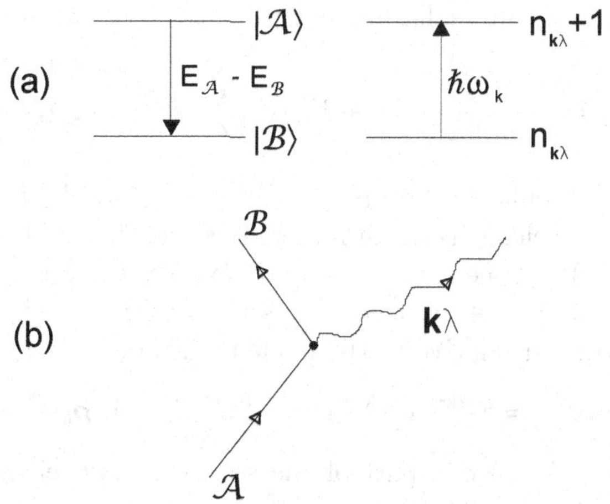

Figure 8.1 Photon emission by atom or molecule. (a) De-excitation of atom (or molecule), accompanied by an excitation of the radiation field. (b) Corresponding Feynman diagram.

and polarization λ is given by Eq. 6.35 for a first-order interaction, namely, Fermi's golden rule:

$$W^{(\mathbf{k}\lambda)}_{\text{emission}} = \frac{2\pi}{\hbar} \left| \langle \mathcal{B} | \langle n_{\mathbf{k}\lambda} + 1 \mid \hat{V} \mid n_{\mathbf{k}\lambda} \rangle | \mathcal{A} \rangle \right|^2 g(E_n). \tag{8.3}$$

Note immediately that only the first-order interaction term, $(-e/mc)\, \hat{\mathbf{A}}_i \cdot \hat{\mathbf{p}}_i$, produces a non-vanishing matrix element between the specified initial and final states (the term $(e^2/2mc^2)\, \hat{A}_i^2$ will contribute to the second-order processes that will be discussed in the next chapter). Thus, the transition probability (per unit time) is

$$W^{(\mathbf{k}\lambda)}_{\text{emission}} = \frac{2\pi}{\hbar} \frac{e^2}{m^2} \frac{2\pi\hbar}{L^3\omega_k} (n_{\mathbf{k}\lambda} + 1) \left| \sum_i \langle \mathcal{B} \mid e^{-i\mathbf{k}\cdot\mathbf{r}_i} \boldsymbol{\epsilon}_{\mathbf{k}\lambda} \cdot \hat{\mathbf{p}}_i \mid \mathcal{A} \rangle \right|^2 g(E_n). \tag{8.4}$$

$g(E_n)dE_n$, the number of final states in an energy interval dE_n, is equivalent to the number of states available to a photon in frequency interval $d\omega_k$, i.e.,

$$g(E_n)dE_n = g(\omega_k)d\omega_k. \tag{8.5}$$

Using Eq. 5.7 and the fact that $dE_n = \hbar d\omega_k$, the density-of-states factor becomes

$$g(E_n) = \frac{1}{c\hbar} \left(\frac{L}{2\pi} \right)^3 \left(\frac{\omega_k}{c} \right)^2 d\Omega. \tag{8.6}$$

This result is a factor of 2 less than expected from Eq. 5.7 because here we are only considering a single polarization. From Eqs. 8.4 and 8.6 we obtain the emission rate of photons of type $(\mathbf{k},\lambda)$ into solid angle $d\Omega$:

$$W^{(\mathbf{k}\lambda)}_{\text{emission}} = \frac{e^2\omega_k}{2\pi\hbar c^3 m^2}(n_{\mathbf{k}\lambda}+1)\left|\langle\mathcal{B}\mid\sum_i e^{-i\mathbf{k}\cdot\mathbf{r}_i}\boldsymbol{\epsilon}_{\mathbf{k}\lambda}\cdot\hat{\mathbf{p}}_i\mid\mathcal{A}\rangle\right|^2 d\Omega. \tag{8.7}$$

The ensuing calculation becomes greatly simplified by focusing our attention on photons having wavelengths much larger than the size of the atom (e.g., photons in the optical regime). To see this, suppose we express the positions of the individual atomic electrons as $\mathbf{r}_i = \mathbf{R} + \boldsymbol{\rho}_i$, where $\mathbf{R}$ is the position of the atomic nucleus and $\boldsymbol{\rho}_i$ is electron i's position relative to the nucleus. Then,

$$e^{-i\mathbf{k}\cdot\mathbf{r}_i} = e^{-i\mathbf{k}\cdot\mathbf{R}}e^{-i\mathbf{k}\cdot\boldsymbol{\rho}_i} = e^{-i\mathbf{k}\cdot\mathbf{R}}\left(1 - i\mathbf{k}\cdot\boldsymbol{\rho}_i + \cdots\right). \tag{8.8}$$

For photons in, say, the visible part of the spectrum (wavelength $\sim$ 400–700 nm), $\mathbf{k}\cdot\boldsymbol{\rho}_i \ll 1$, and we can make the replacement

$$e^{-i\mathbf{k}\cdot\mathbf{r}_i} \simeq e^{-i\mathbf{k}\cdot\mathbf{R}}. \tag{8.9}$$

This is known as the *electric dipole approximation.* In addition, we also derive the following useful identity:

$$\langle\mathcal{B}\mid\hat{\mathbf{p}}_i\mid\mathcal{A}\rangle = -im\omega_k\langle\mathcal{B}\mid\hat{\mathbf{r}}_i\mid\mathcal{A}\rangle. \tag{8.10}$$

Consider the commutator between $\hat{H}_0$, the Hamiltonian of the unperturbed atom, and $\hat{r}_{ik}$, the operator for the kth Cartesian component of the position of electron i. Since the potential energy contribution to $\hat{H}_0$ is purely a function of the electron coordinates, it does not contribute to the commutator, and we have

$$\left[\hat{H}_0, \hat{r}_{ik}\right] = \left[\sum_{i'}\frac{\hat{\mathbf{p}}^2_{i'}}{2m}, \hat{r}_{ik}\right] = \frac{1}{2m}\sum_{i',k'}\left[\hat{p}^2_{i'k'}, \hat{r}_{ik}\right]. \tag{8.11}$$

Reducing the expression still further gives

$$\begin{aligned}\left[\hat{H}_0, \hat{r}_{ik}\right] &= \frac{1}{2m}\sum_{i',k'}\left\{\hat{p}_{i'k'}\left[\hat{p}_{i'k'}, \hat{r}_{ik}\right] + \left[\hat{p}_{i'k'}, \hat{r}_{ik}\right]\hat{p}_{i'k'}\right\} \\ &= -\frac{1}{m}\sum_{i',k'} i\hbar\hat{p}_{i'k'}\delta_{ii'}\delta_{kk'} = -\frac{i\hbar}{m}\hat{p}_{ik}.\end{aligned} \tag{8.12}$$

This now allows us to write

$$\langle\mathcal{B}\mid\hat{p}_{ik}\mid\mathcal{A}\rangle = \frac{im}{\hbar}\langle\mathcal{B}\mid\left[\hat{H}_0, \hat{r}_{ik}\right]\mid\mathcal{A}\rangle = \frac{im}{\hbar}\left(E_{\mathcal{B}} - E_{\mathcal{A}}\right)\langle\mathcal{B}\mid\hat{r}_{ik}\mid\mathcal{A}\rangle. \tag{8.13}$$

Note however that the energy of the emitted photon is $\hbar\omega_k = E_{\mathcal{A}} - E_{\mathcal{B}}$.* Therefore, in vector form, Eq. 8.13 reduces to Eq. 8.10, as desired.

---

*At present, the effects of *linewidth* are being neglected. This subject will be addressed in Sect. 8.2.

Now substitute Eqs. 8.9 and 8.10 into Eq. 8.7 for the transition rate. The expression becomes

$$W^{(\mathbf{k}\lambda)}_{\text{emission}} = \frac{\omega_k^3}{2\pi\hbar c^3}\,(n_{\mathbf{k}\lambda}+1)\,|\langle\mathcal{B}\mid\boldsymbol{\epsilon}_{\mathbf{k}\lambda}\cdot\widehat{\boldsymbol{\mu}}\mid\mathcal{A}\rangle|^2\,d\Omega, \tag{8.14}$$

where we have introduced the operator for the electric dipole moment of the atom, given by

$$\widehat{\boldsymbol{\mu}} = e\sum_i \hat{\mathbf{r}}_i. \tag{8.15}$$

A more compact way of writing Eq. 8.14 is to use the notation $\boldsymbol{\mu}_{\mathcal{BA}} \equiv \langle\mathcal{B}\mid\widehat{\boldsymbol{\mu}}\mid\mathcal{A}\rangle$ to represent the dipole-moment matrix element associated with the atomic transition.† Then,

$$W^{(\mathbf{k}\lambda)}_{\text{emission}} = \frac{\omega_k^3}{2\pi\hbar c^3}\,(n_{\mathbf{k}\lambda}+1)\,|\boldsymbol{\epsilon}_{\mathbf{k}\lambda}\cdot\boldsymbol{\mu}_{\mathcal{BA}}|^2\,d\Omega. \tag{8.16}$$

At this point, it is left to the reader to show that, in fact, Eq. 8.16 could have been obtained in a more direct fashion by using the interaction Hamiltonian $-\widehat{\boldsymbol{\mu}}\cdot\hat{\mathbf{E}}$ instead of $(-e/mc)\sum_i \hat{\mathbf{A}}_i\cdot\hat{\mathbf{p}}_i$.

Equation 8.16 shows that the emission probability is composed of two parts, i.e.,

$$W^{(\mathbf{k}\lambda)}_{\text{emission}} = W^{(\mathbf{k}\lambda)}_{\text{spon}} + W^{(\mathbf{k}\lambda)}_{\text{stim}}, \tag{8.17}$$

where

$$W^{(\mathbf{k}\lambda)}_{\text{spon}} = \frac{\omega_k^3}{2\pi\hbar c^3}\,|\boldsymbol{\epsilon}_{\mathbf{k}\lambda}\cdot\boldsymbol{\mu}_{\mathcal{BA}}|^2\,d\Omega \tag{8.18}$$

and

$$W^{(\mathbf{k}\lambda)}_{\text{stim}} = \frac{\omega_k^3}{2\pi\hbar c^3}n_{\mathbf{k}\lambda}\,|\boldsymbol{\epsilon}_{\mathbf{k}\lambda}\cdot\boldsymbol{\mu}_{\mathcal{BA}}|^2\,d\Omega. \tag{8.19}$$

$W^{(\mathbf{k}\lambda)}_{\text{spon}}$ represents the production of $(\mathbf{k}\lambda)$-photons by *spontaneous emission* and $W^{(\mathbf{k}\lambda)}_{\text{stim}}$ is the contribution by *stimulated emission*. We now discuss each of these contributions in turn:

Spontaneous Emission

This process can occur whether or not an external field is present—that is to say, a photon can be produced by a downward atomic transition even in the total absence of a pre-existing field. Despite the fact that spontaneous emission caused by the atomic transition $|\mathcal{A}\rangle \to |\mathcal{B}\rangle$ produces the single photon frequency $\hbar\omega_k = E_{\mathcal{A}} - E_{\mathcal{B}}$, a variety of field-modes can become excited because the photon's polarization and

†In the initial and final atomic states, the expectation value of the dipole moment vanishes, i.e., $\boldsymbol{\mu}_{\mathcal{AA}} = \boldsymbol{\mu}_{\mathcal{BB}} = 0$, as this is true for any stationary state. However, during a transition between such states, an oscillating non-zero dipole moment $\boldsymbol{\mu}_{\mathcal{BA}}$ develops giving rise to electric dipole radiation from the atom.

emission direction are random. For a given frequency, the transition rate into all possible modes is sometimes referred to as *Einstein's "A_e" coefficient of spontaneous emission*, and can be determined in the following manner:

Take the wavevector of the emitted photon, $\mathbf{k}$, to be along the z-axis, with the two choices for the field-polarization vector, $\boldsymbol{\epsilon}_{\mathbf{k}1}$ and $\boldsymbol{\epsilon}_{\mathbf{k}2}$, directed along the x- and y-axes, respectively, as shown in Fig. 8.2. The orientation of the dipole-moment matrix element $\boldsymbol{\mu}_{BA}$ is given by the angles θ and ϕ in spherical coordinates. With these conventions, Eq. 8.18 becomes

$$W_{\text{spon}}^{(\mathbf{k}\lambda)} = \frac{\omega_k^3}{2\pi\hbar c^3}\mu_{BA}^2 \cos^2\Theta_\lambda \, d\Omega, \tag{8.20}$$

where $\cos\Theta_\lambda$ is the direction cosine of $\boldsymbol{\mu}_{BA}$ with respect to $\boldsymbol{\epsilon}_{\mathbf{k}\lambda}$. Specifically,

$$\cos\Theta_1 = \sin\theta\cos\phi \qquad \text{and} \qquad \cos\Theta_2 = \sin\theta\sin\phi. \tag{8.21}$$

By summing over the two possible polarizations, the spontaneous emission rate takes the form

$$W_{\text{spon}}^{(\mathbf{k})} = \frac{\omega_k^3}{2\pi\hbar c^3}\mu_{BA}^2 \sin^2\theta \, d\Omega. \tag{8.22}$$

Notice the $\sin^2\theta$ angular distribution of the radiation, just as we previously found for radiation from an oscillating classical electric dipole (see Fig. 4.3). If we integrate this expression over all θ and ϕ to account for the random direction of photon emission, the result is Einstein's A_e coefficient for spontaneous emission:

$$A_e = \frac{4\omega_k^3}{3\hbar c^3}\mu_{BA}^2. \tag{8.23}$$

P, the total power spontaneously emitted by the atom, is obtained by simply multiplying the latter expression by the photon energy $\hbar\omega_k$:

$$P = \frac{4\omega_k^4}{3c^3}\mu_{BA}^2. \tag{8.24}$$

Comparing this result to Eq. 4.135, we again see that our quantum-mechanically obtained expression is very nearly identical to the one derived for a classical dipole, i.e., both expressions for the emitted power are proportional to the fourth-power of the radiation frequency and the square of the dipole moment. At a certain level, it is then natural to think of the emission of EM radiation from an atom as being due to the oscillating dipole moment produced by the circulating atomic electrons. Keep in mind, however, that this is a semiclassical point of view having limited utility.

Equation 8.23 allows one to estimate the lifetime τ of an excited atomic state before it spontaneously decays. For an order-of-magnitude calculation, we can replace μ_{BA} by ea, where a is the approximate linear dimension of the atom. Then

$$\tau^{-1} = A_e \sim \alpha\left(\frac{\omega_k a}{c}\right)^2 \omega_k, \tag{8.25}$$

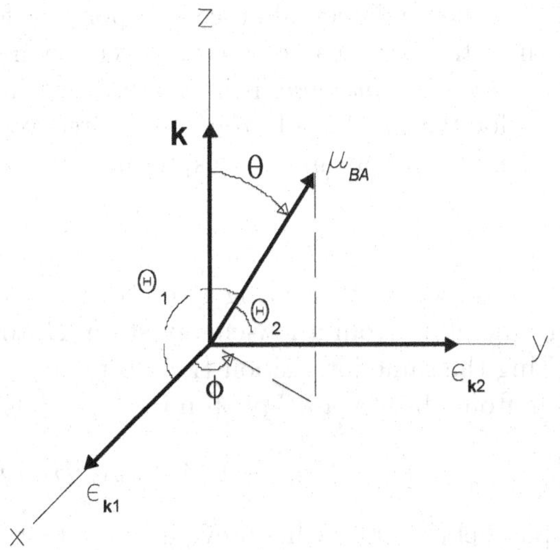

Figure 8.2 Vectors for calculating spontaneous emission probability.

where $\alpha = e^2/\hbar c \approx 1/137$ is the fine structure constant. Furthermore, $\hbar\omega_k = E_{\mathcal{A}} - E_{\mathcal{B}} \sim e^2/a$, or $\omega_k a/c \sim \alpha$, so

$$\tau^{-1} \sim \alpha^3 \omega_k. \tag{8.26}$$

This shows that for excited atomic states leading to transitions in the visible region of the spectrum ($\omega_k \sim 10^{15}$ s^{-1}), the lifetimes are on the order of nanoseconds, while x-ray transitions ($\omega_k \sim 10^{18}$ s^{-1}) correspond to lifetimes of picoseconds.

Stimulated Emission

This process increases the number of photons in an already excited mode of the EM field. As a result, only photons of the same frequency, polarization, and direction-of-travel as those inducing the emission are produced. This is very different from what occurs in spontaneous emission, where the photons produced can populate any number of modes, some previously occupied and some not. Furthermore, as can be seen from Eqs. 8.18 and 8.19, stimulated emission into mode-($\mathbf{k}\lambda$) is $n_{\mathbf{k}\lambda}$-times as likely to occur as spontaneous emission into that mode, i.e.,

$$W_{\text{stim}}^{(\mathbf{k}\lambda)} = n_{\mathbf{k}\lambda} W_{\text{spon}}^{(\mathbf{k}\lambda)}. \tag{8.27}$$

Clearly, as stimulated emission increases the occupation number of a given mode, it becomes even more probable that further emission will occur. The cumulative result is a coherent amplification of that mode. Of course, without additional intervention, the described scenario does not take place because of the more frequent occurrence of photon-absorption by atoms. However, "light amplification by the stimulated emission of radiation" is the fundamental process responsible for the operation of the *laser*—this is where the rate of stimulated emission overwhelms the rate of absorption because a mechanism for *optical pumping* is introduced, which effectively inverts the statistical distribution for the number of atoms in excited versus low-energy states. In lasers, spontaneous emission appears as background noise.

8.1.2 Absorption

Photon absorption is the process where a single photon is removed from the radiation field, thus promoting the atom from a low-energy state, $\mathcal{B}$, to one of higher energy, $\mathcal{A}$, as in Fig. 8.3. Using the same interaction Hamiltonian as before, the probability per unit time that an atom absorbs a $\mathbf{k}\lambda$-photon is

$$W^{(\mathbf{k}\lambda)}_{\text{absorption}} = \frac{2\pi}{\hbar}\left|\langle\mathcal{A}|\,\langle n_{\mathbf{k}\lambda}-1\mid\hat{V}\mid n_{\mathbf{k}\lambda}\rangle\,|\mathcal{B}\rangle\right|^2 g(E_n). \tag{8.28}$$

We can now simply parallel the arguments previously used for obtaining the emission probability. In doing so, it should be noted that now $g(E_n)$ is the density of the final states into which the $(\mathbf{k}\lambda)$-mode spreads as a result of losing a photon. However, this expression is, in fact, equivalent to the one previously used, namely Eq. 8.6. After all is done, the absorption probability per unit time is found to be

$$W^{(\mathbf{k}\lambda)}_{\text{absorption}} = \frac{\omega_k^3}{2\pi\hbar c^3} n_{\mathbf{k}\lambda}\left|\boldsymbol{\epsilon}_{\mathbf{k}\lambda}\cdot\boldsymbol{\mu}_{\mathcal{AB}}\right|^2 d\Omega. \tag{8.29}$$

Example 8.1 Atoms and Radiation in Thermal Equilibrium—the Blackbody Spectrum Revisited. The results of this section can be used to derive the well-known blackbody radiation spectrum for atoms and field exchanging photons under conditions of thermal equilibrium. Previously, in Example 7.1, we considered the specific case of radiation in thermal equilibrium with the walls of an enclosing cavity. At that time recall that the blackbody spectrum was derived without explicit reference to the atomic emission and absorption events occurring in the lining of the cavity. Now, however, the rates for these processes are known. In particular, $W^{(\mathbf{k}\lambda)}_{\text{emission}}$ and $W^{(\mathbf{k}\lambda)}_{\text{absorption}}$ represent the emission and absorption rates *per atom* associated with mode-$(\mathbf{k}\lambda)$ of the field. For radiation confined to a cavity, the polarization and angular-dependence of the photons are random—thus, it is appropriate to consider only $W^{(k)}_{\text{emission}}$ and $W^{(k)}_{\text{absorption}}$, the probabilities for emission and absorption of photons at a particular frequency ω_k. Note that the contribution to $W^{(k)}_{\text{emission}}$ from spontaneous emission was already found to be given by the Einstein coefficient, A_e. The contribution from stimulated emission and the photon absorption rate are both found in an identical manner (refer back to Eqs. 8.20–8.23), i.e., by integrating over all orientations of the photon's wavevector relative to the atomic dipole moment, and adding the contributions from both polarizations. Thus, we quickly find that

$$W^{(k)}_{\text{emission}} = A_e\,(n_k+1) = \frac{4\omega_k^3}{3\hbar c^3}\mu_{\mathcal{BA}}^2\,(n_k+1) \tag{8.30}$$

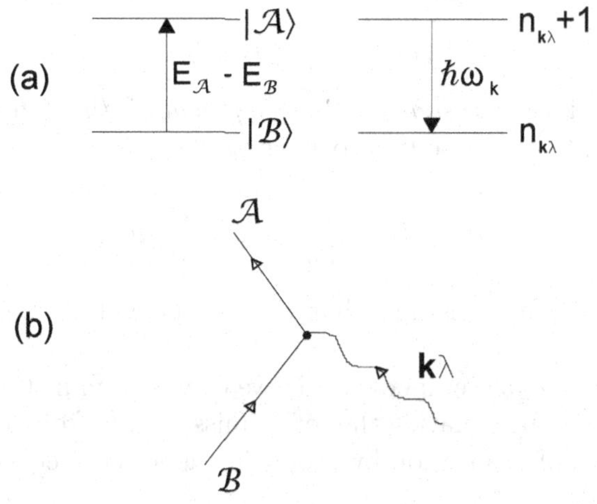

Figure 8.3 Photon absorption by atom or molecule. (a) Excitation of atom (or molecule), accompanied by a de-excitation of the radiation field. (b) Corresponding Feynman diagram.

and

$$W^{(k)}_{\text{absorption}} = n_k \frac{4\omega_k^3}{3\hbar c^3} \mu^2_{\mathcal{AB}}, \tag{8.31}$$

where n_k denotes the total occupation number of all modes belonging to frequency ω_k, irrespective of polarization and propagation direction. The n_k's only depend on the energy density of the field at various frequencies. So, letting $u(\omega_k)\,d\omega_k$ be the energy density in the interval $\omega_k \to \omega_k + d\omega_k$, we write

$$u(\omega_k)\,d\omega_k = \frac{n_k \hbar \omega_k}{L^3} g(\omega_k) d\omega_k. \tag{8.32}$$

Then, by integrating Eq. 5.7 for the density-of-states factor over all solid angles, we see that

$$u(\omega_k)\,d\omega_k = \frac{n_k \hbar \omega_k}{L^3} \left[2 \times 4\pi \times \left(\frac{L}{2\pi c} \right)^3 \omega_k^2 d\omega_k \right] = n_k \hbar \omega_k \left(\frac{\omega_k^2 d\omega_k}{\pi^2 c^3} \right). \tag{8.33}$$

This allows us to replace the photon numbers with an energy density according to

$$n_k = \frac{\pi^2 c^3}{\hbar \omega_k^3} u(\omega_k). \tag{8.34}$$

The emission and absorption rates (per atom) can now be reduced to

$$W^{(k)}_{\text{emission}} = A_{\text{e}} + B_{\text{e}} u(\omega_k) \tag{8.35}$$

and

$$W^{(k)}_{\text{absorption}} = B_{\text{a}} u(\omega_k), \tag{8.36}$$

where we have introduced *Einstein's "B_{e}" coefficient of stimulated emission* and *"B_{a}" coefficient of absorption*. These two coefficients are in fact equal, and given by

$$B_{\text{e}} = B_{\text{a}} = \frac{4\pi^2}{3\hbar^2}\mu^2_{\mathcal{BA}} = \frac{4\pi^2}{3\hbar^2}\mu^2_{\mathcal{AB}}. \tag{8.37}$$

Notice that B_{e} and B_{a} are constants that are independent of frequency and the cavity temperature.

For a cavity at some temperature T, we now determine the radiation spectrum at thermal equilibrium by equating the total emission rate from atoms in excited state $|\mathcal{A}\rangle$ to the total rate of absorption by atoms in the lower energy-state $|\mathcal{B}\rangle$, i.e.,

$$N_{\mathcal{A}} W^{(k)}_{\text{emission}} = N_{\mathcal{B}} W^{(k)}_{\text{absorption}}. \tag{8.38}$$

This procedure is sometimes known as *detailed balancing*. Applying Eqs. 8.35 and 8.36 produces

$$u(\omega_k) = \frac{A_{\text{e}}/B_{\text{a}}}{(N_{\mathcal{B}}/N_{\mathcal{A}}) - 1}. \tag{8.39}$$

From Eqs. 8.23 and 8.37, we have $A_{\text{e}}/B_{\text{a}} = \hbar\omega_k^3/\pi^2c^3$. Furthermore, for atoms in thermal equilibrium, $N_{\mathcal{A}}/N_{\mathcal{B}}$ is given by the Boltzmann factor $\exp[-(E_{\mathcal{A}} - E_{\mathcal{B}})/k_BT] = \exp(-\beta\hbar\omega_k)$. The final result is, as expected, the blackbody spectral distribution (compare with Eq. 7.37):

$$u(\omega_k, T) = \frac{\hbar\omega_k^3/\pi^2c^3}{e^{\beta\hbar\omega_k} - 1}. \tag{8.40}$$

Example 8.2 IR Absorption Spectra: Band Shapes and Molecular Rotations. Each band in an infrared (IR) absorption spectrum appears as the result of transitions between certain vibrational-rotational levels in the molecular system. While the center-frequency of a given band is the signature for a vibrational excitation associated with a particular electronic configuration of the molecules,[‡] the detailed shape of the band is, to a large extent, a reflection of the rotational behavior of the molecules. Because interactions from neighboring molecules affect aspects of the rotational motion, band shapes are highly sensitive to the particular phase of the molecular system and, in the case of molecular solutions, to the choice of background medium (i.e., solvent). In this example we show that, for a given vibrational excitation, the Fourier transform of the absorption band shape produces a rotational time-correlation function for the molecule. The derivation below is an extension of the one presented by Gordon [36].

[‡]Refer to Example 3.5 for a cursory discussion of molecular spectra and, in particular, the Born-Oppenheimer approximation.

Begin by defining an operator for the dipole-moment density at position $\mathbf{r}$ in the absorbing molecular system:

$$\hat{\mathbf{M}}(\mathbf{r}) = \sum_{\ell} \hat{\boldsymbol{\mu}}_{\ell} \delta\left(\mathbf{r} - \mathbf{R}_{\ell}\right). \tag{8.41}$$

$\mathbf{R}_{\ell}$ is the position vector of the ℓth molecule and $\hat{\boldsymbol{\mu}}_{\ell}$ its dipole-moment vector. As alluded to earlier (Sect. 8.1.1), the interaction Hamiltonian can be taken to be

$$\hat{V} = -\sum_{\ell} \hat{\boldsymbol{\mu}}_{\ell} \cdot \hat{\mathbf{E}}\left(\mathbf{R}_{\ell}\right) = -\int \hat{\mathbf{E}}\left(\mathbf{r}\right) \cdot \hat{\mathbf{M}}(\mathbf{r})\, d^3 r. \tag{8.42}$$

For single-photon absorption, the relevant part of the electric-field operator is given by the first term in Eq. 5.38, i.e., $i\sqrt{2\pi\hbar\omega_k/L^3}\hat{a}_{\mathbf{k}\lambda}e^{i\mathbf{k}\cdot\mathbf{r}}\boldsymbol{\epsilon}_{\mathbf{k}\lambda}$, so that the interaction Hamiltonian becomes

$$\begin{aligned} \hat{V} &= -\int i\sqrt{\frac{2\pi\hbar\omega_k}{L^3}}\hat{a}_{\mathbf{k}\lambda}e^{i\mathbf{k}\cdot\mathbf{r}}\boldsymbol{\epsilon}_{\mathbf{k}\lambda}\cdot\hat{\mathbf{M}}(\mathbf{r})\, d^3 r \\ &= -i\sqrt{\frac{2\pi\hbar\omega_k}{L^3}}\hat{a}_{\mathbf{k}\lambda}\boldsymbol{\epsilon}_{\mathbf{k}\lambda}\cdot\hat{\mathbf{M}}(\mathbf{k}), \end{aligned} \tag{8.43}$$

where

$$\hat{\mathbf{M}}(\mathbf{k}) = \int e^{i\mathbf{k}\cdot\mathbf{r}}\hat{\mathbf{M}}(\mathbf{r})d^3 r = \sum_{\ell} \hat{\boldsymbol{\mu}}_{\ell} e^{i\mathbf{k}\cdot\mathbf{R}_{\ell}} \tag{8.44}$$

is the spatial Fourier transform of the molecular system's dipole-moment density. We can now write the square of the matrix element between some initial state $|m\rangle = |E_i; n_{\mathbf{k}}\rangle$ and final state $|n\rangle = |E_f; n_{\mathbf{k}\lambda} - 1\rangle$ as

$$\left|\langle n \mid \hat{V} \mid m\rangle\right|^2 = \frac{2\pi\hbar\omega_k}{L^3} n_{\mathbf{k}\lambda} \left|\langle E_f \mid \boldsymbol{\epsilon}_{\mathbf{k}\lambda}\cdot\hat{\mathbf{M}}(\mathbf{k}) \mid E_i\rangle\right|^2. \tag{8.45}$$

$|E_i\rangle$ and $|E_f\rangle$ denote the unperturbed eigenstates of the molecular system alone. Therefore, according to Fermi's golden rule, the transition rate is

$$W_{fi}\left(\mathbf{k}\lambda\right) = \frac{2\pi}{\hbar}\frac{2\pi\hbar\omega_k}{L^3} n_{\mathbf{k}\lambda} \left|\langle E_f \mid \boldsymbol{\epsilon}_{\mathbf{k}\lambda}\cdot\hat{\mathbf{M}}(\mathbf{k}) \mid E_i\rangle\right|^2 \delta\left(E_f - E_i - \hbar\omega_k\right) \tag{8.46}$$

(the delta-function acts as the density of final states and guarantees that energy is conserved). The quantity that is usually measured is the rate of energy absorption, or the power removed from the incident beam. This is given by

$$P\left(\mathbf{k}\lambda\right) = \frac{2\pi}{\hbar}\frac{2\pi\hbar\omega_k}{L^3} n_{\mathbf{k}\lambda}\hbar\omega_k \frac{1}{\hbar} \left|\langle E_f \mid \boldsymbol{\epsilon}_{\mathbf{k}\lambda}\cdot\hat{\mathbf{M}}(\mathbf{k}) \mid E_i\rangle\right|^2 \delta\left(\frac{E_f - E_i}{\hbar} - \omega_k\right). \tag{8.47}$$

However, from the Fourier representation of the Dirac delta-function

$$\delta\left(\omega\right) = \frac{1}{2\pi}\int_{-\infty}^{+\infty} e^{i\omega t} dt, \tag{8.48}$$

we have

$$\left|\left\langle E_f \mid \boldsymbol{\epsilon}_{\mathbf{k}\lambda}\cdot\hat{\mathbf{M}}(\mathbf{k}) \mid E_i\right\rangle\right|^2 \delta\left(\frac{E_f - E_i}{\hbar} - \omega_k\right)$$

$$= \frac{1}{2\pi}\int_{-\infty}^{+\infty} dt\, e^{-i\omega_k t}\left\langle E_i \mid \boldsymbol{\epsilon}_{\mathbf{k}\lambda}\cdot\hat{\mathbf{M}}(-\mathbf{k}) \mid E_f\right\rangle$$
$$\times\left\langle E_f \mid e^{iE_f t/\hbar}\hat{\mathbf{M}}(\mathbf{k})e^{-E_i t/\hbar}\cdot\boldsymbol{\epsilon}_{\mathbf{k}\lambda} \mid E_i\right\rangle. \tag{8.49}$$

Recall though, that if $\hat{H}_S$ represents the Hamiltonian of the unperturbed molecular system, then $e^{iE_f t/\hbar}\hat{\mathbf{M}}(\mathbf{k})e^{-E_i t/\hbar} = e^{i\hat{H}_S t/\hbar}\hat{\mathbf{M}}(\mathbf{k})e^{-\hat{H}_S t/\hbar} = \hat{\mathbf{M}}(\mathbf{k}, t)$ is the transformation of the dipole-moment operator from the Schrödinger to the Heisenberg picture. Thus, Eq. 8.49 becomes

$$\left|\left\langle E_f \mid \boldsymbol{\epsilon}_{\mathbf{k}\lambda}\cdot\hat{\mathbf{M}}(\mathbf{k}) \mid E_i\right\rangle\right|^2 \delta\left(\frac{E_f - E_i}{\hbar} - \omega_k\right)$$
$$= \frac{1}{2\pi}\int_{-\infty}^{+\infty} dt\, e^{-i\omega_k t}\left\langle E_i \mid \boldsymbol{\epsilon}_{\mathbf{k}\lambda}\cdot\hat{\mathbf{M}}\left(-\mathbf{k}, 0\right) \mid E_f\right\rangle\left\langle E_f \mid \hat{\mathbf{M}}(\mathbf{k}, t)\cdot\boldsymbol{\epsilon}_{\mathbf{k}\lambda} \mid E_i\right\rangle. \tag{8.50}$$

After using this result in Eq. 8.47, we should also average over all initial and final states of the molecular absorber, i.e.,

$$P\left(\mathbf{k}\lambda\right) = \frac{2\pi}{\hbar}\frac{2\pi\hbar\omega_k}{L^3}n_{\mathbf{k}\lambda}\hbar\omega_k\frac{1}{\hbar}\frac{1}{2\pi}\int_{-\infty}^{+\infty} dt\, e^{-i\omega_k t}\sum_{i,f} P(E_i)$$
$$\times\left\langle E_i \mid \boldsymbol{\epsilon}_{\mathbf{k}\lambda}\cdot\hat{\mathbf{M}}\left(-\mathbf{k}, 0\right) \mid E_f\right\rangle\left\langle E_f \mid \hat{\mathbf{M}}(\mathbf{k}, t)\cdot\boldsymbol{\epsilon}_{\mathbf{k}\lambda} \mid E_i\right\rangle. \tag{8.51}$$

The expression is simplified by using the closure relation for the final states, i.e., $\hat{I} = \sum_f |E_f\rangle\langle E_f|$. In addition, we let $P\left(E_i\right)$ be an equilibrium distribution for the initial states. Then the power absorbed from the beam is

$$P\left(\mathbf{k}\lambda\right) = \frac{2\pi}{\hbar}\frac{2\pi\hbar\omega_k}{L^3}n_{\mathbf{k}\lambda}\hbar\omega_k\frac{1}{\hbar}\frac{1}{2\pi}\int_{-\infty}^{+\infty}\left\langle\boldsymbol{\epsilon}_{\mathbf{k}\lambda}\cdot\mathbf{M}(-\mathbf{k}, 0)\mathbf{M}(\mathbf{k}, t)\cdot\boldsymbol{\epsilon}_{\mathbf{k}\lambda}\right\rangle e^{-i\omega_k t}dt. \tag{8.52}$$

In terms of the intensity of the incident beam

$$I\left(\omega_k\right) = \frac{cn_{\mathbf{k}\lambda}\hbar\omega_k}{L^3}, \tag{8.53}$$

the expression is

$$P(\mathbf{k}\lambda) = \frac{4\pi^2}{c\hbar}\omega_k I(\omega_k)\frac{1}{2\pi}\int_{-\infty}^{+\infty}\langle \boldsymbol{\epsilon}_{\mathbf{k}\lambda}\cdot\mathbf{M}(-\mathbf{k},0)\mathbf{M}(\mathbf{k},t)\cdot\boldsymbol{\epsilon}_{\mathbf{k}\lambda}\rangle\, e^{-i\omega_k t}dt. \tag{8.54}$$

We can express the final result as the infrared absorption cross-section, $\sigma_{\mathrm{IR}}(\omega_k)$; it is given by the absorbed power divided by the incident intensity:

$$\sigma_{\mathrm{IR}}(\omega_k) = \frac{P(\mathbf{k}\lambda)}{I(\omega_k)} = \left(\frac{4\pi^2\omega_k}{c\hbar}\right)\frac{1}{2\pi}\int_{-\infty}^{+\infty}\langle \boldsymbol{\epsilon}_{\mathbf{k}\lambda}\cdot\mathbf{M}(-\mathbf{k},0)\mathbf{M}(\mathbf{k},t)\cdot\boldsymbol{\epsilon}_{\mathbf{k}\lambda}\rangle\, e^{-i\omega_k t}dt. \tag{8.55}$$

Consider now the absorption spectrum from an isotropic system like a gas or liquid. The polarization of the incident radiation is then inconsequential, so

$$\sigma_{\mathrm{IR}}(\omega_k) = \left(\frac{4\pi^2\omega_k}{3c\hbar}\right)\frac{1}{2\pi}\int_{-\infty}^{+\infty}\langle \mathbf{M}(-\mathbf{k},0)\cdot\mathbf{M}(\mathbf{k},t)\rangle\, e^{-i\omega_k t}dt. \tag{8.56}$$

At this point, it becomes more convenient to define an *absorption function*,

$$\alpha(\omega_k) = \frac{3c\hbar}{4\pi^2}\omega_k^{-1}\sigma_{\mathrm{IR}}(\omega_k) = \frac{1}{2\pi}\int_{-\infty}^{+\infty}\langle \mathbf{M}(-\mathbf{k},0)\cdot\mathbf{M}(\mathbf{k},t)\rangle\, e^{-i\omega_k t}dt, \tag{8.57}$$

as this is the temporal Fourier transform of the following time-correlation function (see Eq. 8.44):

$$\langle \mathbf{M}(-\mathbf{k},0)\cdot\mathbf{M}(\mathbf{k},t)\rangle = \left\langle \sum_{\ell\ell'} \boldsymbol{\mu}_{\ell'}(0)\cdot\boldsymbol{\mu}_{\ell}(t)\, e^{i\mathbf{k}\cdot[\mathbf{R}_\ell(t)-\mathbf{R}_{\ell'}(0)]}\right\rangle. \tag{8.58}$$

Arguments presented by Gordon [36] show that the cross-terms with $\ell \neq \ell'$ do not contribute to the correlation function. Furthermore, for times probed by IR frequencies, a typical molecular displacement $\mathbf{R}_\ell(t) - \mathbf{R}_\ell(0)$ is much less than the wavelength of the radiation, so $\exp\{i\mathbf{k}\cdot[\mathbf{R}_\ell(t) - \mathbf{R}_\ell(0)]\} \to 1$. The absorption spectrum therefore gives a self-correlation function relating the initial dipole moment of a typical molecule to the dipole moment of the same molecule at a time t later, i.e.,

$$\langle \mathbf{M}(0)\cdot\mathbf{M}(t)\rangle = N\langle \boldsymbol{\mu}(0)\cdot\boldsymbol{\mu}(t)\rangle, \tag{8.59}$$

where N is the number of molecules. The magnitude of $\boldsymbol{\mu}$ is fixed by the transition dipole moment belonging to a particular vibrational excitation—this selects the center frequency, ω_0, of one of the absorption bands. On the other hand, we choose to concentrate solely on the rotational motion of a molecule as given by the time-dependence of the dipole orientation. This is accomplished by relating the absorbed photon frequency ω_k in Eq. 8.57 to the band-center, ω_0, and the frequency relative to the band-center, $\Delta\omega$, i.e.,

$$\omega_k = \omega_0 + \Delta\omega. \tag{8.60}$$

Then, the shape of the absorption function about the band-center, $\alpha(\Delta\omega)$, becomes the spectrum of interest. Fourier transforming $\alpha(\Delta\omega)$ over the extent of the absorption band then produces a purely rotational version of the self-correlation function given by Eq. 8.59:

$$\langle \boldsymbol{\mu}(0) \cdot \boldsymbol{\mu}(t) \rangle \sim \int_{\text{band}} d(\Delta\omega)\, e^{i\Delta\omega t} \alpha(\Delta\omega). \tag{8.61}$$

For example, Fig. 8.4 shows the shape of the fundamental vibration band of the carbon monoxide molecule in two different liquid solvents, along with the resulting rotational correlation function. The latter can be thought of as the mean projection of the vector $\boldsymbol{\mu}$ onto the same vector at a time t earlier (or, for that matter, at a time t later). When dissolved in $CHCl_3$, the correlation function remains positive, indicating that rotation of the CO-molecule past 90^0 relative to the original orientation is highly unlikely because of external torques produced by the solvent. In the case of the n-C_7H_{16} solvent, a small negative correlation exists after a time of about 0.4 picoseconds, so chances are that the molecule will reorient itself past 90^0 after this amount of time. Furthermore, in the case of n-C_7H_{16}, the correlation function settles down to zero after about 1 picosecond. This represents the time it typically takes before the molecular orientation becomes thermally randomized. For the $CHCl_3$ solvent, this time is longer, i.e., it is more like a few picoseconds.

8.2 The Origins of Linewidth

Up to this point it has been assumed that the frequency of emitted or absorbed radiation is determined exactly by knowing the precise energy-difference between the initial and final states of the atom or molecule. However, in actuality, experimentally observed spectral lines exhibit a frequency distribution that is not infinitely sharp. The observed *linewidth* comes about because of a number of physical mechanisms. The most important ones are discussed here.

8.2.1 Natural Linewidth

The mere fact that excited atomic and molecular states have a finite lifetime leads to the existence of a *natural* linewidth for emitted and absorbed radiation. Even though broadening contributions such as those caused by molecular collisions and/or thermal motion usually dominate in a given situation (see Sect. 8.2.2), these effects can often be reduced by decreasing pressure and/or temperature. On the other hand,

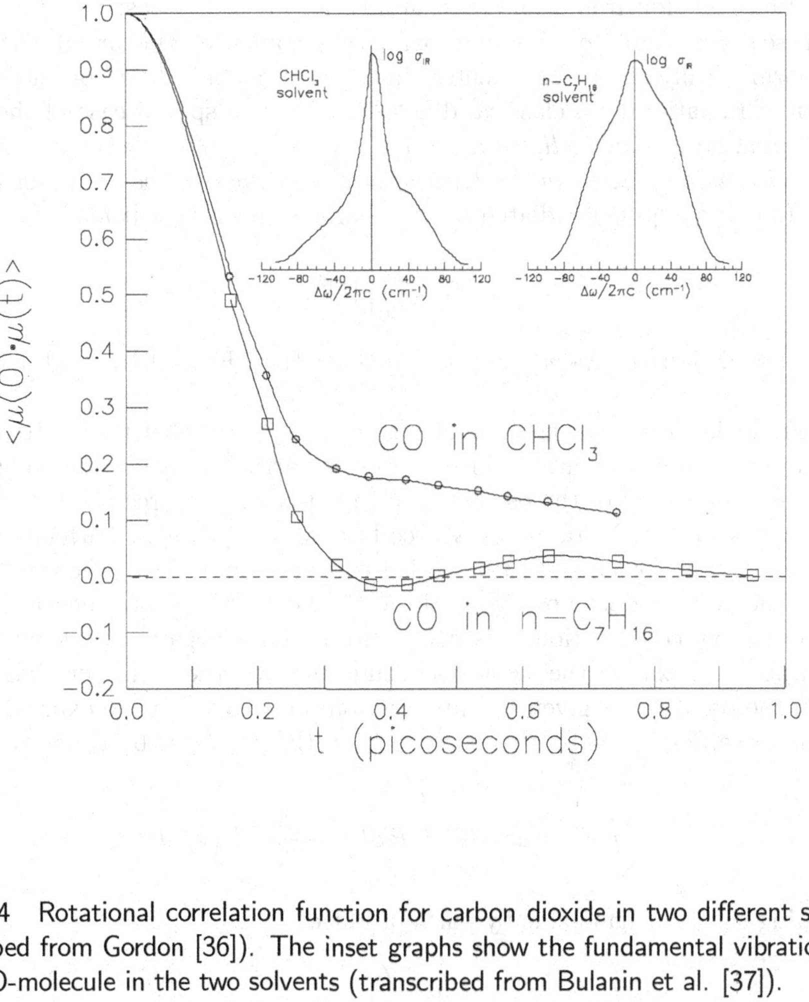

Figure 8.4 Rotational correlation function for carbon dioxide in two different solvents (transcribed from Gordon [36]). The inset graphs show the fundamental vibration band of the CO-molecule in the two solvents (transcribed from Bulanin et al. [37]).

the natural contribution to the linewidth, albeit small, is more fundamental in that it cannot be reduced. As we show below, both the classical and quantum theories of radiation predict the presence of a natural linewidth.

Classical Theory of Radiation Damping

Earlier, in Example 4.4, we learned that a time-varying electric dipole will always radiate away energy in the form of an electromagnetic wave. A semiclassical treatment of radiation emission from molecules, atoms, and nuclei is based on this principle, since these systems are, to a first approximation, composed of charged particles oscillating harmonically about some center. In the next chapter it will be shown that the emission of radiation by a classical dipole is, in fact, a special case of the following more general fact: *Classically, whenever a charged particle experiences an acceleration, it will radiate energy in the form of electromagnetic waves.* Specifically, for an acceleration $\dot{\mathbf{v}}$, the power radiated by the charge is given by *Larmor's formula*,

$$P = \frac{2e^2}{3c^3} |\dot{\mathbf{v}}|^2 . \tag{8.62}$$

For a charged particle undergoing harmonic motion, Eqs. 4.131 and 4.135 clearly follow.

From classical considerations alone, one can predict that the emitted radiation must have a finite linewidth. Since energy is carried away by the radiated field, the mechanical energy of the oscillating charged-particle source must diminish with time—that is to say, the radiation source behaves as a *damped* harmonic oscillator. Said another way, for energy to be added to the radiated field, the oscillator must provide a force that does positive work on the field. As a consequence, in order to maintain energy conservation, the field must perform negative work by exerting a reaction force of exactly the same magnitude back on the oscillator. The damping force on the oscillator is given by this reaction force. To find the expression for the damping force, $\mathbf{F}_D$, consider the work done on the oscillator by the field over some time t:

$$\int_0^t \mathbf{F}_D \cdot \mathbf{v}\, dt = -\int_0^t P\, dt = -\frac{2e^2}{3c^3} \int_0^t |\dot{\mathbf{v}}|^2\, dt. \tag{8.63}$$

Integrating the right-hand side by parts produces

$$\int_0^t \mathbf{F}_D \cdot \mathbf{v}\, dt = -\frac{2e^2}{3c^3} \left(\dot{\mathbf{v}} \cdot \mathbf{v}\, \Big|_0^t - \int_0^t \ddot{\mathbf{v}} \cdot \mathbf{v}\, dt \right) . \tag{8.64}$$

For the time being, let us assume that the damping force is much weaker than the restoring force of the oscillator, i.e., small enough so that the oscillator motion is highly underdamped (see Sect. 2.4). If we then choose t to be the period of the motion with negligible damping, the first term on the right-hand side vanishes because the motion of the particle has come very close to repeating after the specified time. Consequently, by comparing the two integrands, we find that

$$\mathbf{F}_D = \frac{2e^2}{3c^3}\ddot{\mathbf{v}}. \tag{8.65}$$

The resulting motion will be approximately simple harmonic with frequency ω_0, so $\ddot{\mathbf{v}} \simeq -\omega_0^2\mathbf{v}$, and, as must be the case for an underdamped oscillation, the damping force turns out to be proportional to the instantaneous velocity of the charge:

$$\mathbf{F}_D = -\frac{2e^2}{3c^3}\omega_0^2\mathbf{v}. \tag{8.66}$$

The equation of motion for any of the Cartesian components now becomes the standard one for a damped oscillator, i.e.,

$$\ddot{x} + \Gamma\dot{x} + \omega_0^2 x = 0, \tag{8.67}$$

where

$$\Gamma = \frac{2e^2}{3mc^3}\omega_0^2. \tag{8.68}$$

For the above analysis to be correct, the damping constant Γ must satisfy the criterion $\Gamma < 2\omega_0$ (see Eq. 2.82), i.e., underdamped motion only occurs when $\omega_0 < 3mc^3/e^2$. This condition is met as long as the wavelength $\lambda\,(= 2\pi c/\omega_0)$ is greater than the quantity $\frac{2\pi}{3}\,(e^2/mc^2)$. The constant $r_0 \equiv e^2/mc^2$, known as the *classical electron radius*, has the value 2.818×10^{-13}cm. Clearly, except for gamma-radiation of extremely high frequency, all portions of the electromagnetic spectrum contain wavelengths that are much greater than r_0—therefore the criterion is met. The solution to Eq. 8.67 for the underdamped motion is

$$x(t) = x_0 e^{-\Gamma t/2} e^{-i\omega_0 t}. \tag{8.69}$$

The field amplitude of the emitted radiation is proportional to $\ddot{x}$, the acceleration of the charged particle. However, since $\ddot{x} \simeq -\omega_0^2 x$, the time-dependence of the field is simply

$$\mathbf{E}(t) = \mathbf{E}_0 e^{-\Gamma t/2} e^{-i\omega_0 t}. \tag{8.70}$$

Finally, the radiation power-spectrum, $P(\omega)$, which gives the frequency distribution of the emitted energy, is proportional to $|\mathbf{E}(\omega)|^2$, where $\mathbf{E}(\omega)$ is the Fourier-transform representation of the field, given by

$$\mathbf{E}(\omega) = \frac{1}{2\pi}\int_{-\infty}^{+\infty}\mathbf{E}(t)e^{i\omega t}dt \sim \frac{1}{i\,(\omega_0 - \omega) + \frac{1}{2}\Gamma}. \tag{8.71}$$

It then follows that the resulting power spectrum has a Lorentzian shape, i.e.,

$$P(\omega) \sim \frac{1}{(\omega - \omega_0)^2 + \frac{1}{4}\Gamma^2}. \tag{8.72}$$

As shown in Fig. 8.5, the full-width at half-maximum of this power distribution is just the damping factor, Γ. Thus, not only is radiation emitted at the undamped oscillator frequency ω_0, but also at frequencies within Γ, or so, on either side of ω_0. In other words, Γ is the sought after *linewidth*.

Classically, the *lifetime* τ of the oscillator can be taken to be the time for its total energy $\langle \mathcal{E} \rangle$, averaged over a cycle, to decay to a certain fraction of its initial value $\mathcal{E}_0$. For a weakly damped oscillator, note that

$$\langle \mathcal{E}(t) \rangle = \frac{1}{2} m\dot{x}^2 + \frac{1}{2} m\omega_0^2 x^2 \simeq \frac{1}{2} m\omega_0^2 x_0^2 e^{-\Gamma t} = \mathcal{E}_0 e^{-\Gamma t}. \tag{8.73}$$

If we define the lifetime as the time for the oscillator's energy to decay to e^{-1} of the initial value, then $\tau = \Gamma^{-1}$, and a measurement of the radiation linewidth leads to a determination of the lifetime.

Elements of the Weisskopf-Wigner Quantum Theory

Consider an atom in an excited state that decays with some lifetime τ. The excited energy level must have a finite width, ΔE, because of one of the most basic principles of quantum mechanics, namely, the time-energy uncertainty principle (Eq. 3.119). On this basis alone, the minimum width of the excited state must be on the order of

$$\Delta E \sim \hbar/\tau. \tag{8.74}$$

It then follows that the photons emitted upon de-excitation should have a frequency spread, or linewidth, Γ, on the order of

$$\Gamma = \Delta E/\hbar \sim \tau^{-1}. \tag{8.75}$$

Although essentially correct, such simple arguments are lacking when it comes to a more detailed description of emission (and absorption) lines. Fortunately, in 1930, Weisskopf and Wigner [39] put forth a full quantum-mechanical treatment of natural linewidth. Some of the more important elements of that theory are presented here.

The Weisskopf-Wigner theory treats the most general case of an atom having many possible states, and hence many possible transitions. To simplify the theory, we will take the liberty of assuming that the atom has only two possible states. Initially, the atom is in the excited state $|\psi(0)\rangle = |\mathcal{A}\rangle$ and, after some time, it de-excites by making a transition to the ground-state, $|\mathcal{B}\rangle$. In addition, assume that no radiation field is present at the start (denoted by the vacuum state $|0\rangle$), and that the excited atom emits a single $\mathbf{k}\lambda$-photon during the downward transition. In the interaction

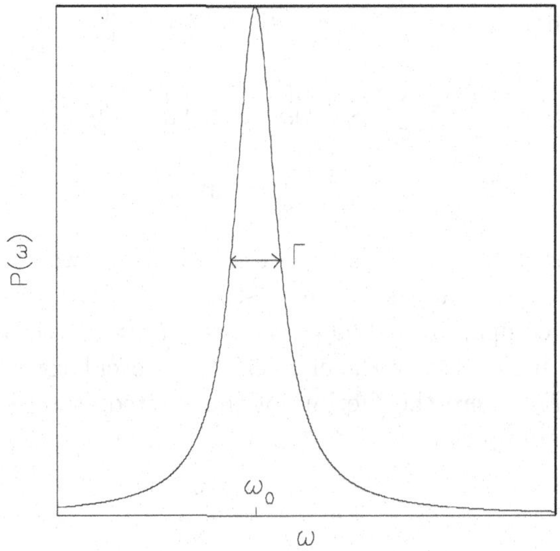

Figure 8.5 Lorentzian shape of radiation spectrum due to natural broadening. Γ represents the natural linewidth.

picture, this situation is represented by the following state vector for the combined system of atom + field:

$$|\psi_I(t)\rangle = c_{\mathcal{A}}(t)\,|\mathcal{A}\rangle\,|0\rangle + \sum_{\mathbf{k}\lambda} c_{\mathcal{B}}^{(\mathbf{k}\lambda)}(t)\,|\mathcal{B}\rangle\,|1_{\mathbf{k}\lambda}\rangle\,. \tag{8.76}$$

The summation over **k** and λ accounts for all possible wavevectors and polarizations of the emitted photon.

Now, recall that the equation of motion for the interaction picture (previously Eq. 6.7) is:

$$i\hbar\frac{d}{dt}\,|\psi_I(t)\rangle = \hat{V}_I(t)\,|\psi_I(t)\rangle\,. \tag{8.77}$$

As before, the relevant part of the atom-field interaction potential is given by the first term of Eq. 8.1. Using the fact that $\hat{V}_I(t) = e^{+i\hat{H}_0t/\hbar}\hat{V}(t)e^{-i\hat{H}_0t/\hbar}$, one can show that the time-evolution of the probability amplitudes is governed by the equations

$$\begin{aligned} i\hbar\frac{dc_{\mathcal{A}}(t)}{dt} &= \sum_{\mathbf{k}\lambda}\langle\mathcal{A}\mid \hat{V}_{\mathbf{k}\lambda}^{(+)}e^{-i\omega_k t}\mid\mathcal{B}\rangle c_{\mathcal{B}}^{(\mathbf{k}\lambda)}(t)e^{i\omega_{\mathcal{AB}}t} \\ i\hbar\frac{dc_{\mathcal{B}}^{(\mathbf{k}\lambda)}(t)}{dt} &= \langle\mathcal{B}\mid \hat{V}_{\mathbf{k}\lambda}^{(-)}e^{+i\omega_k t}\mid\mathcal{A}\rangle c_{\mathcal{A}}(t)e^{i\omega_{\mathcal{BA}}t}, \end{aligned} \tag{8.78}$$

where

$$\hat{V}_{\mathbf{k}\lambda}^{(\pm)} = \sum_i -\frac{e}{m}\sqrt{\frac{2\pi\hbar}{L^3\omega_k}}\boldsymbol{\epsilon}_{\mathbf{k}\lambda}\cdot\hat{\mathbf{p}}_i e^{\pm i\mathbf{k}\cdot\mathbf{r}_i}. \tag{8.79}$$

Notice, however, that $\langle \mathcal{A} \mid \hat{V}^{(+)}_{\mathbf{k}\lambda} e^{-i\omega_k t} \mid \mathcal{B}\rangle = \langle \mathcal{B} \mid \hat{V}^{(-)}_{\mathbf{k}\lambda} e^{+i\omega_k t} \mid \mathcal{A}\rangle^*$, so Eqs. 8.78 become

$$\begin{aligned} i\hbar \frac{dc_{\mathcal{A}}(t)}{dt} &= \sum_{\mathbf{k}\lambda} \langle \mathcal{B} \mid \hat{V}^{(-)}_{\mathbf{k}\lambda} \mid \mathcal{A}\rangle^* c^{(\mathbf{k}\lambda)}_{\mathcal{B}}(t) e^{i(\omega_{\mathcal{AB}}-\omega_k)t} \\ i\hbar \frac{dc^{(\mathbf{k}\lambda)}_{\mathcal{B}}(t)}{dt} &= \langle \mathcal{B} \mid \hat{V}^{(-)}_{\mathbf{k}\lambda} \mid \mathcal{A}\rangle c_{\mathcal{A}}(t) e^{i(\omega_{\mathcal{BA}}+\omega_k)t}. \end{aligned} \tag{8.80}$$

To characterize the decay of the excited state, we focus our attention on $|c_{\mathcal{A}}(t)|^2$, the probability that the atom is still in the initial state $|\mathcal{A}\rangle$ after a time t. For times long compared to $1/\omega_{\mathcal{AB}}$, i.e., $\omega_{\mathcal{AB}} t \gg 1$, we are motivated by the exponentially-decaying energy of the classical atom/oscillator to guess that, likewise, $|c_{\mathcal{A}}(t)|^2 = \exp(-\Gamma t)$, where the lifetime of the excited state is defined as $\tau = \Gamma^{-1}$, just as before. An appropriate trial solution for $c_{\mathcal{A}}(t)$ is therefore

$$c_{\mathcal{A}}(t) = e^{-\gamma t/2}, \tag{8.81}$$

where γ is a complex constant of the form

$$\gamma = \Gamma + 2\Delta i. \tag{8.82}$$

Substituting this form for $c_{\mathcal{A}}(t)$ into Eqs. 8.80 gives

$$\begin{aligned} -i\hbar\gamma &= 2\sum_{\mathbf{k}\lambda} \langle \mathcal{B} \mid \hat{V}^{(-)}_{\mathbf{k}\lambda} \mid \mathcal{A}\rangle^* c^{(\mathbf{k}\lambda)}_{\mathcal{B}}(t) e^{i(\omega_{\mathcal{AB}}-\omega_k-i\gamma/2)t} \\ i\hbar \frac{dc^{(\mathbf{k}\lambda)}_{\mathcal{B}}(t)}{dt} &= \langle \mathcal{B} \mid \hat{V}^{(-)}_{\mathbf{k}\lambda} \mid \mathcal{A}\rangle e^{i(\omega_{\mathcal{BA}}+\omega_k+i\gamma/2)t}. \end{aligned} \tag{8.83}$$

We can now apply the initial condition $c^{(\mathbf{k}\lambda)}_{\mathcal{B}}(0) = 0$ to the last equation to solve for $c^{(\mathbf{k}\lambda)}_{\mathcal{B}}(t)$. The result is

$$c^{(\mathbf{k}\lambda)}_{\mathcal{B}}(t) = \langle \mathcal{B} \mid \hat{V}^{(-)}_{\mathbf{k}\lambda} \mid \mathcal{A}\rangle \frac{1 - e^{i(\omega_k-\omega_{\mathcal{AB}}+i\gamma/2)t}}{\hbar\left(\omega_k - \omega_{\mathcal{AB}} + i\gamma/2\right)}. \tag{8.84}$$

Substituting this back into the first of Eqs. 8.83 produces the following expression for γ:

$$\gamma = \frac{2i}{\hbar^2} \sum_{\mathbf{k}\lambda} \left| \langle \mathcal{B} \mid \hat{V}^{(-)}_{\mathbf{k}\lambda} \mid \mathcal{A}\rangle \right|^2 \frac{e^{-i(\omega_{\mathcal{AB}}-\omega_k-i\gamma/2)t} - 1}{\omega_k - \omega_{\mathcal{AB}} + i\gamma/2}. \tag{8.85}$$

The summation over $\mathbf{k}$ can be replaced by an integral over all possible photon frequencies and emission directions. With $g(\omega_k)\, d\omega_k$ being the number of photon states in the frequency interval $d\omega_k$, Eq. 8.85 can be rewritten as

$$\gamma = \frac{2i}{\hbar^2} \int \frac{e^{-i(\omega_{\mathcal{AB}}-\omega_k-i\gamma/2)t} - 1}{\omega_k - \omega_{\mathcal{AB}} + i\gamma/2} \left| \langle \mathcal{B} \mid \hat{\mathrm{V}} \mid \mathcal{A}\rangle \right|^2 g(\omega_k)\, d\omega_k, \tag{8.86}$$

where the operator $\hat{\mathsf{V}}$ represents

$$\hat{\mathsf{V}} = \sum_{\lambda} \int \hat{V}^{(-)}_{\mathbf{k}\lambda}\, d\Omega. \tag{8.87}$$

The assumed form of the solution for $c_{\mathcal{A}}(t)$ given by Eq. 8.81 cannot be exactly correct since γ does not appear to be a constant independent of t, as required. However, it is reasonable to assume that the lifetime of the atom greatly exceeds $\omega_{\mathcal{AB}}^{-1}$, thus $|\gamma/\omega_{\mathcal{AB}}| \ll 1$, and

$$\frac{e^{-i(\omega_{\mathcal{AB}}-\omega_k-i\gamma/2)t}-1}{\omega_k-\omega_{\mathcal{AB}}+i\gamma/2} \rightarrow \frac{e^{-i(\omega_{\mathcal{AB}}-\omega_k)t}-1}{\omega_k-\omega_{\mathcal{AB}}}. \tag{8.88}$$

For our purposes, we are mainly interested in Γ, which is the real part of γ, or

$$\Gamma \approx \frac{2}{\hbar^2}\int \frac{\sin\left(\omega_{\mathcal{AB}}-\omega_k\right)t}{\omega_{\mathcal{AB}}-\omega_k}\left|\langle\mathcal{B}\mid\hat{\mathsf{V}}\mid\mathcal{A}\rangle\right|^2 g\left(\omega_k\right)d\omega_k. \tag{8.89}$$

Note, however, that for sufficiently large t,

$$\frac{\sin\left(\omega_{\mathcal{AB}}-\omega_k\right)t}{\omega_{\mathcal{AB}}-\omega_k} \rightarrow \pi\delta\left(\omega_{\mathcal{AB}}-\omega_k\right), \tag{8.90}$$

so an excellent approximation for Γ is the time-independent expression

$$\Gamma \approx \frac{2\pi}{\hbar}\left|\langle\mathcal{B}\mid\hat{\mathsf{V}}\mid\mathcal{A}\rangle\right|^2 g\left(\hbar\omega_{\mathcal{AB}}\right), \tag{8.91}$$

where the matrix element $\langle\mathcal{B}\mid\hat{\mathsf{V}}\mid\mathcal{A}\rangle$ is to be evaluated at $\omega_k = \omega_{\mathcal{AB}}$. Comparing with Eq. 6.35, we see that Γ is the same as the transition rate between state $|\mathcal{A}\rangle$ and state $|\mathcal{B}\rangle$. The result agrees with our previous statement that Γ^{-1} is the lifetime of excited state $|\mathcal{A}\rangle$.

Now consider $P\left(\omega_k\right)$, the frequency spectrum of the emitted radiation. $P\left(\omega_k\right)$ is proportional to the probability of emitting a photon into various frequency intervals, $d\omega_k$, regardless of the photon's direction or polarization, i.e.,

$$P\left(\omega_k\right)d\omega_k \sim g\left(\omega_k\right)d\omega_k \sum_{\lambda}\int\left|c_{\mathcal{B}}^{(\mathbf{k}\lambda)}(t\rightarrow\infty)\right|^2 d\Omega. \tag{8.92}$$

It is necessary to evaluate $c_{\mathcal{B}}^{(\mathbf{k}\lambda)}(t)$ at very large time ($t \gg \Gamma^{-1}$) to guarantee that the atom has decayed. Using Eqs. 8.84 and 8.87, and making the replacement $\gamma = \Gamma + 2\Delta i$, gives

$$P\left(\omega_k\right) \sim \frac{1}{2\pi}\frac{\frac{2\pi}{\hbar}\left|\langle\mathcal{B}\mid\hat{\mathsf{V}}\mid\mathcal{A}\rangle\right|^2 g\left(\hbar\omega_k\right)}{\left[\omega_k-\left(\omega_{\mathcal{AB}}+\Delta\right)\right]^2+\frac{1}{4}\Gamma^2}. \tag{8.93}$$

From Eq. 8.91, the numerator is just the constant Γ, so the emitted power spectrum has a Lorentzian form, in agreement with the result from the classical calculation (see Eq. 8.72), i.e.,

$$P(\omega_k) \sim \frac{1}{[\omega_k - (\omega_{AB} + \Delta)]^2 + \frac{1}{4}\Gamma^2}. \tag{8.94}$$

In the present quantum derivation, notice that a small line-shift of an amount Δ comes about because γ contains an imaginary contribution. The calculation of this shift is quite long and tedious, so, for the sake of brevity, it will not be presented here.[§]

8.2.2 Other Broadening Effects

In most experiments, the observed spectral lines do not, in fact, exhibit the natural-linewidth profile just described. This is because other, usually more pronounced, broadening effects are present. Consider, for example, the linewidth associated with the emission of an optical photon. Typically, the spontaneous lifetime τ of the excited state is on the order of nanoseconds, so the natural linewidth is $\Gamma \sim \tau^{-1} \sim 10^9\ \mathrm{s}^{-1}$. Thus, the emitted line will have a Lorentzian shape with a frequency-width of $\Gamma/2\pi$, which is only on the order of 100 MHz. Compared to the broadening effects brought about by other mechanisms, this natural linewidth is very small, as we discuss below.

Collision Broadening

If each emitting molecule undergoes collisions with other molecules in a gas at temperature T and pressure P, then transitions will be induced, effectively shortening their lifetime. The result is that the width of the Lorentzian spectral line is broadened by an amount Γ_c, which is equal to the collision rate experienced by a molecule. The complete theory of collision broadening is quite involved, and is reviewed in a paper by Van Vleck and Huber [40]. Nevertheless, we can make a rough estimate of the collision rate, and hence the collision width, as follows:

τ_c, the average time between collisions, is given by the ratio of l, the *mean free path* in the gas, to $\langle v \rangle$, the average molecular speed:

$$\tau_c = \frac{l}{\langle v \rangle}. \tag{8.95}$$

If there are n molecules per unit volume of the gas, the mean free path, which represents the average distance a molecule travels between collisions, is given by [41]

$$l = \frac{1}{2^{1/2}\pi D^2 n}. \tag{8.96}$$

[§]The interested reader is referred to the original paper by Weisskopf and Wigner [39], or some of the treatments given by other texts (e.g., see Heitler [35, pp. 339-346] or Louisell [21, pp. 285-296]).

D is the molecular diameter ($\sim$ 0.2–0.5 nm). Also, for a Maxwellian distribution of speeds,

$$\langle v\rangle = \sqrt{\frac{8k_BT}{\pi m}}, \tag{8.97}$$

where m is the mass of a molecule. The collision rate, which is the inverse of the collision time, therefore becomes

$$\Gamma_c = \tau_c^{-1} = 4D^2 n\sqrt{\frac{\pi k_BT}{m}}. \tag{8.98}$$

Assuming the gas to be ideal, we can make the substitution $n = P/k_BT$, so that

$$\Gamma_c = 4D^2 P\sqrt{\frac{\pi}{mk_BT}}. \tag{8.99}$$

For a gas at room temperature and a pressure of 1 atm, the collision width is of order 10^{10} s^{-1}, which is much larger than the natural linewidth. The collision width can be reduced, however, by lowering the pressure of the gas.

Doppler Broadening

Because of the thermal motion associated with a gas at temperature T, the frequency of radiation emitted becomes shifted. If one considers a photon emitted in the x-direction, and the emitting molecule has a velocity component v_x along that direction, then, according to the *Doppler effect*, the frequency will be shifted to the value of

$$\omega_k = \frac{1+v_x}{\sqrt{1-v^2/c^2}}\omega_{AB}, \tag{8.100}$$

where $v = \sqrt{v_x^2+v_y^2+v_z^2}$. As before, $\omega_{AB} = (E_A - E_B)/\hbar$ is the frequency emitted had the molecule been stationary. For $v \ll c$, the frequency shift is

$$\omega_k - \omega_{AB} \simeq \frac{v_x}{c}\omega_{AB}. \tag{8.101}$$

At a given temperature T, the probability that a molecule in the gas has a velocity component between v_x and $v_x + dv_x$ is given by the Maxwellian velocity distribution

$$P(v_x)\,dv_x = \left(\frac{m}{2\pi k_BT}\right)^{1/2} e^{-mv_x^2/2k_BT}\,dv_x. \tag{8.102}$$

Then, according to Eq. 8.101, the emitted frequencies also follow this distribution, i.e.,

$$P(\omega_k) = P(v_x)\frac{dv_x}{d\omega_k} = \frac{c}{\omega_{AB}}\left(\frac{m}{2\pi k_BT}\right)^{1/2} e^{-mc^2(\omega_k-\omega_{AB})^2/2\omega_{AB}^2k_BT}. \tag{8.103}$$

The Doppler effect thus gives rise to a broadened line having a Gaussian profile. The Doppler width, Γ_d, can be taken as the full-width at half-maximum of the distribution, or

$$\Gamma_d = 2\omega_{AB}\sqrt{\frac{2k_BT\,\ln 2}{mc^2}}. \tag{8.104}$$

For optical photons emitted by a gas at room temperature, the Doppler width is on the order of 10^{10} s^{-1}, which is comparable to the previously calculated collision width. Clearly, by cooling the gas, the effects of Doppler broadening can be reduced. For the purpose of comparison, Fig. 8.6 shows a Lorentzian and a Gaussian distribution, assuming they have the same width and the same integrated area. Notice that, for frequencies far-removed from the center of the peak, the intensity is dominated by the tail of the Lorentzian, which stems from collision broadening and the natural linewidth.

8.3 The Photoelectric Effect

Up to this point, we have been discussing the emission and absorption of photons that accompany transitions between discrete, or bound, atomic states. In the *photoelectric effect*, a photon is absorbed by an atom, kicking one of the electrons out of the atom. In other words, in the final state the atom is unbound, with a so-called *photoelectron* promoted to the energy continuum. The kinetic energy of the ejected photoelectron is simply

$$\frac{1}{2}mv^2 = \hbar\omega_k - E_b, \tag{8.105}$$

where $\hbar\omega_k$ is the energy of the absorbed photon and E_b represents the binding energy of the electron in the atom. Clearly, the photoelectric process is possible only if the incoming photon energy is at least as large as the binding energy of the most weakly-bound electron in an outer atomic shell. Emission of an electron from a more tightly-bound shell requires the absorption of a higher-energy photon, where $\hbar\omega_k$ exceeds the shell's *absorption edge*. Specifically, the *K*, *L*, *M*, ...-edges refer to the binding energies of the $n = 1, 2, 3, \ldots$ atomic electrons.

It should also be emphasized that whenever a photoelectric event occurs, the emitted electron leaves behind a vacancy in one of the atomic shells. Consequently, the photoelectric effect is always followed by downward transitions made by outer-shell atomic electrons, which in turn are accompanied by the emission of characteristic x-rays from the ionized atom. It is observed, however, that the measured intensities and widths of the resulting x-ray lines are often very different from what one would expect, especially when there is an inner-shell ionization of a low-Z element. This is because, in addition to de-excitation by x-ray emission, there may also be a significant probability for a "radiationless transition" within the atom, known as the *Auger effect* [42][43]. In this process, rather than emitting a photon of some characteristic energy, the ionized atom emits a secondary electron with kinetic energy equal to the aforementioned characteristic energy less the electron's binding energy in the already

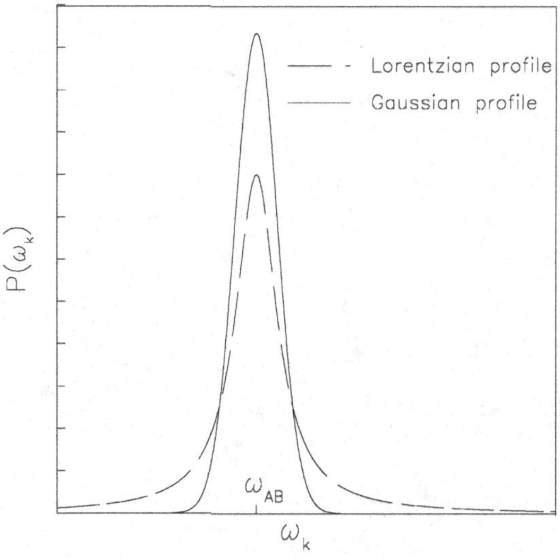

Figure 8.6 Comparison of Lorentzian and Gaussian distributions with same width and integrated area.

singly-ionized atom. Such an electron is readily distinguishable from the primary photoelectron in that the energy of the latter depends on the energy of the incident photon (see Eq. 8.105), whereas the energy of an Auger electron does not. For atoms ionized in a given shell, one is usually interested in knowing the *fluorescence yield*, w, which is the fraction of transitions to the vacant level that produce an x-ray (rather than an Auger electron). Experimentally measured and theoretically calculated fluorescent yields have been compiled by Fink *et al.* [44]. Figure 8.7 shows that the fluorescence yield of a given shell increases with atomic number according to the approximate form [45]

$$w = \frac{1}{1 + \alpha Z^{-4}}. \tag{8.106}$$

For the K-shell fluorescence yield, the experimental data is best fit by $\alpha \simeq 1.12 \times 10^6$, while for the L-shell¶ $\alpha \simeq 6.4 \times 10^7$. A better predictor for the K-shell fluorescence yield is the refined theoretical curve in Fig. 8.7, which is based on calculations made by Callan [47][48] and Listengarten [49][50]. For shells above the L-shell, there is almost no information available in the literature on fluorescent yields; the only noticeable exception is some very limited data reported on the M-shell [46][51][52].

Returning to the photoelectric effect itself, we now derive the cross-section for photon absorption by a K-shell electron under the assumption that the incident

¶The value of α stated here is for the mean L-shell fluorescence yield, obtained by weighting the fluorescent yields of the individual L-subshells by the probability of producing a vacancy in that subshell.

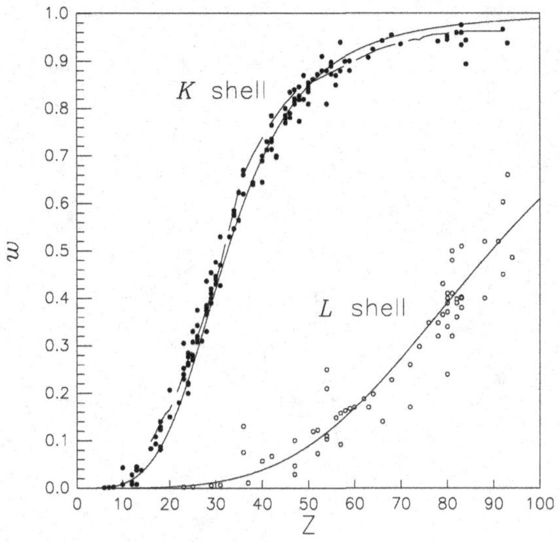

Figure 8.7 K-shell and L-shell fluorescence yields, w, as a function of atomic number, Z. The experimental points for the K-shell are taken from data compiled by Fink [43]; the points for the L-shell come from data tabulated by Lay [45]. The solid curves represent the approximate theoretical form for the fluorescence yield, as given by Eq. 8.106. The dashed K-shell fluorescence curve is based on refined theoretical calculations made by Callan [46][47] and Listergarten [48][49].

energy is far above the absorption edge (i.e., $\hbar\omega_k \gg E_b$). The latter condition guarantees that the kinetic energy of the emitted electron greatly exceeds the electron's binding energy in the atom, i.e.,

$$\frac{1}{2}mv^2 \gg E_b. \tag{8.107}$$

For electrons in the K-shell, one need not worry about screening effects due to other electrons in the atom, so the binding energy is simply given by the Bohr expression

$$E_b = \frac{me^4}{2\hbar^2}Z^2 \approx (13.6\text{ eV})\,Z^2. \tag{8.108}$$

Combining the last two equations produces the condition

$$\hbar v \gg Ze^2. \tag{8.109}$$

Furthermore, our derivation will be valid for cases where the photoelectrons produced are non-relativistic; this means that the kinetic energy of the electron is much less than its rest energy, or

$$\frac{1}{2}mv^2 \approx \hbar\omega_k \ll mc^2. \tag{8.110}$$

Much of our treatment follows the one presented by Heitler [35, pp. 204-211].

The differential cross-section is derived by calculating the ratio of W_{nm}, the photoelectric-event (transition) probability per unit time, to the incident flux. To handle the photon flux, we imagine the system of target-atom + photon to be enclosed in a hypothetical box of volume L^3. Then the flux is just the number density of incident photons (i.e., 1 photon/L^3) times the speed of a photon, c. As a result, the differential cross-section is

$$d\sigma = \frac{W_{nm}}{c/L^3}. \tag{8.111}$$

The photoelectric effect is a first-order process since a single photon is annihilated as a result of the interaction. Therefore, W_{nm} is given by Eq. 6.35, so that

$$d\sigma = \frac{L^3}{c} \cdot \frac{2\pi}{\hbar} \left| \langle n \mid \hat{V} \mid m \rangle \right|^2 g\left(E_n\right). \tag{8.112}$$

Let $|\mathcal{A}\rangle$ be the initial bound-state of the electron in the K-shell; then the initial state of the atom-field system is $|m\rangle = |1_{\mathbf{k}\lambda}\rangle\,|\mathcal{A}\rangle$. After the photon has been absorbed by the atom, the state of the system becomes $|n\rangle = |0_{\mathbf{k}\lambda}\rangle\,|\mathcal{B}\rangle$, where $|\mathcal{B}\rangle$ is the final state of the electron in the continuum. $g\left(E_n\right)$, the final density of states for the whole system, is identical to $g\left(E_{\mathcal{B}}\right)$, the photoelectron's density of states. If ejected into a solid angle $d\Omega$, the free electron will be within some energy range $E_{\mathcal{B}} \rightarrow E_{\mathcal{B}} + dE_{\mathcal{B}}$, or equivalently, within some momentum range $p \rightarrow p + dp$. With $\mathbf{q} = \mathbf{p}/\hbar$ denoting the electron's wavevector, we can write

$$g\left(E_{\mathcal{B}}\right) dE_{\mathcal{B}} = g\left(q\right) dq = \left(\frac{L}{2\pi}\right)^3 q^2 dq\, d\Omega = \left(\frac{L}{2\pi\hbar}\right)^3 p^2 dp\, d\Omega. \tag{8.113}$$

Since $E_{\mathcal{B}} = p^2/2m$, we have $dp/dE_{\mathcal{B}} = m/p$, and

$$g\left(E_{\mathcal{B}}\right) = m\left(\frac{L}{2\pi\hbar}\right)^3 p\, d\Omega. \tag{8.114}$$

This results in the following expression for the angular differential cross-section:

$$\frac{d\sigma}{d\Omega} = \frac{mp}{c}\left(\frac{L^3}{2\pi\hbar^2}\right)^2 \left|\langle n \mid \hat{V} \mid m \rangle\right|^2. \tag{8.115}$$

To continue, we need to calculate the transition matrix element

$$\begin{aligned} \left\langle n \mid \hat{V} \mid m \right\rangle &= -\frac{e}{mc} \langle \mathcal{B}|\, \langle 0_{\mathbf{k}\lambda}|\, \hat{\mathbf{A}}\cdot\hat{\mathbf{p}}\, |1_{\mathbf{k}\lambda}\rangle\, |\mathcal{A}\rangle \\ &= -\frac{e}{m}\sqrt{\frac{2\pi\hbar}{L^3\omega_k}}\, \langle \mathcal{B}|\, e^{i\mathbf{k}\cdot\mathbf{r}} \boldsymbol{\epsilon}_{\mathbf{k}\lambda}\cdot\hat{\mathbf{p}}\, |\mathcal{A}\rangle\,. \end{aligned} \tag{8.116}$$

This is accomplished by transforming to the position-space representation (see Sect. 3.4):

$$\langle n \mid \hat{V} \mid m \rangle = -\frac{e}{m}\sqrt{\frac{2\pi\hbar}{L^3\omega_k}} \int \psi_{\mathcal{B}}^* e^{i\mathbf{k}\cdot\mathbf{r}} \boldsymbol{\epsilon}_{\mathbf{k}\lambda}\cdot\left(-i\hbar\boldsymbol{\nabla}\psi_{\mathcal{A}}\right) d^3r. \tag{8.117}$$

The differential cross-section then becomes

$$\frac{d\sigma}{d\Omega} = \frac{e^2 p L^3}{2\pi\hbar m c \omega_k} \left| \int \psi_{\mathcal{B}}^* (\mathbf{r}) e^{i\mathbf{k}\cdot\mathbf{r}} \boldsymbol{\epsilon}_{\mathbf{k}\lambda}\cdot\boldsymbol{\nabla}\psi_{\mathcal{A}} (\mathbf{r})\, d^3r \right|^2 . \tag{8.118}$$

The wavefunction for the initial state is that of the hydrogen-atom ground-state, i.e.,

$$\psi_{\mathcal{A}} (\mathbf{r}) = \left(\pi a^3\right)^{-1/2} e^{-r/a}, \tag{8.119}$$

where the constant a is the Bohr radius scaled by the atomic number Z:

$$a = \frac{\hbar^2}{m e^2 Z}. \tag{8.120}$$

The final-state wavefunction should represent an electron in the continuum spectrum. However, by applying the Born approximation,‖ we are able to use a simple plane-wave form representing a free electron ejected with some definite momentum $\mathbf{p}$—i.e., a momentum eigenstate:

$$\psi_{\mathcal{B}} (\mathbf{r}) = \frac{1}{L^{3/2}} e^{i\mathbf{q}\cdot\mathbf{r}}. \tag{8.121}$$

Letting $\hbar\mathbf{Q} = \hbar(\mathbf{k} - \mathbf{q})$ be the recoil momentum of the atom, the cross-section reduces to

$$\frac{d\sigma}{d\Omega} = \frac{e^2 q}{2\pi m c \omega_k} \left| \int e^{i\mathbf{Q}\cdot\mathbf{r}} \boldsymbol{\epsilon}_{\mathbf{k}\lambda}\cdot\boldsymbol{\nabla}\psi_{\mathcal{A}} (\mathbf{r})\, d^3r \right|^2 . \tag{8.122}$$

The integral can be evaluated by parts:

$$\begin{aligned}
\int e^{i\mathbf{Q}\cdot\mathbf{r}} \boldsymbol{\epsilon}_{\mathbf{k}\lambda}\cdot\boldsymbol{\nabla}\psi_{\mathcal{A}} (\mathbf{r})\, d^3r &= -i\boldsymbol{\epsilon}_{\mathbf{k}\lambda}\cdot\mathbf{Q} \int e^{i\mathbf{Q}\cdot\mathbf{r}} \psi_{\mathcal{A}} (\mathbf{r})\, d^3r \\
&= i\boldsymbol{\epsilon}_{\mathbf{k}\lambda}\cdot\mathbf{q} \int e^{i\mathbf{Q}\cdot\mathbf{r}} \psi_{\mathcal{A}} (\mathbf{r})\, d^3r \\
&= \frac{4\pi i}{Q} (\boldsymbol{\epsilon}_{\mathbf{k}\lambda}\cdot\mathbf{q}) \int_0^\infty r (\sin Qr)\, \psi_{\mathcal{A}} (r)\, dr \\
&= \frac{8i\sqrt{\pi} a^{3/2}}{(1 + a^2 Q^2)^2} (\boldsymbol{\epsilon}_{\mathbf{k}\lambda}\cdot\mathbf{q}) .
\end{aligned} \tag{8.123}$$

The result is

$$\frac{d\sigma}{d\Omega} = \frac{32 e^2 a^3 q\, (\boldsymbol{\epsilon}_{\mathbf{k}\lambda}\cdot\mathbf{q})^2}{m c \omega_k (1 + a^2 Q^2)^4}. \tag{8.124}$$

‖The Born approximation was previously mentioned in conjunction with the thermal neutron scattering problem (see Example 6.1). In the present case, the use of the Born approximation is justified by the condition given by Eq. 8.109.

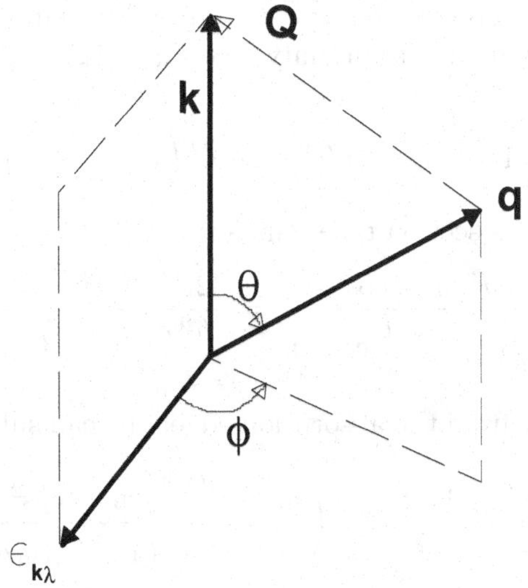

Figure 8.8 Vectors for calculating photoelectric effect cross-section.

Figure 8.8 shows the geometry defined by the various vector quantities in the problem: θ is the angle between **k** and **q**, i.e., the photon and electron wavevectors, and ϕ denotes the angle between the **k**-**q**-plane and the **k**-$\boldsymbol{\epsilon}_{\mathbf{k}\lambda}$-plane. The dot-product in Eq. 8.124, which represents the projection of the electron wavevector onto the direction defined by the photon polarization, is then

$$\boldsymbol{\epsilon}_{\mathbf{k}\lambda}\cdot\mathbf{q} = q\sin\theta\cos\phi. \tag{8.125}$$

To obtain the denominator in the cross-section, start with the law of cosines:

$$a^2Q^2 = a^2k^2 + a^2q^2 - 2kqa^2\cos\theta. \tag{8.126}$$

Far away from the absorption edge (see Eq. 8.109),

$$aq = \frac{\hbar^2}{me^2Z}\cdot\frac{mv}{\hbar} = \frac{\hbar v}{Ze^2} \gg 1, \tag{8.127}$$

so

$$a^2Q^2 \simeq q^2a^2\left(1 - 2\frac{k}{q}\cos\theta\right). \tag{8.128}$$

However, since the photon and electron energies are about equal, i.e., $\hbar\omega_k \simeq mv^2/2$, one has

$$\frac{k}{q} = \frac{\hbar k}{\hbar q} = \frac{\hbar\omega_k/c}{\hbar q} \simeq \frac{mv^2/2c}{mv} = \frac{v}{2c}, \tag{8.129}$$

and Eq. 8.128 becomes

$$a^2Q^2 \simeq q^2a^2\left(1 - \frac{v}{c}\cos\theta\right). \tag{8.130}$$

For a non-relativistic photoelectron, $v/c \ll 1$, and the quantity in parentheses is close to 1. Since aq is much greater than unity (see Eq. 8.127), so is the quantity aQ. We can therefore write

$$1 + a^2Q^2 \simeq a^2Q^2 \simeq q^2a^2\left(1 - \frac{v}{c}\cos\theta\right), \tag{8.131}$$

and the differential cross-section takes the form

$$\frac{d\sigma}{d\Omega} = \frac{32e^2a^3q}{mc\omega_k} \cdot \frac{q^2\sin^2\theta\cos^2\phi}{q^8a^8\left(1 - \frac{v}{c}\cos\theta\right)^4} = \frac{32Z^5m^4e^{12}}{\hbar^{10}c\omega_k q^5} \cdot \frac{\sin^2\theta\cos^2\phi}{\left(1 - \frac{v}{c}\cos\theta\right)^4}. \tag{8.132}$$

Again, since we are far from the absorption edge, one can substitute $\hbar^2q^2/2m \simeq \hbar\omega_k$. The result is

$$\frac{d\sigma}{d\Omega} = \frac{4\sqrt{2}}{137^4}Z^5r_0^2\left(\frac{mc^2}{\hbar\omega_k}\right)^{7/2} \cdot \frac{\sin^2\theta\cos^2\phi}{\left(1 - \frac{v}{c}\cos\theta\right)^4}, \tag{8.133}$$

where we have used the definition of the classical electron radius $r_0 = e^2/mc^2$ and the fine structure constant $e^2/\hbar c = 1/137$.

The expression for $d\sigma/d\Omega$ shows that no photoelectrons are emitted in the direction of the incident photon momentum ($\theta = 0$). For negligibly small values of v/c, the most likely emission direction is given by $\theta = \pi/2$, $\phi = 0$. This means that more electrons are ejected in the direction of the photon polarization than in any other direction. For electrons that are more energetic (yet still non-relativistic, i.e., $v/c \leq 0.1$), the peak differential cross-section shifts slightly toward the forward direction ($\theta < \pi/2$, $\phi = 0$). This effect is illustrated in Fig. 8.9.

Far from the absorption edge, the total atomic cross-section for the photoelectric effect from the K-shell is

$$\sigma_K = 2\int \frac{d\sigma}{d\Omega}d\Omega \simeq \frac{4\sqrt{2}}{137^4}Z^5\left(\frac{8\pi}{3}r_0^2\right)\left(\frac{mc^2}{\hbar\omega_k}\right)^{7/2}. \tag{8.134}$$

The integral is doubled because there are two electrons in an atom's K-shell.** Most importantly, observe that $\sigma_K \propto Z^5/\left(\hbar\omega_k\right)^{7/2}$, i.e., the cross-section increases rapidly with atomic number, while it diminishes quickly with the energy of the incident photon.

If the analysis incorporates the possibility of being near the K-shell absorption edge, then it may be that $\hbar\omega_k \approx E_b$, and many of our previously stated approximations fail. According to Stobbe [53], the correct expression becomes

$$\sigma_K = \frac{4\sqrt{2}}{137^4}Z^5\left(\frac{8\pi}{3}r_0^2\right)\left(\frac{mc^2}{\hbar\omega_k}\right)^{7/2} f(\xi), \tag{8.135}$$

that is, we need to multiply our old expression by the correction factor

**Obviously, for hydrogen this is not the case, and the factor of 2 should not be included.

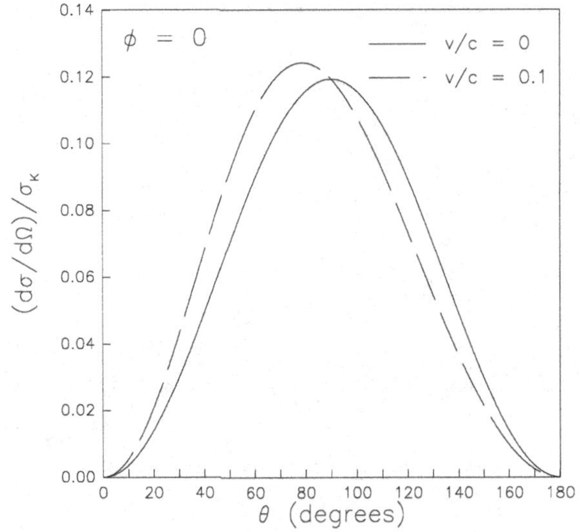

Figure 8.9 Angular distribution of K-shell photoelectron emission (non-relativistic).

$$f(\xi) = 2\pi\sqrt{\frac{E_b}{\hbar\omega_k}}\frac{e^{-4\xi\cot^{-1}\xi}}{1-e^{-2\pi\xi}} \tag{8.136}$$

having the argument

$$\xi \equiv \sqrt{\frac{E_b}{\hbar\omega_k - E_b}} = \frac{Ze^2}{\hbar v}. \tag{8.137}$$

Right at the absorption edge, $\hbar\omega_k = E_b$, and $\xi \to \infty$; this gives $f(\xi) \simeq 0.115$. When considering the other extreme, i.e., cases far from the K-edge, $\hbar\omega_k \gg E_b$, and $\xi \to 0$. Then

$$\frac{e^{-4\xi\cot^{-1}\xi}}{1-e^{-2\pi\xi}} \to \frac{1}{2\pi\xi} \approx \frac{1}{2\pi}\sqrt{\frac{\hbar\omega_k}{E_b}}, \tag{8.138}$$

in which case $f(\xi) \to 1$, and the cross-section reduces to our old result.

K-shell photoelectric cross-sections have also been calculated for the case of relativistic photoelectrons. For example, Sauter [54] [55] gives an analytical expression that is good for light elements, while Hulme *et al.* [56] have numerically calculated cross-section data for heavy elements. One particularly useful result is provided by Hall [57]—it is an approximate analytical expression for σ_K that holds when the emitted electron is highly relativistic, i.e., $\hbar\omega_k \gg mc^2$, and it is appropriate for any value of Z:

$$\sigma_K \simeq \frac{3}{2}\left(\frac{8\pi}{3}r_0^2\right)\frac{Z^5}{137^4}\left(\frac{mc^2}{\hbar\omega_k}\right)e^{-\pi\alpha+2\alpha^2(1-\log\alpha)}. \tag{8.139}$$

Here, $\alpha = Z/137$.

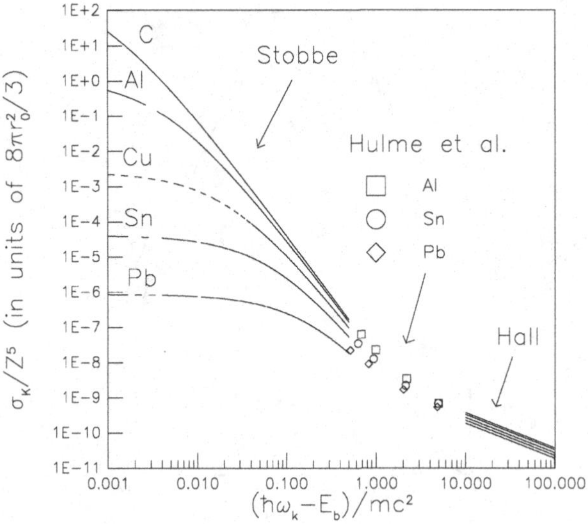

Figure 8.10 K-shell photoelectric cross-sections.

Figure 8.10 shows a plot of σ_K/Z^5 vs. $(\hbar\omega_k - E_b)/mc^2$ computed using Eqs. 8.135 and 8.139. For the region $0.5 < (\hbar\omega_k - E_b)/mc^2 < 10$, points are filled in using the numerical calculations by Hulme *et al.* [56].

Example 8.3 Photodisintegration of the Deuteron. A problem that can be handled in a way very much akin to the photoelectric effect for atoms is the photon-induced break-up of the deuteron, the bound system of neutron + proton. The process is represented by the nuclear photodisintegration process

$$\gamma + \mathrm{H}^2 \rightarrow \mathrm{n} + \mathrm{p}. \tag{8.140}$$

The threshold energy for this reaction is 2.225 MeV, which is the binding energy, E_b, of the deuteron. There is only one bound state of the deuteron, namely the ground state. This is an s-state ($l = 0$) in which the neutron and proton spins are aligned, giving the deuteron a total angular momentum, or nuclear spin, of 1. Since the magnetic quantum number of the system can take on one of three possible values (i.e., -1, 0, $+1$), the deuteron's ground state is sometimes referred to as a *triplet state*. The interaction between the neutron and proton is approximately represented by an attractive square well $V(r)$ of range $r_0 \simeq 2$ F and depth $V_0 \simeq 36$ MeV, as shown in Fig. 8.11. Notice that the ground-state energy of -2.225 MeV occurs very close to the top of the well because the radial wavefunction $u(r) = rR(r)$ needs to have enough curvature to "turn over" within the very short distance r_0 (in order to satisfy the boundary conditions at the edge).

The initial steps in the derivation of the cross-section for the photodisintegration reaction still follow Eqs. 8.111–8.118 used for the photoelectric effect, except for

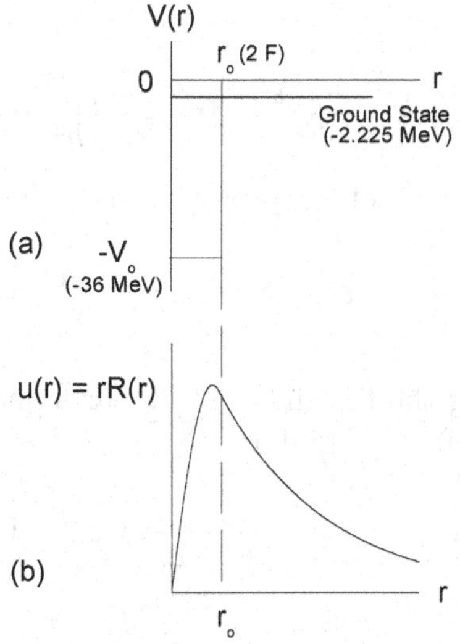

Figure 8.11 (a) Square-well potential representative of neutron-proton interaction in the deuteron. The ground-state energy is -2.225 MeV. (b) Radial wavefunction $u(r) = rR(r)$ for deuteron ground state.

two changes: (1) The potential $\hat{V}$ occurs as a result of the interaction between the proton and EM field, so the proton mass, m_p, replaces the mass of the electron, in the matrix element $\langle n \mid \hat{V} \mid m \rangle$. (2) The differential cross-section has to be calculated in the center-of-mass frame of the n-p system. Consequently, the reduced mass, $m_p/2$, should be used in the density-of-states calculation. Equation 8.121 for the final wavefunction is, of course, still appropriate—it represents the emitted proton as a plane wave with wavevector $\mathbf{q}$. However, to simplify the evaluation of $d\sigma/d\Omega$, we make use of the fact that for incident energies well below about 1 GeV, the photon wavelength is much larger than the effective size of the deuteron—thus $\exp(i\mathbf{k}\cdot\mathbf{r}) \approx 1$. After incorporating these considerations, the expression for the cross-section in the center-of-mass frame becomes

$$\begin{aligned}\left(\frac{d\sigma}{d\Omega}\right)_{\rm cm} &= \frac{e^2 q}{4\pi m_p c \omega_k}\left|\int e^{-i\mathbf{q}\cdot\mathbf{r}}\boldsymbol{\epsilon}_{\mathbf{k}\lambda}\cdot\boldsymbol{\nabla}\psi_{\mathcal{A}}(\mathbf{r})\,d^3r\right|^2 \\ &= \frac{e^2 q}{4\pi m_p c \omega_k}\left(\boldsymbol{\epsilon}_{\mathbf{k}\lambda}\cdot\mathbf{q}\right)^2\left|\int e^{-i\mathbf{q}\cdot\mathbf{r}}\psi_{\mathcal{A}}(\mathbf{r})\,d^3r\right|^2, \end{aligned} \tag{8.141}$$

where, as before, we have integrated by parts. $\psi_{\mathcal{A}}(\mathbf{r})$ is the initial wavefunction, in other words, it is the ground-state wavefunction of the deuteron.

To evaluate the integral in Eq. 8.141 first observe that $\psi_{\mathcal{A}}$ must satisfy the

Schrödinger equation

$$\nabla^2\psi_{\mathcal{A}}(\mathbf{r}) - \frac{m_p E_b}{\hbar^2}\psi_{\mathcal{A}}(\mathbf{r}) = \frac{m_p V(r)}{\hbar^2}\psi_{\mathcal{A}}(\mathbf{r}), \tag{8.142}$$

where $m_p/2$, the reduced mass of the proton, is used. This means that

$$\frac{\hbar^2}{m_p}\int e^{-i\mathbf{q}\cdot\mathbf{r}}\left[\nabla^2\psi_{\mathcal{A}}(\mathbf{r})\right] d^3r - E_b\int e^{-i\mathbf{q}\cdot\mathbf{r}}\psi_{\mathcal{A}}(\mathbf{r})\, d^3r = \int e^{-i\mathbf{q}\cdot\mathbf{r}} V(r)\,\psi_{\mathcal{A}}(\mathbf{r})\, d^3r. \tag{8.143}$$

However, after performing the first integral by parts and rearranging terms, we are able to write our sought after integral as

$$\begin{aligned}\int e^{-i\mathbf{q}\cdot\mathbf{r}}\psi_{\mathcal{A}}(\mathbf{r})\, d^3r &= -\frac{1}{E+E_b}\int e^{-i\mathbf{q}\cdot\mathbf{r}} V(r)\,\psi_{\mathcal{A}}(\mathbf{r})\, d^3r \\ &\simeq -\frac{1}{E+E_b}\int V(r)\,\psi_{\mathcal{A}}(\mathbf{r})\, d^3r.\end{aligned} \tag{8.144}$$

Here, $E = \hbar^2 q^2/m_p$ is the energy of the emitted proton in the center-of-mass reference frame. In the last step, we set $\exp(-i\mathbf{q}\cdot\mathbf{r}) \approx 1$, since the integral is only over the very short range of the potential. It is left to the reader to now show that the Schrödinger equation also leads to

$$\int V(r)\,\psi_{\mathcal{A}}(\mathbf{r})\, d^3r = -E_b\int \psi_{\mathcal{A}}(\mathbf{r})\, d^3r. \tag{8.145}$$

Therefore,

$$\int e^{-i\mathbf{q}\cdot\mathbf{r}}\psi_{\mathcal{A}}(\mathbf{r})\, d^3r = \frac{E_b}{E+E_b}\int \psi_{\mathcal{A}}(\mathbf{r})\, d^3r. \tag{8.146}$$

From the reaction energetics, one should recognize that $E + E_b$ is identical to $\hbar\omega_k$, the energy of the absorbed photon. In addition, since $\psi_{\mathcal{A}}$ is the wavefunction for an s-state, we have

$$\int e^{-i\mathbf{q}\cdot\mathbf{r}}\psi_{\mathcal{A}}(\mathbf{r})\, d^3r = \frac{E_b}{\hbar\omega_k}\int_0^\infty \left[\frac{1}{\sqrt{4\pi}}R(r)\right] 4\pi r^2 dr. \tag{8.147}$$

The radial integral can be approximated by changing the lower limit to r_0. This is justified because the wavefunction decays very slowly in the region $r > r_0$, while r_0 itself is very small. The appropriate (approximately) normalized wavefunction is therefore

$$R(r) = \frac{u(r)}{r} \approx \sqrt{2\kappa}\frac{e^{-\kappa(r-r_0)}}{r}, \tag{8.148}$$

where

$$\kappa = \left(\frac{m_p E_b}{\hbar^2}\right)^{1/2}. \tag{8.149}$$

The needed integral then becomes

$$\int e^{-i\mathbf{q}\cdot\mathbf{r}}\psi_{\mathcal{A}}(\mathbf{r})\, d^3r = \frac{E_b}{\hbar\omega_k}\sqrt{\frac{8\pi}{\kappa^3}}\,(1+\kappa r_0)\,. \qquad (8.150)$$

We can now write the following expression for the differential cross-section:

$$\left(\frac{d\sigma}{d\Omega}\right)_{\mathrm{cm}} = \frac{2e^2\hbar q E_b^2}{m_p c\,(\hbar\omega_k)^3\,\kappa^3}\,(1+\kappa r_0)^2\,(\boldsymbol{\epsilon}_{\mathbf{k}\lambda}\cdot\mathbf{q})^2\,. \qquad (8.151)$$

From Eq. 8.125, this is equivalent to

$$\left(\frac{d\sigma}{d\Omega}\right)_{\mathrm{cm}} = \frac{2E_b^2}{137\hbar m_p \kappa^3}\,(1+\kappa r_0)^2\left(\frac{\hbar q}{\hbar\omega_k}\right)^3 \sin^2\theta_{\mathrm{cm}}\cos^2\phi_{\mathrm{cm}}, \qquad (8.152)$$

where here, of course, the angles are measured in the center-of-mass frame. Finally, one can use $\hbar q = \sqrt{m_p E} = \sqrt{m_p\,(\hbar\omega_k - E_b)}$ and Eq. 8.149 to obtain the result

$$\left(\frac{d\sigma}{d\Omega}\right)_{\mathrm{cm}} = \frac{2}{137\kappa^2}\,(1+\kappa r_0)^2\left(\frac{\sqrt{E_b\,(\hbar\omega_k - E_b)}}{\hbar\omega_k}\right)^3 \sin^2\theta_{\mathrm{cm}}\cos^2\phi_{\mathrm{cm}}. \qquad (8.153)$$

The total cross-section σ is the same whether viewed from the center-of-mass or the laboratory coordinate system. Simply integrating Eq. 8.153 over all solid angles produces

$$\sigma = \frac{8\pi}{3}\frac{(1+\kappa r_0)^2}{137\kappa^2}\left(\frac{\sqrt{E_b\,(\hbar\omega_k - E_b)}}{\hbar\omega_k}\right)^3. \qquad (8.154)$$

Our expression is nearly identical to the one given by Evans [58]. The only difference is that the constant factor $(1+\kappa r_0)^2$ is replaced by the somewhat reduced value of $(1-\kappa r_0)^{-1}$ in that reference. A plot of Eq. 8.154 is shown in Fig. 8.12. Evans discusses the fact that our expression significantly underestimates the cross-section for photon energies within a few MeV above the threshold energy. This is because the calculation we have presented assumes that the n-p system undergoes a pure electric-dipole transition. In actuality, however, it is more likely that a magnetic-dipole transition occurs when in the immediate vicinity of the reaction threshold. The probability for this "photomagnetic disintegration" of the deuteron falls off quite rapidly with photon energy, so that for $\hbar\omega_k$ greater than about 5 or 6 MeV, its contribution to the cross-section is rather negligible, and the measured cross-section is well-represented by Eq. 8.154.

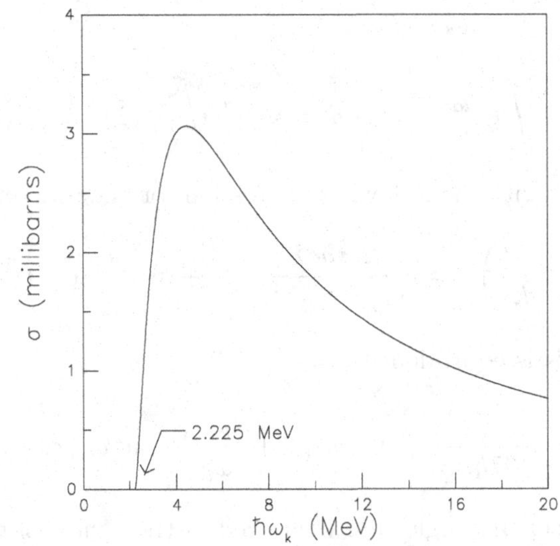

Figure 8.12 Deuteron photodisintegration cross-section vs. photon energy.

Suggested References

A detailed treatment of first-order (and higher) photon-atom interactions, as well as the subject of radiation linewidth, can be found in the following texts:

[a] W. Heitler, *The Quantum Theory of Radiation*, 3rd ed. (Dover Publications, New York, 1984).

[b] R. Loudon, *The Quantum Theory of Light*, 3rd ed. (Oxford University Press, Oxford, 2000).

[c] W. H. Louisell, *Quantum Statistical Properties of Radiation* (John Wiley and Sons, New York, 1973).

[d] D. Marcuse, *Engineering Quantum Electrodynamics* (Harcourt, Brace and World, New York, 1970).

[e] J. Weiner and P.-T. Ho, *Light-Matter Interaction: Fundamentals and Applications* (John Wiley and Sons, Hoboken, NJ, 2003).

Problems

1. In the electric dipole approximation, prove the following famous *selection rules* for transitions between one-electron states in an atom:

$$\Delta l = \pm 1 \qquad \text{and} \qquad \Delta m = 0, \pm 1$$

i.e., transitions are only allowed if the angular momentum quantum number increments or decrements by unity and, in addition, the magnetic quantum number either remains unchanged or changes by unity. (**Hint:** The initial and final wavefunctions are of the form $R(r)\,Y_{l,m}(\theta,\phi)$, where $Y_{l,m}$'s are the spherical harmonics.)

2. Calculate the numerical value of the lifetime τ of the $2p$ -state of the hydrogen atom. (You will need to look up explicit hydrogen-atom wavefunctions in a standard quantum mechanics text.)
   (Ans.: 1.595 ns)

3. For atomic hydrogen at room temperature and atmospheric pressure, calculate the numerical values of the natural linewidth (see previous problem), the Doppler linewidth, and the collision-broadened linewidth for the $2p \rightarrow 1s$ transition. Compare the natural linewidth obtained to the experimentally observed value.

4. The *Lyman series* in the emission spectrum of hydrogen corresponds specifically to transitions between excited states and the ground state. Consider the $2p \rightarrow 1s$ transition, the so-called α*-line*, and the $3p \rightarrow 1s$ transition, referred to as the β*-line.* Calculate the ratio of intensities, I_α/I_β, corresponding to the two lines. (You will need to look up explicit hydrogen-atom wavefunctions in a standard quantum mechanics text.)
   (Ans.: 3.18)

5. Suppose the electron were bound to the nucleus by a square-well potential. Calculate the energy dependence of the cross-section for the photoelectric effect. Assume that the photon energy is much larger than the binding energy of the electron, and that the potential has a short range.

Chapter 9
SECOND-ORDER PROCESSES AND THE SCATTERING OF PHOTONS

In first-order radiation processes, a single photon is either created or annihilated as a result of the interaction Hamiltonian $(-e/mc)\,\hat{\mathbf{A}}_i\cdot\hat{\mathbf{p}}_i$ inducing a direct transition between some initial and final state of the combined atom-field system. We now consider second-order photon-atom processes—these involve two-photons. Recall that the term "second-order" refers to processes that are second-order in the interaction and/or second-order in the perturbation. By "second-order in the interaction" we mean that the term $(e^2/2mc^2)\,\hat{A}_i^2$ in the Hamiltonian is responsible for the process. Because $\hat{\mathbf{A}}_i$ is linear in the raising and lowering operators $\hat{a}_{\mathbf{k}\lambda}$ and $\hat{a}^\dagger_{\mathbf{k}\lambda}$, the operator $\hat{A}_i^2$ can produce a non-vanishing matrix element $\langle n \mid \hat{V} \mid m\rangle$ when the process involves two photons. If the process is second-order in the perturbation, then intermediate states $|l\rangle$ must be involved, and one needs to consider paired matrix elements of the form $\langle n \mid \hat{V} \mid l\rangle\langle l \mid \hat{V} \mid m\rangle$. Since each matrix element applied to the first-order interaction term $(-e/mc)\,\hat{\mathbf{A}}_i\cdot\hat{\mathbf{p}}_i$ creates or annihilates a single photon, the second-order perturbation also involves two photons.

In this chapter, we will focus exclusively on second-order processes as they relate to the scattering of photons. A scattering event involves the annihilation of the incident photon and the creation of the scattered photon, hence it is a two-photon process. Examples include the following:

1. Scattering of a photon by a free electron (e.g., Thomson scattering)
2. Scattering of a photon by an atom, i.e.,
   (a) Compton scattering
   (b) Rayleigh scattering
   (c) Raman scattering

We will examine certain aspects of each of these types of scattering processes.

9.1 Scattering of Electromagnetic Radiation by a Free Electron

Let us begin by considering the scattering of a photon having an energy much greater than the binding energy of an atomic electron. It is then quite reasonable to treat the electron as essentially unbound, or free, and initially at rest. Below we show that for this, the simplest of scattering processes, there is a clear link between the quantum-mechanically derived cross-section and the classical result.

9.1.1 Classical Theory

For a classical treatment of the problem, one considers the incident radiation to be a monochromatic, polarized EM wave:

$$\mathbf{E}_i(\mathbf{r},t) = \boldsymbol{\epsilon} E_0 \cos(\mathbf{k}\cdot\mathbf{r} - \omega t)\,. \tag{9.1}$$

The effect of this oscillating field on a free electron is to produce a harmonic acceleration of

$$\dot{\mathbf{v}} = \frac{\mathbf{F}_i}{m} = \frac{e}{m}\mathbf{E}_i = \boldsymbol{\epsilon}\frac{eE_0}{m}\cos(\mathbf{k}\cdot\mathbf{r} - \omega t)\,. \tag{9.2}$$

The electron, in turn, radiates away energy as a scattered EM wave, also at the frequency ω. Assuming the motion of the electron is non-relativistic, the scattered field at position $\mathbf{R}$ relative to the electron is given by [59]

$$\mathbf{E}_s(\mathbf{R},t) = \frac{e}{c^2}\left[\frac{\mathbf{e}_R\times(\mathbf{e}_R\times\dot{\mathbf{v}})}{R}\right], \tag{9.3}$$

where $\mathbf{e}_R = \mathbf{R}/R$ (Fig. 9.1 shows the scattering geometry). From the Poynting vector (see Eq. 4.129)

$$\mathbf{S}_s = \frac{c}{4\pi}\left|\mathbf{E}_s\right|^2 \mathbf{e}_R, \tag{9.4}$$

the power radiated per steradian becomes

$$\frac{dP}{d\Omega} = (\mathbf{S}_s\cdot\mathbf{e}_R)\,R^2 = \frac{e^2}{4\pi c^3}\left|\mathbf{e}_R\times(\mathbf{e}_R\times\dot{\mathbf{v}})\right|^2 = \frac{e^2}{4\pi c^3}\left|\dot{\mathbf{v}}\right|^2\sin^2\gamma, \tag{9.5}$$

where γ is the angle between $\mathbf{e}_R$ and the polarization vector $\boldsymbol{\epsilon}$. Comparison with Eq. 4.130 reminds one that the $\sin^2\gamma$ angular distribution is characteristic of electric dipole radiation, as previously depicted in Fig. 4.3. *Larmor's formula* for the total instantaneous power emitted by the accelerated charge is obtained by simply integrating Eq. 9.5 over all directions:

$$P = \frac{2e^2}{3c^3}\left|\dot{\mathbf{v}}\right|^2. \tag{9.6}$$

However, for the purpose of finding the angular differential cross-section for the scattering process, we need to return to $dP/d\Omega$, and specifically, its time-averaged value, namely

$$\left\langle\frac{dP}{d\Omega}\right\rangle = \frac{e^2}{4\pi c^3}\left\langle\left|\dot{\mathbf{v}}\right|^2\right\rangle\sin^2\gamma. \tag{9.7}$$

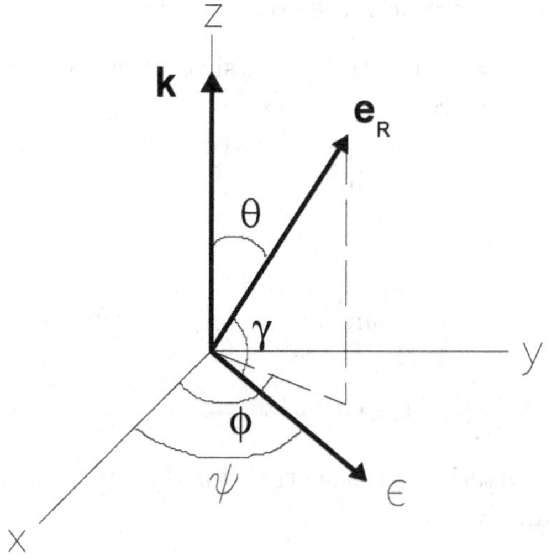

Figure 9.1 Coordinate system for the scattering of EM radiation by a free electron.

We see from Eq. 9.2 that

$$\langle |\dot{\mathbf{v}}|^2 \rangle = \frac{e^2 E_0^2}{m^2} \langle \cos^2 (\mathbf{k}\cdot\mathbf{r} - \omega t) \rangle = \frac{e^2 E_0^2}{2m^2}, \tag{9.8}$$

so

$$\left\langle \frac{dP}{d\Omega} \right\rangle = \frac{e^4 E_0^2}{8\pi m^2 c^3} \sin^2 \gamma. \tag{9.9}$$

The ratio of $\langle dP/d\Omega \rangle$ to the intensity of the incoming field gives the classical differential scattering cross-section, i.e.,

$$\frac{d\sigma}{d\Omega} = \frac{\langle dP/d\Omega \rangle}{\frac{c}{8\pi} E_0^2}. \tag{9.10}$$

Thus, for polarized incident radiation, the cross-section is

$$\left(\frac{d\sigma}{d\Omega} \right)_{\text{pol}} = r_0^2 \sin^2 \gamma, \tag{9.11}$$

where $r_0 = e^2/mc^2$ is the classical electron radius.

To calculate the differential cross-section for unpolarized incident radiation, we refer to Fig. 9.1, and average Eq. 9.11 over all possible values of the angle ψ, i.e.,

$$\left(\frac{d\sigma}{d\Omega} \right)_{\text{unpol}} = r_0^2 \left\langle \sin^2 \gamma \right\rangle_{\psi}. \tag{9.12}$$

In the specified scattering geometry, we observe that

$$\begin{aligned}
\mathbf{e}_R &= \mathbf{e}_x \sin\theta\cos\phi + \mathbf{e}_y \sin\theta\sin\phi + \mathbf{e}_z\cos\theta \\
\boldsymbol{\epsilon} &= \mathbf{e}_x\cos\psi + \mathbf{e}_y\sin\psi \\
\cos\gamma &= \mathbf{e}_R\cdot\boldsymbol{\epsilon} = \sin\theta\cos(\phi-\psi) \\
\sin^2\gamma &= 1-\sin^2\theta\cos^2(\phi-\psi)\,.
\end{aligned} \tag{9.13}$$

Therefore,

$$\begin{aligned}
\left(\frac{d\sigma}{d\Omega}\right)_{\text{unpol}} &= r_0^2\left[1-\sin^2\theta\left\langle\cos^2(\phi-\psi)\right\rangle_\psi\right] = r_0^2\left(1-\frac{1}{2}\sin^2\theta\right) \\
&= \frac{1}{2}r_0^2\left(1+\cos^2\theta\right).
\end{aligned} \tag{9.14}$$

Classically, the total scattering cross-section σ is the same for polarized or unpolarized incident radiation, namely

$$\sigma = \sigma_{\text{pol}} = \sigma_{\text{unpol}} = \int\left(\frac{d\sigma}{d\Omega}\right)d\Omega = \frac{8\pi}{3}r_0^2. \tag{9.15}$$

This result is known as the *Thomson cross-section* for scattering by a free electron.

9.1.2 Quantum Theory

In this case, one treats the incident radiation as a stream of photons, each having energy $\hbar\omega_k$ and momentum $\hbar k$. At the simplest level, one can envision the scattering event to be a classical-like, but relativistic, "collision" between one of these photons and the resting free electron, as illustrated in Fig. 9.2. Because of the recoil of the electron, the scattered photon will have an energy $\hbar\omega_{k'}$ and momentum $\hbar k'$ less than that of the incident photon. Thus, unlike the classical situation, the scattering process causes the radiation frequency to be shifted downward, i.e., $\omega_{k'} < \omega_k$. The precise relationship between the incident and scattered frequencies comes from demanding that both the relativistic energy and momentum of the system be conserved, i.e.,

$$\hbar\omega_k + mc^2 = \hbar\omega_{k'} + E_e \tag{9.16}$$

$$\hbar k = \hbar k'\cos\theta + p_e\cos\phi \tag{9.17}$$

$$0 = \hbar k'\sin\theta - p_e\sin\phi, \tag{9.18}$$

where θ and ϕ are the scattering angles for the photon and electron, respectively. E_e and p_e, the relativistic energy and momentum of the scattered electron, are related by

$$E_e^2 = m^2c^4 + p_e^2c^2. \tag{9.19}$$

Manipulation of Eqs. 9.16–9.19 produces

$$\frac{\omega_{k'}}{\omega_k} = \frac{k'}{k} = \left[1+\alpha(1-\cos\theta)\right]^{-1}, \tag{9.20}$$

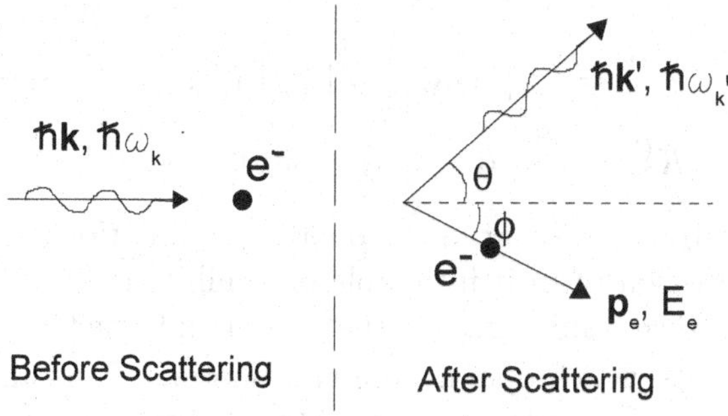

Figure 9.2 The scattering of a photon by a free electron.

where $\alpha = \hbar\omega_k/mc^2$. Equation 9.20, commonly referred to as the *Compton scattering formula*, gives the precise frequency of the radiation scattered into angle θ. One should be aware, however, that more correctly, the true "Compton effect" refers to an event where the photon is, in actuality, scattered by an atomic electron, causing the electron to go from a bound state to the continuum in the process. For photon energies that are non-relativistic, i.e., $\alpha \ll 1$, it is then certainly not appropriate to treat the electron as initially free or at rest. Consequently, Eq. 9.20 is no longer exact, and a small, but measurable, spread of frequencies will appear at a given scattering angle. Later on, in Example 9.2, it will be shown that if the incident photon is non-relativistic, but $\hbar\omega_k$ is much larger than the electron binding energy, then it is appropriate to treat the electron as free, but not initially at rest—this is called the *impulse approximation.* In cases like this, we will see that one can relate the observed frequency spread of the scattered radiation to the electron momentum distribution within the target atom.

We now turn to the quantum-mechanical calculation of the scattering cross-section from a free electron, restricting our derivation to the non-relativistic case. The electron-field Hamiltonian consists of a first and second-order interaction term, i.e.,

$$\hat{V} = \hat{V}_1 + \hat{V}_2, \tag{9.21}$$

where

$$\hat{V}_1 = -\frac{e}{mc}\hat{\mathbf{A}}\cdot\hat{\mathbf{p}} \tag{9.22}$$

$$\hat{V}_2 = \frac{e^2}{2mc^2}\hat{A}^2, \tag{9.23}$$

and the initial and final states are $|m\rangle = |\mathcal{A}\rangle\,|1_{\mathbf{k}\lambda}, 0_{\mathbf{k}'\lambda'}\rangle$ and $|n\rangle = |\mathcal{B}\rangle\,|0_{\mathbf{k}\lambda}, 1_{\mathbf{k}'\lambda'}\rangle$, respectively. Here, $|\mathcal{A}\rangle$ and $|\mathcal{B}\rangle$ represent the initial and final state of the electron alone. From first-order perturbation theory, we have the transition rate

$$W_{nm} = \frac{2\pi}{\hbar}\left|K_1^{(1)} + K_2^{(1)}\right|^2 g\left(E_n\right), \tag{9.24}$$

where

$$K_1^{(1)} = \langle \mathcal{B} \mid \langle 0_{\mathbf{k}\lambda}, 1_{\mathbf{k}'\lambda'} \mid \hat{V}_1 \mid 1_{\mathbf{k}\lambda}, 0_{\mathbf{k}'\lambda'} \rangle \mid \mathcal{A} \rangle \tag{9.25}$$

$$K_2^{(1)} = \langle \mathcal{B} \mid \langle 0_{\mathbf{k}\lambda}, 1_{\mathbf{k}'\lambda'} \mid \hat{V}_2 \mid 1_{\mathbf{k}\lambda}, 0_{\mathbf{k}'\lambda'} \rangle \mid \mathcal{A} \rangle. \tag{9.26}$$

Since the field operator $\hat{\mathbf{A}}$ is linear in the photon annihilation and creation operators, the first-order interaction term is incapable of annihilating one photon and creating another, so the $K_1^{(1)}$-term vanishes.* On the other hand, the term for the second-order interaction involves $\hat{A}^2$, which operates on two photons. This allows $K_2^{(1)}$ to survive, producing the Feynman diagram in Fig. 9.3. Retaining only terms that produce non-zero matrix elements between the initial and final states, we have (recalling Eq. 8.1 for $\hat{\mathbf{A}}$)

$$\begin{aligned} K_2^{(1)} &= \frac{e^2}{2mc^2}\frac{2\pi\hbar c}{L^3}\frac{1}{\sqrt{kk'}}\left(\boldsymbol{\epsilon}_{\mathbf{k}\lambda}\cdot\boldsymbol{\epsilon}_{\mathbf{k}'\lambda'}\right) \\ &\quad\times\left\langle \mathcal{B} \left| \left\langle 0_{\mathbf{k}\lambda}, 1_{\mathbf{k}'\lambda'} \right| \left[\hat{a}_{\mathbf{k}\lambda}\hat{a}^{\dagger}_{\mathbf{k}'\lambda'} + \hat{a}^{\dagger}_{\mathbf{k}'\lambda'}\hat{a}_{\mathbf{k}\lambda}\right] e^{i(\mathbf{k}-\mathbf{k}')\cdot\mathbf{r}} \left| 1_{\mathbf{k}\lambda}, 0_{\mathbf{k}'\lambda'} \right\rangle \right| \mathcal{A} \right\rangle \\ &= 2\left(\frac{e^2}{2mc^2}\right)\left(\frac{2\pi\hbar c}{L^3}\right)\frac{1}{\sqrt{kk'}}\left(\boldsymbol{\epsilon}_{\mathbf{k}\lambda}\cdot\boldsymbol{\epsilon}_{\mathbf{k}'\lambda'}\right) e^{i\mathbf{Q}\cdot\mathbf{r}}\delta_{AB}, \end{aligned} \tag{9.27}$$

where, at this point, we have introduced the wavevector transfer $\mathbf{Q} = \mathbf{k} - \mathbf{k}'$.

The final density of states is

$$g\left(E_n\right) dE_n = \left(\frac{L}{2\pi}\right)^3 k'^2 dk' d\Omega, \tag{9.28}$$

or

$$g\left(E_n\right) = \frac{\left(\frac{L}{2\pi}\right)^3 k'^2 d\Omega}{dE_n/dk'}. \tag{9.29}$$

The latter expression must properly take into account the recoil of the electron, $\hbar\mathbf{Q}$. With the aid of Fig. 9.4, the energy of the final electron-photon state can be written as

$$E_n = \hbar c k' + \frac{\hbar^2 Q^2}{2m} = \hbar c\left[k' + \frac{\hbar}{2mc}\left(k^2 + k'^2 - 2kk'\cos\theta\right)\right]. \tag{9.30}$$

Thus,

$$\frac{dE_n}{dk'} = \hbar c\left[1 + \alpha\left(\frac{k'}{k} - \cos\theta\right)\right]. \tag{9.31}$$

---

*This term vanishes whether the electron is initially stationary or not. However, in order that there not be a scattering contribution from second-order perturbation theory due to the term $\hat{\mathbf{A}}\cdot\hat{\mathbf{p}}$, we must nevertheless demand that the electron be initially motionless.

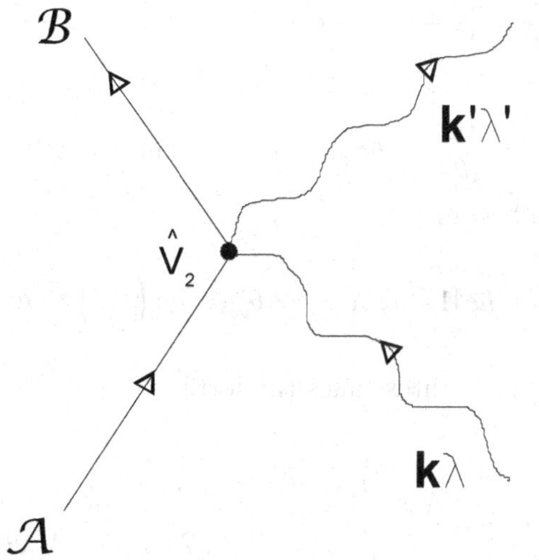

Figure 9.3 Feynman diagram for the scattering of a photon by a free electron.

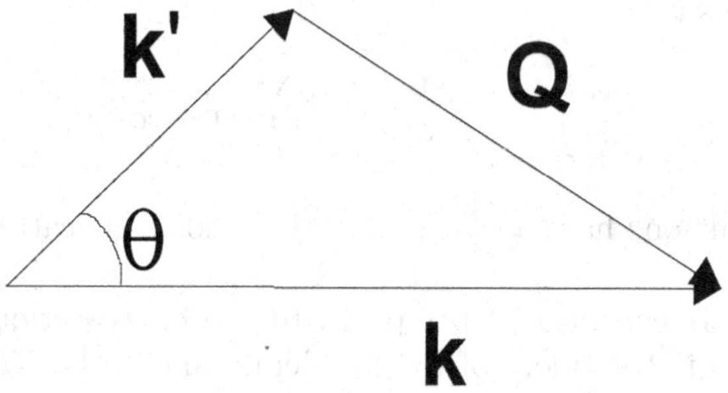

Figure 9.4 Relation between the incident and scattered wavevectors, $\mathbf{k}$ and $\mathbf{k}'$, and the wavevector transfer, $\mathbf{Q}$.

When the basic Compton formula, Eq. 9.20, is applied to the non-relativistic regime where $\alpha \ll 1$, one has

$$\frac{k'}{k} = [1 + \alpha (1 - \cos\theta)]^{-1} \simeq 1 - \alpha (1 - \cos\theta), \tag{9.32}$$

giving

$$\frac{dE_n}{dk'} \simeq \hbar c \left[1 + \alpha (1 - \alpha)(1 - \cos\theta)\right]. \tag{9.33}$$

With $1 - \alpha \simeq 1$, this becomes

$$\frac{dE_n}{dk'} \simeq \hbar c \left[1 + \alpha (1 - \cos\theta)\right] = \hbar c \left(\frac{k}{k'}\right) = \hbar c \left(\frac{\omega_k}{\omega_{k'}}\right). \tag{9.34}$$

Thus, the final density of states takes the form

$$g(E_n) = \frac{\left(\frac{L}{2\pi}\right)^3 k'^2 d\Omega}{\hbar c \left(\frac{\omega_k}{\omega_{k'}}\right)} = \left(\frac{L}{2\pi}\right)^3 \frac{\omega_{k'}^2}{\hbar c^3} \left(\frac{\omega_{k'}}{\omega_k}\right) d\Omega. \tag{9.35}$$

The differential scattering cross-section is obtained by calculating $d\sigma = W_{nm}/\frac{c}{L^3}$. For the case of polarized photons, the expression, in the non-relativistic limit, is

$$\left(\frac{d\sigma}{d\Omega}\right)_{\text{pol}} = r_0^2 \left(\frac{\omega_{k'}}{\omega_k}\right)^2 \left(\boldsymbol{\epsilon}_{\mathbf{k}\lambda} \cdot \boldsymbol{\epsilon}_{\mathbf{k}'\lambda'}\right)^2 = r_0^2 \left(\frac{\omega_{k'}}{\omega_k}\right)^2 \sin^2\gamma. \tag{9.36}$$

Following the same procedure as in the classical calculation, the cross-section for unpolarized photons is

$$\left(\frac{d\sigma}{d\Omega}\right)_{\text{unpol}} = \frac{1}{2} r_0^2 \left(\frac{\omega_{k'}}{\omega_k}\right)^2 \left(1 + \cos^2\theta\right). \tag{9.37}$$

In the classical limit one has $\omega_{k'} = \omega_k$, and the Thomson scattering expression, Eq. 9.14, is recovered.

Equation 9.37 can also be compared to the cross-section expression derived from a full relativistic treatment of the problem, namely the *Klein-Nishina formula* [35, pp. 215-219]:

$$\left(\frac{d\sigma}{d\Omega}\right)_{\text{unpol}} = \frac{1}{2} r_0^2 \left(\frac{\omega_{k'}}{\omega_k}\right)^2 \left(\frac{\omega_k}{\omega_{k'}} + \frac{\omega_{k'}}{\omega_k} - \sin^2\theta\right). \tag{9.38}$$

For $\hbar\omega_k \ll mc^2$ we have $\alpha \to 0$, so according to Eq. 9.20, $\omega_{k'}/\omega_k \to 1$, and this expression clearly reduces to our non-relativistic result. Equation 9.20 can also be

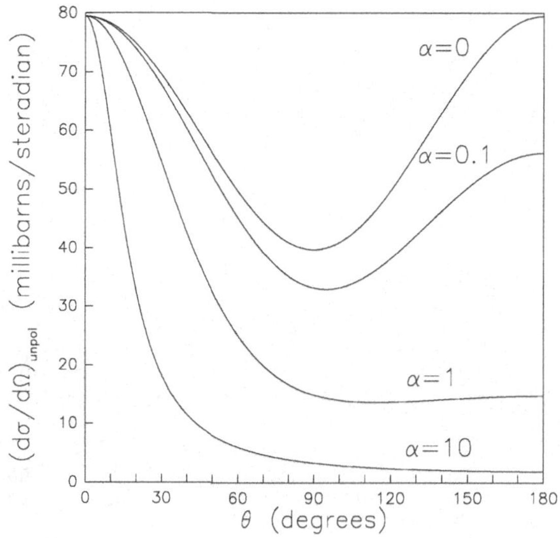

Figure 9.5 Differential cross-section for the scattering of a photon by a free electron, as given by the Klein-Nishina formula ($\alpha = \hbar\omega_k/mc^2$).

used to rewrite the Klein-Nishina formula as

$$\left(\frac{d\sigma}{d\Omega}\right)_{\text{unpol}} = \frac{1}{2} r_0^2 \left(\frac{\omega_{k'}}{\omega_k}\right)^2 \left(1+\cos^2\theta\right)\left\{1 + \frac{\alpha^2\left(1-\cos\theta\right)^2}{\left(1+\cos^2\theta\right)\left[1+\alpha\left(1-\cos\theta\right)\right]}\right\}. \tag{9.39}$$

In this form, the second term in the curly brackets is readily identified as the relativistic correction to the cross-section. As shown in Fig. 9.5, there is a pronounced increase in the fraction of photons scattered in the forward direction with increasing incident photon energy. The total cross-section for the scattering of photons by a free electron is obtained by integrating the Klein-Nishina formula over all solid angles. The resulting expression is

$$\sigma = \frac{8\pi}{3} r_0^2 \left\{\frac{3\left(1+\alpha\right)}{4\alpha^2}\left[\frac{2\left(1+\alpha\right)}{1+2\alpha} - \frac{\ln\left(1+2\alpha\right)}{\alpha}\right] - \frac{3}{4}\left[\frac{1+3\alpha}{\left(1+2\alpha\right)^2} - \frac{\ln\left(1+2\alpha\right)}{2\alpha}\right]\right\}, \tag{9.40}$$

and the corresponding plot appears in Fig. 9.6.

9.2 Scattering of Photons by Atoms

A photon is scattered by an atom because of the interaction of the EM field with the constituent bound electrons of the atom. In order to account for all the interactions, we again have $\hat{V} = \hat{V}_1 + \hat{V}_2$, but now

$$\hat{V}_1 = -\frac{e}{mc}\sum_i \hat{\mathbf{A}}_i \cdot \hat{\mathbf{p}}_i \qquad \text{and} \qquad \hat{V}_2 = \frac{e^2}{2mc^2}\sum_i \hat{A}_i^2, \tag{9.41}$$

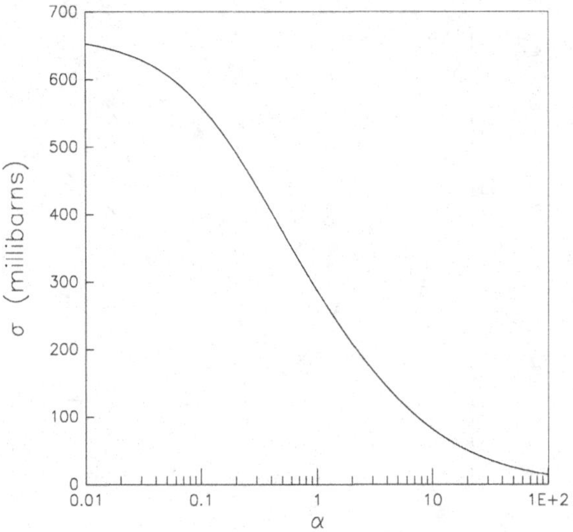

Figure 9.6 Total cross-section for the scattering of a photon by a free electron, as given by the Klein-Nishina formula ($\alpha = \hbar\omega_k/mc^2$).

where the sums account for all the atomic electrons. As before, the initial and final states are $|m\rangle = |\mathcal{A}\rangle\,|1_{\mathbf{k}\lambda}, 0_{\mathbf{k}'\lambda'}\rangle$ and $|n\rangle = |\mathcal{B}\rangle\,|0_{\mathbf{k}\lambda}, 1_{\mathbf{k}'\lambda'}\rangle$, only now $|\mathcal{A}\rangle$ and $|\mathcal{B}\rangle$ are the initial and final states of the entire atom, not just that of a single electron. We have already seen that a first-order perturbation applied to $\hat{V}_2$ gives rise to a diagram as in Fig. 9.3. However, since, in general, the final atomic state is not being limited to the continuum, here one should not interpret $K_2^{(1)}$ as a Compton scattering term. In addition, since the $\hat{\mathbf{p}}_i$'s are not initially zero for electrons that are bound, it is equally important to now consider the second-order perturbation expansion of $\hat{V}_1$. That is to say, the complete expression for the scattering probability to second-order in both the interaction and the perturbation is

$$W_{nm} = \frac{2\pi}{\hbar}\left|K_2^{(1)} + K_1^{(2)}\right|^2 g\left(E_n\right), \tag{9.42}$$

where, generalizing Eq. 9.27, we write

$$K_2^{(1)} = 2\left(\frac{e^2}{2mc^2}\right)\left(\frac{2\pi\hbar c}{L^3}\right)\frac{1}{\sqrt{kk'}}\left(\boldsymbol{\epsilon}_{\mathbf{k}\lambda}\cdot\boldsymbol{\epsilon}_{\mathbf{k}'\lambda'}\right)\langle\mathcal{B}\mid\sum_i e^{i\mathbf{Q}\cdot\mathbf{r}_i}\mid\mathcal{A}\rangle, \tag{9.43}$$

and $K_1^{(2)}$ is given by

$$K_1^{(2)} = \sum_l \frac{\langle n\mid\hat{V}_1\mid l\rangle\langle l\mid\hat{V}_1\mid m\rangle}{E_m - E_l}. \tag{9.44}$$

As we will see, the relative contributions of $K_2^{(1)}$ and $K_1^{(2)}$ to the scattering cross-section depend strongly on the incident photon energy. For this reason, it is best to examine x-ray scattering and light scattering separately.

9.2.1 X-ray Scattering

In this section, we focus solely on the case where $\hbar\omega_k$ is much greater than the binding energy of the atomic electrons. This condition is met in many x-ray scattering experiments where the incident beam is produced by characteristic x-ray emission from a metal target bombarded by electrons. For example, the K_α-line of molybdenum has a wavelength of 0.0711 nm, or a photon energy of 17.4 keV; at least for light elements, this far exceeds the binding energies, which are on the order of 10–100 eV. One should realize that even at these incident photon energies, there are processes that compete with the Compton scattering process previously described, namely, *Rayleigh scattering* and *Raman scattering*, where the atom is left in an unionized state. In general, when the aforementioned condition on the photon energy is satisfied, it has been shown by Eisenberger and Platzman [60] that the term $K_1^{(2)}$ is insignificant relative to $K_2^{(1)}$. Then, as before, one can obtain the cross-section for polarized incident photons:

$$\left(\frac{d\sigma}{d\Omega}\right)_{\text{pol}} = r_0^2\left(\frac{\omega_{k'}}{\omega_k}\right)(\boldsymbol{\epsilon}_{\mathbf{k}\lambda}\cdot\boldsymbol{\epsilon}_{\mathbf{k}'\lambda'})^2\left|\langle\mathcal{B}\mid\sum_i e^{i\mathbf{Q}\cdot\mathbf{r}_i}\mid\mathcal{A}\rangle\right|^2 . \qquad (9.45)$$

Here, the expression only involves $\omega_{k'}/\omega_k$ to the first power rather than its square because in the case of bound electrons, one should not incorporate recoil considerations into the density-of-states factor. That is, one needs to use

$$g(E_n) = \left(\frac{L}{2\pi}\right)^3 k'^2\frac{d\Omega}{\hbar c} = \left(\frac{L}{2\pi}\right)^3\frac{\omega_{k'}^2}{\hbar c^3}d\Omega, \qquad (9.46)$$

rather than Eq. 9.35.

The cross-section expression, Eq. 9.45, can now be applied to three distinct cases:

- Suppose the initial state is the atomic ground-state, i.e., $|\mathcal{A}\rangle = |0\rangle$. Then, if the final state is $|\mathcal{B}\rangle = |n\rangle$, a discrete excited electronic state, the scattering is inelastic and falls under the heading of *electronic Raman scattering*.
- If the final state of the atom $|\mathcal{B}\rangle$ is in the continuum, then we are considering *Compton scattering*, which is also an inelastic process.
- When the scattering is elastic and does not involve any excitation of the atom, that is to say, $|\mathcal{B}\rangle = |\mathcal{A}\rangle = |0\rangle$ and $\omega_{k'} = \omega_k$, then the cross-section corresponds to that of *Rayleigh scattering*.

For the specific case of scattering from hydrogen, Eisenberger and Platzman [60] have calculated the relative magnitudes of the scattering for the three processes as a function of Qa, where a is the Bohr radius. It is seen that for $Qa < 0.5$, Rayleigh scattering is by far the most important process, while for $Qa > 2.0$, both the Rayleigh and Raman contributions diminish as $(Qa)^{-8}$, and Compton scattering dominates. Actually, for the photon energies we are considering, Raman scattering is always small, and only appreciable in the vicinity of $Qa \sim 1.0$.

Of the three processes, only Rayleigh scattering is coherent. It is the important process in x-ray diffraction. In a typical experiment, the x-ray beam is unpolarized, so the relevant cross-section expression is

$$\left(\frac{d\sigma}{d\Omega}\right)_{\text{unpol}} = \frac{1}{2} r_0^2 \left(1 + \cos^2\theta\right) \left|f(\mathbf{Q})\right|^2, \tag{9.47}$$

where we have defined an *atomic form factor*

$$f(\mathbf{Q}) = \langle 0 \mid \sum_i e^{i\mathbf{Q}\cdot\mathbf{r}_i} \mid 0 \rangle. \tag{9.48}$$

Notice that in the limit $Q \to 0$, the form factor approaches the atomic number Z, and the scattering cross-section is proportional to Z^2. Thus, Rayleigh scattering is very weak for light elements. Introducing the average charge-density at position $\mathbf{r}$ in the atom by

$$\rho(\mathbf{r}) = \langle 0 \mid \sum_i \delta(\mathbf{r} - \mathbf{r}_i) \mid 0 \rangle, \tag{9.49}$$

the form factor can be rewritten as the following integral over the atomic volume:

$$f(\mathbf{Q}) = \int_V \rho(\mathbf{r})\, e^{i\mathbf{Q}\cdot\mathbf{r}} d^3r. \tag{9.50}$$

$f(\mathbf{Q})$ is seen to be the Fourier transform of the atomic charge-density distribution.

When scattering from a condensed phase, such as a solid or liquid, the cross-section is written as

$$\left(\frac{d\sigma}{d\Omega}\right)_{\text{unpol}} = \frac{1}{2} r_0^2 \left(1 + \cos^2\theta\right) \left|G(\mathbf{Q})\right|^2, \tag{9.51}$$

where the atomic form factor $f(\mathbf{Q})$ is replaced by the *scattering factor* for the system, defined as

$$G(\mathbf{Q}) = \langle 0 \mid \sum_\ell \sum_i e^{i\mathbf{Q}\cdot(\mathbf{R}_\ell + \mathbf{r}_{\ell i})} \mid 0 \rangle. \tag{9.52}$$

Here, $\mathbf{R}_\ell$ is the center-position of atom ℓ, and $\mathbf{r}_{\ell i}$ is the location of the ith electron in that atom. $G(\mathbf{Q})$ can be decomposed in terms of the form factors of the individual atoms, i.e.,

$$G(\mathbf{Q}) = \sum_\ell f_\ell(\mathbf{Q})\, e^{i\mathbf{Q}\cdot\mathbf{R}_\ell}, \tag{9.53}$$

with

$$f_\ell(\mathbf{Q}) = \langle 0 \mid \sum_i e^{i\mathbf{Q}\cdot\mathbf{r}_{\ell i}} \mid 0 \rangle. \tag{9.54}$$

Example 9.1 X-Ray Diffraction from a Simple Crystal. Consider, as a specific example, x-ray diffraction from a crystalline solid. The scattering pattern is determined by finding the scattering factor, $G(\mathbf{Q})$, defined by Eq. 9.53. The evaluation of $G(\mathbf{Q})$ depends on the fact that a crystal is constructed of repeating blocks, or *unit cells*, each composed of an identical arrangement of atoms. The different types of crystal lattice structures (see Fig. 9.7) are defined by a set of three basic, or *primitive*, vectors $\mathbf{a}_1$, $\mathbf{a}_2$, and $\mathbf{a}_3$. A translation from a particular location in one unit cell to the corresponding location in any other unit cell is accomplished by means of a *lattice vector*

$$\mathbf{d}_{n_1n_2n_3} = n_1\mathbf{a}_1 + n_2\mathbf{a}_2 + n_3\mathbf{a}_3. \tag{9.55}$$

n_1, n_2, and n_3 are a unique triplet of integers for each unit cell. Let us now re-index the atomic positions $\mathbf{R}_\ell$ in the scattering factor as

$$\mathbf{R}_{n_1n_2n_3j} = \mathbf{d}_{n_1n_2n_3} + \mathbf{r}_j, \tag{9.56}$$

where $\mathbf{r}_j$ denotes the location of atom j within a typical unit cell. We then obtain

$$G(\mathbf{Q}) = \sum_{n_1n_2n_3j} f_j(\mathbf{Q})\, e^{i\mathbf{Q}\cdot\mathbf{R}_{n_1n_2n_3j}} = F(\mathbf{Q}) \sum_{n_1n_2n_3} e^{i\mathbf{Q}\cdot\mathbf{d}_{n_1n_2n_3}}, \tag{9.57}$$

where

$$F(\mathbf{Q}) = \sum_j f_j(\mathbf{Q})\, e^{i\mathbf{Q}\cdot\mathbf{r}_j} \tag{9.58}$$

is usually called the *structure factor for the unit cell.* It represents interference arising from a coherent superposition of scattered waves from the various atoms making up a unit cell.

Now let us investigate the summation over the different unit cells appearing in $G(\mathbf{Q})$. If we denote the number of unit cells in each of the three crystal directions by N_1, N_2, and N_3, then

$$\sum_{n_1n_2n_3} e^{i\mathbf{Q}\cdot\mathbf{d}_{n_1n_2n_3}} = \left[\sum_{n_1=0}^{N_1-1} \left(e^{i\mathbf{Q}\cdot\mathbf{a}_1}\right)^{n_1}\right] \left[\sum_{n_2=0}^{N_2-1} \left(e^{i\mathbf{Q}\cdot\mathbf{a}_2}\right)^{n_2}\right] \left[\sum_{n_3=0}^{N_3-1} \left(e^{i\mathbf{Q}\cdot\mathbf{a}_3}\right)^{n_3}\right]. \tag{9.59}$$

Each sum is a geometric series and can be identified as the form

$$\sum_{n_m=0}^{N_m-1} \left(e^{i\mathbf{Q}\cdot\mathbf{a}_m}\right)^{n_m} = \frac{1 - e^{iN_m\alpha_m}}{1 - e^{i\alpha_m}} = \frac{e^{iN_m\alpha_m/2}}{e^{i\alpha_m/2}} \frac{\sin(N_m\alpha_m/2)}{\sin(\alpha_m/2)}, \tag{9.60}$$

where

$$\alpha_m = \mathbf{Q}\cdot\mathbf{a}_m. \tag{9.61}$$

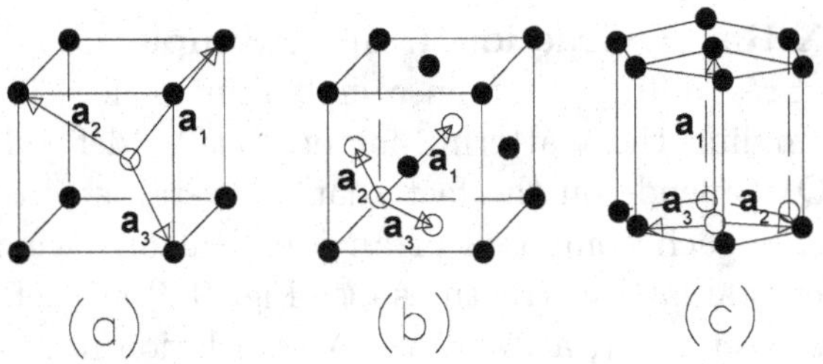

Figure 9.7 Lattice structures for (a) body-centered cubic, (b) face-centered cubic, and (c) hexagonal close-packed crystals.

Replacing all three sums produces

$$\sum_{n_1 n_2 n_3} e^{i\mathbf{Q}\cdot\mathbf{d}_{n_1 n_2 n_3}} = \frac{e^{i(N_1\alpha_1+N_2\alpha_2+N_3\alpha_3)/2}}{e^{i(\alpha_1+\alpha_2+\alpha_3)/2}} \frac{\sin(N_1\alpha_1/2)\sin(N_2\alpha_2/2)\sin(N_3\alpha_3/2)}{\sin(\alpha_1/2)\sin(\alpha_2/2)\sin(\alpha_3/2)}. \tag{9.62}$$

Combining Eqs. 9.51, 9.57, and 9.62, yields the scattering cross-section for the crystal:

$$\begin{aligned}\left(\frac{d\sigma}{d\Omega}\right)_{\text{unpol}} &= \frac{1}{2} r_0^2 \left(1+\cos^2\theta\right) |F(\mathbf{Q})|^2 \\ &\times \left[\frac{\sin(N_1\alpha_1/2)}{\sin(\alpha_1/2)}\right]^2 \left[\frac{\sin(N_2\alpha_2/2)}{\sin(\alpha_2/2)}\right]^2 \left[\frac{\sin(N_3\alpha_3/2)}{\sin(\alpha_3/2)}\right]^2. \end{aligned}\tag{9.63}$$

The result involves the product of three interference factors, one for each set of basic crystal planes. The behavior of one of these factors, having the basic form $\sin^2(N_m\alpha_m/2)/\sin^2(\alpha_m/2)$, is sketched in Fig. 9.8 for two values of N_m. Observe that the scattering is intense when the denominator of each of the interference factors vanishes, i.e., when the α_m's are an integral multiple of 2π, or specifically when the following three conditions are met:

$$\mathbf{Q}\cdot\mathbf{a}_1 = 2\pi h \qquad \mathbf{Q}\cdot\mathbf{a}_2 = 2\pi k \qquad \mathbf{Q}\cdot\mathbf{a}_1 = 2\pi l \qquad (h,k,l = 0,\ \pm1,\ \pm2,\ \ldots). \tag{9.64}$$

This is the well-known *Bragg's law* for x-ray diffraction from a set of crystal planes indexed by hkl. One often encounters Bragg's law in a somewhat more familiar form, namely,

$$2d_{hkl}\sin\phi = m\lambda, \tag{9.65}$$

where λ is the x-ray wavelength, m is an integer representing the diffraction order, and d_{hkl} represents the spacing for one set of planes in the crystal. Figure 9.9 illustrates that ϕ, which is the reflection angle from the planes of interest, is related to our scattering angle θ by $\phi = \theta/2$, and the magnitude of the scattering vector is given by $Q = (4\pi/\lambda)\sin\phi$. It is then worth noting, therefore, that the two forms, Eqs. 9.64 and Eq. 9.65, are really the same law. The values of $\mathbf{Q}$ satisfying the Bragg conditions

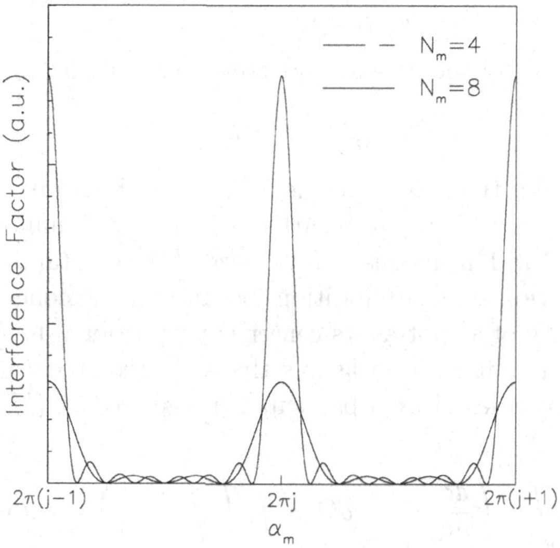

Figure 9.8 Interference factor $\sin^2(N_m\alpha_m/2)/\sin^2(\alpha_m/2)$ for a set of crystal planes ($\alpha_m = \mathbf{Q}\cdot\mathbf{a_m}$).

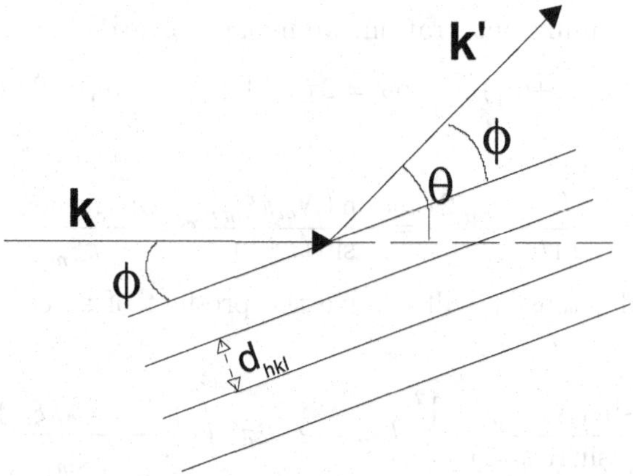

Figure 9.9 Relationship between the scattering angle θ and the reflection angle ϕ for a set of crystal planes, i.e., $\phi = \theta/2$.

are called *reciprocal lattice vectors*—we will denote them by $\mathbf{Q}_{hkl}$. A reciprocal lattice is defined by†

$$\mathbf{Q}_{hkl} = h\mathbf{a}_1^* + k\mathbf{a}_2^* + l\mathbf{a}_3^*, \tag{9.66}$$

where, in order to satisfy the Bragg conditions, i.e., Eqs. 9.64, the $\mathbf{a}_m$'s must be such that

$$\mathbf{a}_m^* \cdot \mathbf{a}_n = 2\pi\delta_{mn}. \tag{9.67}$$

Notice from Fig. 9.8 that the sharpness of the diffraction signal at one of these reciprocal lattice points improves linearly with the total number of unit cells in the crystal, $N = N_1 N_2 N_3$. This means that the size of the crystal can be inferred from the width of the diffraction spot. In addition, because of the coherence of the scattering process, the intensity of a spot at its center is proportional to N^2, not N.

In practice, one usually measures the scattered signal integrated over an entire Bragg reflection indexed by a particular integer-triplet (hkl), i.e., the important quantity is

$$I(hkl) = \int_{\text{reflection}} \left(\frac{d\sigma}{d\Omega}\right)_{\text{unpol}} d\Omega = \int\int\int \left(\frac{d\sigma}{d\Omega}\right)_{\text{unpol}} d\alpha_1 d\alpha_2 d\alpha_3. \tag{9.68}$$

The integrations over the variables $\alpha_1,$ α_2, and α_3 are extended just far enough on each side of $2\pi h$, $2\pi k$, and $2\pi l$, respectively, to cover the entire reflection. To perform the calculation, we use the fact that the expression $(1+\cos^2\theta)\,|F(\mathbf{Q})|^2$ in Eq. 9.63 is approximately constant over the extent of a single reflection; thus, one can pull it out of the angular integral of Eq. 9.68, and evaluate it at the center of the reflection, i.e.,

$$\left(1+\cos^2\theta\right)|F(\mathbf{Q})|^2 \to \left(1+\cos^2\theta_{hkl}\right)|F(\mathbf{Q}_{hkl})|^2. \tag{9.69}$$

To perform the remaining integration, we make the variable transformations

$$\alpha_1 = 2\pi(h+\xi_1) \qquad \alpha_2 = 2\pi(k+\xi_2) \qquad \alpha_3 = 2\pi(l+\xi_3), \tag{9.70}$$

with $\xi_m \ll 1$. Then,

$$\frac{\sin(N_m\alpha_m/2)}{\sin(\alpha_m/2)} = \frac{\sin(N_m\pi\xi_m)}{\sin(\pi\xi_m)} \simeq \frac{\sin(N_m\pi\xi_m)}{\pi\xi_m}, \tag{9.71}$$

and the integrated intensity will involve the product of three integrals, each of the form

$$\begin{aligned}\int\left[\frac{\sin(N_m\alpha_m/2)}{\sin(\alpha_m/2)}\right]^2 d\alpha_m &\simeq 2\pi\int_{-\infty}^{+\infty}\left[\frac{\sin(N_m\pi\xi_m)}{\pi\xi_m}\right]^2 d\xi_m \\ &= 2N_m\int_{-\infty}^{+\infty}\frac{\sin^2 x}{x^2}dx = 2\pi N_m.\end{aligned} \tag{9.72}$$

†In this notation, the superscript $(*)$ does *not* denote a complex conjugate.

As a result, the intensity of a Bragg reflection at a particular hkl is given by

$$I(hkl) = \frac{1}{2}r_0^2\left(1+\cos^2\theta_{hkl}\right)(2\pi)^3 N\left|F\left(\mathbf{Q}_{hkl}\right)\right|^2. \tag{9.73}$$

As a specific example, consider diffraction from a monatomic solid having a body-centered cubic (b.c.c.) structure. These include such elemental solids at room temperature as iron, sodium, chromium, and barium. From Eq. 9.58, the structure factor for the atoms in a unit cell is

$$F\left(\mathbf{Q}_{hkl}\right) = f\left(\mathbf{Q}_{hkl}\right)\sum_j e^{i\mathbf{Q}_{hkl}\cdot\mathbf{r}_j}. \tag{9.74}$$

The atomic form factor $f\left(\mathbf{Q}_{hkl}\right)$, or equivalently $f\left(\sin\phi/\lambda\right)$, has been tabulated for the various elements in the *International Tables for X-Ray Crystallography* [61]. They have been graphed in Fig. 9.10 for the aforementioned solids. The ability to calculate these form factors depends on our knowledge of the electron density, or equivalently, the total ground-state wavefunction, for a given atom. Thus, except in the case of hydrogen, the atomic form factors are approximate. In addition, the tabulated values of f were calculated under our earlier assumption that the energy of the x-ray photon is much larger than the electron binding energy. When this condition is not clearly met, the form factor must be corrected for electron binding effects—this is accomplished by introducing a so-called complex *anomalous dispersion correction* to the form factor, i.e.,

$$f = f_0 + \Delta f' + i\Delta f'', \tag{9.75}$$

where f_0 is the previously tabulated form factor that assumes negligible binding effects (as in Fig. 9.10). For the case of Mo K_α x-rays, the approximate values of $\Delta f' + i\Delta f''$ are listed in Table 9.1 as a function of $\sin\phi/\lambda$.

Return now to Eq. 9.74 for the structure factor. For a b.c.c. crystal, there are two atoms in a unit cell—one atom, in effect, is located at the corner of the cell, or the origin, i.e., $\mathbf{r}_1 = 0$, and the other atom is located at the center of the cell with position $\mathbf{r}_2 = \frac{1}{2}\left(\mathbf{a}_1 + \mathbf{a}_2 + \mathbf{a}_3\right)$. Applying Eqs. 9.66 and 9.67, we then have

$$\begin{aligned}\left|F\left(\mathbf{Q}_{hkl}\right)\right|^2 &= \left|f\left(\mathbf{Q}_{hkl}\right)\right|^2\left|1+e^{i\pi(h+k+l)}\right|^2 \\ &= \left|f\left(\mathbf{Q}_{hkl}\right)\right|^2\left[1+\cos\pi\left(h+k+l\right)\right]^2.\end{aligned} \tag{9.76}$$

This shows that if $h+k+l$ is odd, then that reflection does not appear, whereas if $h+k+l$ is even, then $|F|^2 = 4|f|^2$. As an illustration, consider the case of unpolarized Mo K_α x-rays diffracted by a crystal of iron. Table 9.2 compares the integrated intensity arising from $h\,k\,l = 2\,0\,0$ to that for $4\,0\,0$, along with the factors contributing to the calculation (note: $2d_{100} = 0.2866$ nm). For this particular example, we see that the intensity of the $2\,0\,0$-reflection is about 1.8 times as intense as the $4\,0\,0$-reflection.

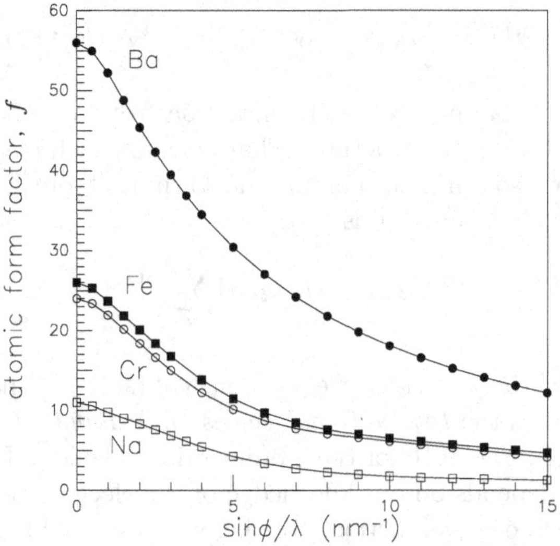

Figure 9.10 Calculated atomic form factors for some monatomic b.c.c. solids (assuming negligible binding effects). Data is taken from the International Tables for X-Ray Crystallography [60].

Table 9.1 Dispersion corrections, $\Delta f' + i\Delta f''$, for some atomic form factors for Mo K_α x-rays ($\lambda = 0.07107$ nm).

$\sin\phi/\lambda$ (nm^{-1})	Na	Cr	Fe	Ba
0	$0.0 + 0.1i$	$0.3 + 0.8i$	$0.4 + 1.0i$	$-0.4 + 3.0i$
9	$0.0 + 0.1i$	$0.3 + 0.7i$	$0.3 + 0.9i$	$-0.6 + 2.5i$
13	$0.0 + 0.0i$	$0.3 + 0.7i$	$0.3 + 0.8i$	$-0.8 + 2.2i$

Table 9.2 Relative intensity calculation for diffraction of Mo K_α x-rays from an Fe crystal.

hkl	$2d_{hkl}$	$\sin\phi/\lambda$ (nm^{-1})	f_0	$\Delta f' + i\Delta f''$	$\lvert f\rvert^2$	$\theta_{hkl} = 2\phi$	$\lvert f\rvert^2(1+\cos^2\theta_{hkl})$
200	0.143	6.98	8.5	$0.3 + 0.9i$	78	59.4^o	98
400	0.0717	14.0	5.0	$0.3 + 0.8i$	29	165.6^o	56

Example 9.2 Compton Scattering Measurement of Electron Momentum Density. The momentum distribution of atomic electrons, a.k.a. the *electron momentum density* (EMD), can be directly related to the spectrum of Compton scattered x-rays. In effect, there is a Doppler shift of the scattered radiation due to the motion of the electrons in the atom. To illustrate the connection between the observed spectrum and the EMD, we shall derive the double-differential cross-section $d^2\sigma/d\Omega d\omega$ for Compton scattering from the simplest atom, hydrogen, in the ground state. To accomplish this, it will be necessary to apply the so-called *impulse approximation* (the specifics of which will be presented shortly). Eisenberger and Platzman [60] have demonstrated that this approximation is valid as long as

$$\frac{E_b}{\hbar^2 Q^2/2m} \ll 1. \tag{9.77}$$

In the case of Compton scattering from hydrogen, the above ratio is $\sim 10^{-3}$, so the impulse approximation certainly holds.

Let a monochromatic x-ray beam having wavevector $\mathbf{k}$, frequency ω_k, and polarization $\boldsymbol{\epsilon}_{\mathbf{k}\lambda}$ be incident upon a hydrogen atom situated at the origin. If the scattered radiation is characterized by $\omega_{k'}$, $\mathbf{k}'$, and $\boldsymbol{\epsilon}_{\mathbf{k}'\lambda'}$, we can then specify the energy and momentum transfers

$$\hbar\omega = \hbar\omega_k - \hbar\omega_{k'} \qquad \text{and} \qquad \hbar\mathbf{Q} = \hbar\mathbf{k} - \hbar\mathbf{k}'. \tag{9.78}$$

Return now to Eq. 9.45 for the angular differential cross-section. For the present case of a single-electron atom, the expression is

$$\frac{d\sigma}{d\Omega} = r_0^2 \left(\frac{\omega_{k'}}{\omega_k}\right) (\boldsymbol{\epsilon}_{\mathbf{k}\lambda} \cdot \boldsymbol{\epsilon}_{\mathbf{k}'\lambda'})^2 \left|\langle \mathcal{B} \mid e^{i\mathbf{Q}\cdot\mathbf{r}} \mid \mathcal{A}\rangle\right|^2. \tag{9.79}$$

Reminiscent of what was done in the case of thermal neutron scattering (see Sect. 6.4), we can now use energy conservation to write a double-differential scattering cross-section:

$$\frac{d^2\sigma}{d\Omega\, d\omega} = r_0^2 \left(\frac{\omega_{k'}}{\omega_k}\right) (\boldsymbol{\epsilon}_{\mathbf{k}\lambda} \cdot \boldsymbol{\epsilon}_{\mathbf{k}'\lambda'})^2 \sum_{\mathcal{A},\mathcal{B}} \left|\langle \mathcal{B} \mid e^{i\mathbf{Q}\cdot\mathbf{r}} \mid \mathcal{A}\rangle\right|^2 \delta\left(\omega - \frac{E_{\mathcal{B}} - E_{\mathcal{A}}}{\hbar}\right). \tag{9.80}$$

This expression can be written in the more compact form

$$\frac{d^2\sigma}{d\Omega\, d\omega} = \left(\frac{d\sigma}{d\Omega}\right)_{\text{free}} \left(\frac{\omega_k}{\omega_{k'}}\right) S(\mathbf{Q}, \omega), \tag{9.81}$$

where

$$\left(\frac{d\sigma}{d\Omega}\right)_{\text{free}} = r_0^2 \left(\frac{\omega_{k'}}{\omega_k}\right)^2 (\boldsymbol{\epsilon}_{\mathbf{k}\lambda} \cdot \boldsymbol{\epsilon}_{\mathbf{k}'\lambda'})^2 \tag{9.82}$$

represents scattering by an initially stationary, free electron (Eq. 9.36) and $S(\mathbf{Q},\omega)$ is the dynamic structure factor

$$S(\mathbf{Q},\omega) = \sum_{\mathcal{A},\mathcal{B}} \left|\langle\mathcal{B}\mid e^{i\mathbf{Q}\cdot\mathbf{r}}\mid\mathcal{A}\rangle\right|^2 \delta\left(\omega - \frac{E_{\mathcal{B}} - E_{\mathcal{A}}}{\hbar}\right). \tag{9.83}$$

It is now left to the reader to show that by replacing the delta-function with its integral representation, and following the same type of arguments as those in Sect. 6.4, the structure factor can be expressed in the same form as in Eq. 6.94, i.e.,

$$S(\mathbf{Q},\omega) = \frac{1}{2\pi}\int_{-\infty}^{+\infty} dt\, e^{-i\omega t}\left\langle e^{-i\mathbf{Q}\cdot\mathbf{r}(0)} e^{i\mathbf{Q}\cdot\mathbf{r}(t)}\right\rangle, \tag{9.84}$$

with the understanding that $\langle\ \rangle \equiv \sum_{\mathcal{A}}\langle\mathcal{A}\mid\ \mid\mathcal{A}\rangle$ and $e^{i\mathbf{Q}\cdot\mathbf{r}(t)} = e^{i\hat{H}_0 t/\hbar} e^{i\mathbf{Q}\cdot\mathbf{r}} e^{-i\hat{H}_0 t/\hbar}$. The Hamiltonian $\hat{H}_0$ is that for an unperturbed hydrogen atom, namely,

$$\hat{H}_0 = \frac{1}{2m}\hat{p}^2 - \frac{Ze^2}{r} = \hat{H}_{00} + \hat{V}(r). \tag{9.85}$$

This is the point at which we introduce the impulse approximation. First, make the expansion (see Eq. 5.128)

$$e^{i\hat{H}_0 t/\hbar} = e^{i\hat{H}_{00} t/\hbar} e^{i\hat{V}t/\hbar} e^{-\left[\hat{H}_{00},\hat{V}\right]t^2/2\hbar^2} \cdots. \tag{9.86}$$

We now only retain terms of order t, or equivalently, of order ω^{-1}, in the exponent—in other words, we set $e^{-\left[\hat{H}_{00},\hat{V}\right]t^2/2\hbar^2}\cdots \simeq 1$; this is the essential step in the impulse approximation. The result is that

$$\begin{aligned} e^{i\mathbf{Q}\cdot\mathbf{r}(t)} &= e^{i\hat{H}_0 t/\hbar} e^{i\mathbf{Q}\cdot\mathbf{r}} e^{-i\hat{H}_0 t/\hbar} = e^{i\hat{H}_{00} t/\hbar} e^{i\hat{V}t/\hbar} e^{i\mathbf{Q}\cdot\mathbf{r}} e^{-i\hat{V}t/\hbar} e^{-i\hat{H}_{00} t/\hbar} \\ &= e^{i\hat{H}_{00} t/\hbar} e^{i\mathbf{Q}\cdot\mathbf{r}} e^{-i\hat{H}_{00} t/\hbar}, \end{aligned} \tag{9.87}$$

and we see that the potential energy $\hat{V}(r)$ drops out of the calculation, i.e., it is treated as essentially constant when the measurement is made over a time that is sufficiently short. One can now apply Eq. 5.128 twice to the above expression and, each time, make use of the commutator relations Eq. 3.26 and 3.72. After a lot of operator algebra, we obtain

$$e^{i\mathbf{Q}\cdot\mathbf{r}(t)} = e^{i\mathbf{Q}\cdot[\mathbf{r}(0)+\mathbf{p}(0)t/m]}. \tag{9.88}$$

Thus, the average in Eq. 9.84 becomes

$$\begin{aligned} \left\langle e^{-i\mathbf{Q}\cdot\mathbf{r}(0)} e^{i\mathbf{Q}\cdot\mathbf{r}(t)}\right\rangle &= \langle\mathcal{A}\mid e^{-i\mathbf{Q}\cdot\mathbf{r}(0)} e^{i\mathbf{Q}\cdot[\mathbf{r}(0)+\mathbf{p}(0)t/m]}\mid\mathcal{A}\rangle \\ &= \langle\mathcal{A}\mid e^{i\left[\hbar Q^2 + 2\mathbf{Q}\cdot\mathbf{p}(0)\right]t/2m}\mid\mathcal{A}\rangle, \end{aligned} \tag{9.89}$$

and, in the impulse approximation, the structure factor is

$$S^I(\mathbf{Q},\omega) = \frac{1}{2\pi}\int_{-\infty}^{+\infty} dt\, e^{-i\omega t}\langle\mathcal{A} \mid e^{i\left[\hbar Q^2 + 2\mathbf{Q}\cdot\mathbf{p}(0)\right]t/2m} \mid \mathcal{A}\rangle. \tag{9.90}$$

On each side of the exponential in Eq. 9.90, insert the complete set of states $|\mathbf{p}\rangle$, which are the momentum eigenstates, or plane waves

$$\psi_{\mathbf{p}}(\mathbf{r}) = \langle\mathbf{r} \mid \mathbf{p}\rangle = \frac{1}{(2\pi)^{3/2}} e^{i\mathbf{p}\cdot\mathbf{r}/\hbar}. \tag{9.91}$$

Then,

$$\begin{aligned} S^I(\mathbf{Q},\omega) &= \frac{1}{2\pi}\int_{-\infty}^{+\infty} dt\, e^{-i\omega t}\int d^3p\int d^3p'\langle\mathcal{A} \mid \mathbf{p}\rangle\langle\mathbf{p} \mid e^{i\left[\hbar Q^2+2\mathbf{Q}\cdot\mathbf{p}(0)\right]t/2m} \mid \mathbf{p}'\rangle\langle\mathbf{p}' \mid \mathcal{A}\rangle \\ &= \frac{1}{2\pi}\int_{-\infty}^{+\infty} dt\, e^{-i\omega t}\int d^3p\, |\langle\mathbf{p} \mid \mathcal{A}\rangle|^2\, e^{i\left[\hbar Q^2+2\mathbf{Q}\cdot\mathbf{p}(0)\right]t/2m}. \end{aligned} \tag{9.92}$$

But the inner product $\langle\mathbf{p} \mid \mathcal{A}\rangle$ is the momentum-space representation (see Sect. 3.5) of the atom's ground state or, from Eq. 3.199, it is the Fourier transform of the ground-state wavefunction:

$$\phi_0(\mathbf{p}) \equiv \langle\mathbf{p} \mid \mathcal{A}\rangle = \frac{1}{(2\pi\hbar)^{3/2}}\int d^3r\, \psi_0(\mathbf{r})\, e^{-i\mathbf{p}\cdot\mathbf{r}/\hbar}. \tag{9.93}$$

As a result,

$$\begin{aligned} S^I(\mathbf{Q},\omega) &= \frac{1}{2\pi}\int d^3p\, |\phi_0(\mathbf{p})|^2 \int_{-\infty}^{+\infty} dt\, e^{-i\omega t} e^{i\left[\hbar Q^2+2\mathbf{Q}\cdot\mathbf{p}(0)\right]t/2m} \\ &= \int d^3p\, |\phi_0(\mathbf{p})|^2\, \delta\left(\omega - \frac{\hbar Q^2}{2m} - \frac{\mathbf{Q}\cdot\mathbf{p}}{m}\right). \end{aligned} \tag{9.94}$$

The delta-function indicates that the frequency of the scattered photon is shifted by two terms— $\hbar Q^2/2m$ is a momentum-transfer term and $\mathbf{Q}\cdot\mathbf{p}/m$ represents a Doppler-shift. To proceed further, take the wavevector transfer $\mathbf{Q}$ to be parallel to the z-axis and define

$$\kappa(Q,\omega) = \frac{m\omega}{Q} - \frac{\hbar Q}{2}. \tag{9.95}$$

Then

$$\begin{aligned} S^I(\mathbf{Q},\omega) &= \int d^3p\, |\phi_0(\mathbf{p})|^2\, \delta\left[\frac{Q}{m}(p_z - \kappa)\right] \\ &= \frac{m}{Q}\int_{-\infty}^{+\infty} dp_x \int_{-\infty}^{+\infty} dp_y\, |\phi(p_x, p_y, \kappa)|^2. \end{aligned} \tag{9.96}$$

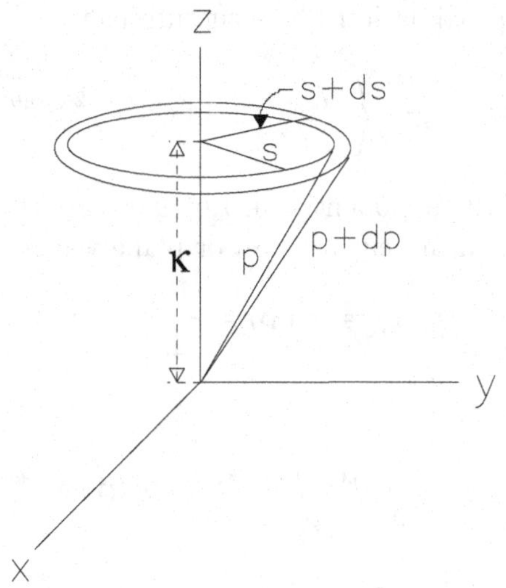

Figure 9.11 Variable transformation to facilitate integration of Eq. 9.96.

Figure 9.11 shows how to change the integration to a more convenient form. This is accomplished by replacing $dp_x dp_y$ with $2\pi s\, ds$, where $s = \sqrt{p^2 - \kappa^2}$; in other words, $dp_x dp_y = 2\pi p dp$, and we integrate over the interval $\kappa \leq p \leq \infty$. The result is

$$S^I(\mathbf{Q}, \omega) = \frac{m}{Q} J(\kappa), \tag{9.97}$$

where

$$J(\kappa) = 2\pi \int_{\kappa}^{\infty} |\phi_0(p)|^2 \, p dp. \tag{9.98}$$

$J(\kappa)$ is known as the *Compton profile*, and it depends on the sought after electron momentum density, or specifically, $4\pi p^2 |\phi_0(p)|^2$.

For hydrogen, the ground-state (1s) wavefunction is

$$\psi_0(\mathbf{r}) = \sqrt{\frac{\alpha^3}{\pi}} e^{-\alpha r}, \tag{9.99}$$

where $\alpha = 1/a$ is the reciprocal Bohr radius ($a = \hbar^2/me^2$). From Eq. 9.93, we can also calculate the ground state in momentum space:

$$\phi_0(p) = \sqrt{\frac{8\alpha^5\hbar^5}{\pi^2}} \left(\hbar^2\alpha^2 + p^2\right)^{-2}. \tag{9.100}$$

Then, from Eq. 9.98, the Compton profile for a hydrogen atom takes the simple form

$$J(\kappa) = \frac{8\alpha^5\hbar^5}{3\pi} \left(\kappa^2 + \hbar^2\alpha^2\right)^{-3}. \tag{9.101}$$

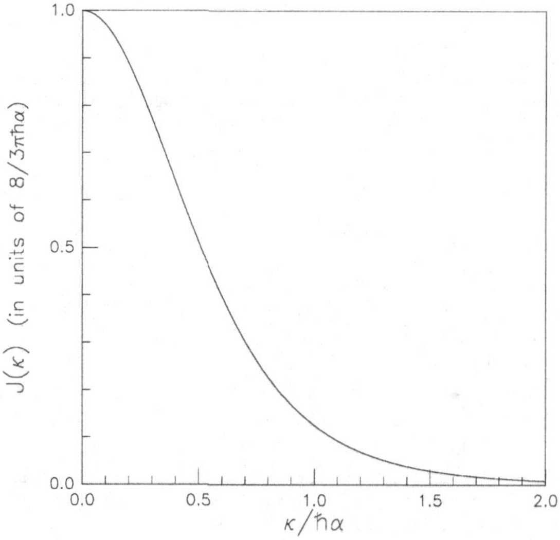

Figure 9.12 Compton profile $J(\kappa)$ for the hydrogen atom calculated using the impulse approximation ($\kappa = \frac{m\omega}{Q} - \frac{\hbar Q}{2}$).

This function is displayed in Fig. 9.12. For the case of hydrogen, an exact analytic expression for $J(\kappa)$ has also been worked out [60]; when it is compared to the profile above, as derived from the impulse approximation, it is found that the agreement is good to order $\left(\frac{E_b}{\hbar^2 Q^2/2m}\right)^2$.

Eisenberger [62] has collected data on Compton scattering from the hydrogen molecule, H_2, and has compared the experimentally measured Compton profile to that calculated theoretically using the impulse approximation. For molecular hydrogen, $J(\kappa)$ is still calculated using Eq. 9.98, however, the exact ground-state wavefunction is not available. Instead, one must make use of the *variational principle*, which is an approximation method for estimating the ground-state energy and $\psi_0(\mathbf{r})$. Henneker (see ref. [62]) has supplied two such calculations of the wavefunction, one specifically based on a Hartree-Fock self-consistent-field calculation, and the other on a multiconfigurational self-consistent field. Figure 9.13 shows a comparison between the theoretical and the measured Compton profiles. Studies of this type have also been performed for more complicated atoms and molecules, such as N_2, O_2, and Ne, with the aim of assessing ground-state wavefunctions calculated using the various approximation techniques [63] [64]. In addition, Compton scattering experiments have been used to examine the momentum density of conduction electrons in metals, e.g., Li and Na [65].

An interesting observation is that when the experimentally measured $\ln J(\kappa)$ is plotted against κ^2 for various atoms, the resulting graph turns out to be quite linear

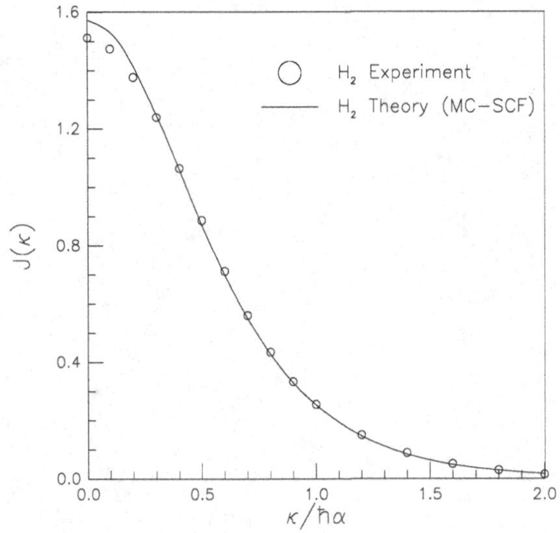

Figure 9.13 Comparison of experimentally and theoretically determined Compton profiles for the hydrogen molecule H_2. The theoretical curve is based on the impulse approximation in conjunction with a multiconfigurational self-consistent field calculation of the ground-state molecular wavefunction (taken from Eisenberger [61]).

for small κ, as seen in Fig. 9.14. In other words, it appears that

$$J(\kappa) \sim e^{-\beta\kappa^2}, \tag{9.102}$$

where $-\beta$ is the slope of the graph. This can be explained if one assumes that $J(\kappa)$ always has the form of Eq. 9.101 in the small-κ region, with the only difference being that the parameter α is replaced by $\alpha^* = 1/a^*$, where a^* represents some effective radius of the scattering atom. Then,

$$J(\kappa) \sim \left[1 + \left(\frac{\kappa}{\hbar\alpha^*}\right)^2\right]^{-3} \sim 1 - 3\left(\frac{\kappa}{\hbar\alpha^*}\right)^2 \sim e^{-3\kappa^2/\hbar^2\alpha^{*2}}, \tag{9.103}$$

so $\beta = 3/\hbar^2\alpha^{*2}$, or

$$\frac{a^*}{a} = \left(\frac{\hbar^2\alpha^2\beta}{3}\right)^{1/2}. \tag{9.104}$$

The insert of Fig. 9.14 shows data for a^*/a vs. the atomic number Z. For comparison, a reasonable theoretical estimate for a^* comes from the Thomas-Fermi statistical model for the ground state of an atom (for example, see Eisberg [66]). This model predicts that, as Z increases, the radial extent of the atomic electron cloud slowly decreases according to

$$\frac{a^*}{a} = \frac{0.885}{Z^{1/3}}. \tag{9.105}$$

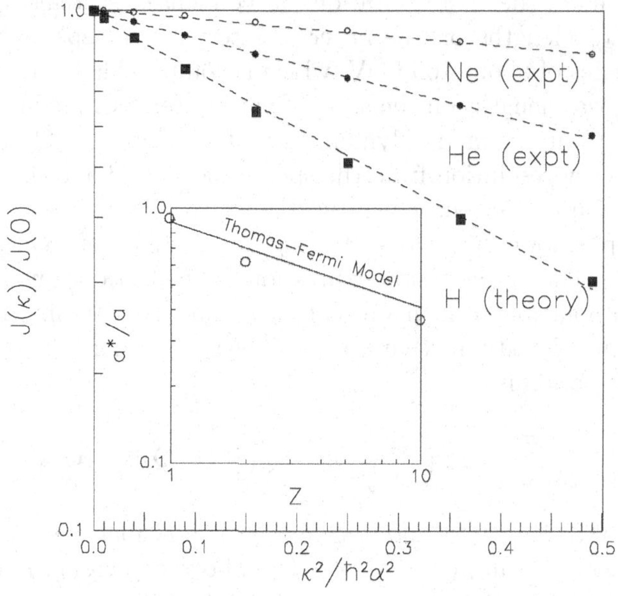

Figure 9.14 Small-κ behavior of the Compton profile $J(\kappa)$ for various atoms, along with the best-fit lines. The inset graph displays a^*/a vs. atomic number Z, where a^* is the effective radius of the scattering atom and a is the Bohr radius. The points come from the slopes of the $J(\kappa)$-curves; the line represents the trend predicted by the Thomas-Fermi model.

The figure insert shows that the atomic sizes obtained from the small-κ behavior of $J(\kappa)$ seem to follow those predicted by the Thomas-Fermi model to a reasonable degree.

Example 9.3 High-Resolution Inelastic X-Ray Scattering from Water at Low Temperature. Traditionally, the study of atomic and molecular density fluctuations in condensed systems is the realm of neutron spectroscopy. The principal reason that inelastic neutron scattering (INS) is particularly suitable for these studies is that the Q, ω-space covered by thermal neutron scattering is well-matched to that of phonon-like collective excitations in condensed systems. More specifically, neutrons with wavelengths on the order of interparticle separations in liquids and solids have energies of about 100 meV, which is comparable to the energies of typical phonons having wavelengths on the scale of nanometers. Consequently, one is able to determine peak positions in the dynamic structure factor, $S(\mathbf{Q}, \omega)$, without requiring excessive energy resolution from the spectrometer and allowing one to utilize the intensity of the source efficiently.

In principle, however, x-ray scattering can also be used to measure the dynamic structure factor. The cross-section expression for inelastic x-ray scattering (IXS) is similar to that applicable to neutron scattering, and the coupling of both x-rays and neutrons to atomic density fluctuations is of the same order of magnitude. The x-ray scattering cross-section is

$$\frac{d^2\sigma}{d\Omega\, d\omega} = r_0^2 \left(\frac{\omega_{k'}}{\omega_k}\right) \left(\boldsymbol{\epsilon}_{\mathbf{k}\lambda} \cdot \boldsymbol{\epsilon}_{\mathbf{k}'\lambda'}\right)^2 \left|f(\mathbf{Q})\right|^2 S(\mathbf{Q}, \omega), \tag{9.106}$$

which is identical to Eqs. 9.81 and 9.82, except it now includes a factorization of the square of the atomic form factor $f(\mathbf{Q})$. The above expression is valid for a system composed of a single atomic species, however, it can easily be generalized to molecular or crystalline systems by replacing $f(\mathbf{Q})$ with either the molecular or the unit-cell form factor. The situation becomes more involved for disordered, multicomponent systems. In such cases, the factorization of the form factor is still justified, but only if some correlated distribution among the different atoms is assumed. In the limiting case where the distribution is completely random, an incoherent contribution appears in the scattering cross-section, as in the case of thermal neutron scattering (see Eq. 6.92). It is worth noting that the coupling strength of x-rays to the atomic electrons in the cross-section is determined by the square of the classical electron radius, $r_0 = 2.82 \times 10^{-13}$cm, and that this strength is comparable to the square of the neutron-nucleus scattering length, which governs the interaction strength in the case of thermal neutron scattering. In spite of the strong similarities between neutron and x-ray inelastic scattering, the development of the IXS technique has, up to now, been limited, mainly because (1) photons with wavelengths near 1 Å have energies of about 10 keV, which means that studies of phonon excitations in the meV region demand an energy resolution of at least $\Delta E/E \approx 10^{-7}$, (2) the dominant contribution to the total attenuation of x-rays with energy near 10 keV is photoelectric absorption, and not the process of scattering, which limits the actual pathlength one can use for the sample, and (3) there is a rapid fall-off of the atomic or molecular form factor with increasing Q which, in turn, is responsible for a drastic reduction in the scattering cross-section, even at relatively small momentum-transfer values.

However, even with the above-mentioned constraints, there are situations where the use of x-ray scattering presents important advantages over neutron scattering. One important example is that in order to study acoustic excitations propagating at the speed of sound, one must make use of a probe particle having a speed exceeding that of the sound wave. This limitation is ordinarily not problematic for INS experiments on crystalline samples because translational invariance in the system allows one to investigate acoustic excitations present in high-order Brillouin zones. This overcomes the difficulty of the above-mentioned kinematic limit for phonon branches characterized by steep dispersions. In contrast, the situation is very different for cases where the system is disordered, where, for the most part, neutron measuements of the dynamic structure factor have been largely unsuccessful. In disordered systems, the absence of periodicity forces one to measure the acoustic excitations at small values of Q. In the case of neutron scattering from water [67] [68], for example, this is an issue that has prevented the measurement, with adequate energy resolution, of $S(Q,\omega)$ over sufficiently extended regions of energy and momentum space. This applies to many other interesting liquids and glasses as well, where the high speed of sound severely limits the utility of existing neutron spectrometers when it comes to measuring acoustic excitations. Inelastic x-ray scattering, on the other hand, is extremely valuable for studying dynamics in disordered systems. IXS does not suffer from the kinematic constraints encountered by neutrons, and can access values of small momentum transfer, provided that the required energy resolution is experimentally achievable. In fact, since energy transfers, $\hbar\omega$, for the case of x-ray scattering are so small compared to the incident and scattered photon energies, the magnitudes of the incident and scattered wavevectors are essentially identical ($k \approx k'$), and a given scattering angle θ completely determines the magnitude of the scattering vector, Q, independent of the energy transfer, i.e., $Q = 2k\sin(\theta/2)$. So, for phonon-like excitations, there is no limitation on the energy transfer accessible at a given value of Q in an inelastic x-ray scattering experiment. Other important advantages of inelastic x-ray scattering are that (1) the cross-section for x-ray scattering is highly coherent, unlike neutron scattering which often requires separating the coherent part, $S(Q,\omega)$, and the incoherent part, $S_s(\mathbf{Q},\omega)$, from the measured signal, (2) multiple scattering is, in general, strongly suppressed by the photoelectric absorption process, reducing the need for sophisticated, and sometimes problematic, procedures for reducing the raw data, and (3) very narrow x-ray beams are attainable at the sample position, a significant advantage when it comes to performing experiments under extreme thermodynamic conditions and/or on systems available in only small quantities.

The power of the IXS technique is nicely illustrated by experiments on the dynamics of low-temperature liquid water performed by Liao, Chen, and Sette [69]. Figure 9.15 displays a series of IXS spectra, collected at various Q-values, for water at a temperature of 273 K. The measurements were carried out at the high-resolution inelastic x-ray scattering beam line (BL21-ID16) at the European Synchrotron Radiation Facility (ESRF) in Grenoble, France. The scattering target was an 18-nm-thick high-purity water sample, and the size of the x-ray beam at the sample was 100 μm $\times$ 300 μm. The energy resolution, i.e., the full-width at half-maximum of the instrument resolution function, was approximately 1.5 meV.‡

The spectra were analyzed using the so-called "three effective eigenmode" (or TEE) model [71], which had previously met with success in describing the dynamic structure factors for a number of model fluids (see, for example, references [72] and [73]). In the TEE model, correlations in the density fluctuations of the system are described by a sum of three exponential functions associated with three slow conserved hydrodynamic eigenmodes of the fluid, the so-called extended heat mode and two extended sound modes. Although this description is an extended hydrodynamic model, it has been shown that it provides a good approximation for $S(Q,\omega)$ over a wide Q-range. In the TEE model, the dynamic structure factor is given by

$$S(Q,\omega)/S(Q) = \frac{1}{\pi}\,\mathrm{Re}\left[\frac{\overleftrightarrow{I}}{i\omega\overleftrightarrow{I} + \overleftrightarrow{H}(Q)}\right]_{1,1} \tag{9.107}$$

where $S(Q) = \int_{-\infty}^{+\infty} S(Q,\omega)\,d\omega$ is the *static structure factor*.§ $\overleftrightarrow{I}$ is the 3×3 identity matrix and the subscript 1, 1 means the (1, 1)-element of the matrix. $\overleftrightarrow{H}(Q)$ is the matrix

$$\overleftrightarrow{H}(Q) = \begin{pmatrix} 0 & if_{un}(Q) & 0 \\ if_{un}(Q) & z_u(Q) & if_{uT}(Q) \\ 0 & if_{uT}(Q) & z_T(Q) \end{pmatrix}, \tag{9.108}$$

with $f_{un}(Q)$ determined by the second moment of the dynamic structure factor to be $Qv_0[S(Q)]^{1/2}$ (where $v_0 = \sqrt{k_B T/m}$ is the thermal speed). The three independent parameters, namely, $z_u(Q)$, $f_{uT}(Q)$, and $z_T(Q)$, are all real.

In the low-Q limit, $\overleftrightarrow{H}(Q)$ tends to the hydrodynamic matrix, where the matrix elements have the following values [71] [74]:

$$\begin{aligned} f_{un}(Q) &= c_s Q/\sqrt{\gamma} && (9.109)\\ z_u(Q) &= \phi Q^2 && (9.110)\\ z_T(Q) &= \gamma D_T Q^2 && (9.111)\\ f_{uT}(Q) &= c_s Q\sqrt{(\gamma-1)/\gamma}. && (9.112) \end{aligned}$$

‡Full details of the spectrometer are given by Ruocco and Sette [70].

§A discussion of the static structure factor, $S(Q)$, appears in Chapter 12.

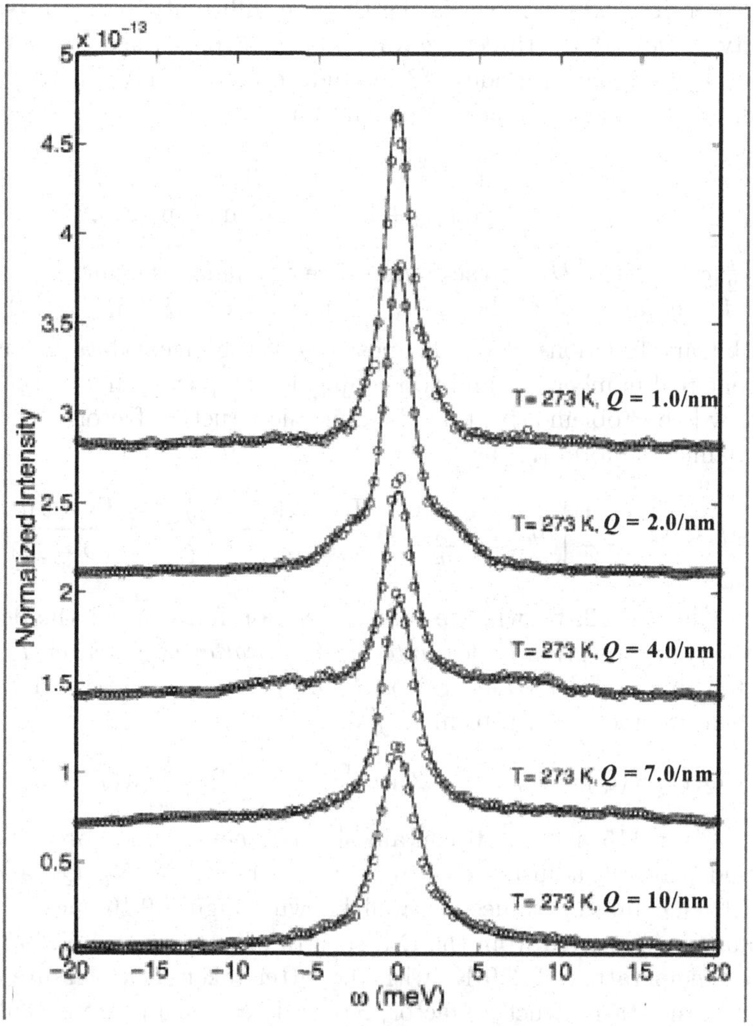

Figure 9.15 Inelastic x-ray scattering (IXS) spectra of H_2O at $T = 273$ K collected at the indicated values of Q (adapted from Liao *et al.* [69]). The solid curves are fits to the experimental data (open circles) using the TEE (three effective eigenmode) model. (The model has been convoluted with the energy resolution of the instrument and the data is normalized to have unit area over the measured range of energy transfer.)

Here, $c_s = v_0 \left[\gamma / S(0)\right]^{1/2}$ is the adiabatic speed of sound; $\gamma = c_p/c_v$ is the ratio of the specific heats at constant pressure and volume; $\phi = \left(\frac{4}{3}\eta + \zeta\right)/\rho$ is the kinematic longitudinal viscosity (η and ζ are the shear and bulk viscosity, respectively, and ρ is the mass-density); $D_T = \lambda/\rho c_p$ is the thermal diffusivity, where λ is the thermal conductivity. At low Q, the three eigenvalues of the hydrodynamic matrix correspond to the three hydrodynamic modes of the fluid. Below, we give the eigenvalues only up to order Q^2 [see end-of-chapter, Problem 2(a)]:

$$z_h(Q) = D_T Q^2 \qquad \text{(heat mode)} \tag{9.113}$$

$$z_\pm(Q) = \pm i c_s Q + \Gamma Q^2 \qquad \text{(sound modes).} \tag{9.114}$$

Here, $\Gamma = \frac{1}{2}\left[\phi + (\gamma - 1) D_T\right]$ is the sound-wave damping constant.

For finite values of Q, however, the parameters $z_u(Q)$, $f_{uT}(Q)$, and $z_T(Q)$ become arbitrary functions of Q. In most cases, the eigenvalues of matrix $\overleftrightarrow{H}(Q)$ consist of one real number, z_h, and two conjugate complex numbers, $\Gamma_s \pm i\omega_s$. Then, one can show [see Problem 2(b)] that the dynamic structure factor may be written in a hydrodynamic-like form [75] as

$$S(Q,\omega)/S(Q) = \frac{1}{\pi}\left[A_h \frac{z_h}{\omega^2 + z_h^2} + A_s \frac{\Gamma_s + b(\omega + \omega_s)}{(\omega + \omega_s)^2 + \Gamma_s^2} + A_s \frac{\Gamma_s - b(\omega - \omega_s)}{(\omega - \omega_s)^2 + \Gamma_s^2}\right]. \tag{9.115}$$

Later on in Chapter 12, we will see that correlation function for the density fluctuations of the system is given by the *intermediate scattering function*, $F(Q,t)$, which is simply the time Fourier transform of $S(Q,\omega)$ (see Eq. 12.65). This correlation function then becomes [see Problem 2(c)]

$$F(Q,t)/S(Q) = A_h e^{-z_h t} + 2A_s e^{-\Gamma_s t}\left[\cos(\omega_s t) + b\sin(\omega_s t)\right]. \tag{9.116}$$

Although Eqs. 9.115 and 9.116 contain six parameters, they are all functions of the three independent, adjustable fitting parameters $z_u(Q)$, $f_{uT}(Q)$, and $z_T(Q)$, for which the low-Q limiting values are well known. Figure 9.16 shows the values of these parameters extracted from the IXS spectra of water at 294 K, 273 K, and at a supercooled temperature of 259 K (with the latter taken at a pressure of 2.0 kbar). Also shown is the static structure factor, $S(Q)$, determined by a molecular dynamics simulation [76], and used as an input. By plotting the sound excitation frequencies, ω_s, extracted from the IXS spectra as a function of Q, one generates the dispersion curve shown in Fig. 9.17. The slope of the curve gives a high-frequency speed of sound equal to approximately $c = 2900$ m/s, which is about twice the adiabatic sound speed of $c_s = 1380$ m/s at low frequency.

9.2.2 Light Scattering

Consider now the scattering of photons in the optical regime, where $\hbar\omega_k$ and $\hbar\omega_{k'}$ are of the same order of magnitude as the spacing of atomic energy levels. Now the term

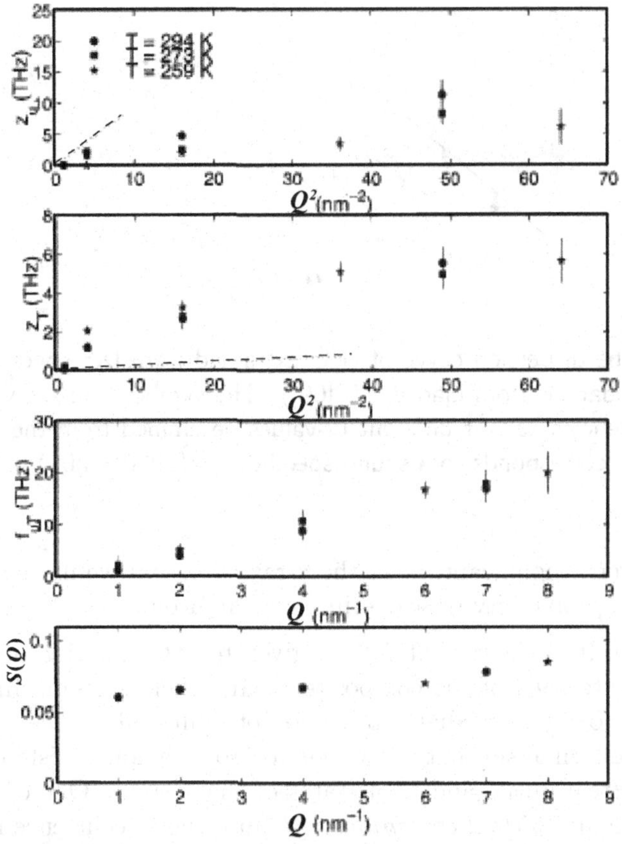

Figure 9.16 The extracted TEE-model fitting parameters, $z_u(Q)$, $z_T(Q)$, and $f_{uT}(Q)$, from IXS spectra of H_2O at the indicated temperatures, plotted along with the input static structure factor, $S(Q)$, from Sciortino [76]. The dashed lines are the expected hydrodynamic behaviors. (Figure adapted from Liao *et al.* [69].)

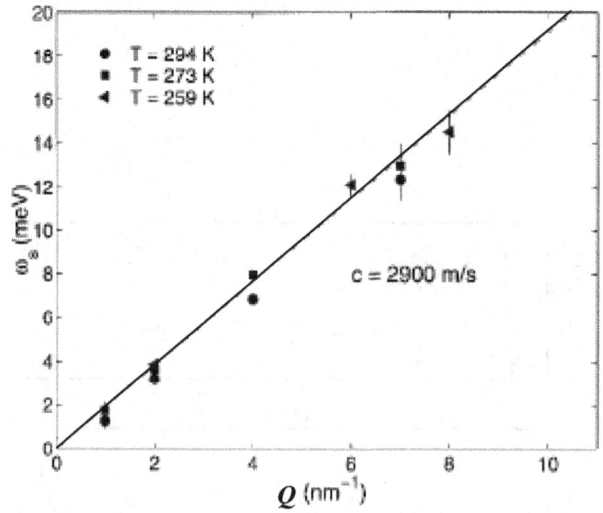

Figure 9.17 The dispersion curve of H_2O extracted from IXS spectra at the indicated temperatures (adapted from Liao *et al.* [69]). The symbols are the values of the sound excitation frequencies, ω_s, at different Q-values detemined from the TEE model. The slope of the line corresponds to a sound speed of $c = (2900 \pm 300)$ m/s

$K_1^{(2)}$ is no longer insignificant, as in the x-ray case, and we must return to Eq. 9.42 to calculate the probability of scattering from an atom.

Equation 9.44 shows that $K_1^{(2)}$ is given by a summation over all the possible intermediate states, $|l\rangle$, of the composite atom + field system. In actuality, transitions to the various intermediate states do not really take place—rather, they are a construct of the formalism that allows one to produce non-vanishing matrix elements connecting the real initial and final states of the system. One often refers to these as *virtual states* and *virtual transitions*. Because their occurrence is not real, virtual transitions do not need to conserve energy. Viewed another way, since the lifetime of a virtual state is infinitesimally small, the time-energy uncertainty principle imposes no restriction on its energy.

For single photon scattering, $K_1^{(2)}$ will have two terms because there are two types of virtual transitions, as illustrated in Fig. 9.18. In the first type, one pictures that an intermediate atomic state, $|\mathcal{I}\rangle$, is created when the incident $\mathbf{k}\lambda$-photon is absorbed; the scattered $\mathbf{k}'\lambda'$-photon is then born, leaving the atom in final state $|\mathcal{B}\rangle$. In this case, the energy of the intermediate state is $E_l = E_{\mathcal{I}}$. In the second type of transition, one imagines that the scattered $\mathbf{k}'\lambda'$-photon appears before the incident $\mathbf{k}\lambda$-photon is absorbed, so the intermediate state has energy $E_l = E_{\mathcal{I}} + \hbar\omega_k + \hbar\omega_{k'}$.

Consider the contribution to $K_1^{(2)}$ from the first type of transition. It is given

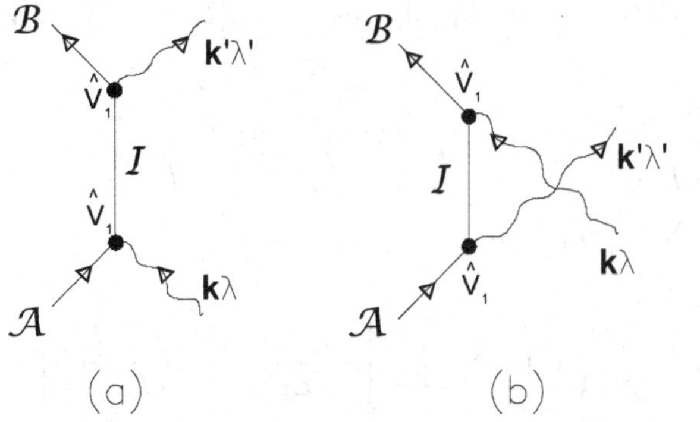

Figure 9.18 Feynman diagrams for the two types of virtual transitions associated with the scattering of a single photon by an atom.

by

$$-\sum_{\mathcal{I}} \frac{\langle\mathcal{B}|\,\langle 0_{\mathbf{k}\lambda}, 1_{\mathbf{k}'\lambda'} \mid \hat{V}_1 \mid 0_{\mathbf{k}\lambda}, 0_{\mathbf{k}'\lambda'}\rangle\,|\mathcal{I}\rangle\,\langle\mathcal{I}|\,\langle 0_{\mathbf{k}\lambda}, 0_{\mathbf{k}'\lambda'} \mid \hat{V}_1 \mid 1_{\mathbf{k}\lambda}, 0_{\mathbf{k}'\lambda'}\rangle\,|\mathcal{A}\rangle}{E_{\mathcal{I}} - E_{\mathcal{A}} - \hbar\omega_k}$$

$$= -\frac{1}{m} r_0 \left(\frac{2\pi\hbar c}{L^3\sqrt{kk'}}\right) \sum_{\mathcal{I}} \left(\frac{1}{E_{\mathcal{I}} - E_{\mathcal{A}} - \hbar\omega_k}\right)$$
$$\times \sum_i \left\langle \mathcal{B} \mid \left\langle 0_{\mathbf{k}\lambda}, 1_{\mathbf{k}'\lambda'} \mid \hat{a}^\dagger_{\mathbf{k}'\lambda'} e^{-i\mathbf{k}'\cdot\mathbf{r}_i} \boldsymbol{\epsilon}_{\mathbf{k}'\lambda'}\cdot\hat{\mathbf{p}}_i \mid 0_{\mathbf{k}\lambda}, 0_{\mathbf{k}'\lambda'} \right\rangle \mid \mathcal{I} \right\rangle$$
$$\times \sum_j \langle \mathcal{I} \mid \langle 0_{\mathbf{k}\lambda}, 0_{\mathbf{k}'\lambda'} \mid \hat{a}_{\mathbf{k}\lambda} e^{i\mathbf{k}\cdot\mathbf{r}_j} \boldsymbol{\epsilon}_{\mathbf{k}\lambda}\cdot\hat{\mathbf{p}}_j \mid 1_{\mathbf{k}\lambda}, 0_{\mathbf{k}'\lambda'} \rangle \mid \mathcal{A} \rangle$$

$$= -\frac{1}{m} r_0 \left(\frac{2\pi\hbar c}{L^3\sqrt{kk'}}\right) \sum_{\mathcal{I}} \frac{\sum_{i,j} \langle \mathcal{B} \mid e^{-i\mathbf{k}'\cdot\mathbf{r}_i} \boldsymbol{\epsilon}_{\mathbf{k}'\lambda'}\cdot\hat{\mathbf{p}}_i \mid \mathcal{I}\rangle\langle \mathcal{I} \mid e^{i\mathbf{k}\cdot\mathbf{r}_j} \boldsymbol{\epsilon}_{\mathbf{k}\lambda}\cdot\hat{\mathbf{p}}_j \mid \mathcal{A}\rangle}{E_{\mathcal{I}} - E_{\mathcal{A}} - \hbar\omega_k}. \tag{9.117}$$

For optical photons, the wavelength is much larger than the size of an atom. This means that we are able to apply the electric dipole approximation (see Sect. 8.1). In other words, if the center of the scattering atom is specified by the vector $\mathbf{R}$, we can write

$$e^{i\mathbf{k}\cdot\mathbf{r}_j} \simeq e^{i\mathbf{k}\cdot\mathbf{R}} \quad \text{and} \quad e^{-i\mathbf{k}'\cdot\mathbf{r}_i} \simeq e^{-i\mathbf{k}'\cdot\mathbf{R}}. \tag{9.118}$$

Also, from Eq. 8.10, we have, for example

$$\langle \mathcal{B} \mid \hat{\mathbf{p}}_i \mid \mathcal{I} \rangle = -i\frac{m}{\hbar}\,(E_I - E_B)\,\langle \mathcal{B} \mid \hat{\mathbf{r}}_i \mid \mathcal{I} \rangle. \tag{9.119}$$

Then the first contribution to $K_1^{(2)}$ becomes

$$
\begin{aligned}
& -\sum_{\mathcal{I}} \frac{\langle\mathcal{B}|\langle 0_{\mathbf{k}\lambda}, 1_{\mathbf{k}'\lambda'} \mid \hat{V}_1 \mid 0_{\mathbf{k}\lambda}, 0_{\mathbf{k}'\lambda'}\rangle |\mathcal{I}\rangle \langle\mathcal{I}| \langle 0_{\mathbf{k}\lambda}, 0_{\mathbf{k}'\lambda'} \mid \hat{V}_1 \mid 1_{\mathbf{k}\lambda}, 0_{\mathbf{k}'\lambda'}\rangle |\mathcal{A}\rangle}{E_{\mathcal{I}} - E_{\mathcal{A}} - \hbar\omega_k} \\
&= \frac{m}{\hbar^2} r_0 \left(\frac{2\pi\hbar c}{L^3\sqrt{kk'}}\right) e^{i\mathbf{Q}\cdot\mathbf{R}} \sum_{\mathcal{I}} \left[\frac{(E_{\mathcal{I}} - E_{\mathcal{B}})(E_{\mathcal{A}} - E_{\mathcal{I}})}{E_{\mathcal{I}} - E_{\mathcal{A}} - \hbar\omega_k}\right] \\
&\quad \times \sum_{i,j} \langle\mathcal{B} \mid \boldsymbol{\epsilon}_{\mathbf{k}'\lambda'}\cdot\hat{\mathbf{r}}_i \mid \mathcal{I}\rangle \langle\mathcal{I} \mid \boldsymbol{\epsilon}_{\mathbf{k}\lambda}\cdot\hat{\mathbf{r}}_j \mid \mathcal{A}\rangle \\
&= \frac{m}{e^2\hbar^2} r_0 \left(\frac{2\pi\hbar c}{L^3\sqrt{kk'}}\right) e^{i\mathbf{Q}\cdot\mathbf{R}} \sum_{\mathcal{I}} \left[\frac{(E_{\mathcal{I}} - E_{\mathcal{B}})(E_{\mathcal{A}} - E_{\mathcal{I}})}{E_{\mathcal{I}} - E_{\mathcal{A}} - \hbar\omega_k}\right] \\
&\quad \times (\boldsymbol{\epsilon}_{\mathbf{k}'\lambda'}\cdot\boldsymbol{\mu}_{\mathcal{B}\mathcal{I}})(\boldsymbol{\epsilon}_{\mathbf{k}\lambda}\cdot\boldsymbol{\mu}_{\mathcal{I}\mathcal{A}}). \qquad (9.120)
\end{aligned}
$$

Here, we have recalled the electric dipole moment operator of the atom, $\widehat{\boldsymbol{\mu}} = e\sum_i \hat{\mathbf{r}}_i$. The contribution to $K_1^{(2)}$ from the second type of transition can be obtained in a similar fashion—the result is

$$
\begin{aligned}
& -\sum_{\mathcal{I}} \frac{\langle\mathcal{B}|\langle 0_{\mathbf{k}\lambda}, 1_{\mathbf{k}'\lambda'} \mid \hat{V}_1 \mid 1_{\mathbf{k}\lambda}, 1_{\mathbf{k}'\lambda'}\rangle |\mathcal{I}\rangle \langle\mathcal{I}| \langle 1_{\mathbf{k}\lambda}, 1_{\mathbf{k}'\lambda'} \mid \hat{V}_1 \mid 1_{\mathbf{k}\lambda}, 0_{\mathbf{k}'\lambda'}\rangle |\mathcal{A}\rangle}{E_{\mathcal{I}} - E_{\mathcal{A}} + \hbar\omega_{k'}} \\
&= \frac{m}{e^2\hbar^2} r_0 \left(\frac{2\pi\hbar c}{L^3\sqrt{kk'}}\right) e^{i\mathbf{Q}\cdot\mathbf{R}} \sum_{\mathcal{I}} \left[\frac{(E_{\mathcal{I}} - E_{\mathcal{B}})(E_{\mathcal{A}} - E_{\mathcal{I}})}{E_{\mathcal{I}} - E_{\mathcal{A}} + \hbar\omega_{k'}}\right] \\
&\quad \times (\boldsymbol{\epsilon}_{\mathbf{k}\lambda}\cdot\boldsymbol{\mu}_{\mathcal{B}\mathcal{I}})(\boldsymbol{\epsilon}_{\mathbf{k}'\lambda'}\cdot\boldsymbol{\mu}_{\mathcal{I}\mathcal{A}}). \qquad (9.121)
\end{aligned}
$$

Upon combining the two contributions, we have

$$
\begin{aligned}
K_1^{(2)} &= \frac{m}{e^2\hbar^2} r_0 \left(\frac{2\pi\hbar c}{L^3\sqrt{kk'}}\right) e^{i\mathbf{Q}\cdot\mathbf{R}} \sum_{\mathcal{I}} (E_{\mathcal{I}} - E_{\mathcal{B}})(E_{\mathcal{A}} - E_{\mathcal{I}}) \\
&\quad \times \left[\frac{(\boldsymbol{\epsilon}_{\mathbf{k}'\lambda'}\cdot\boldsymbol{\mu}_{\mathcal{B}\mathcal{I}})(\boldsymbol{\epsilon}_{\mathbf{k}\lambda}\cdot\boldsymbol{\mu}_{\mathcal{I}\mathcal{A}})}{E_{\mathcal{I}} - E_{\mathcal{A}} - \hbar\omega_k} + \frac{(\boldsymbol{\epsilon}_{\mathbf{k}\lambda}\cdot\boldsymbol{\mu}_{\mathcal{B}\mathcal{I}})(\boldsymbol{\epsilon}_{\mathbf{k}'\lambda'}\cdot\boldsymbol{\mu}_{\mathcal{I}\mathcal{A}})}{E_{\mathcal{I}} - E_{\mathcal{A}} + \hbar\omega_{k'}}\right]. \qquad (9.122)
\end{aligned}
$$

The scattering rate also depends on the term $K_2^{(1)}$. Referring to Eq. 9.43, one can again make use of the electric dipole approximation to reduce it as follows (here index i denotes each of the Z atomic electrons):

$$
\begin{aligned}
K_2^{(1)} &= r_0 \left(\frac{2\pi\hbar c}{L^3\sqrt{kk'}}\right) (\boldsymbol{\epsilon}_{\mathbf{k}\lambda}\cdot\boldsymbol{\epsilon}_{\mathbf{k}'\lambda'}) \langle\mathcal{B} \mid \sum_i e^{i\mathbf{Q}\cdot\mathbf{r}_i} \mid \mathcal{A}\rangle \\
&= r_0 \left(\frac{2\pi\hbar c}{L^3\sqrt{kk'}}\right) (\boldsymbol{\epsilon}_{\mathbf{k}\lambda}\cdot\boldsymbol{\epsilon}_{\mathbf{k}'\lambda'}) \delta_{AB} \sum_i e^{i\mathbf{Q}\cdot\mathbf{R}} \\
&= r_0 Z \left(\frac{2\pi\hbar c}{L^3\sqrt{kk'}}\right) e^{i\mathbf{Q}\cdot\mathbf{R}} (\boldsymbol{\epsilon}_{\mathbf{k}\lambda}\cdot\boldsymbol{\epsilon}_{\mathbf{k}'\lambda'}) \delta_{AB}. \qquad (9.123)
\end{aligned}
$$

Considering all contributions to the scattering probability and inserting the density-of-states-factor¶

$$g(E_n) = \left(\frac{L}{2\pi}\right)^3 k'^2 \frac{d\Omega}{\hbar c}, \tag{9.124}$$

the differential cross-section, $d\sigma = \frac{W_{nm}}{c/L^3}$, becomes

$$\frac{d\sigma}{d\Omega} = r_0^2 \left(\frac{k'}{k}\right) |M_{\mathcal{A}\to\mathcal{B}}|^2, \tag{9.125}$$

where

$$\begin{aligned} M_{\mathcal{A}\to\mathcal{B}} &= Z\left(\boldsymbol{\epsilon}_{\mathbf{k}\lambda}\cdot\boldsymbol{\epsilon}_{\mathbf{k}'\lambda'}\right)\delta_{AB} + \frac{m}{e^2\hbar^2}\sum_{\mathcal{I}}\left(E_{\mathcal{I}} - E_{\mathcal{B}}\right)\left(E_{\mathcal{A}} - E_{\mathcal{I}}\right) \\ &\times \left[\frac{\left(\boldsymbol{\epsilon}_{\mathbf{k}'\lambda'}\cdot\boldsymbol{\mu}_{\mathcal{B}\mathcal{I}}\right)\left(\boldsymbol{\epsilon}_{\mathbf{k}\lambda}\cdot\boldsymbol{\mu}_{\mathcal{I}\mathcal{A}}\right)}{E_{\mathcal{I}} - E_{\mathcal{A}} - \hbar\omega_k} + \frac{\left(\boldsymbol{\epsilon}_{\mathbf{k}\lambda}\cdot\boldsymbol{\mu}_{\mathcal{B}\mathcal{I}}\right)\left(\boldsymbol{\epsilon}_{\mathbf{k}'\lambda'}\cdot\boldsymbol{\mu}_{\mathcal{I}\mathcal{A}}\right)}{E_{\mathcal{I}} - E_{\mathcal{A}} + \hbar\omega_{k'}}\right]. \end{aligned} \tag{9.126}$$

Taken together, Eqs. 9.125 and 9.126 form a general result for the scattering of light by atoms known as the *Kramers-Heisenberg formula.* It is subject to the energy conservation condition

$$E_{\mathcal{A}} + \hbar\omega_k = E_{\mathcal{B}} + \hbar\omega_{k'}. \tag{9.127}$$

The Kramers-Heisenberg formula is the starting point for deriving cross-section expressions for both inelastic Raman scattering and elastic Rayleigh scattering.

Example 9.4 Cross-Section for Electronic Raman Scattering. As a specific application of the Kramers-Heisenberg formula, consider inelastic (Raman) scattering from an atom. In this case, $|\mathcal{A}\rangle$ and $|\mathcal{B}\rangle$ are different discrete electronic states of the atom. Since $\mathcal{A} \neq \mathcal{B}$, we drop the first term in Eq. 9.126 and write

$$\begin{aligned} M_{\mathcal{A}\to\mathcal{B}} &= \frac{m}{e^2\hbar^2}\sum_{\mathcal{I}}\left(E_{\mathcal{I}} - E_{\mathcal{B}}\right)\left(E_{\mathcal{I}} - E_{\mathcal{A}}\right) \\ &\times \left[\frac{\left(\boldsymbol{\epsilon}_{\mathbf{k}'\lambda'}\cdot\boldsymbol{\mu}_{\mathcal{B}\mathcal{I}}\right)\left(\boldsymbol{\epsilon}_{\mathbf{k}\lambda}\cdot\boldsymbol{\mu}_{\mathcal{I}\mathcal{A}}\right)}{E_{\mathcal{A}} - E_{\mathcal{I}} + \hbar\omega_k} + \frac{\left(\boldsymbol{\epsilon}_{\mathbf{k}\lambda}\cdot\boldsymbol{\mu}_{\mathcal{B}\mathcal{I}}\right)\left(\boldsymbol{\epsilon}_{\mathbf{k}'\lambda'}\cdot\boldsymbol{\mu}_{\mathcal{I}\mathcal{A}}\right)}{E_{\mathcal{A}} - E_{\mathcal{I}} - \hbar\omega_{k'}}\right]. \end{aligned} \tag{9.128}$$

In order to proceed any further, it becomes necessary to take a short detour and consider the commutator

$$\left[\sum_i \hat{\mathbf{r}}_i\cdot\boldsymbol{\epsilon}_{\mathbf{k}\lambda}, \sum_j \hat{\mathbf{p}}_j\cdot\boldsymbol{\epsilon}_{\mathbf{k}'\lambda'}\right] = \sum_{i,j}\sum_{m,n}\left[\hat{r}_{im}\epsilon_m, \hat{p}_{jn}\epsilon'_n\right], \tag{9.129}$$

¶Again, in the case of scattering from an atom, recoil effects are neglected.

where the indices i and j refer to the different atomic electrons, and the indices m and n refer to the three Cartesian components of the various vectors. For example, ϵ'_n denotes the nth component of the vector $\boldsymbol{\epsilon}_{\mathbf{k}'\lambda'}$. The commutator can now be reduced to a much simpler form, i.e.,

$$\begin{aligned}\left[\sum_i \mathbf{r}_i\cdot\boldsymbol{\epsilon}_{\mathbf{k}\lambda}, \sum_j \mathbf{p}_j\cdot\boldsymbol{\epsilon}_{\mathbf{k}'\lambda'}\right] &= \sum_{i,j}\sum_{m,n} \epsilon_m\epsilon'_n \left[\hat{r}_{im}, \hat{p}_{jn}\right] = i\hbar\sum_{i,j}\sum_{m,n}\epsilon_m\epsilon'_n\delta_{ij}\delta_{mn} \\ &= i\hbar\sum_i\sum_m \epsilon_m\epsilon'_m = i\hbar Z\boldsymbol{\epsilon}_{\mathbf{k}\lambda}\cdot\boldsymbol{\epsilon}_{\mathbf{k}'\lambda'}. \end{aligned} \tag{9.130}$$

Furthermore, observe that

$$\begin{aligned} 0 &= \langle\mathcal{B}\mid Z\boldsymbol{\epsilon}_{\mathbf{k}\lambda}\cdot\boldsymbol{\epsilon}_{\mathbf{k}'\lambda'}\mid\mathcal{A}\rangle = \frac{1}{i\hbar}\langle\mathcal{B}\mid\left[\sum_i \hat{\mathbf{r}}_i\cdot\boldsymbol{\epsilon}_{\mathbf{k}\lambda}, \sum_j \hat{\mathbf{p}}_j\cdot\boldsymbol{\epsilon}_{\mathbf{k}'\lambda'}\right]\mid\mathcal{A}\rangle \\ &= \frac{1}{i\hbar}\sum_{\mathcal{I}}\sum_{i,j}\langle\mathcal{B}\mid\hat{\mathbf{r}}_i\cdot\boldsymbol{\epsilon}_{\mathbf{k}\lambda}\mid\mathcal{I}\rangle\langle\mathcal{I}\mid\hat{\mathbf{p}}_j\cdot\boldsymbol{\epsilon}_{\mathbf{k}'\lambda'}\mid\mathcal{A}\rangle \\ &\quad -\frac{1}{i\hbar}\sum_{\mathcal{I}}\sum_{i,j}\langle\mathcal{B}\mid\hat{\mathbf{p}}_j\cdot\boldsymbol{\epsilon}_{\mathbf{k}'\lambda'}\mid\mathcal{I}\rangle\langle\mathcal{I}\mid\hat{\mathbf{r}}_i\cdot\boldsymbol{\epsilon}_{\mathbf{k}\lambda}\mid\mathcal{A}\rangle . \end{aligned} \tag{9.131}$$

If we now apply Eq. 9.119, we produce the following useful identity:

$$0 = \sum_{\mathcal{I}}\left[\left(E_{\mathcal{I}}-E_{\mathcal{B}}\right)\left(\boldsymbol{\epsilon}_{\mathbf{k}'\lambda'}\cdot\boldsymbol{\mu}_{\mathcal{B}\mathcal{I}}\right)\left(\boldsymbol{\epsilon}_{\mathbf{k}\lambda}\cdot\boldsymbol{\mu}_{\mathcal{I}\mathcal{A}}\right) - \left(E_{\mathcal{A}}-E_{\mathcal{I}}\right)\left(\boldsymbol{\epsilon}_{\mathbf{k}\lambda}\cdot\boldsymbol{\mu}_{\mathcal{B}\mathcal{I}}\right)\left(\boldsymbol{\epsilon}_{\mathbf{k}'\lambda'}\cdot\boldsymbol{\mu}_{\mathcal{I}\mathcal{A}}\right)\right]. \tag{9.132}$$

The latter zero-term can be added to the sum in Eq. 9.128 to produce

$$\begin{aligned} M_{\mathcal{A}\to\mathcal{B}} &= \frac{m\omega_k}{e^2\hbar}\sum_{\mathcal{I}}\left(E_{\mathcal{I}}-E_{\mathcal{B}}\right)\frac{\left(\boldsymbol{\epsilon}_{\mathbf{k}'\lambda'}\cdot\boldsymbol{\mu}_{\mathcal{B}\mathcal{I}}\right)\left(\boldsymbol{\epsilon}_{\mathbf{k}\lambda}\cdot\boldsymbol{\mu}_{\mathcal{I}\mathcal{A}}\right)}{E_{\mathcal{A}}-E_{\mathcal{I}}+\hbar\omega_k} \\ &\quad -\frac{m\omega_k}{e^2\hbar}\sum_{\mathcal{I}}\left(E_{\mathcal{I}}-E_{\mathcal{A}}\right)\frac{\left(\boldsymbol{\epsilon}_{\mathbf{k}\lambda}\cdot\boldsymbol{\mu}_{\mathcal{B}\mathcal{I}}\right)\left(\boldsymbol{\epsilon}_{\mathbf{k}'\lambda'}\cdot\boldsymbol{\mu}_{\mathcal{I}\mathcal{A}}\right)}{E_{\mathcal{A}}-E_{\mathcal{I}}-\hbar\omega_{k'}}. \end{aligned} \tag{9.133}$$

To obtain this expression we also made use of the energy conservation condition, Eq. 9.127. Finally, add the following zero-term to the latter sum:

$$\begin{aligned} 0 &= \langle\mathcal{B}\mid\left[\left(\boldsymbol{\epsilon}_{\mathbf{k}'\lambda'}\cdot\boldsymbol{\mu}\right)\left(\boldsymbol{\epsilon}_{\mathbf{k}\lambda}\cdot\boldsymbol{\mu}\right)-\left(\boldsymbol{\epsilon}_{\mathbf{k}\lambda}\cdot\boldsymbol{\mu}\right)\left(\boldsymbol{\epsilon}_{\mathbf{k}'\lambda'}\cdot\boldsymbol{\mu}\right)\right]\mid\mathcal{A}\rangle \\ &= \sum_{\mathcal{I}}\left[\left(\boldsymbol{\epsilon}_{\mathbf{k}'\lambda'}\cdot\boldsymbol{\mu}_{\mathcal{B}\mathcal{I}}\right)\left(\boldsymbol{\epsilon}_{\mathbf{k}\lambda}\cdot\boldsymbol{\mu}_{\mathcal{I}\mathcal{A}}\right)-\left(\boldsymbol{\epsilon}_{\mathbf{k}\lambda}\cdot\boldsymbol{\mu}_{\mathcal{B}\mathcal{I}}\right)\left(\boldsymbol{\epsilon}_{\mathbf{k}'\lambda'}\cdot\boldsymbol{\mu}_{\mathcal{I}\mathcal{A}}\right)\right]. \end{aligned} \tag{9.134}$$

The result becomes

$$M_{\mathcal{A}\to\mathcal{B}} = \frac{m\omega_k\omega_{k'}}{e^2}\sum_{\mathcal{I}}\left[\frac{\left(\boldsymbol{\epsilon}_{\mathbf{k}'\lambda'}\cdot\boldsymbol{\mu}_{\mathcal{B}\mathcal{I}}\right)\left(\boldsymbol{\epsilon}_{\mathbf{k}\lambda}\cdot\boldsymbol{\mu}_{\mathcal{I}\mathcal{A}}\right)}{E_{\mathcal{A}}-E_{\mathcal{I}}+\hbar\omega_k}+\frac{\left(\boldsymbol{\epsilon}_{\mathbf{k}\lambda}\cdot\boldsymbol{\mu}_{\mathcal{B}\mathcal{I}}\right)\left(\boldsymbol{\epsilon}_{\mathbf{k}'\lambda'}\cdot\boldsymbol{\mu}_{\mathcal{I}\mathcal{A}}\right)}{E_{\mathcal{A}}-E_{\mathcal{I}}-\hbar\omega_{k'}}\right]. \tag{9.135}$$

Thus, from Eq. 9.125, the cross-section for Raman scattering is

$$\left(\frac{d\sigma}{d\Omega}\right)_{\text{Raman}} = kk'^3 \left| \sum_{\mathcal{I}} \left[\frac{(\boldsymbol{\epsilon}_{\mathbf{k}'\lambda'} \cdot \boldsymbol{\mu}_{\mathcal{BI}})(\boldsymbol{\epsilon}_{\mathbf{k}\lambda} \cdot \boldsymbol{\mu}_{\mathcal{IA}})}{E_{\mathcal{A}} - E_{\mathcal{I}} + \hbar\omega_k} + \frac{(\boldsymbol{\epsilon}_{\mathbf{k}\lambda} \cdot \boldsymbol{\mu}_{\mathcal{BI}})(\boldsymbol{\epsilon}_{\mathbf{k}'\lambda'} \cdot \boldsymbol{\mu}_{\mathcal{IA}})}{E_{\mathcal{A}} - E_{\mathcal{I}} - \hbar\omega_{k'}} \right] \right|^2 . \quad (9.136)$$

This can be written more compactly by defining the *Raman tensor*

$$\tilde{\tilde{\mathbf{R}}} = \sum_{\mathcal{I}} \left[\frac{\boldsymbol{\mu}_{\mathcal{IA}} \boldsymbol{\mu}_{\mathcal{BI}}}{E_{\mathcal{A}} - E_{\mathcal{I}} + \hbar\omega_k} + \frac{\boldsymbol{\mu}_{\mathcal{BI}} \boldsymbol{\mu}_{\mathcal{IA}}}{E_{\mathcal{A}} - E_{\mathcal{I}} - \hbar\omega_{k'}} \right], \quad (9.137)$$

so that

$$\left(\frac{d\sigma}{d\Omega}\right)_{\text{Raman}} = kk'^3 \left| \boldsymbol{\epsilon}_{\mathbf{k}\lambda} \cdot \tilde{\tilde{\mathbf{R}}} \cdot \boldsymbol{\epsilon}_{\mathbf{k}'\lambda'} \right|^2 . \quad (9.138)$$

In Raman scattering, the frequency of the scattered photon can either be less than or greater than the frequency of the incident radiation, depending on which is larger, the energy of the initial or final state. In a case where $E_{\mathcal{B}} > E_{\mathcal{A}}$, the frequency of the scattered radiation is down-shifted relative to the incident beam, i.e., $\omega_{k'} = \omega_k - (E_{\mathcal{B}} - E_{\mathcal{A}})/\hbar$, and is commonly referred to as the *Stokes line.* On the other hand, if $E_{\mathcal{A}} > E_{\mathcal{B}}$, then $\omega_{k'} = \omega_k + (E_{\mathcal{A}} - E_{\mathcal{B}})/\hbar$, and the frequency is up-shifted producing an *anti-Stokes line.*

Of special interest is the scattering at resonance, where the incident photon energy matches the energy difference between some pair of atomic levels, i.e., $\hbar\omega_k = E_{\mathcal{I}} - E_{\mathcal{A}}$. Then, according to our result for the scattering cross-section, one term clearly dominates, and

$$\left(\frac{d\sigma}{d\Omega}\right)^{\text{res}}_{\text{Raman}} = kk'^3 \left| \frac{(\boldsymbol{\epsilon}_{\mathbf{k}'\lambda'} \cdot \boldsymbol{\mu}_{\mathcal{BI}})(\boldsymbol{\epsilon}_{\mathbf{k}\lambda} \cdot \boldsymbol{\mu}_{\mathcal{IA}})}{E_{\mathcal{I}} - E_{\mathcal{A}} - \hbar\omega_k} \right|^2_{\hbar\omega_k \to E_{\mathcal{I}} - E_{\mathcal{A}}} . \quad (9.139)$$

In fact, Eq. 9.139 predicts the cross-section to be infinite! The reason for this non-physical result is that, in our derivation, we have assumed the linewidths of the various atomic levels to be zero. As discussed in Chapter 8, this is certainly not the case, since there is always at least a natural linewidth associated with each state. To account for the non-zero width, we add a small imaginary component to the denominator, so that

$$\begin{aligned} \left(\frac{d\sigma}{d\Omega}\right)^{\text{res}}_{\text{Raman}} &= kk'^3 \left| \frac{(\boldsymbol{\epsilon}_{\mathbf{k}'\lambda'} \cdot \boldsymbol{\mu}_{\mathcal{BI}})(\boldsymbol{\epsilon}_{\mathbf{k}\lambda} \cdot \boldsymbol{\mu}_{\mathcal{IA}})}{E_{\mathcal{I}} - E_{\mathcal{A}} - \hbar\omega_k - i\hbar\Gamma/2} \right|^2_{\hbar\omega_k \approx E_{\mathcal{I}} - E_{\mathcal{A}}} \\ &= kk'^3 \left[\frac{\left|(\boldsymbol{\epsilon}_{\mathbf{k}'\lambda'} \cdot \boldsymbol{\mu}_{\mathcal{BI}})(\boldsymbol{\epsilon}_{\mathbf{k}\lambda} \cdot \boldsymbol{\mu}_{\mathcal{IA}})\right|^2}{(E_{\mathcal{I}} - E_{\mathcal{A}} - \hbar\omega_k)^2 + \frac{1}{4}\hbar^2\Gamma^2} \right]_{\hbar\omega_k \approx E_{\mathcal{I}} - E_{\mathcal{A}}} , \end{aligned} \quad (9.140)$$

and the resonance cross-section becomes very large, but not infinite.

Example 9.5 Rayleigh Scattering of Light. Now consider an elastic scattering event in which the atom returns to its original state, i.e., $|\mathcal{B}\rangle = |\mathcal{A}\rangle$ and $\omega_{k'} = \omega_k$. This corresponds to the case of coherent or Rayleigh scattering. It now becomes necessary to include the term $Z\boldsymbol{\epsilon}_{\mathbf{k}\lambda}\cdot\boldsymbol{\epsilon}_{\mathbf{k}'\lambda'}$ in Eq. 9.126, and we have

$$\begin{aligned} M_{\mathcal{A}\to\mathcal{A}} &= Z\boldsymbol{\epsilon}_{\mathbf{k}\lambda}\cdot\boldsymbol{\epsilon}_{\mathbf{k}'\lambda'} - \frac{m}{e^2\hbar^2}\sum_{\mathcal{I}}(E_{\mathcal{I}} - E_{\mathcal{A}})^2 \\ &\quad\times\left[\frac{(\boldsymbol{\epsilon}_{\mathbf{k}'\lambda'}\cdot\boldsymbol{\mu}_{\mathcal{AI}})(\boldsymbol{\epsilon}_{\mathbf{k}\lambda}\cdot\boldsymbol{\mu}_{\mathcal{IA}})}{E_{\mathcal{I}} - E_{\mathcal{A}} - \hbar\omega_k} + \frac{(\boldsymbol{\epsilon}_{\mathbf{k}\lambda}\cdot\boldsymbol{\mu}_{\mathcal{AI}})(\boldsymbol{\epsilon}_{\mathbf{k}'\lambda'}\cdot\boldsymbol{\mu}_{\mathcal{IA}})}{E_{\mathcal{I}} - E_{\mathcal{A}} + \hbar\omega_k}\right]. \end{aligned} \tag{9.141}$$

For the first term, we write $Z\boldsymbol{\epsilon}_{\mathbf{k}\lambda}\cdot\boldsymbol{\epsilon}_{\mathbf{k}'\lambda'} = \langle\mathcal{A} \mid Z\boldsymbol{\epsilon}_{\mathbf{k}\lambda}\cdot\boldsymbol{\epsilon}_{\mathbf{k}'\lambda'} \mid \mathcal{A}\rangle$, and proceed as in Eq. 9.131 to obtain

$$Z\boldsymbol{\epsilon}_{\mathbf{k}\lambda}\cdot\boldsymbol{\epsilon}_{\mathbf{k}'\lambda'} = \frac{m}{e^2\hbar^2}\sum_{\mathcal{I}}(E_{\mathcal{I}} - E_{\mathcal{A}})\left[(\boldsymbol{\epsilon}_{\mathbf{k}\lambda}\cdot\boldsymbol{\mu}_{\mathcal{AI}})(\boldsymbol{\epsilon}_{\mathbf{k}'\lambda'}\cdot\boldsymbol{\mu}_{\mathcal{IA}}) + (\boldsymbol{\epsilon}_{\mathbf{k}'\lambda'}\cdot\boldsymbol{\mu}_{\mathcal{AI}})(\boldsymbol{\epsilon}_{\mathbf{k}\lambda}\cdot\boldsymbol{\mu}_{\mathcal{IA}})\right]. \tag{9.142}$$

Inserting this expression into Eq. 9.141 and rearranging terms gives

$$M_{\mathcal{A}\to\mathcal{A}} = \frac{m\omega_k}{e^2\hbar}\sum_{\mathcal{I}}(E_{\mathcal{I}} - E_{\mathcal{A}})\left[\frac{(\boldsymbol{\epsilon}_{\mathbf{k}\lambda}\cdot\boldsymbol{\mu}_{\mathcal{AI}})(\boldsymbol{\epsilon}_{\mathbf{k}'\lambda'}\cdot\boldsymbol{\mu}_{\mathcal{IA}})}{E_{\mathcal{I}} - E_{\mathcal{A}} + \hbar\omega_k} - \frac{(\boldsymbol{\epsilon}_{\mathbf{k}'\lambda'}\cdot\boldsymbol{\mu}_{\mathcal{AI}})(\boldsymbol{\epsilon}_{\mathbf{k}\lambda}\cdot\boldsymbol{\mu}_{\mathcal{IA}})}{E_{\mathcal{I}} - E_{\mathcal{A}} - \hbar\omega_k}\right]. \tag{9.143}$$

Now incorporate the following zero-term:

$$\begin{aligned} 0 &= \langle\mathcal{A} \mid [(\boldsymbol{\epsilon}_{\mathbf{k}'\lambda'}\cdot\boldsymbol{\mu})(\boldsymbol{\epsilon}_{\mathbf{k}\lambda}\cdot\boldsymbol{\mu}) - (\boldsymbol{\epsilon}_{\mathbf{k}\lambda}\cdot\boldsymbol{\mu})(\boldsymbol{\epsilon}_{\mathbf{k}'\lambda'}\cdot\boldsymbol{\mu})] \mid \mathcal{A}\rangle \\ &= \sum_{\mathcal{I}}\left[(\boldsymbol{\epsilon}_{\mathbf{k}'\lambda'}\cdot\boldsymbol{\mu}_{\mathcal{AI}})(\boldsymbol{\epsilon}_{\mathbf{k}\lambda}\cdot\boldsymbol{\mu}_{\mathcal{IA}}) - (\boldsymbol{\epsilon}_{\mathbf{k}\lambda}\cdot\boldsymbol{\mu}_{\mathcal{AI}})(\boldsymbol{\epsilon}_{\mathbf{k}'\lambda'}\cdot\boldsymbol{\mu}_{\mathcal{IA}})\right]. \end{aligned} \tag{9.144}$$

The result is

$$M_{\mathcal{A}\to\mathcal{A}} = -\frac{m\omega_k^2}{e^2}\sum_{\mathcal{I}}\left[\frac{(\boldsymbol{\epsilon}_{\mathbf{k}\lambda}\cdot\boldsymbol{\mu}_{\mathcal{AI}})(\boldsymbol{\epsilon}_{\mathbf{k}'\lambda'}\cdot\boldsymbol{\mu}_{\mathcal{IA}})}{E_{\mathcal{I}} - E_{\mathcal{A}} + \hbar\omega_k} + \frac{(\boldsymbol{\epsilon}_{\mathbf{k}'\lambda'}\cdot\boldsymbol{\mu}_{\mathcal{AI}})(\boldsymbol{\epsilon}_{\mathbf{k}\lambda}\cdot\boldsymbol{\mu}_{\mathcal{IA}})}{E_{\mathcal{I}} - E_{\mathcal{A}} - \hbar\omega_k}\right], \tag{9.145}$$

and the Rayleigh scattering cross-section takes the form

$$\left(\frac{d\sigma}{d\Omega}\right)_{\text{Rayleigh}} = k^4\left|\sum_{\mathcal{I}}\left[\frac{(\boldsymbol{\epsilon}_{\mathbf{k}\lambda}\cdot\boldsymbol{\mu}_{\mathcal{AI}})(\boldsymbol{\epsilon}_{\mathbf{k}'\lambda'}\cdot\boldsymbol{\mu}_{\mathcal{IA}})}{E_{\mathcal{I}} - E_{\mathcal{A}} + \hbar\omega_k} + \frac{(\boldsymbol{\epsilon}_{\mathbf{k}'\lambda'}\cdot\boldsymbol{\mu}_{\mathcal{AI}})(\boldsymbol{\epsilon}_{\mathbf{k}\lambda}\cdot\boldsymbol{\mu}_{\mathcal{IA}})}{E_{\mathcal{I}} - E_{\mathcal{A}} - \hbar\omega_k}\right]\right|^2, \tag{9.146}$$

or

$$\left(\frac{d\sigma}{d\Omega}\right)_{\text{Rayleigh}} = k^4\left|\boldsymbol{\epsilon}_{\mathbf{k}\lambda}\cdot\tilde{\tilde{\boldsymbol{\alpha}}}(\omega_k)\cdot\boldsymbol{\epsilon}_{\mathbf{k}'\lambda'}\right|^2, \tag{9.147}$$

where we have introduced the *Rayleigh tensor*

$$\tilde{\tilde{\boldsymbol{\alpha}}}(\omega_k) = \sum_{\mathcal{I}}\left[\frac{\boldsymbol{\mu}_{\mathcal{AI}}\boldsymbol{\mu}_{\mathcal{IA}}}{E_{\mathcal{I}} - E_{\mathcal{A}} + \hbar\omega_k} + \frac{\boldsymbol{\mu}_{\mathcal{IA}}\boldsymbol{\mu}_{\mathcal{AI}}}{E_{\mathcal{I}} - E_{\mathcal{A}} - \hbar\omega_k}\right]. \tag{9.148}$$

In Chapter 13 we will show, by using linear response theory, that $\widetilde{\widetilde{\boldsymbol{\alpha}}}(\omega_k)$ can, in fact, be identified as the *atomic polarizability tensor*.

The reader should notice that the expression for the Raman scattering cross-section developed in the previous example reduces to the above Rayleigh scattering result simply by letting $\mathcal{B} \to \mathcal{A}$ and $\omega_{k'} \to \omega_k$. The fact that this should be true is not at all obvious from the formalism as we have presented it since the calculation of the Rayleigh scattering cross-section required the additional $Z\boldsymbol{\epsilon}_{\mathbf{k}\lambda}\boldsymbol{\cdot}\boldsymbol{\epsilon}_{\mathbf{k}'\lambda'}$-term. To better understand why the Rayleigh cross-section is just a special case of the Raman expression, it becomes advantageous to rework the entire light-scattering formalism with an alternate, but equivalent, interaction Hamiltonian. According to a discussion by Power and Thirunamachandran [77], the sum of the interaction energies, $\hat{V}_1$ and $\hat{V}_2$, can be transformed into an equivalent series of multipole interactions. For optical photons, where the wavelength is large compared to the atomic size, it is only necessary to retain the leading term in the resulting multipole expansion, namely the electric dipole term. It is then permissible to make the replacement

$$\hat{V}_1 + \hat{V}_2 \to -\widehat{\boldsymbol{\mu}}\boldsymbol{\cdot}\hat{\mathbf{E}}, \tag{9.149}$$

where the electric-field operator is given by (recall Eq. 5.38)

$$\hat{\mathbf{E}} = -\frac{1}{c}\frac{\partial \hat{\mathbf{A}}}{\partial t} = \sum_{\mathbf{k},\lambda} i\sqrt{\frac{2\pi\hbar\omega_k}{L^3}}\left[\hat{a}_{\mathbf{k}\lambda}e^{i\mathbf{k}\boldsymbol{\cdot}\mathbf{R}} - \hat{a}^{\dagger}_{\mathbf{k}\lambda}e^{-i\mathbf{k}\boldsymbol{\cdot}\mathbf{R}}\right]\boldsymbol{\epsilon}_{\mathbf{k}\lambda}. \tag{9.150}$$

From a calculational standpoint, this interaction energy is much simpler to work with than the form involving the vector potential $\hat{\mathbf{A}}$. As before, it is still necessary to annihilate a photon in one mode and create a photon in a second mode, but because $-\widehat{\boldsymbol{\mu}}\boldsymbol{\cdot}\hat{\mathbf{E}}$ is purely linear in $\hat{a}_{\mathbf{k}\lambda}$ and $\hat{a}^{\dagger}_{\mathbf{k}\lambda}$, the transition probability is determined solely by second-order perturbation theory—in other words, a first-order perturbation term no longer appears in the cross-section calculation. This easier way of doing things produces Eq. 9.136 directly, and it is now equally valid for both elastic and inelastic scattering. All this is accomplished without having to explicitly transform the $\hat{\mathbf{p}}_i$'s into $\boldsymbol{\mu}$, or having to go through any of the other tedious manipulations associated with the derivation and use of the Kramers-Heisenberg formula!

To conclude, let us now refocus our attention on Rayleigh scattering, and examine three special cases:

1. $\hbar\omega_k \gg E_{\mathcal{I}} - E_{\mathcal{A}}$
   This condition is met in the soft x-ray region of the EM spectrum ($\hbar\omega_k \sim$ a few keV) where the wavelengths are still large compared to the atomic size and the electric dipole approximation, used to derive Eq. 9.146, is still justified. Then we can expand the frequency factors as

$$\frac{1}{E_{\mathcal{I}} - E_{\mathcal{A}} + \hbar\omega_k} = \frac{1}{\hbar\omega_k}\left(1 - \frac{E_{\mathcal{I}} - E_{\mathcal{A}}}{\hbar\omega_k} + \cdots\right) \tag{9.151}$$

$$\frac{1}{E_{\mathcal{I}} - E_{\mathcal{A}} - \hbar\omega_k} = -\frac{1}{\hbar\omega_k}\left(1 + \frac{E_{\mathcal{I}} - E_{\mathcal{A}}}{\hbar\omega_k} + \cdots\right), \tag{9.152}$$

so that

$$\left(\frac{d\sigma}{d\Omega}\right)^{\text{short }\lambda}_{\text{Rayleigh}} = \frac{1}{\hbar^4 c^4} \times \left|\sum_{\mathcal{I}} (E_{\mathcal{I}} - E_{\mathcal{A}}) \left[(\boldsymbol{\epsilon}_{\mathbf{k}\lambda}\cdot\boldsymbol{\mu}_{\mathcal{AI}})(\boldsymbol{\epsilon}_{\mathbf{k}'\lambda'}\cdot\boldsymbol{\mu}_{\mathcal{IA}}) + (\boldsymbol{\epsilon}_{\mathbf{k}'\lambda'}\cdot\boldsymbol{\mu}_{\mathcal{AI}})(\boldsymbol{\epsilon}_{\mathbf{k}\lambda}\cdot\boldsymbol{\mu}_{\mathcal{IA}})\right]\right|^2. \tag{9.153}$$

From Eq. 9.142, this simplifies to

$$\left(\frac{d\sigma}{d\Omega}\right)^{\text{short }\lambda}_{\text{Rayleigh}} = Z^2 r_0^2 \left(\boldsymbol{\epsilon}_{\mathbf{k}\lambda}\cdot\boldsymbol{\epsilon}_{\mathbf{k}'\lambda'}\right)^2. \tag{9.154}$$

In this limit, one sees that the result becomes Z^2 times the Thomson scattering cross-section from a single free electron! The fact the multiplier is Z^2, rather than Z, means that as long as the electric dipole approximation is valid, the atomic electrons all scatter coherently with the same phase.

2. $\hbar\omega_k \simeq E_{\mathcal{I}} - E_{\mathcal{A}}$
This represents the case of elastic scattering at resonance. In analogy with Eq. 9.140, we write

$$\left(\frac{d\sigma}{d\Omega}\right)^{\text{res}}_{\text{Rayleigh}} = k^4 \left[\frac{\left|(\boldsymbol{\epsilon}_{\mathbf{k}'\lambda'}\cdot\boldsymbol{\mu}_{\mathcal{AI}})(\boldsymbol{\epsilon}_{\mathbf{k}\lambda}\cdot\boldsymbol{\mu}_{\mathcal{IA}})\right|^2}{(E_{\mathcal{I}} - E_{\mathcal{A}} - \hbar\omega_k)^2 + \frac{1}{4}\hbar^2\Gamma^2}\right]_{\hbar\omega_k \approx E_{\mathcal{I}} - E_{\mathcal{A}}}. \tag{9.155}$$

The resulting cross-section is extremely large, and explains the phenomenon of *resonance fluorescence*. This effect occurs, for example, when a well-defined beam of monochromatic yellow light originating from a sodium lamp enters a transparent cell containing sodium vapor. When the incoming light undergoes elastic scattering by a gas atom, the process can be thought of as absorption of an incident photon accompanied by the immediate emission of a photon at precisely the same frequency, but in a different direction. Because the probability for such an event is so large at resonance, and because the light frequency does not get degraded by the process, the light scattered by any one atom will be rescattered by another. This sequence of events will occur over and over within the gas, causing the light beam to be strongly diffused. The result is that one observes a pronounced yellow glow from the sodium cell.

3. $\hbar\omega_k \ll E_{\mathcal{I}} - E_{\mathcal{A}}$
Here, we are considering the very long-wavelength limit, which corresponds to the scattering of optical photons, and Eq. 9.146 readily simplifies to

$$\left(\frac{d\sigma}{d\Omega}\right)^{\text{long }\lambda}_{\text{Rayleigh}} = k^4 \left|\sum_{\mathcal{I}} \frac{(\boldsymbol{\epsilon}_{\mathbf{k}\lambda}\cdot\boldsymbol{\mu}_{\mathcal{AI}})(\boldsymbol{\epsilon}_{\mathbf{k}'\lambda'}\cdot\boldsymbol{\mu}_{\mathcal{IA}}) + (\boldsymbol{\epsilon}_{\mathbf{k}'\lambda'}\cdot\boldsymbol{\mu}_{\mathcal{AI}})(\boldsymbol{\epsilon}_{\mathbf{k}\lambda}\cdot\boldsymbol{\mu}_{\mathcal{IA}})}{E_{\mathcal{I}} - E_{\mathcal{A}}}\right|^2. \tag{9.156}$$

This quantum-mechanically derived result exhibits the same characteristic k^4-dependence as was found from the classical formulation of the Rayleigh scattering problem (see Example 4.7). The factor $|\sum_{\mathcal{I}} \cdots|^2$ depends solely on the properties of the scattering atom and its orientation relative to the incident and scattered field polarizations. In particular, the atomic polarizability tensor becomes frequency independent, and simplifies to

$$\tilde{\tilde{\boldsymbol{\alpha}}} \rightarrow \sum_{\mathcal{I}} \left[\frac{\boldsymbol{\mu}_{A\mathcal{I}}\boldsymbol{\mu}_{\mathcal{I}A} + \boldsymbol{\mu}_{\mathcal{I}A}\boldsymbol{\mu}_{A\mathcal{I}}}{E_{\mathcal{I}} - E_A} \right]. \tag{9.157}$$

Furthermore, if the atomic charge distribution is spherically symmetric, then the atomic polarizability becomes a scalar, i.e., $\tilde{\tilde{\boldsymbol{\alpha}}} = \alpha \tilde{\tilde{\mathbf{I}}}$, and

$$\left(\frac{d\sigma}{d\Omega}\right)_{\text{Rayleigh}}^{\text{long } \lambda} \rightarrow k^4 \alpha^2 \left|\boldsymbol{\epsilon}_{\mathbf{k}\lambda} \cdot \boldsymbol{\epsilon}_{\mathbf{k}'\lambda'}\right|^2. \tag{9.158}$$

This result is identical to the classical expression for scattering from a dielectric sphere of polarizability α (see Eq. 4.185).

Suggested References

For further discussions on the scattering of photons by electrons and atoms, as well as other second-order interactions, see the references previously listed at the end of Chapter 8.
Also, for the interested reader, we list below some very readable books on the subject of x-ray diffraction:

[a] G. B. Carpenter, *Principles of Crystal Structure Determination* (W. A. Benjamin, New York, 1969).

[b] D. W. L. Hukins, *X-Ray Diffraction by Disordered and Ordered Systems* (Pergamon Press, New York, 1981).

[c] A. G. Michette and C. J. Buckley, eds., *X-Ray Science and Technology* (Institute of Physics, Bristol, 1993).

[d] B. E. Warren, *X-Ray Diffraction* (Dover Publications, New York, 1990).

A very good review article on the inelastic x-ray scattering (IXS) technique is

[e] G. Ruocco and F. Sette, "*The High-Frequency Dynamics of Liquid Water*", *J. Phys.: Condens. Matter* **11** (1999) R259–R293.

Problems

1. Equation 9.76 gives the unit-cell structure factor for the case of a body-centered cubic (b.c.c.) crystal. Here, investigate two other cases:

   (a) Copper is an example of a metal having a face-centered cubic (f.c.c.) crystal structure. Taking the atomic form factor to be f, write out the expression for F, the structure factor for the unit cell. Determine the combinations of $h\ k\ l$ that produce a vanishing structure factor and state the value of F for the cases that do not vanish.

   (b) Zinc is an example of a metal having a hexagonal close-packed (h.c.p.) crystal structure. Again, write out the structure factor and give the combinations of $h\ k\ l$ that correspond to a vanishing result? Determine $|F|^2$ for the non-vanishing cases.

2. In this problem, you are asked to work out some of the details of the TEE (three effective eigenmode) model used to analyze the inelastic x-ray scattering spectra from water in Example 9.3.

(a) Show that, up to order Q^2, the three eigenvalues of the hydrodynamic matrix (i.e., the low-Q limit of the matrix $\overleftrightarrow{H}(Q)$) are given by Eqs. 9.113 and 9.114.

(b) Given that the eigenvalues of the $\overleftrightarrow{H}(Q)$-matrix consist of one real number, z_h, and two conjugate complex numbers, $\Gamma_s \pm i\omega_s$, show that the dynamic structure factor in the TEE model can written in the form of Eq. 9.115. Work out the expression for the parameter b appearing in that equation.

(c) Show that the expression for $S(Q,\omega)$ appearing in Eq. 9.115 produces the intermediate scattering function given in Eq. 9.116.

(d) In the TEE model, show that one can also cast the dynamic structure factor in the form $S(Q,\omega) = \mathrm{Re}\left[S(Q,z)\right]$, where

$$S(Q,z) = \left[z + \frac{f_{un}^2(Q)}{z + z_u(Q) + \dfrac{f_{uT}(Q)}{z + z_T(Q)}}\right]^{-1}$$

and $z = i\omega$.

3. Define a *generalized oscillator strength* for an atom by

$$\tilde{f}_n(\mathbf{k}) = \frac{2m}{\hbar^2 k^2}(E_n - E_0)\left|\langle n|\sum_{i=1}^{Z} e^{i\mathbf{k}\cdot\mathbf{r}_i}|0\rangle\right|^2 .$$

The sum is over the Z atomic electrons, each of mass m, and $|0\rangle$ denotes the ground state of the atom (with corresponding energy E_0). By following the steps outlined below, prove the Bethe sum rule

$$\sum_n \tilde{f}_n(\mathbf{k}) = Z$$

where the sum over n extends to all states of the atom, discrete or continuum.

(a) Take the Hamiltonian for the atom as

$$\hat{H} = \sum_{i=1}^{Z}\frac{\hat{p}_i^2}{2m} + \hat{V}(\mathbf{r}_1,\ldots,\mathbf{r}_Z)$$

where $\hat{p}_i^2 = -\hbar^2\nabla_i^2$. Now, define an operator $\hat{B}$ by

$$\hat{B} \equiv \sum_{i=1}^{Z} e^{i\mathbf{k}\cdot\mathbf{r}_i}$$

and show that

$$\sum_n |B_{n0}|^2 (E_n - E_0) = \langle 0| \hat{B}^\dagger \left[\hat{H}, \hat{B}\right] |0\rangle .$$

(b) Next, show that

$$\left[\hat{H}, \hat{B}\right] = \frac{\hbar^2}{2m} \sum_{i=1}^{Z} e^{i\mathbf{k}\cdot\mathbf{r}_i} \left(k^2 - 2i\mathbf{k}\cdot\boldsymbol{\nabla}_i\right) .$$

(c) Finally, derive the following result

$$\langle 0| \hat{B}^\dagger \left[\hat{H}, \hat{B}\right] |0\rangle = \frac{\hbar^2 k^2}{2m} Z$$

and show this leads to the Bethe sum rule.

4. In the long-wavelength limit ($k \to 0$), show that the Bethe sum rule for generalized oscillator strengths (see previous problem) reduces to the Thomas–Reiche-Kuhn sum rule

$$\sum_n f_n = Z$$

for the more commonly encountered *oscillator strengths*

$$f_n = \frac{2m}{3\hbar^2 e^2} (E_n - E_0) \mu_{n0}^2$$

involving matrix elements of the atomic dipole moment $\boldsymbol{\mu}$.

5. (a) Consider the classical scattering of a monochromatic electromagnetic wave by a harmonically bound atomic electron having resonant frequency ω_0. For an incident wave of frequency ω, show that the total classical scattering cross-section is given by

$$\sigma_{\rm cl} = \frac{8\pi}{3} r_0^2 \left(\frac{\omega}{\omega_0}\right)^4$$

where r_0 is the classical electron radius.

(b) Let us now compare this classical expression to the quantum-mechanical result for the Rayleigh scattering of light by an atom in the limit where the energy $\hbar\omega$ of the incident photon is very small compared with the excitation energies $E_n - E_0$ of the system (here, E_0 is the ground-state energy and E_n denotes the energy of an excited atomic level). For simplicity, you should assume an atom with only a single bound electron. In particular, first show that the atomic polarizability α (assumed to be scalar) is related to the oscillator strengths f_n (defined in the previous problem) by

$$\alpha = \frac{3e^2}{m} \sum_n \frac{f_n}{\omega_{n0}^2}$$

where $\omega_{no} = (E_n - E_0)/\hbar$. Then show that the quantum-mechanical result for the total Rayleigh scattering cross-section corresponds to the classical result if one makes the replacement

$$\frac{1}{\omega_0^2} \to \sum_n \frac{3f_n}{\omega_{n0}^2}.$$

Since a single-electron atom obeys the Thomas-Reiche-Kuhn sum rule $\sum_n f_n = 1$, this result says that $1/\omega_0^2$ is determined by averaging the quantities $3/\omega_{n0}^2$, corresponding to the various transition frequencies, weighted according to the respective oscillator strengths.

Chapter 10
PRINCIPLES OF NUCLEAR MAGNETIC RESONANCE

The technique of *nuclear magnetic resonance* (NMR) was first developed by E. M. Purcell and F. Bloch *et al.* [78][79] in the mid-1940's. During the early stages of its development, NMR was used to make accurate determinations of nuclear magnetic moments. With the discovery of the chemical shift effect [80][81] in the early 1950's, however, a wealth of chemical applications became possible using NMR spectroscopy. Pulse NMR and spin-echo techniques soon followed. The most recent development has been that of *magnetic resonance imaging* (MRI), which uses spatial encoding of the NMR signal to produce high-resolution images of the human anatomy (see, for example, Ref. [82]).

Figure 10.1 is a crude illustration of the basic apparatus for a nuclear magnetic resonance experiment. A sample is immersed in an applied uniform, static, magnetic field,* $\mathbf{B}_0$. Superimposed on the static field is a weaker radiofrequency (RF) field, $\mathbf{B}(t)$, which is oriented in a direction transverse to $\mathbf{B}_0$. In NMR, one is interested in understanding the spin dynamics caused by the interaction between these applied fields and the atomic nuclei in the sample. This chapter will present the central features of the classical and quantum theories of spin dynamics, including a discussion of spin echo experiments. We will postpone addressing a related topic, i.e., that of *NMR susceptibility*, until a discussion of linear response theory in Chapter 13.

10.1 Energy of a Nuclear Spin in an Applied Magnetic Field

Consider a nucleus which possesses a magnetic moment $\boldsymbol{\mu}$ and an intrinsic angular momentum, or spin, $\mathbf{J} = \hbar\mathbf{I}$. The two are proportional to each other with a proportionality constant γ called the *gyromagnetic ratio*, namely,

$$\boldsymbol{\mu} = \gamma\mathbf{J} = \gamma\hbar\mathbf{I}. \tag{10.1}$$

To estimate the order of magnitude of γ, we imagine a point mass m carrying a charge e moving around a circular orbit of radius r around a center. The angular momentum about the center is

$$\mathbf{J} = \mathbf{r}\times m\mathbf{v} = mvr\mathbf{n} = m\omega r^2\mathbf{n}, \tag{10.2}$$

*Strictly speaking, one should use $\mathbf{H}_0$, i.e., the magnetic intensity, rather than $\mathbf{B}_0$. However, because of notational conflicts that arise with H, the symbol for the Hamiltonian, we choose to use $\mathbf{B}_0$.

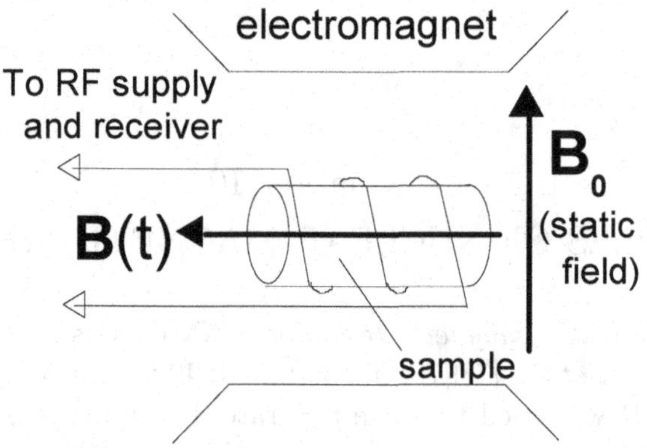

Figure 10.1 Simplified diagram of nuclear magnetic resonance apparatus.

where $\mathbf{n}$ is a unit-vector normal to the plane of the orbit. On the other hand, the magnetic moment $\boldsymbol{\mu}$ generated by a circular loop of area A carrying current i is $(iA/c)\,\mathbf{n}$. Therefore,

$$\boldsymbol{\mu} = \frac{i}{c}\left(\pi r^2\right)\mathbf{n} = \frac{1}{c}\frac{e\omega}{2\pi}\pi r^2\mathbf{n} = \frac{e}{2c}\omega r^2\mathbf{n}. \tag{10.3}$$

Combining Eqs. 10.2 and 10.3, we get

$$\boldsymbol{\mu} = \frac{e}{2mc}\mathbf{J} = \frac{e\hbar}{2mc}\mathbf{I}. \tag{10.4}$$

In this simple case, $\gamma = e/2mc$. For an actual nucleus, it is customary to introduce a g_I-factor and write $\gamma = g_I\,(e/2m_pc)$, where m_p is the proton mass. We then define a *nuclear magneton* by

$$\mu_N = \frac{|e|\,\hbar}{2m_pc} = 5.05 \times 10^{-24}\ \text{erg/G}, \tag{10.5}$$

from which it follows that

$$\boldsymbol{\mu} = \gamma\hbar\mathbf{I} = g_I\mu_N\mathbf{I}. \tag{10.6}$$

Nuclear magnetic moments are normally quoted in units of μ_N. In this convention, $g_I I$ is the magnetic moment. Values are tabulated for various nuclei in Table 10.1.

The energy of the interaction of the magnetic moment $\boldsymbol{\mu}$ with an applied magnetic field $\mathbf{B}_0$ is given by

$$H = -\boldsymbol{\mu}\cdot\mathbf{B}_0. \tag{10.7}$$

Take $\mathbf{B}_0 = B_0\mathbf{e}_z$ to be a DC (i.e., static) field. Then,

$$H = -\mu_z B_0 = -\gamma\hbar B_0 I_z = -\hbar\omega_0 I_z, \tag{10.8}$$

Table 10.1 Spin, magnetic moment, and resonance frequency of some common nuclides. (Data taken from CRC Handbook of Chemistry and Physics [72].)

nuclide	natural abundance (%)	I	$\mu/\mu_N = g_I I$	ν/B_0 or $\gamma/2\pi$ (MHz/tesla)
H^1	99.984	1/2	2.793	42.576
H^2	0.016	1	0.857	6.536
Li^7	92.57	3/2	3.356	16.55
C^{13}	1.11	1/2	0.702	10.71
F^{19}	100	1/2	2.63	40.06
Na^{23}	100	3/2	2.216	11.26
Cl^{35}	75.4	3/2	0.821	4.17
K^{39}	93.08	3/2	0.391	1.99
K^{41}	6.91	3/2	0.215	1.09
I^{127}	100	5/2	2.79	8.52
Si^{29}	4.70	1/2	−0.555	8.46
P^{31}	100	1/2	1.131	17.235
H^3	radioactive	1/2	2.9788	45.414

where we introduce the *Larmor frequency* $\omega_0 = \gamma B_0$. Previously, in Sect. 3.6, it was shown that the general quantum-mechanical angular momentum operators $\hat{J}^2$ and $\hat{J}_z$ commute. In the present case, therefore, we have $\left[\hat{I}^2, \hat{I}_z\right] = 0$ for the spin operators. Also, both $\hat{I}^2$ and $\hat{I}_z$ commute with the Hamiltonian, $\hat{H} = -\hbar\omega_0\hat{I}_z$. As a result, eigenvalues of $\hat{I}^2$ and $\hat{I}_z$ can be simultaneously specified for the energy eigenstates. We can therefore follow the formalism of Sect. 3.6, and take the representation $|I, m\rangle$ for which

$$\hat{I}^2 |I, m\rangle = I(I+1)|I, m\rangle \qquad \text{and} \qquad \hat{I}_z |I, m\rangle = m|I, m\rangle . \tag{10.9}$$

If we introduce the step-up and step-down operators by $\hat{I}_\pm = \hat{I}_x \pm i\hat{I}_y$, then it follows that (see Eq. 3.242)

$$\hat{I}_\pm |I, m\rangle = \sqrt{I(I+1) - m(m \pm 1)}\, |I, m \pm 1\rangle . \tag{10.10}$$

For a given I-value, m can take on a set of $2I + 1$ values, and so does the energy:

$$m = I,\ I-1,\ \ldots,\ -I \tag{10.11}$$

$$E_m = \langle I, m \mid \hat{H} \mid I, m\rangle = -\gamma\hbar B_0 m. \tag{10.12}$$

A set of energy levels E_m with m given by Eq. 10.11 are called *Zeeman levels*. For example, for Na^{23} or Cu^{63}, $I = 3/2$, so one has four Zeeman levels. The Zeeman levels are equally spaced with

$$\Delta E = \gamma\hbar B_0 = \hbar\omega_0. \tag{10.13}$$

In an NMR experiment, one induces transitions between the different Zeeman levels. This is accomplished by introducing an RF field $(2B_1 \cos\omega t)\,\mathbf{e}_x$ in a direction perpendicular to the main static field, $B_0\mathbf{e}_z$. The effect of the RF field is to introduce the sinusoidal perturbation

$$\hat{V} = 2\gamma\hbar B_1 \hat{I}_x \cos\omega t \tag{10.14}$$

into the Hamiltonian. Since the matrix element

$$\langle m' \mid \hat{V} \mid m\rangle \sim \langle m' \mid 2\hat{I}_x \mid m\rangle = \langle m' \mid \hat{I}_+ \mid m\rangle + \langle m' \mid \hat{I}_- \mid m\rangle \tag{10.15}$$

is non-vanishing only when $m' = m \pm 1$, the condition for resonance between the RF field and the nuclear spins is that the energy of an absorbed photon must match the spacing between adjacent Zeeman levels, i.e.,

$$|\Delta E| = \hbar\omega_0 = \hbar\omega. \tag{10.16}$$

Said another way, the frequency ω of the RF field must be identical to the Larmor frequency, $\omega_0 = \gamma B_0$, that is characteristic of the nucleus in the static field.

The resonance frequencies of various nuclei in a 10 kG (or 1 tesla) DC field are listed in Table 10.1. In particular, the gyromagnetic ratio for a proton is $\gamma_p = 2.68 \times 10^4\ \mathrm{s}^{-1}\mathrm{G}^{-1}$, so the resonance frequency is

$$\nu_p = \frac{1}{2\pi}\gamma_p B_0 = 4.26 B_0(\mathrm{G})\ \mathrm{kHz}. \tag{10.17}$$

Compare this value to the resonance frequency of an electron, which is

$$\nu_e = \frac{1}{2\pi}\gamma_e B_0 = 2.80 B_0(\mathrm{G})\ \mathrm{MHz}. \tag{10.18}$$

In a 1 tesla field, the resonance frequencies have the values of $\nu_p = 42.6$ MHz and $\nu_e = 28.0$ GHz.

Since the resonance frequencies of all nuclei are already known to great accuracy, the resonance absorption in itself is of little experimental interest at the present time. Rather, most modern NMR experiments are concerned with studying the dynamics of spin in the presence of different sample environments.

10.2 Quantum Mechanical Description of Motion of a Nuclear Spin in a Static Magnetic Field

The general state vector of a nuclear spin in a static magnetic field can be expanded in terms of the eigenvectors of $\hat{I}_z$. At $t = 0$ it may be written as

$$|\psi(0)\rangle = \sum_{m=-I}^{I} c_m |m\rangle\,. \tag{10.19}$$

The evolution of the state vector is obtained by solving the time-dependent Schrödinger equation (see Sect. 3.3). The Hamiltonian is time-independent, so the state vector at time t is

$$|\psi(t)\rangle = \sum_{m=-I}^{I} c_m e^{-i\hat{H}t/\hbar}|m\rangle = \sum_{m=-I}^{I} c_m e^{-iE_m t/\hbar}|m\rangle . \tag{10.20}$$

The expectation value of any observable can now be calculated using this state vector. The expectation value of the x-component of the nuclear magnetic moment is

$$\langle \mu_x(t)\rangle = \langle \psi(t) \mid \hat{\mu}_x \mid \psi(t)\rangle. \tag{10.21}$$

Considering that $\hat{\mu}_x = \gamma\hbar\hat{I}_x$, we have

$$\langle \mu_x(t)\rangle = \gamma\hbar \sum_{m,m'=-I}^{I} c^*_{m'} c_m \langle m' \mid \hat{I}_x \mid m\rangle e^{-i(E_m - E_{m'})t/\hbar}. \tag{10.22}$$

Since the operator $\hat{I}_x = \frac{1}{2}\left(\hat{I}_+ + \hat{I}_-\right)$, the matrix elements $\langle m' \mid \hat{I}_x \mid m\rangle$ vanish unless $m' = m \pm 1$.

It is instructive to consider a single spin with $I = 1/2$ in order to provide a clear physical insight into the motion of the magnetic moment in a static magnetic field. Denoting the spin-up ($m = +1/2$) and spin-down ($m = -1/2$) states using "+" and "−," respectively, the state vector is given by

$$|\psi(t)\rangle = c_+ e^{-iE_+ t/\hbar}|+\rangle + c_- e^{-iE_- t/\hbar}|-\rangle . \tag{10.23}$$

Then, from Eq. 10.22, we can easily see that the expectation value of μ_x is

$$\langle \mu_x(t)\rangle = \gamma\hbar\left(c^*_+ c_- \langle + \mid \hat{I}_x \mid -\rangle e^{-i\omega_0 t} + c_+ c^*_- \langle - \mid \hat{I}_x \mid +\rangle e^{i\omega_0 t}\right), \tag{10.24}$$

where $\omega_0 = (E_- - E_+)/\hbar = \gamma B_0$ (see Eq. 10.13). Since $\langle + \mid \hat{I}_x \mid -\rangle$ and $\langle - \mid \hat{I}_x \mid +\rangle$ are complex conjugates, Eq. 10.24 can be written as

$$\langle \mu_x(t)\rangle = 2\gamma\hbar \,\mathrm{Re}\left(c^*_+ c_- \langle + \mid \hat{I}_x \mid -\rangle e^{-i\omega_0 t}\right). \tag{10.25}$$

Making use of Eq. 10.10, the matrix element is calculated to be $\langle + \mid \hat{I}_x \mid -\rangle = 1/2$. The two complex coefficients, c_+ and c_-, can be expressed as

$$c_+ = ae^{i\alpha} \qquad \text{and} \qquad c_- = be^{i\beta}, \tag{10.26}$$

where the normalization condition for the c's demands that $a^2 + b^2 = 1$. The resulting expectation value is then

$$\langle \mu_x(t)\rangle = \gamma\hbar ab\cos(\alpha - \beta + \omega_0 t). \tag{10.27}$$

Similarly, the expectation values for the y and z components of the magnetic moment can also be obtained:

$$\begin{aligned} \langle \mu_y (t) \rangle &= -\gamma\hbar ab \sin(\alpha - \beta + \omega_0 t) \\ \langle \mu_z (t) \rangle &= \gamma\hbar (a^2 - b^2)/2. \end{aligned} \tag{10.28}$$

By defining the angles θ and ϕ as

$$\cos\theta = a^2 - b^2, \qquad \sin\theta = 2ab, \qquad \phi = \beta - \alpha - \omega_0 t, \tag{10.29}$$

Eqs. 10.27 and 10.28 reduce to

$$\begin{aligned} \langle \mu_x (t) \rangle &= \tfrac{1}{2}\gamma\hbar \sin\theta \cos\phi \\ \langle \mu_y (t) \rangle &= \tfrac{1}{2}\gamma\hbar \sin\theta \sin\phi \\ \langle \mu_z (t) \rangle &= \tfrac{1}{2}\gamma\hbar \cos\theta. \end{aligned} \tag{10.30}$$

This set of equations tells us that the expectation value of the magnetic moment appears as a vector of length $\gamma\hbar/2$, with direction given by the spherical coordinates θ and ϕ. Since the z-component of the magnetic moment commutes with the Hamiltonian, it is a constant of the motion—hence, the angle θ is fixed. The transverse components of the magnetic moment, however, do not commute with the Hamiltonian and are therefore time-dependent. We can see from Eq. 10.29 that the angle ϕ decreases linearly with time. Thus, Eqs. 10.30 show that the expectation value of the magnetic moment precesses clockwise about the z-axis with a Larmor frequency of ω_0. This is illustrated in Fig. 10.2.

By differentiating Eqs. 10.30 or, alternatively, by invoking the Heisenberg equation of motion, one can show that the expectation value of the magnetic moment obeys the vector equation

$$\frac{d\langle \boldsymbol{\mu} \rangle}{dt} = \langle \boldsymbol{\mu} \rangle \times \gamma \mathbf{B}_0. \tag{10.31}$$

This exactly matches the classical equation of motion of a magnetic moment in a static magnetic field. In fact, Eq. 10.31 holds for a time-dependent magnetic field as well.

10.3 Nuclear Spins in Thermal Equilibrium Under a Static Magnetic Field

Let us now consider a system containing a total of N spins per unit volume. Since there is no coherence between spins, i.e., the phase of spins, ϕ, is randomly distributed, the first two of Eqs. 10.30 show that the net transverse magnetization is zero. However, from the third equation, the net magnetization along the direction of the static magnetic field is given by $N\langle \mu_z \rangle = \frac{1}{2}N\gamma\hbar(a^2 - b^2)$, so it is, in general, not

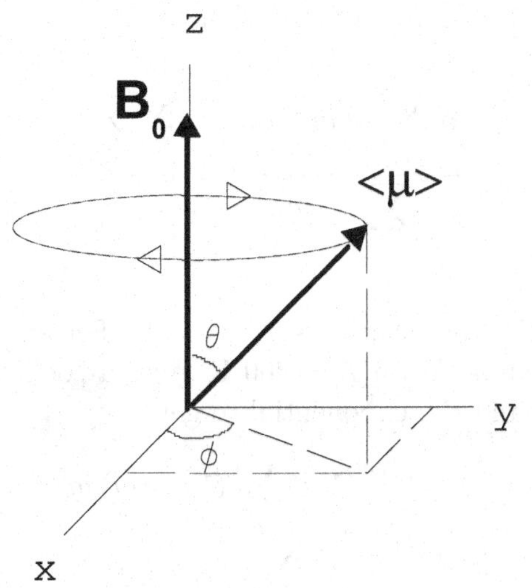

Figure 10.2 Precession of a single spin in a static magnetic field $\mathbf{B}_0 = B_0\mathbf{e}_z$.

zero. We now derive an expression for the net magnetization for a sample in thermal equilibrium.

Denote the number of nuclear spins per unit volume in the state m as $N(m)$. Then for any two Zeeman levels m and m', the Boltzmann principle gives

$$\frac{N(m)}{N(m')} = e^{-(E_m - E_{m'})/k_B T} \tag{10.32}$$

for a system in thermal equilibrium at temperature T. Recall from Eq. 10.12 that the spin energy levels take on the values $E_m = -\gamma\hbar B_0 m$. Noting that the total number of spins per unit volume is

$$N = \sum_m N(m), \tag{10.33}$$

we can rewrite Eq. 10.32 as

$$N(m) = N\frac{e^{-E_m/k_B T}}{Z}, \tag{10.34}$$

where Z is the partition function given by

$$Z = \sum_m e^{-E_m/k_B T}. \tag{10.35}$$

The magnetization, which is the total magnetic moment per unit volume, can now be expressed as

$$M_0 = \langle M_z \rangle = \sum_m N(m) \cdot \gamma\hbar m. \tag{10.36}$$

From Eqs. 10.33, 10.34, and 10.36, we then have

$$M_0 = \frac{N\sum\limits_m \gamma\hbar m N(m)}{\sum\limits_m N(m)} = \frac{N\sum\limits_{m=-I}^{I} \gamma\hbar m e^{\alpha m}}{\sum\limits_{m=-I}^{I} e^{\alpha m}}, \tag{10.37}$$

where $\alpha \equiv \gamma\hbar B_0/k_BT$. Note that α is very small. For example, a proton in a field of $B_0 = 10^4$ G at a temperature $T = 300$ K gives a coefficient of $\alpha = 6.83 \times 10^{-6}$. Therefore, we can expand the exponential to get

$$M_0 = \frac{N\gamma\hbar\sum\limits_m (1+\alpha m)\, m}{\sum\limits_m (1+\alpha m)}. \tag{10.38}$$

This expression can now be reduced using the relations $\sum 1 = 2I+1$, $\sum m^2 = I(I+1)(2I+1)/3$, and $\sum m = 0$, with the result

$$M_0 = \frac{\alpha N\gamma\hbar I(I+1)(2I+1)}{3(2I+1)} = \frac{N\gamma^2\hbar^2 I(I+1)}{3k_BT} B_0 = \chi_0 B_0, \tag{10.39}$$

where

$$\chi_0 = \frac{N\gamma^2\hbar^2 I(I+1)}{3k_BT} = \frac{N\langle\mu^2\rangle}{3k_BT} = \frac{C}{T}. \tag{10.40}$$

χ_0 is called the *static susceptibility* and $C = N\langle\mu^2\rangle/3k_B$ is known as the *Curie constant.* Equation 10.40 was first observed experimentally by P. Curie in 1895, and hence is called Curie's law. It was later derived by P. Langevin in 1905. The fact that χ_0 is positive means that the collection of nuclei is paramagnetic. The susceptibility is inversely proportional to the temperature due to the increased randomization of the nuclear-spin orientations at elevated temperatures.

10.4 Effect of Alternating Transverse Magnetic Field on Spin Dynamics

The energy state of a spin in a static field can be changed if there is a mechanism for inducing transitions between the spin energy levels. Since the energy eigenstates of the spin are also the eigenstates of the z-component of the angular momentum, only $\hat{I}_x$ and $\hat{I}_y$ produce non-vanishing matrix elements between the eigenstates. Thus, transitions between the energy levels can be made to occur by introducing a time-dependent perturbation containing $\hat{I}_x$ or $\hat{I}_y$ into the Hamiltonian. Experimentally, this is accomplished by applying a transverse magnetic field alternating at some angular frequency ω set close to, or at, the Larmor frequency ω_0. Doing so allows for energy exchange between the spin and the alternating transverse field. It is important to realize that only a field oscillating at a frequency close to the Larmor frequency can cause such transitions. In other words, one can only control the spin dynamics within a certain narrow frequency window centered on the Larmor frequency.

Before launching into the discussion of the spin dynamics, it is first useful to derive the identity

$$e^{-i\varphi\hat{C}}\hat{A}e^{i\varphi\hat{C}} = \hat{A}\cos\varphi + \hat{B}\sin\varphi, \tag{10.41}$$

where $\hat{A}$, $\hat{B}$, and $\hat{C}$ are any three operators satisfying the cyclic commutation relationship

$$\left[\hat{A},\hat{B}\right] = i\hat{C} \tag{10.42}$$

$$\left[\hat{C},\hat{A}\right] = i\hat{B} \tag{10.43}$$

$$\left[\hat{B},\hat{C}\right] = i\hat{A}. \tag{10.44}$$

Begin by considering the function $\hat{f}(\varphi)$ defined as

$$\hat{f}(\varphi) = e^{-i\varphi\hat{C}}\hat{A}e^{i\varphi\hat{C}}. \tag{10.45}$$

Observe that the derivative with respect to φ is

$$\frac{d\hat{f}}{d\varphi} = e^{-i\varphi\hat{C}}\left(-i\hat{C}\hat{A} + i\hat{A}\hat{C}\right)e^{i\varphi\hat{C}} = -ie^{-i\varphi\hat{C}}\left[\hat{C},\hat{A}\right]e^{i\varphi\hat{C}}, \tag{10.46}$$

or, using Eq. 10.43,

$$\frac{d\hat{f}}{d\varphi} = e^{-i\varphi\hat{C}}\hat{B}e^{i\varphi\hat{C}}. \tag{10.47}$$

Differentiating a second time produces

$$\frac{d^2\hat{f}}{d\varphi^2} = e^{-i\varphi\hat{C}}\left(-i\hat{C}\hat{B} + i\hat{B}\hat{C}\right)e^{i\varphi\hat{C}}, \tag{10.48}$$

and using Eq. 10.44,

$$\frac{d^2\hat{f}}{d\varphi^2} = -e^{-i\varphi\hat{C}}\hat{A}e^{i\varphi\hat{C}} = -\hat{f}. \tag{10.49}$$

In other words, $\hat{f}(\varphi)$ satisfies the simple second-order differential equation

$$\frac{d^2\hat{f}}{d\varphi^2} + \hat{f} = 0. \tag{10.50}$$

The general solution is

$$\hat{f}(\varphi) = \hat{a}\cos\varphi + \hat{b}\sin\varphi. \tag{10.51}$$

The coefficients $\hat{a}$ and $\hat{b}$ are easily determined by the two boundary conditions at $\varphi = 0$ obtained from Eqs. 10.45 and 10.47, i.e.,

$$\hat{f}(0) = \hat{a} = \hat{A} \qquad \hat{f}'(0) = \hat{b} = \hat{B}, \tag{10.52}$$

and thus Eq. 10.41 is now proven. Because $\hat{A}$, $\hat{B}$, and $\hat{C}$ satisfy a cyclic commutation relationship, the operators in Eq. 10.41 can also be cyclically permuted.

Looking back at Eq. 3.217, one notices that the components of the spin angular momentum operator follow the above cyclic commutation relationship. As a result, we can use Eq. 10.41 to transform the components as follows:

$$\begin{aligned} e^{-i\varphi\hat{I}_z}\hat{I}_x e^{i\varphi\hat{I}_z} &= \hat{I}_x\cos\varphi + \hat{I}_y\sin\varphi \\ e^{-i\varphi\hat{I}_z}\hat{I}_y e^{i\varphi\hat{I}_z} &= -\hat{I}_x\sin\varphi + \hat{I}_y\cos\varphi \\ e^{-i\varphi\hat{I}_z}\hat{I}_z e^{i\varphi\hat{I}_z} &= \hat{I}_z. \end{aligned} \tag{10.53}$$

One can easily recognize that this trio has the effect of rotating the spin angular momentum vector clockwise through an angle φ about the z-axis. This is equivalent to rotating the x-y coordinate axes counterclockwise through the same angle. In a similar fashion, the rotation of spin angular momentum operators about the x or y axis satisfy

$$\begin{aligned} e^{-i\varphi\hat{I}_x}\hat{I}_x e^{i\varphi\hat{I}_x} &= \hat{I}_x \\ e^{-i\varphi\hat{I}_x}\hat{I}_y e^{i\varphi\hat{I}_x} &= \hat{I}_y\cos\varphi + \hat{I}_z\sin\varphi \\ e^{-i\varphi\hat{I}_x}\hat{I}_z e^{i\varphi\hat{I}_x} &= -\hat{I}_y\sin\varphi + \hat{I}_z\cos\varphi \end{aligned} \tag{10.54}$$

and

$$\begin{aligned} e^{-i\varphi\hat{I}_y}\hat{I}_x e^{i\varphi\hat{I}_y} &= \hat{I}_x\cos\varphi - \hat{I}_z\sin\varphi \\ e^{-i\varphi\hat{I}_y}\hat{I}_y e^{i\varphi\hat{I}_y} &= \hat{I}_y \\ e^{-i\varphi\hat{I}_y}\hat{I}_z e^{i\varphi\hat{I}_y} &= \hat{I}_x\sin\varphi + \hat{I}_z\cos\varphi. \end{aligned} \tag{10.55}$$

Let us now go back and consider the effects of a transverse magnetic field at angular frequency ω:

$$\mathbf{B}_x(t) = (2B_1\cos\omega t)\,\mathbf{e}_x. \tag{10.56}$$

This field is a superposition of a right-handed and a left-handed rotating field, i.e.,

$$\mathbf{B}_x(t) = \mathbf{B}_{RH}(t) + \mathbf{B}_{LH}(t), \tag{10.57}$$

where

$$\begin{aligned} \mathbf{B}_{RH}(t) &= B_1(\mathbf{e}_x\cos\omega t + \mathbf{e}_y\sin\omega t) \\ \mathbf{B}_{LH}(t) &= B_1(\mathbf{e}_x\cos\omega t - \mathbf{e}_y\sin\omega t). \end{aligned} \tag{10.58}$$

The left-handed field contribution, $\mathbf{B}_{LH}$, rotates in the same sense as the precession of the magnetic moment in the static field, while the right-handed contribution, $\mathbf{B}_{RH}$, rotates in the opposite sense. At resonance ($\omega = \omega_0$) the rotation frequency of $\mathbf{B}_{LH}$ relative to the spin vector is zero. At the same time, $\mathbf{B}_{RH}$ rotates at a frequency of $2\omega_0$ relative to the spin vector. In other words, the effect of $\mathbf{B}_{LH}$ is to introduce a static perturbation, whereas $\mathbf{B}_{RH}$ produces a high-frequency perturbation. Consequently, we can neglect the counter-rotating field $\mathbf{B}_{RH}$ in a perturbation calculation.

The total magnetic field is that of the strong longitudinal field, $\mathbf{B}_0 = B_0\mathbf{e}_z$, plus the transverse magnetic field rotating at frequency ω:

$$\mathbf{B}(t) = B_0\mathbf{e}_z + B_1(\mathbf{e}_x \cos\omega t - \mathbf{e}_y \sin\omega t). \tag{10.59}$$

The Hamiltonian for the spin-field system is time-dependent, and is given by

$$\hat{H} = -\gamma\hbar\left[B_0\hat{I}_z + B_1\left(\hat{I}_x \cos\omega t - \hat{I}_y \sin\omega t\right)\right]. \tag{10.60}$$

Making the replacement $\varphi \to -\omega t$ in the first of Eqs. 10.53, the Hamiltonian can be rewritten as

$$\hat{H} = -\gamma\hbar\left(B_0\hat{I}_z + B_1 e^{i\omega t\hat{I}_z}\hat{I}_x e^{-i\omega t\hat{I}_z}\right). \tag{10.61}$$

Then the time-dependent Schrödinger equation is

$$i\hbar\frac{\partial}{\partial t}|\psi(t)\rangle = -\gamma\hbar\left(B_0\hat{I}_z + B_1 e^{i\omega t\hat{I}_z}\hat{I}_x e^{-i\omega t\hat{I}_z}\right)|\psi(t)\rangle. \tag{10.62}$$

In order to remove the time-dependence of the Hamiltonian, consider the transformation

$$|\psi_R(t)\rangle = e^{-i\omega t\hat{I}_z}|\psi(t)\rangle. \tag{10.63}$$

Treating the rotating field as a perturbation to the strong static field, at resonance ($\omega = \omega_0$) Eq. 10.63 is equivalent to going from the Schrödinger picture to the interaction picture, $|\psi_I(t)\rangle = e^{i\hat{H}_0 t/\hbar}|\psi_s(t)\rangle$. In more physical terms, the operator $e^{-i\omega t\hat{I}_z}$ (or $e^{-i\omega t\hat{J}_z/\hbar}$) rotates the state vector, $|\psi(t)\rangle$, counterclockwise about the z-axis through an angle ωt; this is demonstrated in the example presented below. $|\psi_R(t)\rangle$ is therefore the evolution of the state vector in a coordinate system rotating clockwise about the z-axis at angular frequency ω. Hence, Eq. 10.63 is appropriately referred to as a *rotating coordinate transformation.*

Example 10.1 Angular Momentum Operators as Generators of Rotations. Angular momenta are intimately connected with rotations. This example demonstrates the connection between them. Consider an ordinary vector $\mathbf{V} = (V_x, V_y, V_z)$ in a coordinate system (x, y, z). If one physically rotates the vector around the z-axis counterclockwise through an angle φ, the vector will have a different set of components along the three coordinate axes, i.e., $\mathbf{V}' = \left(V_x', V_y', V_z'\right)$. The new components are related to the old ones via the transformation

$$\mathbf{V}' = \mathbb{R}_z(\varphi)\mathbf{V}, \tag{10.64}$$

where $\mathbb{R}_z(\varphi)$ is the *rotation matrix*

$$\mathbb{R}_z(\varphi) = \begin{pmatrix} \cos\varphi & -\sin\varphi & 0 \\ \sin\varphi & \cos\varphi & 0 \\ 0 & 0 & 1 \end{pmatrix}. \tag{10.65}$$

Similarly, a rotation about an arbitrary axis n would be produced by a rotation matrix denoted by $\mathbb{R}_n(\varphi)$.

Now consider an analogous situation in quantum mechanics: We wish to determine the form of a *rotation operator* which has the effect of rotating a state vector in ket space. Our approach will be to work, at least initially, in position space. In this case, we require that the rotation operator, $\hat{R}_n(\varphi)$, rotates the wavefunction $\psi(\mathbf{r})$ through an angle φ about axis n. Mathematically, $\hat{R}_n(\varphi)$ must produce the following action:

$$\hat{R}_n(\varphi)\,\psi(\mathbf{r}) = \psi\left(\mathbb{R}_n^{-1}(\varphi)\,\mathbf{r}\right), \tag{10.66}$$

where $\mathbb{R}_n^{-1}(\varphi)$, the inverse of the rotation matrix, is given by

$$\mathbb{R}_n^{-1}(\varphi) = \mathbb{R}_n(-\varphi). \tag{10.67}$$

First consider a rotation through an infinitesimal angle, $d\varphi$, around the z-axis. To second-order in $d\varphi$, the rotation matrix of Eq. 10.65 becomes

$$\mathbb{R}_z(d\varphi) = \begin{pmatrix} 1-\frac{1}{2}d\varphi^2 & -d\varphi & 0 \\ d\varphi & 1-\frac{1}{2}d\varphi^2 & 0 \\ 0 & 0 & 1 \end{pmatrix}. \tag{10.68}$$

Then,

$$\begin{aligned}\hat{R}_z(d\varphi)\,\psi(\mathbf{r}) &= \psi\left(\mathbb{R}_z^{-1}(d\varphi)\,\mathbf{r}\right) = \psi\left(\mathbb{R}_z(-d\varphi)\,\mathbf{r}\right) \\ &= \psi(x + y\,d\varphi, y - x\,d\varphi, z).\end{aligned} \tag{10.69}$$

The resulting right-hand side can be expanded in a Taylor series. To first-order in $d\varphi$, this gives

$$\hat{R}_z(d\varphi)\,\psi(\mathbf{r}) = \left[1 - d\varphi\left(x\frac{\partial}{\partial y} - y\frac{\partial}{\partial x}\right)\right]\psi(\mathbf{r}). \tag{10.70}$$

Recall, however (see Eq. 3.244), that the position-space representation of the orbital angular momentum operator $\hat{L}_z$ is

$$\hat{L}_z = -i\hbar\left(x\frac{\partial}{\partial y} - y\frac{\partial}{\partial x}\right). \tag{10.71}$$

This means that, in the case of an infinitesimal rotation, the rotation operator is

$$\hat{R}_z(d\varphi) = 1 - \frac{i}{\hbar}\hat{L}_z d\varphi, \tag{10.72}$$

or, extending the result to the n-component of a general angular momentum $\mathbf{J}$, we write

$$\hat{R}_n(d\varphi) = 1 - \frac{i}{\hbar}\hat{J}_n d\varphi. \tag{10.73}$$

It is now straightforward to determine the form of $\hat{R}_n$ for a finite rotation. Just think of a finite rotation as an infinite number of small rotations, each with

$$d\varphi = \lim_{N\to\infty} \frac{\varphi}{N}. \tag{10.74}$$

In other words,

$$\hat{R}_n(\varphi) = \lim_{N\to\infty} \left[\hat{R}_n(d\varphi)\right]^N = \lim_{N\to\infty} \left[1 - \frac{i}{\hbar}\hat{J}_n\left(\frac{\varphi}{N}\right)\right]^N. \tag{10.75}$$

Recognize the last expression to be identical to an exponential form. Thus, rotations about a given axis are generated by the angular momentum operator associated with that axis according to

$$\hat{R}_n(\varphi) = e^{-i\varphi\hat{J}_n/\hbar}. \tag{10.76}$$

Return now to our investigation of the spin in a magnetic field. To effect the rotating coordinate transformation, rewrite Eq. 10.63 as

$$|\psi(t)\rangle = e^{i\omega t\hat{I}_z} |\psi_R(t)\rangle \tag{10.77}$$

and substitute into Eq. 10.62. Then,

$$-\hbar\omega\hat{I}_z e^{i\omega t\hat{I}_z} |\psi_R(t)\rangle + i\hbar e^{i\omega t\hat{I}_z}\frac{\partial}{\partial t}|\psi_R(t)\rangle = -\gamma\hbar\left(B_0\hat{I}_z e^{i\omega t\hat{I}_z} + B_1 e^{i\omega t\hat{I}_z}\hat{I}_x\right)|\psi_R(t)\rangle. \tag{10.78}$$

Multiplying both sides from the left by $e^{-i\omega t\hat{I}_z}$, we obtain

$$i\hbar\frac{\partial}{\partial t}|\psi_R(t)\rangle = -\gamma\hbar\left[\left(B_0 - \frac{\omega}{\gamma}\right)\hat{I}_z + B_1\hat{I}_x\right]|\psi_R(t)\rangle. \tag{10.79}$$

Note that the time dependence of the Hamiltonian has been removed, i.e., the transformed Hamiltonian in the rotating frame is

$$\hat{H}_R = -\gamma\hbar\left[\left(B_0 - \frac{\omega}{\gamma}\right)\hat{I}_z + B_1\hat{I}_x\right]. \tag{10.80}$$

This Hamiltonian represents the coupling of the spins with an effective static field

$$\mathbf{B}_{\text{eff}} = \left(B_0 - \frac{\omega}{\gamma}\right)\mathbf{e}_z + B_1\mathbf{e}_x \tag{10.81}$$

which is equivalent to the effective field that appears in the classical equation of motion for a magnetic moment in a rotating frame. The formal solution to the time-dependent Schrödinger equation, Eq. 10.79, is

$$|\psi_R(t)\rangle = e^{-i\hat{H}_R t/\hbar}|\psi_R(0)\rangle \tag{10.82}$$

or, using Eq. 10.63,

$$|\psi(t)\rangle = e^{i\omega t\hat{I}_z} e^{-i\hat{H}_R t/\hbar} |\psi_R(0)\rangle, \tag{10.83}$$

where $|\psi_R(0)\rangle = |\psi(0)\rangle$.

The effect of the rotating field can be understood by calculating the expectation value of the magnetic moment in the rotating coordinate system. For simplicity we focus on a system meeting the resonance condition, $\omega = \omega_0 = \gamma B_0$. In this case, the transformed Hamiltonian is

$$\hat{H}_R = -\gamma\hbar B_1 \hat{I}_x = -\hbar\omega_1 \hat{I}_x, \tag{10.84}$$

where $\omega_1 = \gamma B_1$. The spin dynamics as viewed from the laboratory frame are obtained by calculating the expectation value of the magnetic moment using either $|\psi(t)\rangle$ or $|\psi_R(t)\rangle$ with the corresponding operators. For example, the full motion of the z-component of the magnetic moment is determined by

$$\langle\mu_z(t)\rangle = \langle\psi(t) \mid \hat{\mu}_z \mid \psi(t)\rangle = \langle\psi_R(t) \mid \hat{\mu}_z^R \mid \psi_R(t)\rangle, \tag{10.85}$$

where

$$\hat{\mu}_z^R = e^{-i\omega t\hat{I}_z} \hat{\mu}_z e^{i\omega t\hat{I}_z}. \tag{10.86}$$

However, we are not really interested in the full dynamics, but rather with the spin's motion in the rotating frame. For this, the expectation value of the z-component of the magnetic moment is calculated as

$$\begin{aligned} \langle\mu_z(t)\rangle_R &= \langle\psi_R(t) \mid \hat{\mu}_z \mid \psi_R(t)\rangle \\ &= \gamma\hbar\langle\psi_R(0) \mid e^{-i\omega_1 t\hat{I}_x} \hat{I}_z e^{i\omega_1 t\hat{I}_x} \mid \psi_R(0)\rangle. \end{aligned} \tag{10.87}$$

Using the last of Eqs. 10.54, one learns that

$$\begin{aligned} \langle\mu_z(t)\rangle_R &= \gamma\hbar\langle\psi_R(0) \mid \left(\hat{I}_z \cos\omega_1 t - \hat{I}_y \sin\omega_1 t\right) \mid \psi_R(0)\rangle \\ &= \langle\mu_z(0)\rangle \cos\omega_1 t - \langle\mu_y(0)\rangle \sin\omega_1 t. \end{aligned} \tag{10.88}$$

Similarly, one finds

$$\langle\mu_x(t)\rangle_R = \langle\mu_x(0)\rangle \tag{10.89}$$

$$\langle\mu_y(t)\rangle_R = \langle\mu_y(0)\rangle \cos\omega_1 t + \langle\mu_z(0)\rangle \sin\omega_1 t. \tag{10.90}$$

If the magnetic moment is initially aligned along the positive z-axis at $t = 0$, i.e., $\langle\mu_x(0)\rangle = \langle\mu_y(0)\rangle = 0$, the expectation values are of the form

$$\begin{aligned} &\langle\mu_x(t)\rangle_R = 0 \\ &\langle\mu_y(t)\rangle_R = \langle\mu_z(0)\rangle \sin\omega_1 t \\ &\langle\mu_z(t)\rangle_R = \langle\mu_z(0)\rangle \cos\omega_1 t. \end{aligned} \tag{10.91}$$

Thus, from the vantage point of the rotating frame, the expectation value of the magnetic moment precesses about the B_1-field (i.e., the x-axis of the rotating coordinate system) at frequency ω_1. This is illustrated in Fig. 10.3. Although it is true that the orientation of the magnetic moment of an individual spin can only take on discrete, quantum values, the expectation value $\langle \boldsymbol{\mu} \rangle$, which is under discussion here, can, in fact, precess through a continuum of angles. This expectation value, recall, reflects the total magnetization of a sample containing many nuclear spins, i.e., $\mathbf{M} = N \langle \boldsymbol{\mu} \rangle$.

Since the Larmor frequency is in the radiofrequency region, $\mathbf{B}_1$ must, in fact, be an RF field. By applying the field in the form of a pulse having a controlled width, the experimenter is able to tilt the magnetization vector away from the z-axis by any amount desired. Consider an RF pulse of duration $\omega_1 t = \pi/2$, commonly referred to as a "$\pi/2$-pulse." It would be used to tilt the equilibrium longitudinal magnetization onto the positive y-axis in the rotating frame. When viewed from the laboratory frame, this magnetization will then precess in the transverse (x-y) plane at the Larmor frequency. According to Faraday's law (see Eq. 4.4), the oscillating magnetization induces an NMR signal in the RF coil in the form of an oscillating emf. Although the signal is a high-frequency oscillation at ω_0, its envelope decays exponentially. The signal exhibits what is known as a *free induction decay*, or FID. The rate of decay is characterized by a time-constant, T_2, known as the *transverse* or *spin-spin relaxation time*. In general, relaxations that occur in an NMR experiment are caused by random magnetic-field fluctuations originating from the thermal motion of neighboring spins in the sample. In the case of T_2 relaxation in liquids, fluctuating magnetic dipole-dipole interactions between spins due to the tumbling of molecules lead to a broadening of the Zeeman energy levels. That is to say, the magnetic transitions develop a linewidth, $\Gamma = 1/T_2$, as discussed in Sect. 8.2. This is called a *homogeneous broadening* of the resonance line; it is the natural linewidth due to spin-spin interactions. A second source of broadening comes from any spatial inhomogeneities that may exist in the static (longitudinal) magnetic field. In this case, spins in different regions of the sample experience slightly different values of the field, and hence produce a spread of resonance frequencies. One then talks about an *inhomogeneous broadening* contribution to the linewidth. The overall effect of static field inhomogeneities, as well as spin-spin interactions, is to produce a dephasing of spins in the transverse plane. This will be discussed further in Sect. 10.6. In an actual experiment, what one really observes is an effective time-constant, T_2^*, due to both sources of transverse relaxation.

Another commonly used pulse is one of duration $\omega_1 t = \pi$, known as a "π-pulse." It has the effect of inverting the equilibrium longitudinal magnetization onto the negative z-axis. This magnetization will then exponentially return to the equilibrium value via the so-called T_1 relaxation pathway. The time-constant T_1 is called the *longitudinal* or *spin-lattice relaxation time*. As before, the relaxation is produced by the presence of randomly fluctuating magnetic fields. In this case, however, it is specifically the Fourier component of these fluctuations at the Larmor frequency

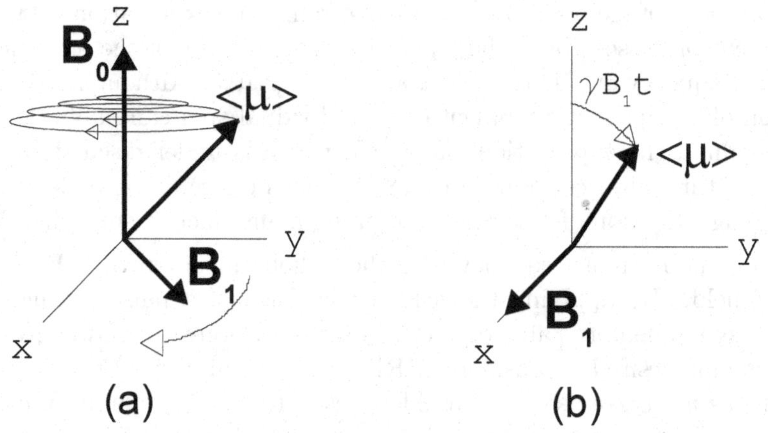

Figure 10.3 Evolution of $\langle \boldsymbol{\mu} \rangle$, the expectation value of the nuclear-spin magnetic moment in the presence of a static, longitudinal field, $\mathbf{B}_0$, and a transverse rotating field, $\mathbf{B}_1$. (a) In the laboratory frame, $\langle \boldsymbol{\mu} \rangle$ follows a rather complicated path. (b) In the rotating frame, $\mathbf{B}_1$ is stationary and the effective longitudinal field vanishes at resonance. Only a precession about $\mathbf{B}_1$ is apparent.

that is responsible for the T_2 relaxation. When a fluctuating field contribution at frequency ω_0 has a component perpendicular to the static field, it acts as a mini-RF signal in resonance with the system. Its effect is to induce transitions between the spin states, causing the magnetization to relax back toward equilibrium.

The NMR signal is detected by the same coil that is used to apply the RF field. The signal representing the emf induced in the coil can be expressed as

$$S_{\text{emf}}(t) = -\frac{\partial}{\partial t}\int_V \left[\frac{\mathbf{B}_1(\mathbf{r},t)}{i}\right] \cdot \mathbf{M}(\mathbf{r},t)\, d^3r, \tag{10.92}$$

where $(\mathbf{B}_1/i)$ is the magnetic field produced by the coil per unit current. The magnetization after a $\pi/2$-pulse can be written as

$$\mathbf{M}(t) = M_0 (\mathbf{e}_x \cos\omega_0 t + \mathbf{e}_y \sin\omega_0 t), \tag{10.93}$$

where, as before, M_0 is the equilibrium magnetization. Assuming that both $(\mathbf{B}_1/i)$ and the magnetization are homogeneous throughout the integration volume V, the signal is of the form

$$S_{\text{emf}}(t) = \omega_0 (\mathbf{B}_1/i)\, M_0 V \cos\omega_0 t. \tag{10.94}$$

Since $M_0 = \chi_0 B_0$ and $\omega_0 = \gamma B_0$, the emf can be rewritten as

$$S_{\text{emf}}(t) = \gamma\chi_0 B_0^2 (\mathbf{B}_1/i)\, V \cos\omega_0 t. \tag{10.95}$$

This shows that the signal is proportional to the square of the strength of the static magnetic field.

10.5 The Bloch Equations—T_1 and T_2 Relaxations

In this section, we derive the classical equations of motion of the magnetization, which are known as the *Bloch equations* [84]. To begin, consider the equation of motion of a free magnetic moment $\boldsymbol{\mu}$ in a general magnetic field $\mathbf{B}$. The rate of change of angular momentum $\hbar\mathbf{I}$ of a nucleus is equal to the torque applied by the field, namely, $\boldsymbol{\mu}\times\mathbf{B}$, so that

$$\hbar\frac{d\mathbf{I}}{dt} = \boldsymbol{\mu}\times\mathbf{B}. \tag{10.96}$$

But $\boldsymbol{\mu} = \gamma\hbar\mathbf{I}$, which produces

$$\frac{d\boldsymbol{\mu}}{dt} = \gamma\boldsymbol{\mu}\times\mathbf{B}. \tag{10.97}$$

Since the nuclear magnetization $\mathbf{M}$ of a sample is defined as the sum $\sum\boldsymbol{\mu}_i$ over all the nuclei in a unit volume, we can sum up both sides of Eq. 10.97 to get

$$\frac{d\mathbf{M}}{dt} = \gamma\mathbf{M}\times\mathbf{B}. \tag{10.98}$$

This is the equation of motion for $\mathbf{M}$ in the absence of any mechanisms for relaxation.

If one places a collection of nuclei in a static field $\mathbf{B} = B_0\mathbf{e}_z$, then

$$\frac{d\mathbf{M}}{dt} = \gamma B_0\left(-M_x\mathbf{e}_y + M_y\mathbf{e}_x\right), \tag{10.99}$$

or

$$\frac{dM_x}{dt} = \gamma B_0 M_y, \qquad \frac{dM_y}{dt} = -\gamma B_0 M_x, \qquad \frac{dM_z}{dt} = 0. \tag{10.100}$$

These three equations show that M_z is a constant and confirm that, in the laboratory frame, the magnetization precesses freely around the z-axis at the Larmor frequency, $\omega_0 = \gamma B_0$, in a clockwise sense. For example, if $M_y(0) = 0$, then $M_x = M_x(0)\cos\omega_0 t$ and $M_y = -M_x(0)\sin\omega_0 t$. For the case of a sample in thermal equilibrium at temperature T, we also have $M_x(0) = 0$, so the net magnetization will be purely along the z-axis, or

$$M_x = M_y = 0 \qquad \text{and} \qquad M_z = M_0 = \frac{C}{T}B_0, \tag{10.101}$$

where, as discussed in Sect. 10.3, C is the Curie constant.

On the other hand, if the magnetization component M_z is initially other than the equilibrium value M_0, it is reasonable to assume that it will relax back toward the equilibrium value at a rate proportional to the instantaneous departure from equilibrium, i.e.,

$$\frac{dM_z}{dt} = \frac{M_0 - M_z}{T_1}, \tag{10.102}$$

where T_1 is the longitudinal (spin-lattice) relaxation time referred to in Sect. 10.4. Equation 10.102, of course, predicts a simple exponential relaxation. Under an applied static field $\mathbf{B}$, Eq. 10.102 is modified to

$$\frac{dM_z}{dt} = \gamma\left(\mathbf{M}\times\mathbf{B}\right)_z + \frac{M_0 - M_z}{T_1}. \tag{10.103}$$

This means that, besides precessing about the static field, M_z will relax to its equilibrium value M_0 at the rate $1/T_1$. Likewise, if the transverse magnetization components, M_x and M_y, are initially non-zero, they will subsequently decay towards the vanishing value characteristic of thermal equilibrium. To provide for this transverse relaxation, the x and y components of Eq. 10.98 are modified to

$$\frac{dM_x}{dt} = \gamma\left(\mathbf{M}\times\mathbf{B}\right)_x - \frac{M_x}{T_2} \tag{10.104}$$

$$\frac{dM_y}{dt} = \gamma\left(\mathbf{M}\times\mathbf{B}\right)_y - \frac{M_y}{T_2}. \tag{10.105}$$

T_2 is the aforementioned transverse (spin-spin) relaxation time.† Equations 10.103–10.105 are, in fact, the desired Bloch equations. Although not exact, the Bloch equations are plausible, and are fairly good for liquid samples.

It is instructive to consider the relaxation of different components of the magnetization (in the laboratory frame) following the application of a single RF pulse. Once such a pulse is turned off, the field is purely longitudinal, i.e., $\mathbf{B} = B_0\mathbf{e}_z$, and the Bloch equations reduce to

$$\begin{aligned} \frac{dM_x}{dt} + \frac{M_x}{T_2} &= \omega_0 M_y \\ \frac{dM_y}{dt} + \frac{M_y}{T_2} &= -\omega_0 M_x \\ \frac{dM_z}{dt} &= \frac{M_0 - M_z}{T_1}. \end{aligned} \tag{10.106}$$

Let us consider two special cases:

1. Following a $\pi/2$-pulse, the initial magnetization is $\mathbf{M} = [0, M_0, 0]$, and the solution to the Bloch equations is

$$\begin{aligned} M_x(t) &= M_0 e^{-t/T_2}\sin\omega_0 t \\ M_y(t) &= M_0 e^{-t/T_2}\cos\omega_0 t \\ M_z(t) &= M_0\left(1 - e^{-t/T_1}\right). \end{aligned} \tag{10.107}$$

   By combining the first two equations, one sees that the magnitude of the transverse magnetization component is given by $M_0 e^{-t/T_2}$. Notice that the longitudinal and transverse components relax with two different time constants.

2. After applying a π-pulse, the initial magnetization is $\mathbf{M} = [0, 0, -M_0]$, and the solution to the Bloch equations is

$$\begin{aligned} M_x(t) &= 0 \\ M_y(t) &= 0 \\ M_z(t) &= M_0\left(1 - 2e^{-t/T_1}\right). \end{aligned} \tag{10.108}$$

   Here, the transverse components remain zero at all times, but the longitudinal component relaxes with time constant T_1.

---

†As pointed out in the last section, the observed transverse relaxation time is actually T_2^*.

10.6 The Principle of Spin Echo

The phenomenon of *spin echo* was first discovered by Hahn [85] and was the driving force behind the development of pulse methods in NMR. The specific pulse sequence we describe here was invented by Carr and Purcell [86]. The spin echo experiment consists of applying two RF pulses, namely a $\pi/2$-pulse and a π-pulse, followed by the observation of an echo.‡ The sequence of events is depicted in Fig. 10.4.

Before proving the occurrence of an echo quantum mechanically, it is helpful to have a physical picture of the spin dynamics resulting in an echo. The $\pi/2$-pulse rotates the equilibrium magnetization from the z-axis onto the positive y-axis in the transverse plane. The resulting transverse magnetization then evolves under a certain Hamiltonian. Specifically, consider the effect of static magnetic-field inhomogeneities within the sample.§ The Larmor frequency of each spin will depend on its position and the precession rate of each spin will become randomized. The net result is that the spin-phases lose coherence, causing a free induction decay of the NMR signal. If a π-pulse is applied at time τ after the initial $\pi/2$-pulse, it will rotate the individual magnetic moments about the x-axis, inverting the phase of all spins. A continued evolution under the same Hamiltonian will then cancel out the various phases accumulated prior to the π-pulse. At time $t = 2\tau$, all spins will again be in phase, and a spin echo will be formed. This is schematically illustrated in Fig. 10.5. A quantum mechanical treatment of spin echo follows below:

The Hamiltonian in the rotating coordinate system was previously given in Eq. 10.80. We rewrite it here as

$$\hat{H}_R = -\gamma\hbar\left(\Delta B_0 \hat{I}_z + B_1 \hat{I}_x\right), \tag{10.109}$$

where $\Delta B_0 = B_0 - \omega/\gamma$. Since we want to illustrate the rephasing of dephased spins due to a time-independent field inhomogeneity, we consider a case where ΔB_0 is not zero. During the RF pulses, we may assume $B_1 \gg \Delta B_0$ for all spins; this is the usual case in most NMR experiments. Then the Hamiltonian during the RF pulses is

$$\hat{H}_R = -\gamma\hbar B_1 \hat{I}_x. \tag{10.110}$$

During the time interval between the RF pulses, $B_1 = 0$ and therefore we have

$$\hat{H}_R = -\gamma\hbar\Delta B_0 \hat{I}_z. \tag{10.111}$$

Since the Hamiltonian is time-independent within each interval $(0, t_1)$, (t_1, t_2), (t_2, t_3), and (t_3, t) displayed in Fig. 10.4, the evolution of the state vector in the rotating

‡Hahn used a $\pi/2$-$\pi/2$ pulse sequence.

§The effect of inherent spin-spin interactions will be neglected since they produce a much slower contribution to the evolution of the transverse magnetization.

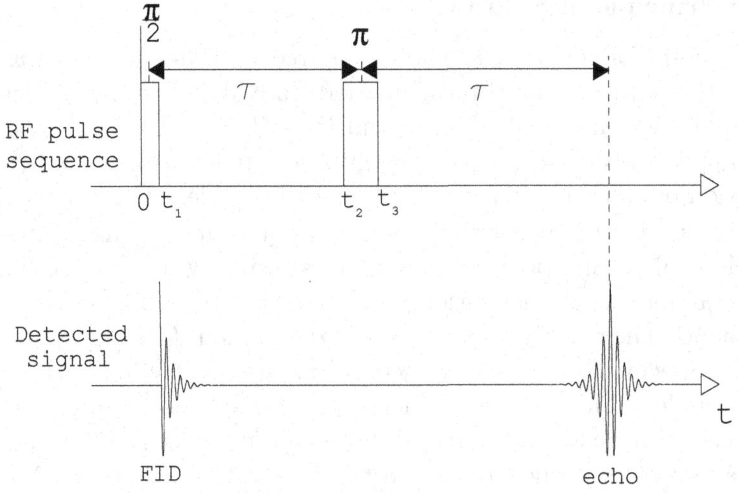

Figure 10.4 Spin-echo pulse sequence and the corresponding free induction decay (FID) and echo formation.

coordinate system can be obtained as follows:

$$\begin{aligned}
|\psi_R(t_1)\rangle &= e^{i\gamma B_1 t_1 \hat{I}_x}\,|\psi_R(0)\rangle \\
|\psi_R(t_2)\rangle &= e^{i\gamma \Delta B_0 (t_2-t_1) \hat{I}_z}\,|\psi_R(t_1)\rangle \\
|\psi_R(t_3)\rangle &= e^{i\gamma B_1 (t_3-t_2) \hat{I}_x}\,|\psi_R(t_2)\rangle \\
|\psi_R(t)\rangle &= e^{i\gamma \Delta B_0 (t-t_3) \hat{I}_z}\,|\psi_R(t_3)\rangle,
\end{aligned} \tag{10.112}$$

where $\gamma B_1 t_1 = \pi/2$ and $\gamma B_1 (t_3 - t_2) = \pi$. For convenience, let us define

$$\begin{aligned}
\hat{T}(t, \Delta B_0) &= e^{i\gamma \Delta B_0 t \hat{I}_z} \\
\hat{X}(\phi) &= e^{i\phi \hat{I}_x}.
\end{aligned} \tag{10.113}$$

Furthermore, we assume that the duration of an RF pulse is negligibly small compared to τ. Then the state vector at time t is

$$|\psi_R(t)\rangle = \hat{T}(t-\tau, \Delta B_0)\,\hat{X}(\pi)\,\hat{T}(\tau, \Delta B_0)\,\hat{X}(\pi/2)\,|\psi(0)\rangle. \tag{10.114}$$

The expectation value of the magnetic moment of each individual spin in the rotating frame is then calculated by

$$\langle \mu(t, \Delta B_0)\rangle_R = \langle \psi_R(t) \mid \hat{\mu} \mid \psi_R(t)\rangle. \tag{10.115}$$

Remembering that each spin sees a different ΔB_0, the total magnetic moment of all spins is

$$\langle \mathbf{M}(t)\rangle_R = N \int P(\Delta B_0)\,\langle \boldsymbol{\mu}(t, \Delta B_0)\rangle_R\, d(\Delta B_0), \tag{10.116}$$

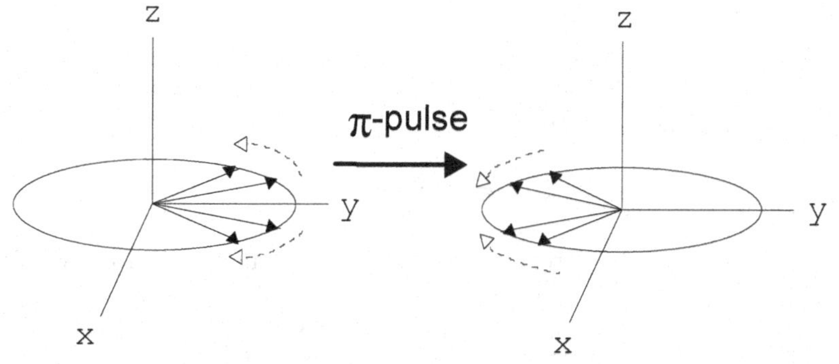

Figure 10.5 The formation of a spin echo as viewed from the rotating frame. Following a $\pi/2$-pulse, which aligns the magnetization vector along the positive y-axis, inhomogeneities in the local magnetic field cause the individual spin magnetic moments to precess with their own Larmor frequencies, leading to a dephasing of spins. A π-pulse applied at $t = \tau$ then rotates all the magnetic moments about the x-axis, inverting the phases of all spins. Upon continuing their precession, the dephasing of spins is reversed until, at $t = 2\tau$, they are completely in phase again, only this time along the negative y-axis. This occurence is signalled by an echo.

where, as before, N is the number of spins per unit volume and $P(\Delta B_0)\, d(\Delta B_0)$ is the probability that a spin experiences a static field between ΔB_0 and $\Delta B_0 + d(\Delta B_0)$. In particular, the y-component of the magnetization is

$$\langle M_y(t)\rangle_R = \gamma\hbar N \int P(\Delta B_0)\, \langle I_y(t, \Delta B_0)\rangle_R\, d(\Delta B_0)\,. \tag{10.117}$$

Using Eq. 10.114 gives

$$\begin{aligned}\langle I_y(t, \Delta B_0)\rangle_R &= \langle \psi_R(t) \mid \hat{I}_y \mid \psi_R(t)\rangle \\ &= \langle \psi(0) \mid \hat{X}^{-1}(\pi/2)\, \hat{T}^{-1}(\tau, \Delta B_0)\, \hat{X}^{-1}(\pi) \\ &\quad \times \hat{T}^{-1}(t-\tau, \Delta B_0)\, \hat{I}_y \hat{T}(t-\tau, \Delta B_0) \\ &\quad \times \hat{X}(\pi)\, \hat{T}(\tau, \Delta B_0)\, \hat{X}(\pi/2) \mid \psi(0)\rangle,\end{aligned} \tag{10.118}$$

where $\hat{X}^{-1}$ and $\hat{T}^{-1}$ are the inverses of operators $\hat{X}$ and $\hat{T}$. Let us now insert an identity operator $\hat{X}(\pi)\, \hat{X}^{-1}(\pi)$ into a portion of the above operator sequence, i.e.,

$$\begin{aligned}&\hat{X}^{-1}(\pi)\, \hat{T}^{-1}(t-\tau, \Delta B_0)\, \hat{I}_y \hat{T}(t-\tau, \Delta B_0)\, \hat{X}(\pi) \\ = \;& \hat{X}^{-1}(\pi)\, \hat{T}^{-1}(t-\tau, \Delta B_0)\, \hat{X}(\pi)\, \hat{X}^{-1}(\pi)\, \hat{I}_y \\ & \times \hat{X}(\pi)\, \hat{X}^{-1}(\pi)\, \hat{T}(t-\tau, \Delta B_0)\, \hat{X}(\pi)\,.\end{aligned} \tag{10.119}$$

The right-hand side of Eq. 10.119 can be broken into the following three terms:

$$\hat{X}^{-1}(\pi)\, \hat{T}^{-1}(t-\tau, \Delta B_0)\, \hat{X}(\pi)$$

$$\hat{X}^{-1}(\pi)\,\hat{I}_y\hat{X}(\pi)$$
$$\hat{X}^{-1}(\pi)\,\hat{T}(t-\tau,\Delta B_0)\,\hat{X}(\pi)\,.$$

From Eq. 10.54, the second term is

$$\hat{X}^{-1}(\pi)\,\hat{I}_y\hat{X}(\pi) = -\hat{I}_y. \tag{10.120}$$

To handle the first and third terms, we consider the fact that for two operators $\hat{A}$ and $\hat{B}$,

$$A^{-1}e^{i\hat{B}}\hat{A} = e^{i\hat{A}^{-1}\hat{B}\hat{A}}. \tag{10.121}$$

This can easily be proven using the series expansion of the exponential operator and repeatedly inserting $\hat{A}\hat{A}^{-1}$. Using Eqs. 10.121 and 10.54, the first term can be transformed as follows:

$$\begin{aligned}\hat{X}^{-1}(\pi)\,\hat{T}^{-1}(t-\tau,\Delta B_0)\,\hat{X}(\pi) &= \hat{X}^{-1}(\pi)\,e^{-i\gamma\Delta B_0(t-\tau)\hat{I}_z}\hat{X}(\pi)\\ &= e^{-i\gamma\Delta B_0(t-\tau)\hat{X}^{-1}(\pi)\hat{I}_z\hat{X}(\pi)}\\ &= e^{i\gamma\Delta B_0(t-\tau)\hat{I}_z}\\ &= \hat{T}(t-\tau,\Delta B_0)\,.\end{aligned} \tag{10.122}$$

Similarly, the third term is

$$\hat{X}^{-1}(\pi)\,\hat{T}(t-\tau,\Delta B_0)\,\hat{X}(\pi) = \hat{T}^{-1}(t-\tau,\Delta B_0)\,. \tag{10.123}$$

Therefore, using Eqs. 10.120, 10.122, and 10.123, we have

$$\begin{aligned}\langle I_y(t,\Delta B_0)\rangle_R &= -\langle\psi(0)\mid\hat{X}^{-1}(\pi/2)\,\hat{T}^{-1}(\tau,\Delta B_0)\\ &\quad\times\hat{T}(t-\tau,\Delta B_0)\,\hat{I}_y\hat{T}^{-1}(t-\tau,\Delta B_0)\\ &\quad\times\hat{T}(\tau,\Delta B_0)\,\hat{X}(\pi/2)\mid\psi(0)\rangle.\end{aligned} \tag{10.124}$$

When $t = 2\tau$, i.e., $t-\tau = \tau$, the expectation value simply becomes

$$\begin{aligned}\langle I_y(2\tau,\Delta B_0)\rangle_R &= -\langle\psi(0)\mid\hat{X}^{-1}(\pi/2)\,\hat{I}_y\hat{X}(\pi/2)\mid\psi(0)\rangle\\ &= -\langle\psi(0)\mid\hat{I}_z\mid\psi(0)\rangle\\ &= -\left\langle\hat{I}_z(0)\right\rangle.\end{aligned} \tag{10.125}$$

This proves that the dephasing of spins due to magnetic field inhomogeneities, ΔB_0, is refocused (rephased) and a spin echo is produced. The time $t = 2\tau$ is the time at which the echo occurs, independent of the particular ΔB_0 seen by any one spin.

From Eq. 10.117, the total magnetization along the y-direction at the time of the echo is

$$\langle M_y(2\tau)\rangle_R = -\gamma\hbar N\int P(\Delta B_0)\,\langle I_z(0)\rangle\,d(\Delta B_0)\,. \tag{10.126}$$

Since any dependence on ΔB_0 has been removed, the integration simply gives

$$\langle M_y (2\tau)\rangle_R = -\gamma\hbar N \langle I_z (0^-)\rangle = -\langle M_z (0^-)\rangle , \tag{10.127}$$

where $t = 0^-$ stands for the instant of time immediately prior to turning on the $\pi/2$-pulse. Assuming that the magnetization was initially aligned along the z-axis, one can also show that the other components of magnetization vanish at the echo center, i.e.,

$$\begin{aligned} \langle M_x (2\tau)\rangle_R &= \langle M_x (0)\rangle = 0 \\ \langle M_z (2\tau)\rangle_R &= \langle M_y (0)\rangle = 0. \end{aligned} \tag{10.128}$$

The results show that the occurrence of the echo signals that the magnetization is rephased along the negative y-direction and its magnitude is the same as that at $t = 0^-$. The analysis presented neglects T_1 relaxations and the inherent T_2 relaxations due to spin-spin interactions. If the latter are included, then the magnitude of the magnetization at the echo center will be attenuated by a factor of $e^{-2\tau/T_2}$.

Example 10.2 Pulsed Gradient Spin Echo—the NMR Equivalent of Incoherent Scattering. The Larmor frequency is proportional to the external static magnetic field $\mathbf{B}_0$. If, in addition to the strong static magnetic field $\mathbf{B}_0$, a magnetic field which varies linearly across a sample, or a magnetic field gradient, is applied, the Larmor frequencies of the spins will depend on their spatial locations according to

$$\omega (\mathbf{r}) = \gamma B_0 + \gamma \mathbf{G}\cdot\mathbf{r}, \tag{10.129}$$

where $\mathbf{G}$ is the gradient of the additional field parallel to $\mathbf{B}_0$:

$$\mathbf{G} = \frac{\partial B_z}{\partial x}\mathbf{e}_x + \frac{\partial B_z}{\partial y}\mathbf{e}_y + \frac{\partial B_z}{\partial z}\mathbf{e}_z. \tag{10.130}$$

Consider what happens if this gradient field is applied just after rotating the equilibrium magnetization into the transverse plane using a $\pi/2$-pulse. The gradient then acts as a static field inhomogeneity. This causes the spins to evolve so that their phases vary linearly with location, creating a sinusoidally varying transverse magnetization across the sample. This is illustrated in Fig. 10.6 for a gradient applied along the z-direction.

The linear relationship between the Larmor frequency and the spatial coordinate of the spins allows one to investigate the location, and hence the translational motion, of spins along the direction of the gradient. This principle is the basis for the *pulsed gradient spin echo* (PGSE) technique. The pulse sequence of PGSE is shown in Fig. 10.7. The gradient field is applied in pulses and it is assumed that the applied linear field variation is much larger than any other inhomogeneities that might be present in the static magnetic field. This allows us to treat $\mathbf{B}_0$ as essentially homogeneous.

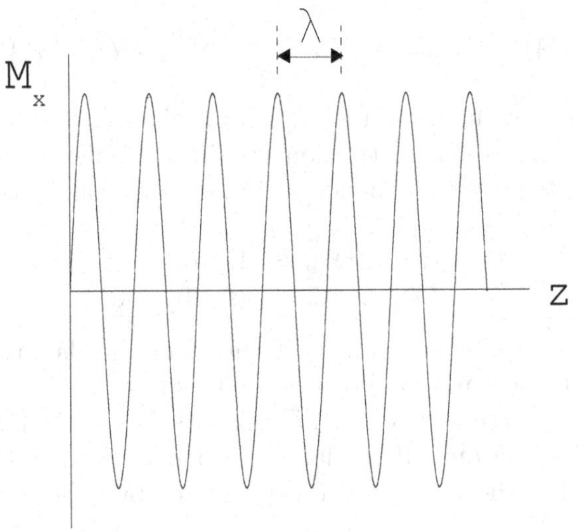

Figure 10.6 Sinusoidal transverse magnetization grating produced by a linear magnetic field gradient **G** applied along the z-direction of a sample. The wavelength of the grating, λ, is inversely related to the wavevector defined as $Q_z = \gamma G_z t$.

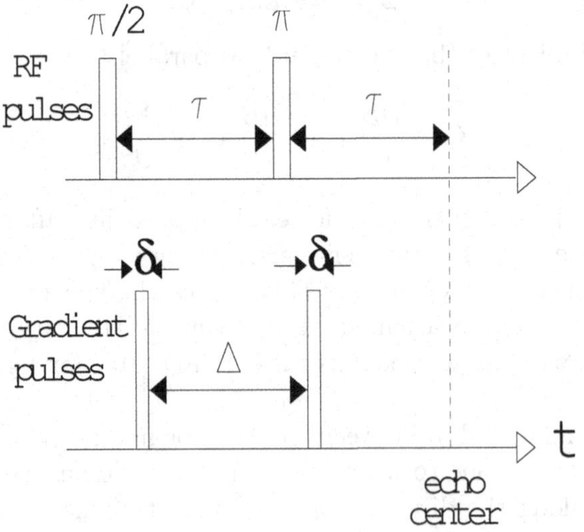

Figure 10.7 Pulsed gradient spin echo sequence. The first and second gradients have the same magnitude, direction, and duration.

At resonance the Hamiltonian in the rotating coordinate system is

$$\hat{H}_R = -\gamma\hbar\left(\mathbf{G}\cdot\mathbf{r}\hat{I}_z + B_1\hat{I}_x\right). \tag{10.131}$$

During the RF pulses, when no gradient field is present, we have $\mathbf{G} = 0$, so the Hamiltonian is

$$\hat{H}_R = -\gamma\hbar B_1\hat{I}_x. \tag{10.132}$$

During the gradient pulses, the RF field has been turned off, so $B_1 = 0$, and

$$\hat{H}_R = -\gamma\hbar\mathbf{G}\cdot\mathbf{r}\hat{I}_z. \tag{10.133}$$

For all other times, the Hamiltonian vanishes. Compared with Eqs. 10.110 and 10.111, the only difference is that the field inhomogeneity term ΔB_0 is replaced by $\mathbf{G}\cdot\mathbf{r}$. Therefore the evolution of the spin state vector can be followed in the same way as before, without repeating all the previous calculations. Suffice it to say, that the dephasing of spins which, in this case, is due to the action of the gradient pulses, will be refocused at the echo center at time $t = 2\tau$ after the $\pi/2$-pulse.

If the gradient pulses are sufficiently narrow, we may neglect the translational motion of a spin during these pulses. Thus, the duration, δ, of a gradient pulse is made much smaller than the time, Δ, between pulses. Now we focus our attention on the expectation value of the transverse component of the magnetic moment of a spin. First, the phase factor accumulated during the first gradient pulse for a spin at $\mathbf{r}(0)$ is $e^{i\gamma\delta\mathbf{G}\cdot\mathbf{r}(0)}$. Next, the sign of the phase is inverted by the π-pulse, so the phase factor right after the π-pulse is $e^{-i\gamma\delta\mathbf{G}\cdot\mathbf{r}(0)}$. During the second gradient pulse, the spin obtains an additional phase factor of $e^{i\gamma\delta\mathbf{G}\cdot\mathbf{r}(\Delta)}$. Thus, the total phase of the spin after the second gradient is $e^{i\gamma\delta\mathbf{G}\cdot[\mathbf{r}(\Delta)-\mathbf{r}(0)]}$. Should the spin not move during time Δ, then $\mathbf{r}(\Delta) = \mathbf{r}(0)$, and the phase evolution due to the gradient pulses are completely rephased. However, if the spin does move during this time, then $\mathbf{r}(\Delta) \neq \mathbf{r}(0)$ and the rephasing is incomplete, with a remaining phase factor of $e^{i\gamma\delta\mathbf{G}\cdot[\mathbf{r}(\Delta)-\mathbf{r}(0)]}$ at the echo center.

The signal at the echo center, denoted by $E_\Delta(\mathbf{G})$, is a superposition of transverse magnetizations, i.e., an ensemble average in which each phase term $e^{i\gamma\delta\mathbf{G}\cdot[\mathbf{r}(\Delta)-\mathbf{r}(0)]}$ is weighted by the probability that a spin moves from $\mathbf{r}(0)$ to $\mathbf{r}(\Delta)$. This probability is given by $\rho(\mathbf{r})\, P_s(\mathbf{r} \mid \mathbf{r}', \Delta)$, where $\rho(\mathbf{r})$ is a normalized spin density at position $\mathbf{r}$ and $P_s(\mathbf{r} \mid \mathbf{r}', \Delta)$ is the conditional (self) probability for a spin at $\mathbf{r}$ to move to $\mathbf{r}'$ in time Δ. Thus,

$$E_\Delta(\mathbf{G}) = \int \rho(\mathbf{r}) \int P_s(\mathbf{r} \mid \mathbf{r}', \Delta)\, e^{-i\gamma\delta\mathbf{G}\cdot(\mathbf{r}'-\mathbf{r})} d^3r' d^3r, \tag{10.134}$$

where we have omitted any constant prefactor. Let us now define a reciprocal wavevector $\mathbf{Q}$ as

$$\mathbf{Q} = \gamma\mathbf{G}\delta. \tag{10.135}$$

Then the echo signal can be rewritten as

$$E_\Delta(\mathbf{Q}) = \int \rho(\mathbf{r}) \int P_s(\mathbf{r} \mid \mathbf{r}', \Delta)\, e^{i\mathbf{Q}\cdot(\mathbf{r}-\mathbf{r}')} d^3r' d^3r. \tag{10.136}$$

Notice that this integral is very reminiscent of the expression $\left\langle e^{i\mathbf{Q}\cdot[\mathbf{r}(t)-\mathbf{r}(0)]}\right\rangle$ that previously appeared in the context of incoherent neutron scattering (see Eq. 6.96).¶ As a result, pulse gradient spin echo also goes by the name *incoherent NMR scattering*.

If the spin density in the sample is uniform and the conditional probability is independent of the initial position, i.e., it depends only on the net displacement **R**, then Eq. 10.136 can be rewritten as

$$E_\Delta(\mathbf{Q}) = \int P_s(\mathbf{R}, \Delta)\, e^{i\mathbf{Q}\cdot\mathbf{R}} d\mathbf{R}. \tag{10.137}$$

This shows that the echo signal and the net displacement probability function are related through a simple Fourier transform relationship. Therefore, $P_s(\mathbf{R}, \Delta)$ can be reconstructed from the echo signal in **Q**-space.

For example, in the case of free self-diffusion, the net displacement probability function is gaussian in the displacement. If the gradient pulses are applied along the z-direction, i.e., $\mathbf{G} = G\mathbf{e}_z$, the echo signal becomes‖

$$\begin{aligned} E_\Delta(\mathbf{Q}) &= (4\pi D\Delta)^{-1/2} \int e^{-z^2/4D\Delta} e^{iQz} dz \\ &= e^{-\gamma^2\delta^2 G^2 D\Delta}. \end{aligned} \tag{10.138}$$

In other words, the acquisition of the echo signal along a particular direction in **Q**-space allows one to measure the diffusion coefficient D for that direction.

¶In Chapter 12, we will identify $\left\langle e^{i\mathbf{Q}\cdot[\mathbf{r}(t)-\mathbf{r}(0)]}\right\rangle$ as the so-called *self intermediate scattering function*, $F_s(\mathbf{Q}, t)$.

‖This is equivalent to the incoherent neutron scattering result for the intermediate scattering function, i.e., $F_s(Q, t) = e^{-DQ^2 t}$.

Suggested References

Authoritative general references on NMR include

[a] A. Abragam, *The Principles of Nuclear Magnetism* (Oxford University Press, New York, 1983).

[b] C. P. Slichter, *Principles of Magnetic Resonance*, 3rd ed. (Springer-Verlag, Berlin, 1996).

[c] P. T. Callaghan, *Principles of Nuclear Magnetic Resonance Microscopy* (Oxford University Press, New York, 1994).

[d] P. J. Hore, *Nuclear Magnetic Resonance* (Oxford University Press, New York, 1995).

[e] M. H. Levitt, *Spin Dynamics: Basics of Nuclear Magnetic Resonance* (John Wiley and Sons, Chichester, UK, 2001).

Another very readable text with the chemist in mind is

[f] R. K. Harris, *Nuclear Magnetic Resonance Spectroscopy: A Physicochemical View* (Pitman Publishing, London, 1983).

For more details on pulse methods, consider reading

[g] T. C. Farrar and E. D. Becker, *Pulse and Fourier Transform NMR: Introduction to Theory and Methods* (Academic Press, New York, 1971).

[h] E. Fukushima and S. B. W. Roeder, *Experimental Pulse NMR: A Nuts and Bolts Approach* (Addison-Wesley, Reading, MA, 1981).

Problems

1. Equations 10.27 and 10.28 give expressions for the expectation values of the various components of the nuclear magnetic moment for the case of a spin-1/2 nucleus immersed in a static magnetic field. Generalize these expressions for the case of a spin I.

2. In this problem, we examine matrix representations for the various angular momentum operators:

   (a) For a spin-1/2 system, there are two eigenkets $|j, m\rangle$ for the angular momentum operators $\hat{J}^2$ and $\hat{J}_z$. These can be represented as two-element column vectors, i.e.,

$$\left|\frac{1}{2}, +\frac{1}{2}\right\rangle \to \begin{pmatrix} 1 \\ 0 \end{pmatrix} \qquad \text{and} \qquad \left|\frac{1}{2}, -\frac{1}{2}\right\rangle \to \begin{pmatrix} 0 \\ 1 \end{pmatrix}.$$

By considering the action of $\hat{J}_z$, $\hat{J}_+$, and $\hat{J}_-$ on each of the above states, show that the various angular momentum operators can be written as

$$\hat{J}_x = \frac{\hbar}{2}\hat{\sigma}_x \qquad \hat{J}_y = \frac{\hbar}{2}\hat{\sigma}_y \qquad \hat{J}_z = \frac{\hbar}{2}\hat{\sigma}_z$$

where $\hat{\sigma}_x$, $\hat{\sigma}_y$, and $\hat{\sigma}_z$ are the *Pauli spin matrices*

$$\hat{\sigma}_x \to \begin{pmatrix} 0 & 1 \\ 1 & 0 \end{pmatrix} \qquad \hat{\sigma}_y \to \begin{pmatrix} 0 & -i \\ i & 0 \end{pmatrix} \qquad \hat{\sigma}_z \to \begin{pmatrix} 1 & 0 \\ 0 & -1 \end{pmatrix}.$$

More compactly, one can also define a *Pauli spin operator* as $\hat{\mathbf{J}} = \frac{\hbar}{2}\hat{\boldsymbol{\sigma}}$.

(b) Construct the matrices $\hat{J}_x$, $\hat{J}_y$, and $\hat{J}_z$ for a spin-1 system, where the three eigenkets are represented as

$$|1,1\rangle = \begin{pmatrix} 1 \\ 0 \\ 0 \end{pmatrix} \qquad |1,0\rangle = \begin{pmatrix} 0 \\ 1 \\ 0 \end{pmatrix} \qquad |1,-1\rangle = \begin{pmatrix} 0 \\ 0 \\ 1 \end{pmatrix}.$$

3. Consider a standard NMR experiment with a nucleus of spin $I = 1/2$ and gyromagnetic ratio γ. The Hamiltonian for the system is given by Eq. 10.60, i.e.,

$$\hat{H} = -\gamma\hbar\left[B_0\hat{I}_z + B_1\left(\hat{I}_x\cos\omega t - \hat{I}_y\sin\omega t\right)\right],$$

where B_0 is the strength of the strong DC magnetic field in the $+z$-direction and B_1 is the strength of the weak transverse RF field rotating at frequency ω.

(a) By connecting $\hat{I}_x$, $\hat{I}_y$, and $\hat{I}_z$ to Pauli spin matrices (see previous problem), construct the 2×2 Hamiltonian matrix for this system. Write the matrix elements in terms of the RF frequency ω, the Larmor frequency $\omega_0 = \gamma B_0$, and the frequency $\omega_1 = \gamma B_1$.

(b) Consider the case where the RF frequency is set to match the Larmor frequency, i.e., $\omega = \omega_0$. You are now asked to exactly solve the time-dependent Schrödinger equation

$$i\hbar\frac{\partial}{\partial t}|\psi(t)\rangle = \hat{H}|\psi(t)\rangle$$

by using the following expansion in the Schrödinger picture

$$|\psi(t)\rangle = c_+(t)\,e^{-iE_+t/\hbar}|+\rangle + c_-(t)\,e^{-iE_-t/\hbar}|-\rangle,$$

where $|+\rangle$ and $|-\rangle$ are the eigenstates for spin-up and spin-down, respectively, and $E_+ = -\hbar\omega_0/2$ and $E_- = +\hbar\omega_0/2$ are the corresponding energy eigenvalues. Alternatively, this can be represented as the column vector

$$|\psi(t)\rangle \to \begin{pmatrix} c_+(t)\,e^{+i\omega_0 t/2} \\ c_-(t)\,e^{-i\omega_0 t/2} \end{pmatrix}.$$

In particular, show that the Schrödinger equation leads to two coupled first-order differential equations for the amplitudes $c_+(t)$ and $c_-(t)$. Solve for $c_+(t)$ and $c_-(t)$ given that the nucleus is initially in the spin-up state. Find $P(t)$, the probability that the nucleus has made a transition to the spin-down state as a function of time.

(c) Now consider the more general case of an off-resonant RF driving frequency, i.e., $\omega \neq \omega_0$. Again, given that the nucleus starts out in the spin-up state at $t = 0$, derive the following result for the probability that the nucleus has flipped to spin down at time t:

$$P(t) = \left(\frac{\omega_1}{\omega'}\right)^2 \sin^2\left(\frac{\omega' t}{2}\right)$$

where

$$\omega' \equiv \sqrt{(\omega - \omega_0)^2 + \omega_1^2}.$$

The result is known as *Rabi's formula.*

4. The previous problem examines the time dependence of the spin-flip probability for a spin-1/2 nucleus in a standard NMR experiment. Here, we consider a spin-1 nucleus, where the eigenstates $|+\rangle$, $|0\rangle$, and $|-\rangle$ now correspond to the eigenvalues of +1, 0, and -1, respectively, for the operator $\hat{I}_z$. As in the last problem, construct the Hamiltonian matrix (use the results of Problem 2b). Then, for the case where the RF frequency ω is precisely tuned to resonance (i.e., $\omega = \omega_0 = \gamma B_0$), solve the time-dependent Schrödinger equation for

$$|\psi(t)\rangle = c_+(t)\, e^{-iE_+t/\hbar}\, |+\rangle + c_0(t)\, e^{-iE_0t/\hbar}\, |0\rangle + c_-(t)\, e^{-iE_-t/\hbar}\, |-\rangle,$$

where E_+, E_0, and E_- are the corresponding energies of the three eigenstates. Given that the nucleus is initially in state $|+\rangle$, find the probabilities (in terms of $\omega_1 = \gamma B_1$) for making transitions to each of the states $|0\rangle$ and $|-\rangle$ as a function of time.

5. In a pulsed gradient spin echo (PGSE) experiment, assuming a very narrow pulse-width, show that the echo signal for the case of self-diffusion superimposed on a flow of velocity $\mathbf{v}$ is given by

$$E_\Delta(Q) = \exp\left(-\gamma^2\delta^2 G^2 D\Delta + i\gamma\delta \mathbf{G}\cdot\mathbf{v}\Delta\right).$$

(More generally, it can be shown [87] that when the pulse-width, δ, is not narrow, the echo takes the revised form

$$E_\Delta(Q) = \exp\left[-\gamma^2\delta^2 G^2 D\left(\Delta - \frac{1}{3}\delta\right) + i\gamma\delta \mathbf{G}\cdot\mathbf{v}\Delta\right],$$

which is an exact result.)

Chapter 11
THEORY OF PHOTON COUNTING STATISTICS

A practical method for detecting light is by means of a photoelectric detector. This is a device that generates an electronic pulse, or photocount, as the result of a photoelectric absorption event (see Sect. 8.3) occurring in some photosensitive surface. A common example is the photomultiplier tube, where photoelectrons ejected from the sensitive surface (referred to as the *photocathode*) are amplified in number by means of an electron multiplication scheme, thereby producing a usable pulse, and hence, a recorded photocount [88]. The goal of the present chapter is to show how the data collected in a photoelectron counting experiment is related to the statistical properties of the particular type of light field being measured. Fluctuations in the intensity of the observed light, which show up as fluctuations in the number of photocounts recorded over some specified time interval, are governed by the quantum state of the radiation field, i.e., whether it is mixed, as in the case of light from a conventional lamp, or whether it is pure, such as in the case of a coherent state or squeezed state. The results, as presented in this chapter, however, are most easily derived from semiclassical considerations.

We will find that the subject of photocount statistical fluctuations naturally leads to a discussion of intensity correlation functions and photon correlation measurements. The technique of *photon correlation spectroscopy* will be discussed at some length in this chapter because it provides the basis for quasi-elastic light scattering experiments, which are used for measuring slow fluctuations in fluids.

11.1 Statistical Distribution of Photoelectron Counts

Let us consider the statistical fluctuations observed in a photon counting experiment where a photoelectric detector is used. The fluctuations arise because (1) the radiation field itself contains fluctuations that are quantum-mechanical or classical in origin, and (2) light detection by the photoelectric effect is a quantum process which is probabilistic in nature. Our aim is to derive an expression for the statistical distribution of photoelectron counts, $p(n, T)$, which is the probability of registering n counts during a time interval T. We begin by making the following statement: The probability of registering a single pulse over a very short counting time $(T, T + \Delta T)$ is proportional to the instantaneous light intensity $I(T)$ and the counting interval

ΔT. In other words,

$$p(1,\Delta T) = \epsilon I(T)\Delta T \qquad \text{and} \qquad p(0,\Delta T) = 1 - \epsilon I(T)\Delta T. \tag{11.1}$$

ϵ is a proportionality constant that includes both the quantum efficiency of the photocathode and any factors related to detection geometry. It has been shown by Mandel *et al.* [89] that the above assertion holds for quasi-monochromatic light as long as $\Delta\omega^{-1} \gg \Delta T \gg \langle\omega\rangle^{-1}$, where $\langle\omega\rangle$ is the mean frequency of the light ($\sim 10^{15}\mathrm{s}^{-1}$) and $\Delta\omega$ is its bandwidth. Another assumption we make is that the probability of registering or not registering a count is independent of previous photoelectron emissions. Said another way, there is no *dead time* associated with the detection process. This allows us to write

$$\begin{aligned} p(n,T+\Delta T) &= p(n-1,T)\,p(1,\Delta T) + p(n,T)\,p(0,\Delta T) \\ &= p(n-1,T)\,\epsilon I(T)\Delta T + p(n,T)\left[1-\epsilon I(T)\Delta T\right]. \end{aligned} \tag{11.2}$$

or

$$\frac{\partial}{\partial T}p(n,T) = \epsilon I(T)\left[p(n-1,T) - p(n,T)\right]. \tag{11.3}$$

Now introduce a generating function [90]

$$G(\lambda,T) = \sum_{m=0}^{\infty}(1-\lambda)^m p(m,T) \tag{11.4}$$

from which $p(n,T)$ can be obtained by the procedure

$$p(n,T) = \left[\frac{1}{n!}\left(-\frac{\partial}{\partial\lambda}\right)^n G(\lambda,T)\right]_{\lambda=1}. \tag{11.5}$$

For example, to find the probability for obtaining three photon counts by using the generating function, we write

$$\left[\frac{1}{3!}\left(-\frac{\partial}{\partial\lambda}\right)^3 G(\lambda,T)\right]_{\lambda=1} = \left[\frac{1}{6}\sum_{m=0}^{\infty} m(m-1)(m-2)(1-\lambda)^{m-3}p(m,T)\right]_{\lambda=1}. \tag{11.6}$$

Notice that the only non-vanishing term on the right is for $m=3$, so that

$$\left[\frac{1}{3!}\left(-\frac{\partial}{\partial\lambda}\right)^3 G(\lambda,T)\right]_{\lambda=1} = p(3,T), \tag{11.7}$$

as it should. Obviously, Eq. 11.5 is useful only if $G(\lambda,T)$ can be obtained by some independent means. To see how this is done, consider the derivative of the generating function with respect to T:

$$\frac{\partial}{\partial T}G(\lambda,T) = \sum_{m=0}^{\infty}(1-\lambda)^m\frac{\partial}{\partial T}p(m,T). \tag{11.8}$$

Then, using Eq. 11.3, this can be rewritten as

$$\begin{aligned}\frac{\partial}{\partial T}G(\lambda,T) &= \epsilon I(T)\left[\sum_{m=0}^{\infty}(1-\lambda)^m p(m-1,T)-\sum_{m=0}^{\infty}(1-\lambda)^m p(m,T)\right]\\ &= \epsilon I(T)\left[\sum_{m=0}^{\infty}(1-\lambda)^{m+1} p(m,T)-\sum_{m=0}^{\infty}(1-\lambda)^m p(m,T)\right]\\ &= -\epsilon I(T)\lambda G(\lambda,T).\end{aligned} \tag{11.9}$$

Therefore, the generating function becomes

$$G(\lambda,T) = e^{-\lambda\epsilon W(T)}, \tag{11.10}$$

where $W(T)$ is the light intensity integrated over the counting interval, i.e.,

$$W(T) = \int_0^T I(t)\,dt. \tag{11.11}$$

Returning to Eq. 11.5, we now see that the counting distribution is given by

$$p(n,T) = \frac{1}{n!}\left[\epsilon W(T)\right]^n e^{-\epsilon W(T)}. \tag{11.12}$$

Because of fluctuations, the instantaneous intensity $I(T)$ is, in fact, a random variable; that is to say, $I(T)$ represents a sample of the fluctuating intensity. It then naturally follows that the integrated intensity is a random variable as well. Consequently, in order for the calculation of $p(n,T)$ to make sense, it is necessary to perform an ensemble average over the distribution function $P(W)$ for the random variable W. We interpret $P(W)\,dW$ as the probability of measuring the integrated intensity to be between W and $W+dW$. The appropriate expression for the photocount distribution function is then

$$p(n,T) = \int_0^\infty \frac{1}{n!}(\epsilon W)^n e^{-\epsilon W} P(W)\,dW. \tag{11.13}$$

This result is equivalent to taking the generating function to be

$$G(\lambda,T) = \left\langle e^{-\lambda\epsilon W(T)}\right\rangle, \tag{11.14}$$

where the average is over the distribution $P(W)$. Equation 11.13 is a fundamental result known as *Mandel's formula*.

One can also define the so-called *factorial moments* of n by

$$n^{(k)} = \langle n(n-1)(n-2)\cdots(n-k+1)\rangle. \tag{11.15}$$

The average $\langle\ \rangle$ depends on the specific photocount distribution, so we write

$$\begin{aligned} n^{(k)} &= \sum_{m=0}^{\infty} m\,(m-1)\,(m-2)\cdots(m-k+1)\,p\,(m,T) \\ &= \left[\left(-\frac{\partial}{\partial\lambda}\right)^{k} G\,(\lambda,T)\right]_{\lambda=0} . \end{aligned} \tag{11.16}$$

The last line follows from the original definition of the generating function, Eq. 11.4. Furthermore, by inserting $G\,(\lambda,T)$ from Eq. 11.14, we learn that the kth factorial moment of n is determined by the kth moment of $W\,(T)$, i.e.,

$$n^{(k)} = \epsilon^{k} \left\langle W^{k} \right\rangle . \tag{11.17}$$

11.2 Intensity Fluctuations and Correlations

The first and second factorial moments are particularly useful because, as we will show in Sect. 11.3, they are derived from experimentally determined quantities. $n^{(1)}$ is equivalent to the first moment, or mean, of the count distribution. It is proportional to the average light intensity $\langle I \rangle$ and the counting period T, as shown below:

$$n^{(1)} = \langle n \rangle = \epsilon \langle W \rangle = \epsilon \int_{0}^{T} \langle I\,(t) \rangle \; dt = \epsilon T \langle I \rangle . \tag{11.18}$$

$n^{(2)}$, on the other hand, gives useful information about the nature of the intensity fluctuations of the light field. Specifically,

$$n^{(2)} = \langle n\,(n-1) \rangle = \epsilon^{2} \left\langle W^{2} \right\rangle = \epsilon^{2} \int_{0}^{T} dt_1 \int_{0}^{T} dt_2 \,\langle I\,(t_1)\, I\,(t_2) \rangle . \tag{11.19}$$

We can also write the second factorial moment in a normalized form, i.e.,

$$\frac{\langle n\,(n-1) \rangle}{\langle n \rangle^{2}} = \frac{1}{T^{2}} \int_{0}^{T} dt_1 \int_{0}^{T} dt_2 \, g^{(2)}\,(t_1,t_2) , \tag{11.20}$$

where

$$g^{(2)}\,(t_1,t_2) = \frac{\langle I\,(t_1)\, I\,(t_2) \rangle}{\langle I \rangle^{2}} . \tag{11.21}$$

Now let us introduce a *delay time* $\tau = t_2 - t_1$, so that $g^{(2)}\,(t_1,t_2) = \langle I\,(t_1)\, I\,(t_1+\tau) \rangle / \langle I \rangle^{2}$. By treating the fluctuating light field as a stationary random process,* $\langle I\,(t_1)\, I\,(t_1+\tau) \rangle$

*See the short discussion on stationary processes in Example 7.3.

should be independent of t_1, and should only be a function of τ. Therefore, $g^{(2)}(t_1, t_2)$ is replaced by

$$g^{(2)}(\tau) = \frac{\langle I(0)\, I(\tau)\rangle}{\langle I\rangle^2}, \tag{11.22}$$

which is the normalized *intensity autocorrelation function* of the light.† Some of the general properties of correlation functions produced by fluctuating signals were presented previously in Example 7.3. Recall that in the case of random fluctuations, the autocorrelation function will decay with some finite correlation time τ_0. Also, according to the Wiener-Khintchine theorem, $g^{(2)}(\tau)$ is the Fourier transform of the intensity signal's power spectrum (see Eq. 7.75).

Finally, Eq. 11.20 shows how the second factorial moment of n can be obtained from the intensity autocorrelation function $g^{(2)}(\tau)$:

$$\begin{aligned}
\frac{\langle n(n-1)\rangle}{\langle n\rangle^2} &= \frac{1}{T^2}\int_0^T dt_1 \int_{-t_1}^{-t_1+T} d\tau\, g^{(2)}(\tau) \\
&= \frac{1}{T^2}\left[\int_{-T}^{0} d\tau \int_{-\tau}^{T} dt_1\, g^{(2)}(\tau) + \int_0^T d\tau \int_0^{T-\tau} dt_1\, g^{(2)}(-\tau)\right] \\
&= \frac{2}{T^2}\int_0^T d\tau\, (T-\tau)\, g^{(2)}(\tau).
\end{aligned} \tag{11.23}$$

This indicates that the fluctuations, and hence the observed statistics of the light field, strongly depend on the length of the counting time T compared to the characteristic correlation time τ_0 of the measured light. Counting statistics for cases where T is either much less than or much greater than τ_0 are examined below.

11.2.1 Statistics for Short Counting Time

First we shall consider the case where $T \ll \tau_0$. Under this condition, the intensity can essentially be treated as a constant during the time interval T, so

$$W(T) = \int_0^T I(t)\, dt = TI. \tag{11.24}$$

Thus, instead of integrating over the probability density for the integrated intensity $P(W)$ in Mandel's formula (Eq. 11.13), we integrate over $P(I)$, which is the

†In Sect. 11.3, we will see that, in practice, $g^{(2)}(\tau)$ is determined experimentally by using a digital correlator, which measures the so-called *photocount correlation function*, $\langle n(t_1)\, n(t_2)\rangle$.

probability density for the instantaneous intensity. The photocount distribution then becomes

$$p(n,T) = \frac{(\epsilon T)^n}{n!} \int_0^\infty I^n e^{-\epsilon T I} P(I)\, dI. \tag{11.25}$$

Example 11.1 Intensity-Stabilized Light. Here we consider the radiation emitted by a stabilized, single-mode laser well above threshold. In the ideal case, the amplitude of the field is perfectly constant (no fluctuations), so

$$P(I) = \delta(I - \langle I \rangle). \tag{11.26}$$

Said another way, the intensity signal is correlated over an infinitely long time, so $\tau_0 \to \infty$. Consequently, the counting time T is always much less than the correlation time, and Eq. 11.25 is applicable for all T. The immediate result is that of Poisson counting statistics, i.e.,

$$p(n,T) = \frac{(\epsilon T)^n}{n!} \langle I \rangle^n e^{-\epsilon T \langle I \rangle} = \frac{e^{-\langle n \rangle}}{n!} \langle n \rangle^n, \tag{11.27}$$

where in the last step we used $\langle I \rangle = \langle n \rangle / \epsilon T$ (see Eq. 11.18). This is precisely what we previously found for a pure coherent state (see Sect. 5.3) or for a coherent state with a mixed phase (see Example 7.2). For Poisson statistics, the photocount distribution function is shown in Fig. 11.1.

One general method for obtaining the second factorial moment is to determine the generating function $G(\lambda, T)$, and then make use of Eq. 11.16. For the Poisson distribution, the generating function reduces to a simple form:

$$\begin{aligned} G(\lambda,T) &= \sum_{m=0}^{\infty} (1-\lambda)^m p(m,T) = e^{-\langle n \rangle} \sum_{m=0}^{\infty} \frac{[\langle n \rangle (1-\lambda)]^m}{m!} \\ &= e^{-\langle n \rangle} e^{\langle n \rangle (1-\lambda)} = e^{-\lambda \langle n \rangle}. \end{aligned} \tag{11.28}$$

Therefore,

$$\frac{\langle n(n-1) \rangle}{\langle n \rangle^2} = \frac{1}{\langle n \rangle^2} \left(\frac{\partial^2 G}{\partial \lambda^2} \right)_{\lambda=0} = 1. \tag{11.29}$$

From this, the variance of the distribution is $(\Delta n)^2 = \langle n^2 \rangle - \langle n \rangle^2 = \langle n \rangle$, as expected (see Eq. 5.99).

The second factorial moment also follows from the intensity autocorrelation function of the light. Since τ_0 is infinite, $g^{(2)}(\tau)$ does not decay, and must be a constant in time. Furthermore, since

$$g^{(2)}(0) = \frac{\langle I(0) I(0) \rangle}{\langle I \rangle^2} = \frac{\langle I^2 \rangle}{\langle I \rangle^2} = 1 \tag{11.30}$$

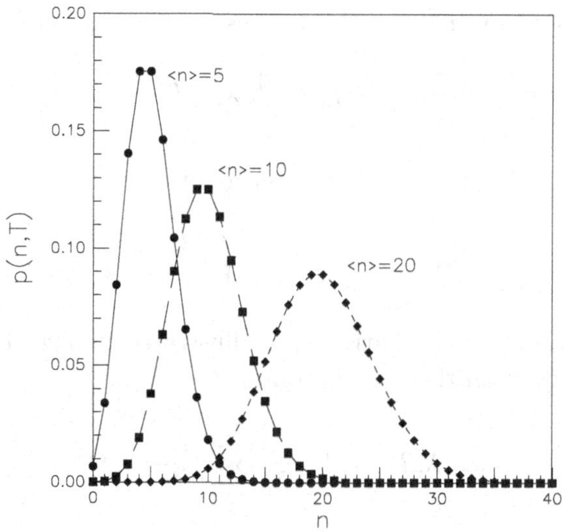

Figure 11.1 Photocount distribution for Poisson statistics.

for constant-intensity light, the correlation function in this case can only be

$$g^{(2)}(\tau) = 1. \tag{11.31}$$

Substituting into Eq. 11.23 again produces the result $\langle n(n-1)\rangle / \langle n\rangle^2 = 1$.

Example 11.2 Narrow-Band Chaotic Light. The term *chaotic light* applies to any light field that originates from a large number of independently emitting atoms. Some sources of broad-band chaotic light are incandescent lamps, thermal (or black-body) cavities, and lasers operating well-below threshold. In the present example, however, we will be focusing specifically on narrow-band chaotic light, such as the light associated with the spectral lines of gas discharge lamps. Of course, broad-band chaotic light that has been spectrally filtered also falls into this category. Later on in this chapter, we will find that another related example is laser light scattered by a randomly fluctuating medium, such as a liquid containing particles in Brownian motion. In this case, a narrow-band spectrum is produced because the amplitude of the laser light is randomly modulated by fluctuations in the scattering medium which are slow relative to the laser frequency.

Sometimes chaotic light is called *Gaussian light* because the complex electric-field amplitude can be represented by a *Gaussian random process*.‡ This makes the instantaneous value of the field modulus, $|E(t)|$, a Gaussian-distributed random variable, i.e., $P(|E|) \sim \exp\left(-|E|^2 / \langle |E|^2 \rangle\right)$. Since $I \propto |E|^2$, the normalized probability

‡For more of a discussion on Gaussian light, see Loudon [91].

density for the instantaneous intensity follows an exponential form, i.e.,

$$P(I) = \frac{1}{\langle I \rangle} e^{-I/\langle I \rangle}. \tag{11.32}$$

Inserting into Eq. 11.25 produces the counting distribution for $T \ll \tau_0$:

$$p(n,T) = \frac{1}{1+\langle n \rangle} \left(\frac{\langle n \rangle}{1+\langle n \rangle} \right)^n. \tag{11.33}$$

This represents Bose-Einstein statistics, as illustrated in Fig. 11.2. From Eq. 11.4, the generating function in this case becomes

$$G(\lambda,T) = (1-\xi) \sum_{m=0}^{\infty} [(1-\lambda)\xi]^m, \tag{11.34}$$

where $\xi(T) = \langle n \rangle / (1+\langle n \rangle)$. The summation corresponds to a geometric series converging to $[1-(1-\lambda)\xi]^{-1}$, so

$$G(\lambda,T) = \frac{1-\xi}{1-\xi+\lambda\xi} = \frac{1}{1+\lambda\langle n \rangle}. \tag{11.35}$$

Again applying Eq. 11.16, we obtain the normalized second factorial moment:

$$\frac{\langle n(n-1) \rangle}{\langle n \rangle^2} = 2. \tag{11.36}$$

Thus, for Bose-Einstein statistics, the variance of the distribution function is

$$(\Delta n)^2 = \langle n^2 \rangle - \langle n \rangle^2 = \langle n \rangle (1+\langle n \rangle). \tag{11.37}$$

Clearly, the counting fluctuations are enhanced relative to what is seen for Poisson statistics, and we say that the light-field exhibits *photon bunching.*

The phenomenon of bunching has its origins in the non-zero spectral width of the detected light.[§] This point can be illustrated by computing the intensity fluctuations directly from the spectral make-up of the light-field. To do this, we first need to be able to connect the power spectral density of the electric field, $|E(\omega)|^2$, to the intensity fluctuations. Here, $E(\omega)$ represents the frequency components of the electric field:

$$E(\omega) = \frac{1}{2\pi} \int_{-\infty}^{+\infty} E(t) e^{i\omega t} dt. \tag{11.38}$$

[§]Bunching is in no way related to the fact that photons are bosons, as one might be tempted to believe.

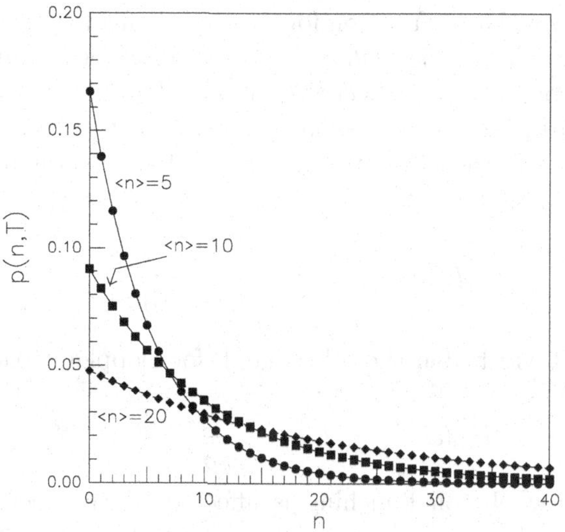

Figure 11.2 Photocount distribution for Bose-Einstein statistics.

For a randomly fluctuating field amplitude, the Wiener-Khintchine theorem relates $|E(\omega)|^2$ to the electric-field amplitude autocorrelation function, $\langle E(0)\,E(\tau)\rangle$, via a Fourier transform:

$$\langle E(0)\,E(\tau)\rangle = \int_{-\infty}^{+\infty} |E(\omega)|^2 e^{-i\omega\tau} d\tau. \tag{11.39}$$

In turn, $\langle E(0)\,E(\tau)\rangle$ can be related to the intensity autocorrelation function, $\langle I(0)\,I(\tau)\rangle$. We write

$$\begin{aligned}\langle I(0)\,I(\tau)\rangle &= \langle E^*(0)\,E(0)\,E^*(\tau)\,E(\tau)\rangle \\ &= \left\langle |E(0)|^2\right\rangle \left\langle |E(\tau)|^2\right\rangle + \langle E^*(0)\,E(\tau)\rangle \langle E(0)\,E^*(\tau)\rangle \\ &\qquad + \langle E^*(0)\,E^*(\tau)\rangle \langle E(0)\,E(\tau)\rangle \\ &= \langle I\rangle^2 + |\langle E^*(0)\,E(\tau)\rangle|^2 + \langle E^*(0)\,E^*(\tau)\rangle \langle E(0)\,E(\tau)\rangle\,. \end{aligned} \tag{11.40}$$

It is shown in the literature that the second line above follows when the field is a Gaussian random process ¶. In that case, it can also be shown that the term $\langle E^*(0)\,E^*(\tau)\rangle \langle E(0)\,E(\tau)\rangle$ vanishes. As a result, the normalized intensity correlation function becomes

$$g^{(2)}(\tau) = \frac{\langle I(0)\,I(\tau)\rangle}{\langle I\rangle^2} = 1 + \left|g^{(1)}(\tau)\right|^2, \tag{11.41}$$

where

$$g^{(1)}(\tau) = \frac{\langle E^*(0)\,E(\tau)\rangle}{\left\langle |E|^2\right\rangle} \tag{11.42}$$

¶See, for example, Mandel [92][93].

is the normalized correlation function for the electric-field amplitude. Equation 11.41 is sometimes referred to as the *factorization condition* for a Gaussian field.

The shape of the power-spectrum profile, $|E(\omega)|^2$, is determined by the particular mechanism in the source responsible for frequency broadening. For example, in Sect. 8.2, it was learned that the signature of collision-induced broadening is a Lorentzian power spectrum

$$|E_{\rm L}(\omega)|^2 = \left\langle |E|^2 \right\rangle \frac{\gamma/\pi}{(\omega-\omega_0)^2+\gamma^2}, \tag{11.43}$$

where γ is the half-width. On the other hand, for Doppler broadening, the profile is Gaussian:

$$|E_{\rm G}(\omega)|^2 = \left\langle |E|^2 \right\rangle \frac{1}{\sqrt{2\pi\sigma^2}} e^{-(\omega-\omega_0)^2/2\sigma^2}. \tag{11.44}$$

In order to see how photon bunching is affected by the specific spectrum, let us calculate the factorial moment $\langle n(n-1)\rangle / \langle n\rangle^2$ for these two types of profiles. First, from Eqs. 11.39 and 11.42, the respective field autocorrelation functions become

$$g_{\rm L}^{(1)}(\tau) = e^{-i\omega_0\tau} e^{-\gamma|\tau|} \tag{11.45}$$

and

$$g_{\rm G}^{(1)}(\tau) = e^{-i\omega_0\tau} e^{-\sigma^2\tau^2/2}. \tag{11.46}$$

We say that $\tau_c^{(\rm L)} = \gamma^{-1}$ is the *coherence time* of the Lorentzian frequency spectrum and $\tau_c^{(\rm G)} = \sigma^{-1}$ is the coherence time for the Gaussian spectrum. In general, the coherence time is inversely related to the spectral width. The intensity autocorrelation function is now obtained from Eq. 11.41, which gives

$$g_{\rm L}^{(2)}(\tau) = 1 + e^{-2\gamma|\tau|} \tag{11.47}$$

and

$$g_{\rm G}^{(2)}(\tau) = 1 + e^{-\sigma^2\tau^2}. \tag{11.48}$$

We see that the correlation time for the intensity signal is $\tau_0^{(\rm L)} = \tau_c^{(\rm L)}/2$ and $\tau_0^{(\rm G)} = \tau_c^{(\rm G)}/\sqrt{2}$ for the two spectra. Finally, applying Eq. 11.23, we have

$$\left[\frac{\langle n(n-1)\rangle}{\langle n\rangle^2}\right]_{\rm L} = 1 + \frac{1}{x_{\rm L}} - \frac{1}{2x_{\rm L}^2}\left(1 - e^{-2x_{\rm L}}\right) \tag{11.49}$$

and

$$\left[\frac{\langle n(n-1)\rangle}{\langle n\rangle^2}\right]_{\rm G} = 1 + \frac{\sqrt{\pi}}{x_{\rm G}}\,\mathrm{erf}\,x_{\rm G} - \frac{1}{x_{\rm G}^2}\left(1 - e^{-x_{\rm G}^2}\right), \tag{11.50}$$

where $x_{\rm L} = \gamma T = T/\tau_c^{(\rm L)}$ and $x_{\rm G} = \sigma T = T/\tau_c^{(\rm G)}$. Figure 11.3 shows the variation of $\langle n(n-1)\rangle / \langle n\rangle^2$ as a function of counting time for the two cases. Observe that

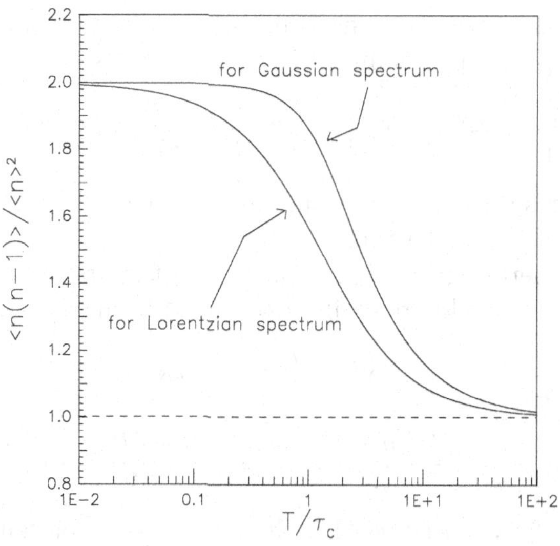

Figure 11.3 Dependence of second factorial moment on the counting time T for both Gaussian and Lorentzian broadened light.

$\langle n(n-1)\rangle / \langle n\rangle^2 \to 2$ for $T/\tau_c \ll 1$, so bunching (or Bose-Einstein statistics) occurs independent of the specific frequency spectrum (this was previously predicted by Eq. 11.36). For $T/\tau_c \gg 1$, we find that $\langle n(n-1)\rangle / \langle n\rangle^2 \to 1$, and no bunching, (i.e., Poisson statistics) is observed. A bit later on we will see that a Poisson counting distribution is characteristically measured in the limit of very long counting time, independent of the light spectrum. It is interesting to note, however, that when the duration of the counting interval is comparable to τ_c (or, for that matter, τ_0), there is an intermediate degree of photon bunching. This indicates that the counting distribution is neither Poisson nor Bose-Einstein in character. In addition, the distribution function depends on the particular shape of the frequency spectrum.

Example 11.3 Superposition of Intensity-Stabilized and Narrow-Band Chaotic Light. This case applies to light emitted by a laser operating close to, but above, threshold. A coherent, or intensity-stabilized, field is accompanied by a certain level of background noise in the form of Gaussian light. Let $\langle n\rangle_S$ represent the mean number of photocounts produced by the stabilized field in the absence of any background signal. Likewise, let $\langle n\rangle_N$ be the mean number of counts registered by the noise field alone. Then, it has been shown by Glauber [94] that the generating function for short counting time ($T \ll \tau_0$) is of the form

$$G(\lambda, T) = \frac{1}{1+\lambda\langle n\rangle_N} e^{-\lambda\langle n\rangle_S/\left(1+\lambda\langle n\rangle_N\right)}. \tag{11.51}$$

Using Eq. 11.5, one finds that the counting distribution involves the generating function for the Laguerre polynomials, L_n. The result is

$$p(n,T) = \frac{\langle n\rangle_N^n}{\left(1+\langle n\rangle_N\right)^{n+1}} L_n\left[-\frac{\langle n\rangle_S}{\langle n\rangle_N\left(1+\langle n\rangle_N\right)}\right] e^{-\langle n\rangle_S/\left(1+\langle n\rangle_N\right)}. \tag{11.52}$$

Figure 11.4 compares the count distributions for a pure coherent field, a pure chaotic field, and a combination of the two.

From the generating function (Eq. 11.51), the first and second factorial moments for the combined field are easily determined from Eq. 11.16 to be

$$\langle n\rangle = \langle n\rangle_S + \langle n\rangle_N \tag{11.53}$$

and

$$\frac{\langle n(n-1)\rangle}{\langle n\rangle^2} = \frac{1+4R+2R^2}{(1+R)^2}, \tag{11.54}$$

where $R = \langle n\rangle_N / \langle n\rangle_S$. Figure 11.5 shows how the normalized second factorial moment varies with the ratio R. Notice that for a pure coherent field ($R = 0$), $\langle n(n-1)\rangle / \langle n\rangle^2 = 1$, and Poisson statistics is recovered, whereas for pure chaotic light ($R \to \infty$), $\langle n(n-1)\rangle / \langle n\rangle^2 = 2$, and the distribution is Bose-Einstein. Equation 11.54 also gives the variance of the distribution as

$$\langle n^2\rangle - \langle n\rangle^2 = \langle n\rangle_S + \langle n\rangle_N\left(1+\langle n\rangle_N\right) + 2\langle n\rangle_S\langle n\rangle_N. \tag{11.55}$$

Clearly, the last term represents an interference effect.

Example 11.4 Squeezed Light. The concept of a squeezed state of the radiation field was introduced in Sect. 5.4, where it was learned that such states are classified as being either scaling squeezed states or coherent squeezed states. The counting statistics ($T \ll \tau_0$) for each of the two squeezing processes appear in Ref. [95], and the results are presented here:

A scaling squeezed state, recall, is generated by transforming a coherent state, $|\alpha_{\rm in}\rangle$, according to $|\alpha_{\rm ss};\mu,\nu\rangle = \hat{S}(\mu,\nu)|\alpha_{\rm in}\rangle$, where $\hat{S}(\mu,\nu)$ is the squeezing operator. The generating function for the state $|\alpha_{\rm ss};\mu,\nu\rangle$ is found to be

$$G_{\rm ss}(\lambda,T) = \sqrt{\vartheta}\exp\left\{\left[(1-\lambda)\vartheta - 1\right]|\alpha_{\rm in}|^2 - \frac{\nu}{2\mu}\left[1-(1-\lambda)^2\vartheta\right]\left(e^{-i\theta}\alpha_{\rm in}^2 + e^{i\theta}\alpha_{\rm in}^{*2}\right)\right\}, \tag{11.56}$$

where

$$\vartheta = \left[\mu^2 - (1-\lambda)^2\nu^2\right]^{-1}. \tag{11.57}$$

θ represents the angle of squeezing; it is related to the squeezing parameters μ and ν through Eqs. 5.153. If we now let $\alpha_{\rm in} = |\alpha_{\rm in}|e^{i\phi_{\rm in}}$, then the form of the corresponding counting distribution becomes

$$p_{\rm ss}(n,T) = \frac{1}{\mu n!}\left(\frac{\nu}{2\mu}\right)^n |H_n(Z_{\rm ss})|^2 \exp\left\{-\langle n\rangle_{\rm in}\left[1+\frac{\nu}{\mu}\cos\left(2\phi_{\rm in}-\theta\right)\right]\right\}, \tag{11.58}$$

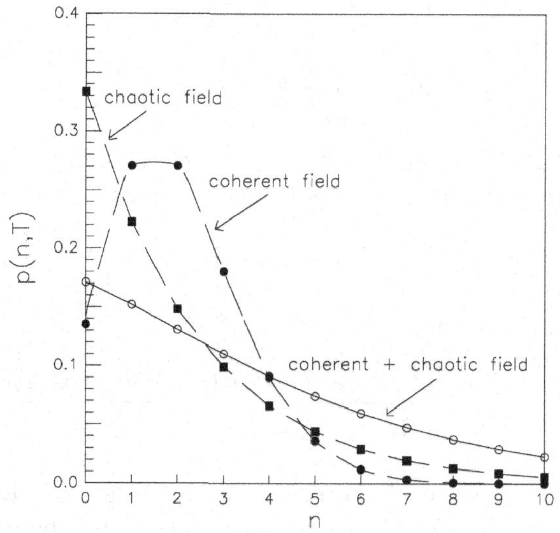

Figure 11.4 Photocount distribution for pure coherent light, pure chaotic light, and a superposition of the two.

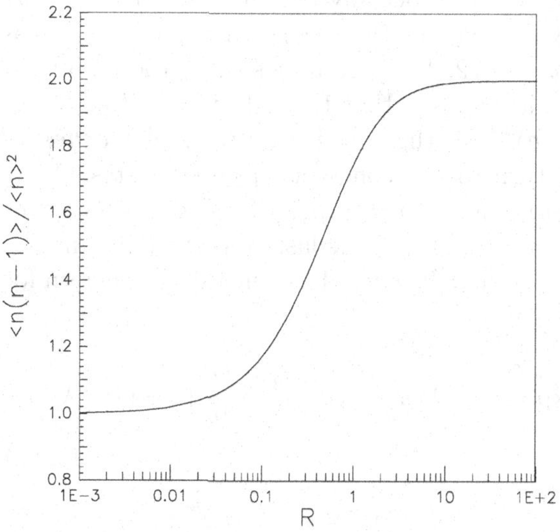

Figure 11.5 Second factorial moment vs. $R = \langle n \rangle_N / \langle n \rangle_S$, where $\langle n \rangle_S$ is the mean number of photocounts produced by the intensity-stabilized field alone. $\langle n \rangle_N$ represents the mean number of counts arising from Gaussian background light.

where $\langle n \rangle_{\text{in}} = |\alpha_{\text{in}}|^2$. The H_n's are nth-order Hermite polynomials involving the complex argument

$$Z_{\text{ss}} = \sqrt{\frac{\langle n \rangle_{\text{in}}}{2\mu\nu}}\, e^{i\left[\phi_{\text{in}} - \frac{1}{2}(\theta - \pi)\right]}. \tag{11.59}$$

The first and second factorial moments also follow from the generating function. Aided by the identity $\mu^2 - \nu^2 = 1$, it is rather straightforward to show that

$$\langle n \rangle_{\text{ss}} = \left[\mu^2 + \nu^2 + 2\mu\nu\cos\left(2\phi_{\text{in}} - \theta\right)\right] \langle n \rangle_{\text{in}} + \nu^2 \tag{11.60}$$

and

$$\left[\frac{\langle n(n-1)\rangle}{\langle n \rangle^2}\right]_{\text{ss}} = 1 + \frac{2\langle n \rangle_{\text{in}}\left[\nu^2\left(4\mu^2 - 1\right) + \mu\nu\left(\mu^2 + 3\nu^2\right)\cos\left(2\phi_{\text{in}} - \theta\right)\right] + \nu^2\left(2\mu^2 - 1\right)}{\langle n \rangle_{\text{ss}}^2}. \tag{11.61}$$

An interesting feature of the squeezed states is that the counting statistics depend on the phase of the input state, not just $|\alpha_{\text{in}}|$ and the squeezing parameters. In addition, observe from Eq. 11.60 that the squeezing process causes amplification. Figures 11.6 and 11.7 show representative counting distributions $p_{\text{ss}}(n, T)$ for different values of the squeezing parameter ν and the angle $\phi_{\text{in}} - \theta/2$. We find that the distribution functions are oscillatory in nature. This behavior is very different from what one finds for Poisson statistics which characterize the unsqueezed coherent states. The oscillations are especially pronounced when $\phi_{\text{in}} - \theta/2 = \pi/2$.

Figure 11.8 is a plot of the second factorial moment as a function of ν for different values of $\phi_{\text{in}} - \theta/2$. For parameters that produce a value of $\langle n(n-1)\rangle / \langle n \rangle^2$ that is less than unity, we say that the field is *anti-bunching*. On the other hand, when $\langle n(n-1)\rangle / \langle n \rangle^2 \geq 2$, the light is said to exhibit *enhanced-bunching*.

Now let us turn to the coherent squeezed states. In this case, one takes a squeezed vacuum state, $|0; \mu, \nu\rangle = \hat{S}(\mu, \nu)|0\rangle$, and injects a coherent state with complex amplitude $\alpha_{\text{cs}} = |\alpha_{\text{cs}}| e^{i\phi_{\text{cs}}}$. This last step is equivalent to the displacement operation $|\alpha_{\text{cs}}; \mu, \nu\rangle = \hat{D}(\alpha_{\text{cs}})|0; \mu, \nu\rangle$. The generating function for a coherent squeezed state is given by

$$G_{\text{cs}}(\lambda, T) = \sqrt{\vartheta}\exp\left\{\left[(1-\lambda)\vartheta - 1\right]|\beta_{\text{cs}}|^2 + \frac{\nu}{2\mu}\left[1 - (1-\lambda)^2\vartheta\right]\left(e^{-i\theta}\beta_{\text{cs}}^2 + e^{i\theta}\beta_{\text{cs}}^{*2}\right)\right\}, \tag{11.62}$$

where

$$\beta_{\text{cs}} = \mu\alpha_{\text{cs}} + \nu e^{i\theta}\alpha_{\text{cs}}^*. \tag{11.63}$$

The corresponding statistical distribution is

$$p_{\text{cs}}(n, T) = \frac{1}{\mu n!}\left(\frac{\nu}{2\mu}\right)^n |H_n(Z_{\text{cs}})|^2 \exp\left\{-\left(\langle n \rangle_{\text{cs}} - \nu^2\right)\left[1 + \frac{\nu}{\mu}\cos\left(2\phi_{\text{cs}} - \theta\right)\right]\right\} \tag{11.64}$$

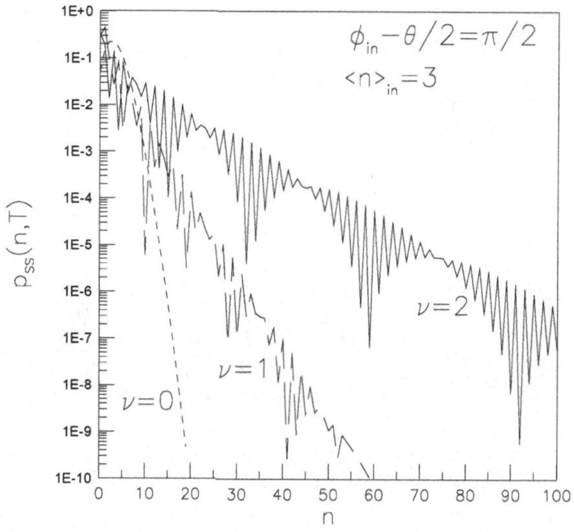

Figure 11.6 Photocount distribution for scaling squeezed states for different values of the squeezing parameter ν. The case $\nu = 0$ corresponds to the unsqueezed coherent state.

$$Z_{cs} = \sqrt{\frac{\langle n\rangle_{cs} - \nu^2}{2\mu\nu}} \left[(\mu+\nu)\cos\left(\phi_{cs} - \frac{\theta}{2}\right) + i(\mu-\nu)\sin\left(\phi_{cs} - \frac{\theta}{2}\right)\right], \tag{11.65}$$

having the first two factorial moments

$$\langle n\rangle_{cs} = |\alpha_{cs}|^2 + \nu^2 \tag{11.66}$$

$$\left[\frac{\langle n(n-1)\rangle}{\langle n\rangle^2}\right]_{cs} = 1 + \frac{2\left(\langle n\rangle_{cs} - \nu^2\right)\left[\nu^2 - \mu\nu\cos\left(2\phi_{cs} - \theta\right)\right] + \nu^2\left(2\mu^2 - 1\right)}{\langle n\rangle_{cs}^2}. \tag{11.67}$$

The statistics for the coherent squeezed states depend on the phase, not just the magnitude of the injected state, α_{cs}. As in the case of the scaling squeezed states, Figs. 11.9 and 11.10 show that there are oscillations in the counting distribution. The second factorial moment, which gives the degree of photon bunching, is a strong function of the squeezing parameters and the state α_{cs}—this is illustrated in Fig. 11.11. For the special case of a squeezed vacuum state, i.e., the case where $\alpha_{cs} = 0$, the photocount distribution reduces to

$$p(n,T) = \frac{|\nu|^n\left[(n-1)(n-3)\cdots 3\cdot 1\right]^2}{|\mu|^{n+1}\, n!} \qquad \text{for even } n, \tag{11.68}$$

and vanishes for odd n. This is shown in Fig. 11.12. The mean of the distribution becomes ν^2 and the second factorial moment is $\langle n(n-1)\rangle / \langle n\rangle^2 = 3 + \nu^{-2}$. A squeezed vacuum state, therefore, always has enhanced bunching.

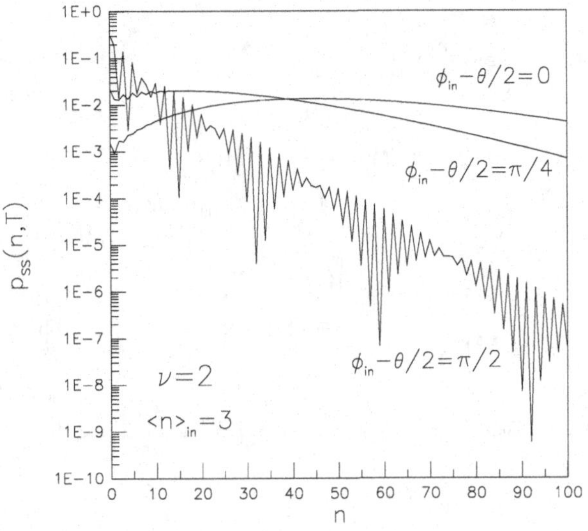

Figure 11.7 Photocount distribution for scaling squeezed states for different values of the angle $\phi_{in} - \theta/2$.

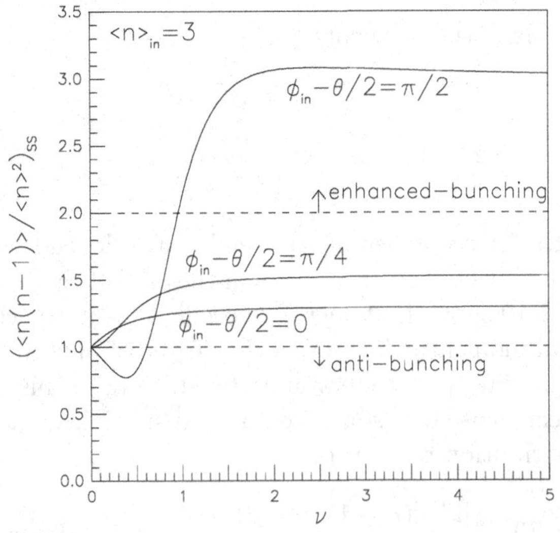

Figure 11.8 Second factorial moment of the scaling squeezed states as a function of the squeezing parameter ν for different values of the angle $\phi_{in} - \theta/2$.

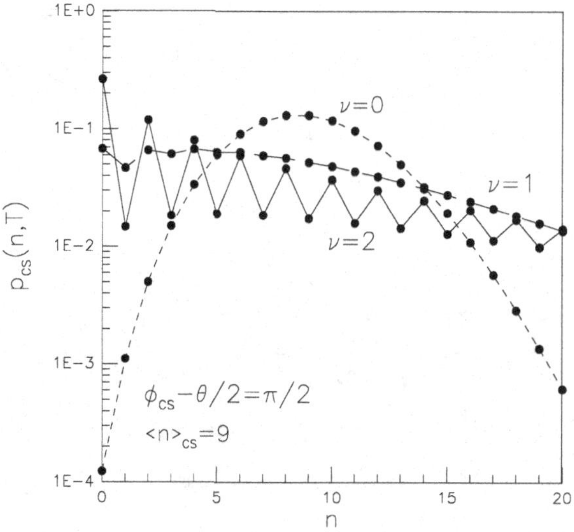

Figure 11.9 Photocount distribution for coherent squeezed states for different values of the squeezing parameter ν. The case $\nu = 0$ corresponds to the unsqueezed coherent state.

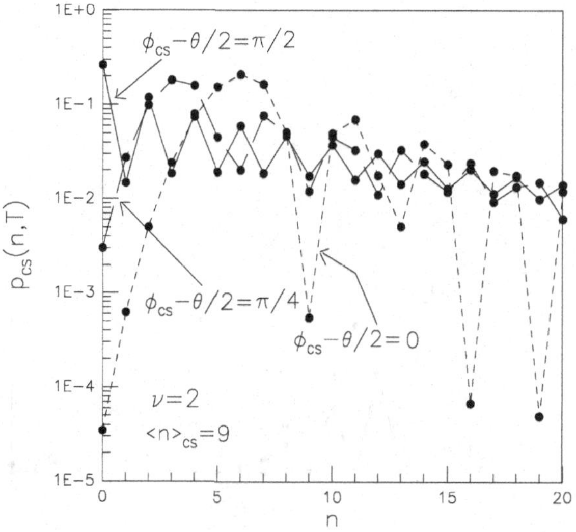

Figure 11.10 Photocount distribution for coherent squeezed states for different values of the angle $\phi_{cs} - \theta/2$.

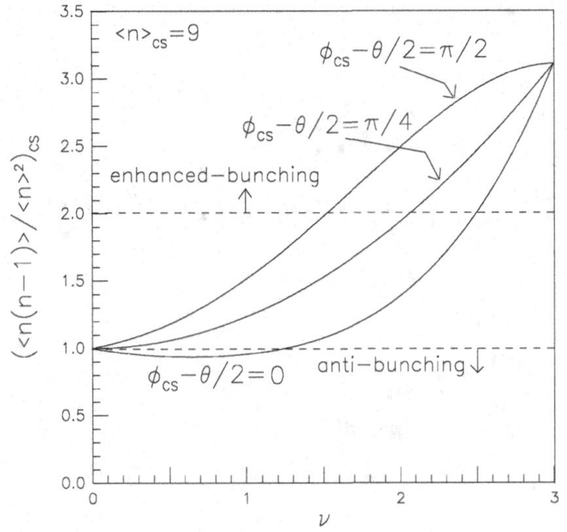

Figure 11.11 Second factorial moment of the coherent squeezed states as a function of the squeezing parameter ν for different values of the angle $\phi_{cs} - \theta/2$.

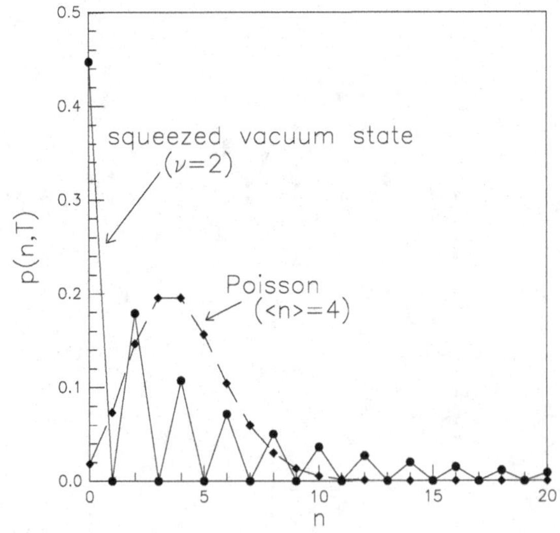

Figure 11.12 Photocount distribution for squeezed vacuum state with $\nu = 2$. For comparison, a Poisson distribution having the same mean ($\langle n \rangle = 4$) is also shown.

11.2.2 Statistics for Long Counting Time

When $T \gg \tau_0$, the integral $W(T) = \int_0^T I(t)\,dt$ averages out any intensity fluctuations, so that

$$W(T) = \langle I \rangle T. \tag{11.69}$$

Since W is a constant (for a specified counting time T), the distribution function for W is simply $P(W) = \delta(W - \langle I \rangle T)$. Then, according to Eq. 11.13, the counting distribution becomes

$$p(n,T) = \frac{1}{n!}\left[\epsilon \langle I \rangle T\right]^n e^{-\epsilon \langle I \rangle T} = \frac{e^{-\langle n \rangle}}{n!}\langle n \rangle^n, \tag{11.70}$$

which represents Poisson statistics. We find that the statistical distribution of photoelectron counts is, in fact, identical to that found for intensity-stabilized laser light (see Eq. 11.27)! In both cases, a Poisson distribution is observed because a quantum counter makes the detection process itself probabilistic in nature, not because of statistical fluctuations in the light field. For laser light well-above threshold, we have assumed that intensity fluctuations are truly non-existent, whereas for the case under current consideration, any existing fluctuations are smoothed out due to the long duration of the counting time relative to the correlation time of the intensity signal. For example, light detected from an incandescent lamp produces Poisson statistics—this is because the fluctuations in light intensity are so rapid that $T \gg \tau_0$ for any real detector.

The statistics for long counting time have been studied in greater detail by Glauber for the specific case of narrow-band chaotic light with a Lorentzian spectrum [90][94]. It is found that the precise photocount distribution is not solely determined by the duration of the counting interval, as we have suggested, but also by the mean counting rate, $r = \langle n \rangle / T$, which depends on the brightness of the source. For counting intervals T much longer than the coherence time, i.e., $T/\tau_c = \gamma T \gg 1$, Glauber shows that the generating function is

$$G(\lambda, T) = e^{\left(\gamma - \sqrt{\gamma^2 + 2\gamma r \lambda}\right)T}. \tag{11.71}$$

When the counting rate r is much less than the linewidth γ, the exponent can be expanded to lowest order giving $G(\lambda, T) \rightarrow e^{-\lambda r T} = e^{-\lambda \langle n \rangle}$; this gives rise to Poisson counting statistics, as previously predicted. However, for higher counting rates, some photon bunching occurs. The increased fluctuations appear as a broadening of the distribution function, and a deviation from Poisson statistics. The generating function of Eq. 11.71 leads to the following general expression for the photocount distribution [90][94]:

$$p(n,T) = \frac{1}{n!}\left(\frac{\gamma r T}{\Gamma}\right)^n s_n(\Gamma T)\, e^{-(\Gamma - \gamma)T}. \tag{11.72}$$

The parameter Γ is defined as

$$\Gamma = \sqrt{\gamma^2 + 2r\gamma} \tag{11.73}$$

and we introduce the functions

$$s_n(\xi) = e^{\xi}\sqrt{\frac{2\xi}{\pi}}K_{n-\frac{1}{2}}(\xi), \tag{11.74}$$

where the $K_{n-\frac{1}{2}}$'s denote modified Hankel functions of half-integral order.‖ The mean and variance of the Eq. 11.72 distribution are most easily obtained by using the generating function, Eq. 11.71, to determine the first and second factorial moments (see Eq. 11.16). We then find that the mean is $\langle n\rangle = rT$, as expected, and the normalized second factorial moment is $\langle n(n-1)\rangle / \langle n\rangle^2 = 1 + 1/\gamma T$ (notice that this is in agreement with Eq. 11.49 in the limit when γT is large). The resulting variance is then

$$\langle n^2\rangle - \langle n\rangle^2 = \langle n\rangle(1+v), \tag{11.75}$$

where $v = r/\gamma = \langle n\rangle/\gamma T$. This parameter represents the enhanced width of the distribution function brought about by intensity fluctuations, or photon bunching. For a case where $\langle n\rangle \gg 1$ and v is sufficiently large such that $n \gg \langle n\rangle\sqrt{2/v}$, Glauber shows that the distribution function, Eq. 11.72, reduces to the following asymptotic form:

$$p(n,T) = \frac{\langle n\rangle}{n^{3/2}\sqrt{2\pi v}}\exp\left[-\frac{1}{2v}\left(\sqrt{n} - \frac{\langle n\rangle}{\sqrt{n}}\right)^2\right]. \tag{11.76}$$

The criteria stated above for the validity of this expression are easily met for large values of n. However, in order for the expression to also hold for n in the vicinity of and well-below $\langle n\rangle$, the count rate r must be very high (i.e., the light source must be very intense) and/or the linewidth of the field needs to be extremely narrow. Figure 11.13 shows $p(n,T)$ for an extreme case with $\langle n\rangle = 10,000$ and $v = 2,000$. It is displayed along with the Poisson distribution having the same mean in order to illustrate that the distribution is broadened.

11.3 Photon Correlation Measurements

In this section we show that, for $T \ll \tau_0$, the experimental determination of the intensity correlation function, $g^{(2)}(\tau)$, is accomplished by using a data-processing

‖More explicitly, the functions $s_n(\xi)$ are polynomials in ξ^{-1}. The first few functions are

$$\begin{aligned} s_0(\xi) &= s_1(\xi) = 1 \\ s_2(\xi) &= 1 + \xi^{-1} \\ s_3(\xi) &= 1 + 3\xi^{-1} + 3\xi^{-2} \\ s_4(\xi) &= 1 + 6\xi^{-1} + 15\xi^{-2} + 15\xi^{-3}. \end{aligned}$$

They obey the recursion relation

$$s_{n+1}(\xi) = -s_n'(\xi) + \left(1 + n\xi^{-1}\right)s_n(\xi).$$

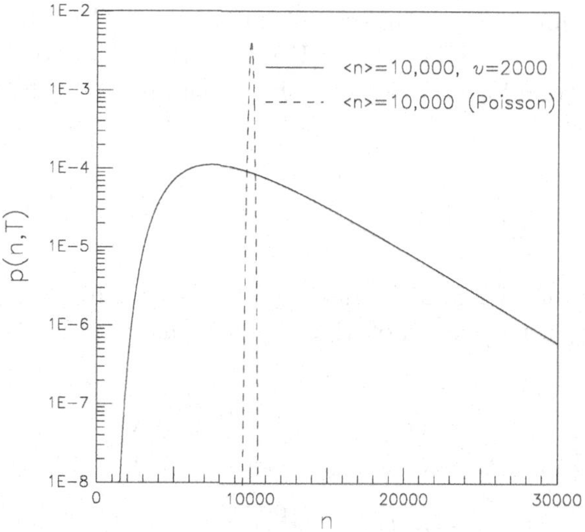

Figure 11.13 Effect of mean counting rate, $r = \langle n \rangle / T$, on the photocount distribution for Lorentzian narrow-band light in the limit of long counting time ($T \gg \tau_c$). The distribution for the case $v = r\tau_c = 2000$ is compared to a Poisson distribution having the same mean ($\langle n \rangle = 10,000$).

instrument known as a *digital correlator*, which converts the incoming train of photoelectron pulses into a so-called *photocount correlation function* $\langle n(t_1, T)\, n(t_2, T) \rangle$. The quantity $n(t, T)$ represents the number of photocounts recorded during the time-interval $(t, t+T)$. For the sake of brevity, let us write the photocount correlation function using the shorthand notation $\langle n(t_1)\, n(t_2) \rangle$, where the *sampling time* T is treated implicitly. The discussion presented below will show that the normalized intensity and photocount correlation functions are, in fact, identical when the sampling time is sufficiently less than τ_0.

Begin by defining the joint photocount distribution $p(n_1, n_2, T)$, which is the probability of recording n_1 counts during the interval $(t_1, t_1 + T)$ and n_2 counts during the interval $(t_2, t_2 + T)$. Then, by analogy with Eq. 11.4, we introduce a joint generating function

$$G(\lambda_1, \lambda_2, T) = \sum_{m_1=0}^{\infty} \sum_{m_2=0}^{\infty} (1-\lambda_1)^{m_1} (1-\lambda_2)^{m_2}\, p(m_1, m_2, T) . \qquad (11.77)$$

Using arguments that parallel those in Sect. 11.1, leads to the following alternate expression:

$$G(\lambda_1, \lambda_2, T) = \left\langle e^{-\epsilon\lambda_1 W_1(T)} e^{-\epsilon\lambda_2 W_2(T)} \right\rangle . \qquad (11.78)$$

This is a generalization of Eq. 11.14. The brackets represent an ensemble average over the distribution function $P(W_1, W_2)$, where $P(W_1, W_2)\, dW_1 dW_2$ is interpreted as the probability of finding the two integrated intensities $\int_{t_1}^{t_1+T} I(t)\,dt$ and $\int_{t_2}^{t_2+T} I(t)\,dt$ lying within the intervals (W_1, W_1+dW_1) and (W_2, W_2+dW_2), respectively. It is now left as an exercise for the reader to show that the joint photocount distribution is given by

$$p(n_1,n_2,T) = \left[\frac{1}{n_1!n_2!}\left(-\frac{\partial}{\partial\lambda_1}\right)^{n_1}\left(-\frac{\partial}{\partial\lambda_2}\right)^{n_2} G(\lambda_1,\lambda_2,T)\right]_{\lambda_1=\lambda_2=1}, \tag{11.79}$$

which is an extension of Eq. 11.5. Inserting the generating function from Eq. 11.78 produces a more general version of Mandel's formula (Eq. 11.13):

$$\begin{aligned} p(n_1,n_2,T) &= \left\langle \frac{(\epsilon W_1)^{n_1}}{n_1!}e^{-\epsilon W_1}\frac{(\epsilon W_2)^{n_2}}{n_2!}e^{-\epsilon W_2}\right\rangle \\ &= \int_0^\infty dW_1 \int_0^\infty dW_2 \frac{(\epsilon W_1)^{n_1}}{n_1!}e^{-\epsilon W_1}\frac{(\epsilon W_2)^{n_2}}{n_2!}e^{-\epsilon W_2} P(W_1,W_2). \end{aligned} \tag{11.80}$$

The photocount correlation function can now be obtained directly from the joint distribution of photocounts by using

$$\langle n(t_1)\,n(t_2)\rangle = \sum_{m_1=0}^{\infty}\sum_{m_2=0}^{\infty} m_1 m_2\, p(m_1,m_2,T). \tag{11.81}$$

From the form of the generating function given by Eq. 11.77, it then becomes a straightforward matter to verify that

$$\langle n(t_1)\,n(t_2)\rangle = \left[\left(\frac{\partial}{\partial\lambda_1}\right)\left(\frac{\partial}{\partial\lambda_2}\right) G(\lambda_1,\lambda_2,T)\right]_{\lambda_1=\lambda_2=0}. \tag{11.82}$$

Finally, from Eq. 11.78, the correlation function takes the form

$$\langle n(t_1)\,n(t_2)\rangle = \epsilon^2 \langle W_1(T)\,W_2(T)\rangle. \tag{11.83}$$

For $T \ll \tau_0$, one has $W_1(T) = I(t_1)\,T$ and $W_2(T) = I(t_2)\,T$, so that $\langle n(t_1)\,n(t_2)\rangle = \epsilon^2 T^2 \langle I(t_1)\,I(t_2)\rangle$. By treating the fluctuating light field as a stationary random process, we are able to write the correlation functions in terms of the delay time $\tau = t_2 - t_1$, i.e.,

$$\langle n(0)\,n(\tau)\rangle = \epsilon^2 T^2 \langle I(0)\,I(\tau)\rangle. \tag{11.84}$$

Hence, for short sampling time,

$$g^{(2)}(\tau) \equiv \frac{\langle I(0)\,I(\tau)\rangle}{\langle I\rangle^2} = \frac{\langle n(0)\,n(\tau)\rangle}{\langle n\rangle^2}. \tag{11.85}$$

Let us estimate how short T must be in order for this equivalence between the photocount and intensity correlation functions to hold. From Eq. 11.83, the photocount correlation function is

$$\langle n(t_1)\, n(t_2)\rangle = \epsilon^2 \int_{t_1}^{t_1+T} dt' \int_{t_2}^{t_2+T} dt'' \langle I(t')\, I(t'')\rangle\,, \tag{11.86}$$

or in normalized form,

$$\frac{\langle n(t_1)\, n(t_2)\rangle}{\langle n\rangle^2} = \frac{1}{T^2} \int_{t_1}^{t_1+T} dt' \int_{t_2}^{t_2+T} dt'' g^{(2)}(|t'' - t'|)\,. \tag{11.87}$$

As an example, consider the case of Gaussian light with a Lorentzian spectral profile such that (see Eq. 11.47)

$$g^{(2)}(|t'' - t'|) = 1 + e^{-2\gamma|t''-t'|}. \tag{11.88}$$

Then the integral produces the expression

$$\frac{\langle n(t_1)\, n(t_2)\rangle}{\langle n\rangle^2} = 1 + \left(\frac{\sinh \gamma T}{\gamma T}\right)^2 e^{-2\gamma|t_2-t_1|} \qquad (t_2 > t_1)\,. \tag{11.89}$$

Clearly, the photocount and intensity correlation functions are identical when $\left(\frac{\sinh \gamma T}{\gamma T}\right)^2$ is very close to unity. Notice, however, that $\left(\frac{\sinh \gamma T}{\gamma T}\right)^2 = 1 + \frac{1}{3}\gamma^2 T^2 + \cdots$. Thus, it is reasonable to require that $\frac{1}{3}\gamma^2 T^2 \leq 0.001$, or $\gamma T = T/\tau_c \leq 0.055$. Since $\tau_c = 2\tau_0$ (see Example 11.2), a good rule of thumb is that $g^{(2)}(\tau) = \langle n(0)\, n(\tau)\rangle / \langle n\rangle^2$, as long as $T \leq 0.1\tau_0$.

In an actual experiment, the photocount correlation function is obtained as follows: Each detected photon pulse is first amplified, then fed to a discriminator that creates a standardized narrow ($\sim$ 25 nsec) pulse with a 5-volt amplitude. The correlation function is then formed by sending the pulse sequence to a digital autocorrelator. An "ideal" digital correlator constructs a time-averaged correlation function

$$C(t_i) = \frac{1}{M} \sum_{j=1}^{M} n(t_j)\, n(t_{j+i})\,, \tag{11.90}$$

based on collecting, say, M samples, each of duration $T = t_{i+1} - t_i$. For a stationary random process, this time average is assumed to be identical to the sought after ensemble average required for $g^{(2)}(\tau)$,** and

$$C(t_i) = C(\tau) = \langle n(0)\, n(\tau)\rangle\,. \tag{11.91}$$

**The equivalence between the time average and ensemble average is known as the *ergodic hypothesis*.

In the next section, we will see that for the purposes of light scattering spectroscopy, correlation functions can also be measured using a "clipped" correlator, which is faster and slightly simpler in design than the ideal one described above.

11.4 Quasi-Elastic Light Scattering

An important application of the photon correlation technique is the scattering of laser light from fluid media. In general, the presence of any local density fluctuations in a fluid will produce the Rayleigh scattering of incident light. As we saw in Chapter 9, Rayleigh scattering is nominally an elastic process, characterized by a frequency shift of $\omega = 0$. In the present case, however, the temporal properties of the fluctuations present in the scattering medium will produce a broadening of the observed Rayleigh line about $\omega = 0$. The profile of this line mirrors the power spectrum of the fluctuations, and is directly related to the dynamic structure factor $S(\mathbf{Q}, \omega)$ previously encountered in the analysis of neutron and x-ray scattering experiments (see Examples 6.1 and 9.2). This can be seen by starting with the basic result for Rayleigh scattering from a single atom (i.e., Eq. 9.147)

$$\frac{d\sigma}{d\Omega} = k^4 \left| \boldsymbol{\epsilon}_{\mathbf{k}\lambda} \cdot \tilde{\tilde{\boldsymbol{\alpha}}} \cdot \boldsymbol{\epsilon}_{\mathbf{k}'\lambda'} \right|^2 , \tag{11.92}$$

and then applying energy conservation, as in the case of thermal neutron scattering. It is left to the reader to fill in the intermediate steps which lead to the following double-differential cross-section for scattering by a fluid containing N molecules:

$$\frac{d\sigma}{d\Omega\, d\omega} = k^4 \frac{1}{2\pi} \int_{-\infty}^{+\infty} dt\, e^{-i\omega t} \\ \times \left\langle \sum_{\ell\ell'}^{N} \left[\boldsymbol{\epsilon}_{\mathbf{k}'\lambda'} \cdot \tilde{\tilde{\boldsymbol{\alpha}}}_\ell (0) \cdot \boldsymbol{\epsilon}_{\mathbf{k}\lambda} \right] \left[\boldsymbol{\epsilon}_{\mathbf{k}'\lambda'} \cdot \tilde{\tilde{\boldsymbol{\alpha}}}_{\ell'} (t) \cdot \boldsymbol{\epsilon}_{\mathbf{k}\lambda} \right] e^{-i\mathbf{Q}\cdot\mathbf{R}_\ell(0)} e^{i\mathbf{Q}\cdot\mathbf{R}_{\ell'}(t)} \right\rangle . \tag{11.93}$$

Here, $\tilde{\tilde{\boldsymbol{\alpha}}}_\ell (t)$ represents the effective molecular polarizability of the ℓth molecule. Its time dependence arises from changes in the instantaneous orientation and/or conformation of the molecule. For a system of optically isotropic molecules, the polarizability is a scalar quantity, so that

$$\frac{d\sigma}{d\Omega\, d\omega} = N k^4 \alpha^2 \left| \boldsymbol{\epsilon}_{\mathbf{k}'\lambda'} \cdot \boldsymbol{\epsilon}_{\mathbf{k}\lambda} \right|^2 \left[\frac{1}{2\pi} \int_{-\infty}^{+\infty} dt\, e^{-i\omega t} \left\langle \frac{1}{N} \sum_{\ell\ell'}^{N} e^{-i\mathbf{Q}\cdot\mathbf{R}_\ell(0)} e^{i\mathbf{Q}\cdot\mathbf{R}_{\ell'}(t)} \right\rangle \right] . \tag{11.94}$$

In other words, the scattering is proportional to the dynamic structure factor:

$$\frac{d\sigma}{d\Omega\, d\omega} = N k^4 \alpha^2 \left| \boldsymbol{\epsilon}_{\mathbf{k}'\lambda'} \cdot \boldsymbol{\epsilon}_{\mathbf{k}\lambda} \right|^2 S(\mathbf{Q}, \omega) . \tag{11.95}$$

For relatively high-frequency phenomena, the power spectrum of the scattered light is conventionally measured by using an optical filter followed by a square-law detector. This technique, known as *amplitude correlation spectroscopy*, is appropriate for frequencies above $\omega \sim 10^7$ s^{-1}. However, many important phenomena of physical, chemical, and biological interest are characterized by slow fluctuations in the range 10^7 down to 1 s^{-1}. For such cases, where the spectral broadening of the Rayleigh-scattered line is small relative to the laser frequency ($\sim 3 \times 10^{15}$ s^{-1}), we call the process *quasi-elastic light scattering* (QELS). Amplitude correlation spectroscopy cannot be used for QELS because, in practice, there are lower limits on both the optical filter bandwidth and the linewidth of the incident laser light. Instead of carrying out the measurements in frequency space, one uses the technique of *photon correlation spectroscopy* which operates in the time domain.

Before discussing the specifics of photon correlation spectroscopy and how it is used to measure the properties of the scatterer, let us first mention some of the processes in fluids that give rise to quasi-elastic scattering:

- **Density fluctuations and concentration fluctuations in one and two-component fluids, respectively, near the critical point.** The characteristic decay time for the thermal fluctuations is given by $1/DQ^2$, where D is the thermal or concentration diffusion coefficient, and the magnitude of the wavevector transfer, Q, is on the order of 10^5 cm^{-1} for light scattering experiments. For temperatures within 10 to 10^{-3} K of the critical point, D is typically in the range 5×10^{-6}–10^{-8} cm^2/s. This means that one expects to observe a quasi-elastic linewidth, $\Gamma = DQ^2$, on the order of 5×10^4–10^2 s^{-1}.

- **Diffusion of macromolecules and molecular aggregates in solution.** Consider particles having sizes in the range $a \sim 10$–10^3 nm. In an aqueous solution of viscosity $\eta \sim 10^{-3}$ poise, an estimate of the diffusion coefficient is obtained from the *Stokes-Einstein relation*

$$D = \frac{k_B T}{6\pi\eta a} \tag{11.96}$$

  to be 2×10^{-6}–2×10^{-8} cm^2/s. So one finds that $\Gamma \sim 2 \times 10^4$–$2 \times 10^2$ s^{-1}, i.e., the measured linewidths are on the same order as in the previous example.

- **Motile particles in solution.** One is sometimes interested in examining particles moving in straight-line trajectories, either under the influence of external forces (such as carried by a flow or under an electric field) or by their self-propulsion (such as bacteria with flagella). The difference between this type of motion and diffusion is that the mean free path between events that change the direction of the particle motion is much larger in the present situation.[††] Diffusive motion can be described as a random walk process, where each step is much

[††]Clearly, if there is no change of direction, the mean free path is infinite.

shorter than the observational wavelength of Q^{-1}, whereas for motile particles the steps cover a distance much larger than this value. As an example, consider the motion of *E. coli* bacteria in vitro.‡‡ Observations of certain strains through a microscope suggest that in between directional changes, a typical bacterium moves approximately in a straight line at constant speed for a distance on the order of ten times its body length of about 1.5 μm. If no gradients or chemical agents are present which might bias the direction of movement, the motion is found to be isotropic, and the average speed of a bacterium during a step is $\langle v \rangle \sim 15$ μm/s (see Example 11.5). Furthermore, the Doppler-broadened Rayleigh linewidth for motile particles is given approximately by $\Gamma \sim Q \langle v \rangle$, so we see that $\Gamma \sim 150$ s^{-1}.

The basic setup for performing light scattering experiments using photon correlation spectroscopy is shown in Fig. 11.14. A light beam from a well-stabilized gas laser is sent into a sample via a focusing lens. Scattered light is collimated by two pinholes which serve to define the scattering angle θ and to reduce the light incident on the photocathode surface of the photomultiplier tube to a spot size of approximately one *coherence area.* The scattered field vectors are spatially correlated within a coherence area, thus a correlated signal can only be measured if this condition is satisfied. As described in the last section, after being passed through an amplifier and discriminator, the train of photon pulses is fed into a digital correlator, and the photon correlation function is formed. In actuality, the digital correlator used in most experiments is not an ideal one. Rather than measuring $\langle n(0)\, n(\tau) \rangle$, one ordinarily uses a so-called *clipped digital correlator* which constructs the time correlation function $\langle n_K(0)\, n(\tau) \rangle$. The signal $n_K(t)$ is a clipped version of the original $n(t)$—it has the property that

$$n_K(t) = \begin{cases} 1 \text{ if } n(t) > K \\ 0 \text{ if } n(t) \leq K, \end{cases} \tag{11.97}$$

where K is an integer called the *clipping level.* The effect of this clipping is that the time consuming process of digital multiplication required for an ideal correlator is reduced to a much simpler and faster gating process [99]. However, unlike the full photocount correlation function, which is related simply to the intensity correlation function $\langle I(0)\, I(\tau) \rangle$ (see Eq. 11.84), the clipped correlation function depends, in a quite complicated way, on all higher-order intensity correlation functions $\langle I(0)\, I^j(\tau) \rangle$ (where j is any positive integer) [100]. Fortunately though, if the scattered field is due to the random motion of many independently moving particles, one can show [98] that if K is chosen to have a value close to $\langle n \rangle$, the clipped function $\langle n_K(0)\, n(\tau) \rangle$ reduces to the full correlation function $\langle n(0)\, n(\tau) \rangle$.

---

‡‡Nossal *et al.* [96] were the first to use quasi-elastic light scattering to quantitatively analyze the motility of bacteria (*E. coli*) in solution. For reviews, and some extensions of this work, also see Refs. [97] and [98].

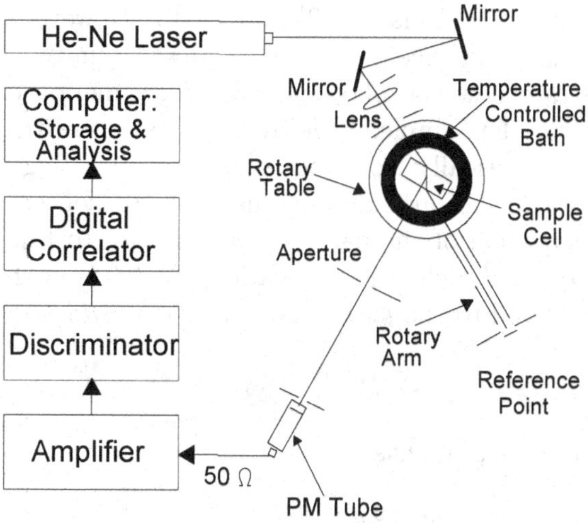

Figure 11.14 Basic setup for doing laser light-scattering using the photon correlation technique.

As indicated by Eq. 11.85, the photocount correlation function can be equated to the normalized intensity correlation function $g^{(2)}(\tau) = \langle I(0) I(\tau)\rangle / \langle I\rangle^2$ provided that T, the sampling time chosen for the experiment, is much less than the characteristic correlation time τ_0, i.e.,

$$\langle n(0) n(\tau)\rangle = \langle n\rangle^2 g^{(2)}(\tau). \tag{11.98}$$

Furthermore, under the proper conditions, one is also justified in applying the factorization condition which connects the scattered-field amplitude correlation function, $g^{(1)}(\tau) = \langle E_s^*(0) E_s(\tau)\rangle / \langle |E_s|^2\rangle$, to the intensity correlation function, as in Eq. 11.41. Recall that the derivation of the factorization condition is normally based on the assumption that the scattered field amplitude is a Gaussian random process. However, Mandel [101] has shown that because of the phase fluctuations exhibited by a real gas laser, the scattered field is, in fact, not a Gaussian random process at all. Yet, in spite of this, it has been shown that Eq. 11.41 is still valid for most types of quasi-elastic light scattering experiments. In particular, Tartaglia *et al.* [102] [103] have demonstrated that when the scattering medium contains either independently moving (e.g., diffusing or motile) particles or critical fluctuations, the Gaussian assumption is unnecessary for deriving the factorization condition. This means that the photon correlation function for the scattered light can be written as

$$\langle n(0) n(\tau)\rangle = \langle n\rangle^2 \left[1 + f(A) \left|g^{(1)}(\tau)\right|^2\right]. \tag{11.99}$$

$f(A)$, known as a *spatial coherence factor*, accounts for the finite area A of the photocathode surface. The specific form and numerical values of $f(A)$ for a commonly used geometry are given in Refs. [102] and [104]. However, if the photocathode subtends less than one coherence area, $f(A)$ is close to unity.

Let us illustrate how the photon correlation function can be related to the motion of particles in the sample. Consider a case where the illuminated volume contains N spherical (optically isotropic) scatterers; in doing so, the only relevant particle motions are translational, and one need not be concerned with internal motions, such as reorientation of the particle. Now return to Eq. 4.188, which relates the scattered field to local dielectric fluctuations $\Delta\varepsilon(\mathbf{r}', t)$ in the sample. The latter quantity is simply related to the *local polarizability density* $\alpha(\mathbf{r}', t)$ by

$$\Delta\varepsilon(\mathbf{r}', t) = 4\pi\alpha(\mathbf{r}', t). \tag{11.100}$$

With the aid of Eq. 4.189, one has

$$E_s(\mathbf{r}, t) = \left(\frac{\omega_L}{c}\right)^2 \frac{e^{ikr}}{r} E_0 e^{-i\omega_L t} \int_V d^3r'\, \alpha(\mathbf{r}', t)\, e^{i\mathbf{Q}\cdot\mathbf{r}'} \tag{11.101}$$

for the case of scattering in the VV-geometry (ω_L represents the laser frequency). We can now make the substitution

$$\alpha(\mathbf{r}', t) = \sum_{i=1}^{N} \alpha_i\, \delta\left[\mathbf{r}' - \mathbf{R}_i(t)\right], \tag{11.102}$$

where α_i and $\mathbf{R}_i(t)$ are the effective polarizability and instantaneous position of the ith particle. The time dependence of the scattered field then becomes

$$E_s(t) = \sum_{i=1}^{N} A_i e^{i[\mathbf{Q}\cdot\mathbf{R}_i(t) - \omega_L t]}. \tag{11.103}$$

A_i, or the *scattering amplitude* of the ith particle, is

$$A_i = \left(\frac{\omega_L}{c}\right)^2 \frac{e^{ikr}}{r} E_0 \alpha_i. \tag{11.104}$$

Implicit in Eq. 11.103 is the assumption that the Rayleigh-Gans-Debye (RGD) approximation is valid for the scattering system. This is because the above simple expression for the scattered field is derived from Eq. 4.188, which (as was discussed in Sect. 4.4) is based on the RGD approximation. Let us now further assume that the particles are identical (i.e., no polydispersity) and that they move independently. It then follows that the scattered-field correlation function is given by

$$g^{(1)}(\tau) \equiv \frac{\langle E_s^*(0)\, E_s(\tau)\rangle}{\langle |E_s|^2 \rangle} = e^{-i\omega_L \tau} F_s(\mathbf{Q}, \tau), \tag{11.105}$$

where

$$F_s(\mathbf{Q},\tau) = \left\langle e^{i\mathbf{Q}\cdot\mathbf{R}(\tau)} e^{-i\mathbf{Q}\cdot\mathbf{R}(0)} \right\rangle . \tag{11.106}$$

Thus, from Eq. 11.99, the photon correlation function becomes

$$\langle n(0)\, n(\tau)\rangle = \langle n\rangle^2 \left[1 + f(A)\left|F_s(\mathbf{Q},\tau)\right|^2\right] . \tag{11.107}$$

In Chapter 12 we will see that the function $F_s(\mathbf{Q},\tau)$ plays a central role in scattering problems in general, and can be identified as the so-called *self intermediate scattering function*. Observe, from Eq. 6.96, that the temporal Fourier transform of $F_s(\mathbf{Q},\tau)$ is what we have previously referred to as the *self dynamic structure factor*, $S_s(\mathbf{Q},\omega)$, and, for the case of independently moving particles, it is proportional to the power spectrum of the quasi-elastically scattered light (refer back to Eq. 11.94). The precise functional form of the scattering function is determined by the type of motion exhibited by an individual particle, for example, whether the motion is diffusive or motile. The full significance of $F_s(\mathbf{Q},\tau)$ will be discussed in Chapter 12. For now, however, we illustrate an application in the example below:

Example 11.5 Light Scattering from Motile Bacteria. As mentioned previously, photon correlation spectroscopy has been used to measure the motility of *E. coli* bacteria in solution. Here one has a situation where the validity of the RGD approximation comes into question because the size of a bacterial cell and the wavelength of light are both on the order of a micron. Fortunately, however, numerical calculations [105] have demonstrated that the use of the RGD approximation to calculate the scattered field autocorrelation function produces results that fall within 10% of the true correlation function values, making the use of Eq. 11.106 for the scattering function justified. For simplicity, this example will only address the case where no chemoattractants are present, and the motion is isotropic.

Begin by rewriting Equation 11.106 for the self intermediate scattering function in the form

$$F_s(\mathbf{Q},\tau) = \left\langle e^{i\mathbf{Q}\cdot[\mathbf{R}(\tau)-\mathbf{R}(0)]} \right\rangle . \tag{11.108}$$

Although this step may appear to be obvious, it is, in fact, not an automatic one. The reason is that, rigorously speaking, $\mathbf{R}(\tau)$ and $\mathbf{R}(0)$ are non-commuting quantum-mechanical operators. If, between changes in direction, a bacterium (of mass M) swims with momentum $\mathbf{P}(0)$, then $\mathbf{R}(\tau) = \mathbf{R}(0) + \tau\mathbf{P}(0)/M$; since $\mathbf{R}(0)$ and $\mathbf{P}(0)$ do not commute, then neither do $\mathbf{R}(\tau)$ and $\mathbf{R}(0)$. As a consequence, there is a quantum-mechanical factor that should multiply the expression in Eq. 11.108. We will see this factor show up in Chapter 12. In the present situation, however, the mass M of a bacterial cell is large enough to make the quantum factor essentially unity.

Now, if $\mathbf{V}$ represents the velocity of a swimming cell between direction changes, then $\mathbf{R}(\tau) - \mathbf{R}(0) = \mathbf{V}\tau$, and

$$F_s(\mathbf{Q},\tau) = \left\langle e^{i\mathbf{Q}\cdot\mathbf{V}\tau} \right\rangle = \int e^{i\mathbf{Q}\cdot\mathbf{V}\tau} P(\mathbf{V})\, d\mathbf{V}, \tag{11.109}$$

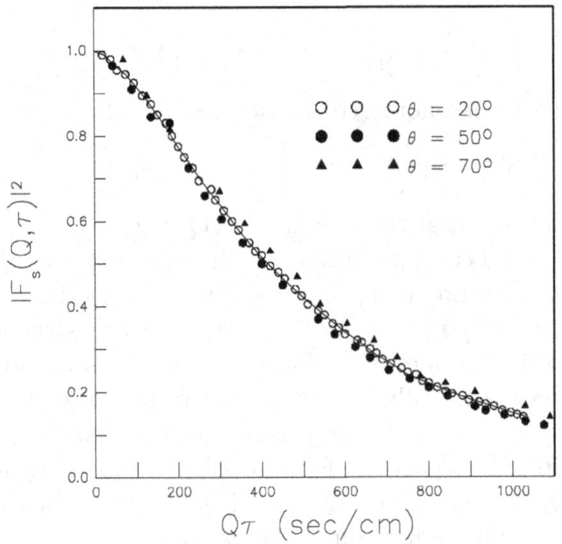

Figure 11.15 Square of the self intermediate scattering function vs. $Q\tau$ for isotropically moving E. coli bacteria. Data from three different scattering angles collapses onto a single curve.

where the integration is over the velocity distribution $P(\mathbf{V})$ of the bacteria. For an isotropic distribution where $P(\mathbf{V}) = P(V)$, the angular integration is easily performed and we focus on the speed distribution

$$P_s(V) = 4\pi V^2 P(V). \tag{11.110}$$

The result is that $F_s(\mathbf{Q},\tau)$ is found to depend only on the single scaled variable $x = Q\tau$ according to

$$F_s(x) = \int_0^\infty P_s(V)\left(\frac{\sin xV}{xV}\right) dV. \tag{11.111}$$

Equation 11.111 can now be inverted [106] to give the distribution of bacterial speeds:

$$P_s(V) = \frac{2V}{\pi}\int_0^\infty xF_s(x)\sin xV\,dx. \tag{11.112}$$

For *E. coli* undergoing isotropic motion, Nossal *et al.* [96] measured the (clipped) photocount correlation function at three different scattering angles. Equation 11.107 predicts that aside from a constant background, the correlation function in each case is proportional to $|F_s(\mathbf{Q},\tau)|^2$. The three curves appear different when plotted against the delay time τ, however, as shown in Fig. 11.15, they collapse to

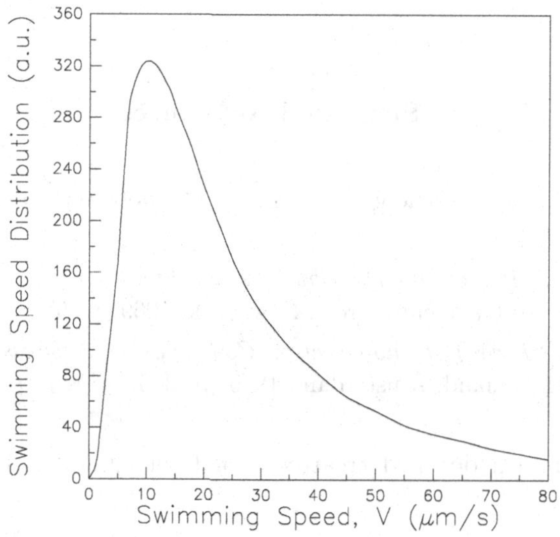

Figure 11.16 Distribution of swimming speeds for isotropically moving E. coli bacteria.

a single curve when plotted against the scaled variable x. This fact confirms that the bacteria are indeed moving isotropically in the solution. The swimming speed distribution can now be extracted by using Eq. 11.112, producing the profile shown in Fig. 11.16.

Suggested References

These two references review the general theory of photoelectron counting statistics:

[a] F. T. Arecchi, *"Photocount Distributions and Field Statistics"*, in *Quantum Optics*, ed. R. J. Glauber (Academic Press, New York, 1969) p. 57.

[b] C. L. Mehta, *"Theory of Photoelectron Counting"*, in *Progress in Optics*, Vol. 8, ed. E. Wolf (North-Holland, Amsterdam, 1970) p. 373.

Readers interested in a detailed treatment of laser-light statistics near and above threshold are referred to

[c] J. A. Armstrong and A. W. Smith, *"Experimental Studies of Intensity Fluctuations in Lasers"*, in *Progress in Optics*, Vol. 6, ed. E. Wolf (North-Holland, Amsterdam, 1967) p. 211.

[d] H. Risken, *"Statistical Properties of Laser Light"*, in *Progress in Optics*, Vol. 8, ed. E. Wolf (North-Holland, Amsterdam, 1970) p. 239.

The statistics of squeezed light is presented in

[e] O. Hirota, ed., *Squeezed Light* (Elsevier Science Publishers, Amsterdam, 1992).

The following references detail the basic principles and instrumentation associated with photon correlation spectroscopy as applied to quasi-elastic light scattering experiments. Included are descriptions of specific experiments on macromolecular and biological systems:

[f] B. J. Berne and R. Pecora, *Dynamic Light Scattering: With Applications to Chemistry, Biology, and Physics* (John Wiley and Sons, New York, 1976).

[g] B. Chu, *Laser Light Scattering: Basic Principles and Practice*, 2nd ed. (Academic Press, San Diego, 1991).

[h] R. Pecora, ed., *Dynamic Light Scattering: Applications of Photon Correlation Spectroscopy* (Plenum Press, New York, 1985).

[i] W. Brown, ed., *Dynamic Light Scattering: The Method and Some Applications* (Oxford University Press, Oxford, 1993).

Problems

1. Statistical considerations enter into photon counting both because of fluctuations inherent in the light field and because light detection is itself a quantum process that is probabilistic in nature. Consider a beam of coherent light characterized by fluctuations obeying a Poisson distribution, i.e.,

$$p(n) = \frac{e^{-\langle n \rangle}}{n!} \langle n \rangle^n .$$

Here, $p(n)$ represents the probability for counting n photons (during some specified time interval), with the assumption that there are no counting fluctuations due to the actual detection process. In other words, the quantum efficiency of the detector is unity. $\langle n \rangle$ is the average number of counts for this ideal detector. Let us now consider the effect of having a real photoelectric detector with quantum efficiency ϵ:

   (a) One can represent the process of detection by a binomial distribution (why?). Suppose one assumes that exactly n photons strike the detector face. If the detector has quantum efficiency ϵ, express as a binomial distribution the probability, $p_n(C)$, of registering exactly C counts.

   (b) What is the joint probability $p(C, n)$ of having both n photons strike the detector and registering C counts?

   (c) Show that $\langle C \rangle$, the average number of counts registered by the detector reduces to simply $\langle C \rangle = \epsilon \langle n \rangle$.

2. To a good approximation, one can consider the light emitted by an incandescent lamp to be thermal light (i.e., blackbody radiation). Suppose that one passes this light through a monochromator, producing a beam of nearly monochromatic light, and detects it with a photomultiplier tube. One repeatedly counts the number of photons detected per second, getting a series of counts n_1, n_2, n_3, ... One can then compute the average counts $\langle n \rangle$ and plot out the photocount distribution $p(n)$, i.e., the probability of getting n counts in a measurement. One can predict these quantities theoretically by assuming that one can regard the monochromatic light to be a single mode of radiation of frequency ω in equilibrium with a thermal reservoir at temperature T.

   (a) Write down the density operator for this beam of light.

   (b) Compute $\langle n \rangle$, and express in terms of ω and T.

   (c) Give the explicit expression for $p(n)$, and sketch as a function of n.

3. For a light signal containing low-frequency noise, such as a linear drift, the photon correlation function $C(t_i) = \langle n(t_j)\, n(t_{j+i}) \rangle$ is ill-defined, i.e., it becomes not only a function of t_i, but a function of t_j as well. A quantity that remains well-defined in the presence of drift is the "*photon structure function*," defined by

$$D(t_i) = \left\langle \left[n(t_j) - n(t_{j+i})\right]^2 \right\rangle .$$

   (a) For a stationary signal, determine the general relationship between the photon structure function and the photon correlation function.

   (b) Quasielastic light scattering by a diffusing particle produces a self intermediate scattering function, $F_s(Q,\tau)$, that is a simple decaying exponential. Sketch the forms of both the photon correlation and the photon structure functions.

4. Equation 11.111 gives the self intermediate scattering function for motile bacteria with an isotropic speed distribution, $P_s(V)$. The power spectrum of the scattered light is proportional to the self dynamic structure factor, $S_s(Q,\omega)$, which is simply the Fourier transform of the self intermediate scattering function, i.e.,

$$S_s(Q,\omega) = \frac{1}{2\pi} \int_{-\infty}^{+\infty} d\tau\, e^{-i\omega\tau} F_s(Q,\tau) .$$

   Show that

$$S_s(Q,\omega) = \frac{1}{2Q} \int_{\omega/Q}^{\infty} \frac{P_s(V)}{V} dV$$

   and hence that

$$Q^2 \frac{d}{d\omega} S_s(Q,\omega) = -\frac{1}{2} \left. \frac{P_s(V)}{V} \right|_{V=\omega/Q} .$$

   This result shows that, in the case of isotropic motion, the slope of the self dynamic structure factor at each frequency uniquely gives the value of $P_s(V)/V$ at the speed $V = \omega/Q$.

Chapter 12
DYNAMIC STRUCTURE FACTORS

The concept of a *dynamic structure factor* was first introduced in Chapter 6 during our discussion of thermal neutron scattering. At that time, we defined the so-called *self* dynamic structure factor, $S_s(\mathbf{Q},\omega)$, and *full* dynamic structure factor, $S(\mathbf{Q},\omega)$. In Chapters 9–11, we have also seen these structure factors and related functions appear in the context of light, x-ray, and NMR scattering. They are, in fact, the basic signature of scattering experiments in general.

The structure factors contain information pertaining to spatial and temporal correlations in the scattering medium. The specific form of the self dynamic structure factor is governed by the characteristic motion of a single particle in the medium, and can be determined from incoherent (neutron) scattering measurements. On the other hand, the full dynamic structure factor reflects the interparticle correlations set up between particles in the system. It can be obtained from measurements of the coherent scattering.

The intent of this chapter is to calculate the dynamic structure factors for a few basic types of systems, namely, a perfect gas, a simple liquid, and a harmonic oscillator. We will also derive some general properties common to all dynamic structure factors for systems in thermal equilibrium.

12.1 Dynamic Structure Factors for Simple Fluid Systems

In this section, the calculation of $S_s(\mathbf{Q},\omega)$ and $S(\mathbf{Q},\omega)$ for a simple homogeneous, isotropic fluid will be performed. Consider the fluid to contain N identical particles in thermal equilibrium.

12.1.1 The Self Dynamic Structure Factor

The basic form for $S_s(\mathbf{Q},\omega)$ was given previously as Eq. 6.94. We rewrite it here, however, in a somewhat more convenient way, namely,

$$S_s(\mathbf{Q},\omega) = \frac{1}{2\pi}\int_{-\infty}^{+\infty} dt\, e^{-i\omega t} F_s(\mathbf{Q},t)\,, \tag{12.1}$$

where we introduce the *self intermediate scattering function*

$$F_s(\mathbf{Q},t) = \left\langle \frac{1}{N}\sum_{\ell=1}^{N} e^{-i\mathbf{Q}\cdot\mathbf{R}_\ell(0)} e^{i\mathbf{Q}\cdot\mathbf{R}_\ell(t)} \right\rangle . \tag{12.2}$$

$\mathbf{R}_\ell(t)$ is the position vector of the ℓth particle* at time t and the brackets $\langle\ \rangle$ represent an ensemble average over the system. Equation 12.1 shows that the dynamic structure factor is the temporal Fourier transform of the intermediate scattering function.

In the case of a simple fluid, the various particles are statistically equivalent, and one can make the simplification

$$F_s(\mathbf{Q},t) = \left\langle e^{-i\mathbf{Q}\cdot\mathbf{R}(0)} e^{i\mathbf{Q}\cdot\mathbf{R}(t)} \right\rangle . \tag{12.3}$$

Here, $\mathbf{R}(t)$ represents the position vector of a typical particle at time t. It now becomes convenient to introduce a *test particle density*, $n_t(\mathbf{r},t)$, and its spatial Fourier component, $n_t(\mathbf{Q},t)$, i.e.,

$$n_t(\mathbf{r},t) = \delta\left[\mathbf{r} - \mathbf{R}(t)\right] \tag{12.4}$$

and

$$n_t(\mathbf{Q},t) = \int d^3r\, e^{i\mathbf{Q}\cdot\mathbf{r}} n_t(\mathbf{r},t) = e^{i\mathbf{Q}\cdot\mathbf{R}(t)} . \tag{12.5}$$

Then, Eq. 12.5 together with Eq. 12.3 gives

$$F_s(\mathbf{Q},t) = \langle n_t(-\mathbf{Q},0)\, n_t(\mathbf{Q},t)\rangle , \tag{12.6}$$

and we see that the self intermediate scattering function is a $\mathbf{Q}$-dependent test particle time-correlation function. It satisfies the property $F_s(\mathbf{Q},t=0) = 1$.

If we wish to pursue a description of the test particle's propagation in space and time, we need to introduce the so-called *van-Hove space-time self correlation function*, $G_s(\mathbf{r},t)$, defined as

$$\begin{aligned} G_s(\mathbf{r},t) &= \int d^3r' \langle n_t(\mathbf{r}',0)\, n_t(\mathbf{r}+\mathbf{r}',t)\rangle \\ &= \int d^3r' \langle \delta\left[\mathbf{r}' - \mathbf{R}(0)\right] \delta\left[\mathbf{r}+\mathbf{r}' - \mathbf{R}(t)\right]\rangle . \end{aligned} \tag{12.7}$$

*A word about notation: In Chapter 6, the symbol r_ℓ was used to denote the position vector of a typical atomic nucleus in the scattering medium. In the present discussion, we want to be more general, and consider the motion of a "particle." This could mean the motion of an atom, a molecule, a molecular aggregate, or, for that matter, even a microorganism. The symbol R_ℓ is used to denote the center-of-mass of such a particle. In the general case, the observed differential scattering cross-section will depend not only on the structure factor, but also on an intraparticle form factor which is a function of the particle size, shape, composition, and orientation relative to the scattering vector (see, for example, Ref. [107]).

In general, the specified integration with respect to $\mathbf{r}'$ cannot be carried out because the position vectors, $\mathbf{R}(0)$ and $\mathbf{R}(t)$, are quantum-mechanical operators that do not commute. However, in the classical limit, this problem does not occur, and the latter integration can be carried out to give

$$G_s^{\mathrm{cl}}(\mathbf{r},t) = \langle \delta[\mathbf{r} + \mathbf{R}(0) - \mathbf{R}(t)] \rangle . \tag{12.8}$$

This expression is valid for all fluids except liquid helium and liquid hydrogen.

From the general definition of $G_s(\mathbf{r},t)$ given by Eq. 12.7, it follows that

$$G_s(\mathbf{r},t=0) = \delta(\mathbf{r}) . \tag{12.9}$$

In addition, by integrating both sides of Eq. 12.7 with respect to the space coordinate $\mathbf{r}$, we have the relation

$$\int d^3r\, G_s(\mathbf{r},t) = 1. \tag{12.10}$$

Although $G_s(\mathbf{r},t)$ is in general a quantum-mechanical complex function, in the classical limit it reduces to a function that is real. In that case, the above two relations taken together allows us to attach a physical meaning to the classical van Hove self correlation function:

> $G_s^{\mathrm{cl}}(\mathbf{r},t)\, d^3r$ is the conditional probability of finding the test particle at position $\mathbf{r}$ (within volume d^3r) at time t, given that the particle was located at the origin at $t = 0$.

At $t = 0$, $G_s^{\mathrm{cl}}(\mathbf{r},t)$ starts out as a delta-function centered at the origin. As time goes on, it spreads out in a way that reflects the particle's characteristic movement away from the origin. This function is particularly useful for studying the self-diffusion of a particle.

Classically, the self intermediate scattering function can be written as

$$F_s^{\mathrm{cl}}(\mathbf{Q},t) = \left\langle e^{-i\mathbf{Q}\cdot[\mathbf{R}(0)-\mathbf{R}(t)]} \right\rangle . \tag{12.11}$$

Therefore, using Eq. 12.8, it is easy to see that

$$F_s^{\mathrm{cl}}(\mathbf{Q},t) = \int d^3r\, e^{i\mathbf{Q}\cdot\mathbf{r}} G_s^{\mathrm{cl}}(\mathbf{r},t) . \tag{12.12}$$

Hence, $F_s^{\mathrm{cl}}(\mathbf{Q},t)$ and $G_s^{\mathrm{cl}}(\mathbf{r},t)$ form a spatial Fourier-transform pair. Consequently, we also have

$$G_s^{\mathrm{cl}}(\mathbf{r},t) = \frac{1}{(2\pi)^3} \int d^3Q\, e^{-\mathbf{Q}\cdot\mathbf{r}} F_s^{\mathrm{cl}}(\mathbf{Q},t) . \tag{12.13}$$

Even though the validity of Eqs. 12.12 and 12.13 appears to hinge on the assumption that we are using classical functions, these relations are in fact universally true and hold for quantum-mechanical systems as well [108]. Hence, in their most general form, the latter pair of relations becomes

$$F_s(\mathbf{Q},t) = \int d^3r\, e^{i\mathbf{Q}\cdot\mathbf{r}} G_s(\mathbf{r},t) \tag{12.14}$$

and

$$G_s(\mathbf{r},t) = \frac{1}{(2\pi)^3} \int d^3Q\, e^{-\mathbf{Q}\cdot\mathbf{r}} F_s(\mathbf{Q},t). \tag{12.15}$$

Let us now calculate $F_s(\mathbf{Q},t)$and $G_s(\mathbf{r},t)$, along with the self dynamic structure factor, $S_s(\mathbf{Q},\omega)$, in the limit of both short time and long time:

Perfect Gas (or short time) Limit

We begin by performing a quantum-mechanical calculation of $F_s(\mathbf{Q},t)$, as given by Eq. 12.3, and treat the position vectors, $\mathbf{R}(0)$ and $\mathbf{R}(t)$, as quantum-mechanical operators. The product of exponentials is handled by recalling the Baker-Hausdorff theorem (previously Eq. 5.128)

$$e^{\hat{A}}e^{\hat{B}} = e^{\hat{A}+\hat{B}+\frac{1}{2}[\hat{A},\hat{B}]} \tag{12.16}$$

with

$$\hat{A} = -i\mathbf{Q}\cdot\hat{\mathbf{R}}(0) \tag{12.17}$$

$$\hat{B} = i\mathbf{Q}\cdot\hat{\mathbf{R}}(t) \tag{12.18}$$

$$\left[\hat{A},\hat{B}\right] = \left[-i\mathbf{Q}\cdot\hat{\mathbf{R}}(0), i\mathbf{Q}\cdot\hat{\mathbf{R}}(t)\right]. \tag{12.19}$$

For t much shorter than the characteristic collision time in the fluid, the test particle has essentially no chance of colliding with neighboring particles, and can therefore be treated as free. It is then valid to use the linear relation

$$\hat{\mathbf{R}}(t) = \hat{\mathbf{R}}(0) + \frac{\hat{\mathbf{P}}(0)}{M}t, \tag{12.20}$$

where $\hat{\mathbf{P}}$ is the linear momentum of the particle and M is its mass. The commutator then simplifies to

$$\left[\hat{A},\hat{B}\right] = \left[-i\mathbf{Q}\cdot\hat{\mathbf{R}}(0), i\frac{t}{M}\mathbf{Q}\cdot\hat{\mathbf{P}}(0)\right] = i\frac{\hbar t}{M}Q^2. \tag{12.21}$$

Then, from Eqs. 12.3 and 12.16, the resulting scattering function is

$$F_s(\mathbf{Q},t) = \left\langle e^{it\mathbf{Q}\cdot\mathbf{P}/M} e^{i\hbar t Q^2/2M}\right\rangle = e^{i\hbar t Q^2/2M}\left\langle e^{i\mathbf{Q}\cdot\mathbf{V}t}\right\rangle. \tag{12.22}$$

To compute the equilibrium statistical average represented by the bracket, we use the Maxwellian distribution function

$$P(\mathbf{V}) = \left(2\pi V_0^2\right)^{-3/2} e^{-V^2/2V_0^2}, \tag{12.23}$$

where $V_0^2 = k_B T/M$. We find

$$\begin{aligned}\left\langle e^{i\mathbf{Q}\cdot\mathbf{V}t}\right\rangle &= \int d^3V\, P(\mathbf{V})\, e^{i\mathbf{Q}\cdot\mathbf{V}t} \\ &= 4\pi \int_0^\infty dV\, P(\mathbf{V})\, V^2 \frac{\sin(QVt)}{QVt} \\ &= e^{-Q^2V_0^2t^2/2}.\end{aligned} \tag{12.24}$$

The result of combining Eqs. 12.22 and 12.24 gives

$$F_s(\mathbf{Q},t) = e^{i\hbar t Q^2/2M} e^{-Q^2V_0^2t^2/2} = e^{-\frac{1}{2}Q^2 W(t)}, \tag{12.25}$$

where we have introduced a complex *width function*

$$W(t) = V_0^2 t\left(t - \frac{i\hbar}{k_B T}\right) \tag{12.26}$$

that is quadratic in time. According to Eqs. 12.1 and 12.15, $S_s(\mathbf{Q},\omega)$ and $G_s(\mathbf{r},t)$ are obtained by Fourier transforming the intermediate scattering function. The results are

$$S_s(\mathbf{Q},\omega) = \left(2\pi Q^2V_0^2\right)^{-1/2} \exp\left[-\frac{\left(\omega - \dfrac{\hbar Q^2}{2M}\right)^2}{2Q^2V_0^2}\right] \tag{12.27}$$

and

$$G_s(\mathbf{r},t) = \left[2\pi W(t)\right]^{-3/2} e^{-r^2/2W(t)}. \tag{12.28}$$

Compare the quantum-mechanical result in the limit of short time to the one obtained from a purely classical calculation. From Eq. 12.11, the classical scattering function takes on the simple Gaussian form

$$F_s^{\mathrm{cl}}(\mathbf{Q},t) = \left\langle e^{-i\mathbf{Q}\cdot[\mathbf{R}(0)-\mathbf{R}(t)]}\right\rangle = \left\langle e^{i\mathbf{Q}\cdot\mathbf{V}t}\right\rangle = e^{-Q^2V_0^2t^2/2}. \tag{12.29}$$

The corresponding dynamic structure factor is then

$$S_s^{\mathrm{cl}}(\mathbf{Q},\omega) = \left(2\pi Q^2V_0^2\right)^{-1/2} e^{-\omega^2/2V_0^2Q^2}. \tag{12.30}$$

This represents a Doppler-broadened Gaussian line. The corresponding van Hove correlation function is also Gaussian, and is given by

$$G_s^{\mathrm{cl}}(\mathbf{r},t) = \left(2\pi V_0^2 t^2\right)^{-3/2} e^{-r^2/2V_0^2t^2}. \tag{12.31}$$

We see that the width function is now

$$W(t) = V_0^2 t^2, \tag{12.32}$$

which tells us that the test particle is moving away from the origin with speed V_0.

A direct comparison of Eqs. 12.27 and 12.30 gives the following interesting relationship:

$$S_s(\mathbf{Q},\omega) = e^{\hbar\omega/2k_BT} e^{-\hbar^2Q^2/8Mk_BT} S_s^{\rm cl}(\mathbf{Q},\omega). \tag{12.33}$$

This shows that the quantum-mechanical expression for $S_s(\mathbf{Q},\omega)$ is obtained from the corresponding classical one simply by multiplying by two exponential factors. The first factor, $\exp(\hbar\omega/2k_BT)$, is known as the *detailed balance factor*—it produces an asymmetry in the quantum-mechanical structure factor, whereas the classical one is an even function of ω, as will be shown in Section 12.3. The second factor, $\exp(-\hbar^2Q^2/8Mk_BT)$, can also be written as $\exp(-E_R/4k_BT)$, where $E_R = \hbar^2Q^2/2M$ is the recoil energy of the target particle (see Example 12.3). Hence, this exponential is known as the *recoil factor*. Equation 12.33 is exactly true only in the perfect gas (or short time) limit, however, it is also approximately valid, in general, to first order in $\hbar$.

Example 12.1 Neutron Scattering from a Resting, Free Nucleus. When dealing with the basic problem of neutron thermalization, one often encounters a quantity denoted by $\sigma(E,E',\theta)$, which is the differential cross-section for a neutron (mass m_n) with incoming energy E to be scattered into angle θ with a final energy E' following a collision with some target nucleus (mass M). Actually, this quantity is identical to the double-differential cross-section defined earlier in Section 6.4. Inspection of Eq. 6.85 shows that the neutron scattering cross-section from a single nucleus can be written as

$$\sigma(E,E',\theta) \equiv \frac{d^2\sigma}{d\Omega\, dE'} = \frac{\sigma_b}{4\pi\hbar}\left(\frac{k'}{k}\right) S_s(\mathbf{Q},\omega), \tag{12.34}$$

where $\mathbf{Q} = \mathbf{k} - \mathbf{k}'$, $\hbar\omega = E - E'$, and $Q^2 = k^2 + k'^2 - 2kk'\cos\theta$. The quantity σ_b is the total cross-section that would be presented by a fixed target nucleus, also known as the *bound-atom cross-section*. It is given by $\sigma_b = 4\pi\overline{b^2}$, where b is the bound scattering length of the nucleus.

Assume that the target atom has a mass number $A = M/m_n$, and it is at rest prior to the collision. The intermediate scattering function for a resting particle is given by Eq. 12.25 to be

$$F_s(\mathbf{Q},t) = e^{i\hbar t Q^2/2M}, \tag{12.35}$$

so the dynamic structure factor is simply

$$S_s(\mathbf{Q},\omega) = \frac{1}{2\pi}\int_{-\infty}^{+\infty} e^{-i\omega t} e^{i\hbar t Q^2/2M} = \delta\left(\omega - \frac{\hbar Q^2}{2M}\right). \tag{12.36}$$

Substituting into Eq. 12.34, we obtain

$$\begin{aligned}\sigma(E,E',\theta) &= \frac{\sigma_b}{4\pi\hbar}\left(\frac{k'}{k}\right)\delta\left(\omega-\frac{\hbar Q^2}{2M}\right)\\ &= \frac{\sigma_b}{4\pi}\left(\frac{k'}{k}\right)\delta\left(E-E'-\frac{E+E'-2\sqrt{EE'}\cos\theta}{A}\right).\end{aligned} \tag{12.37}$$

We shall now calculate the energy transfer cross-section $\sigma(E,E')$ by integrating out the angular dependence of $\sigma(E,E',\theta)$. Rather than integrating over all solid angles, $d\Omega = 2\pi\sin\theta\, d\theta$, from $\theta = 0$ to π, it is better to transform to the variable x defined as

$$x = E - E' - \frac{E+E'-2\sqrt{EE'}\cos\theta}{A}. \tag{12.38}$$

The integral then takes the form

$$\sigma(E,E') \equiv \int_0^\pi \sigma(E,E',\theta)\, 2\pi\sin\theta\, d\theta = \frac{A\sigma_b}{4E}\int_{x_-}^{x_+} dx\,\delta(x) \tag{12.39}$$

with integration limits

$$x_\pm = E - E' - \frac{\left(\sqrt{E}\mp\sqrt{E'}\right)^2}{A}. \tag{12.40}$$

The cross-section is non-zero only for a certain range of final neutron energies. This occurs when the lower limit of the integral in Eq. 12.39 is less than zero. The result, which is displayed in Fig. 12.1, is

$$\sigma(E,E') = \begin{cases} \sigma_f\dfrac{(A+1)^2}{4AE}, & \text{for } \left(\dfrac{A-1}{A+1}\right)^2 E \le E' \le E \\ 0, & \text{for all other } E', \end{cases} \tag{12.41}$$

where we have introduced the *free-atom cross-section*, σ_f, given by

$$\sigma_b = \left(1+\frac{1}{A}\right)^2 \sigma_f. \tag{12.42}$$

Our calculation shows that within the range of the final energies allowed by the kinematics, the probability of scattering into energy E' is constant and equal to $(A+1)^2/4AE$. When $\theta = 0$, the neutron is forward scattered, and $E' = E$. When $\theta = \pi$, the neutron is backscattered, and $E' = \left(\dfrac{A-1}{A+1}\right)^2 E$. Notice also that the neutron slowing-down cross-section is inversely proportional to the incident energy E.

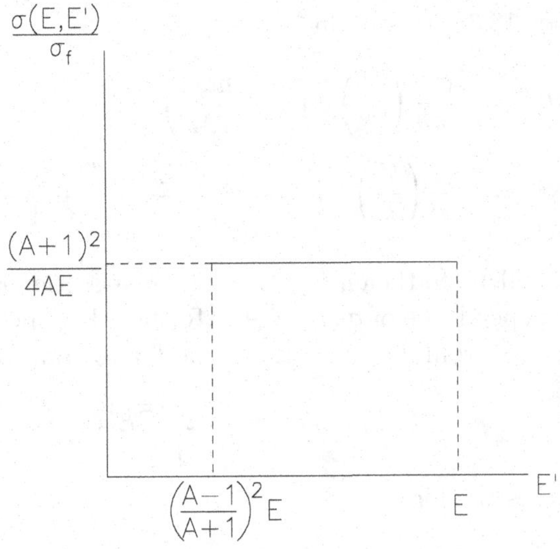

Figure 12.1 Cross-section for scattering from neutron energy E to energy E' off of a free target nucleus (mass number A) initially at rest. σ_f is the free-atom cross-section of the target nucleus.

Consider next the calculation of the angular differential cross-section:

$$\frac{d\sigma}{d\Omega} \equiv \int \sigma\left(E, E', \theta\right) dE' = \frac{\sigma_b}{4\pi} \int \left(\frac{k'}{k}\right) \delta\left(\hbar\omega - \frac{\hbar^2 Q^2}{2M}\right) dE'. \tag{12.43}$$

To integrate the right hand side, we change to the variable $x = k'/k$ so that

$$dE' = \frac{\hbar^2}{m_n} k'\, dk' = 2Ex\, dx \tag{12.44}$$

and

$$\hbar\omega - \frac{\hbar^2 Q^2}{2M} = E\left(1 - x^2 - \frac{1 + x^2 - 2x\mu}{A}\right), \tag{12.45}$$

where $\mu = \cos\theta$. The integral then reduces to

$$\frac{d\sigma}{d\Omega} = \frac{\sigma_b}{4\pi} \int 2x^2 \delta\left(1 - x^2 - \frac{1 + x^2 - 2x\mu}{A}\right) dx. \tag{12.46}$$

We now make use of the following property of the delta-function:

$$\delta\left[f\left(x\right)\right] = \frac{\delta\left(x - x_0\right)}{\left|f'\left(x_0\right)\right|}. \tag{12.47}$$

Here, x_0 is the positive root of the equation $f(x_0) = 0$. After going through some algebra, the final result for the angular differential cross-section is

$$\frac{d\sigma}{d\Omega} = \frac{\sigma_b}{4\pi\left(1+\frac{1}{A}\right)^2}\left[\frac{2\mu}{A} + \frac{\left(1-\frac{1}{A^2}+\frac{2\mu^2}{A^2}\right)}{\left(1-\frac{1}{A^2}+\frac{\mu^2}{A^2}\right)^{1/2}}\right]. \tag{12.48}$$

We now make a number of observations: First of all, in the limit of an infinite-mass target nucleus ($A \to \infty$), the differential cross-section reduces to the bound-atom value of $\sigma_b/4\pi$, as it should. Secondly, the cross-section is biased toward the forward direction, as illustrated in Fig. 12.2. The effect is most dramatic in the case of hydrogen ($A = 1$), for which the cross-section formula reduces to

$$\frac{d\sigma}{d\Omega} = \begin{cases} \frac{\sigma_b}{4\pi}\cos\theta, & \text{for } \theta \leq \pi/2 \\ 0, & \text{for } \theta \geq \pi/2. \end{cases} \tag{12.49}$$

The average cosine of the scattering angle can be calculated as

$$\bar{\mu} = \frac{1}{\sigma_f}\int_{-1}^{+1} \mu \frac{d\sigma}{d\Omega} 2\pi\, d\mu = \frac{2}{3A}. \tag{12.50}$$

The total scattering cross-section is given by

$$\sigma(E) = 2\pi \int \frac{d\sigma}{d\Omega}\, d\mu. \tag{12.51}$$

The integration is most readily performed by using the expression provided by Eq. 12.46. Then,

$$\begin{aligned} \sigma(E) &= 2\pi\left(\frac{\sigma_b}{4\pi}\right)\int 2x^2 dx \int \delta\left(1 - x^2 - \frac{1+x^2-2x\mu}{A}\right) d\mu \\ &= 2\pi\left(\frac{\sigma_b}{4\pi}\right)\int 2x^2 dx \left(\frac{A}{2x}\right) = \frac{\sigma_b A}{2}\int_{x_1}^{x_2} x\, dx. \end{aligned} \tag{12.52}$$

From Eq. 12.41, the lower and upper integration limits are seen to be

$$x_1 = \frac{A-1}{A+1} \quad \text{and} \quad x_2 = 1, \tag{12.53}$$

so the total cross-section becomes

$$\sigma(E) = \left(1 + \frac{1}{A}\right)^{-2} \sigma_b. \tag{12.54}$$

Since we assumed from the beginning that the nucleus is free and at rest, $\sigma(E)$ should be identified as the free-atom scattering cross-section, σ_f. This result is then identical to that of Eq. 12.42.

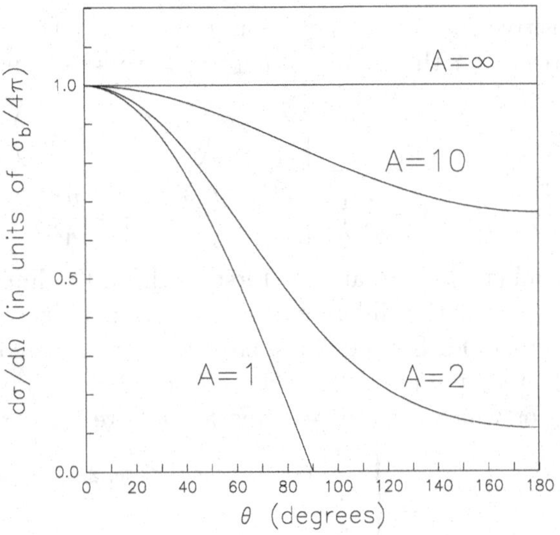

Figure 12.2 Angular differential cross-section for neutron scattering from a free nucleus initially at rest for different values of the mass number (A) of the target nucleus. Cross-sections are plotted in units of $\sigma_b/4\pi$, where σ_b is the bound-atom cross-section of the target nucleus.

Diffusion (or long time) Limit

In the limit of long time, the test particle has a chance to collide many times with the surrounding particles, and hence the net displacement is describable by a stochastic random walk or diffusion. The classical part of the intermediate scattering function may then be written as

$$F_s^{\mathrm{cl}}(\mathbf{Q},t) = \left\langle e^{i\mathbf{Q}\cdot[\mathbf{R}(t)-\mathbf{R}(0)]}\right\rangle = \left\langle e^{iQ\int_0^t V_x(t')dt'}\right\rangle, \tag{12.55}$$

where we have taken the vector $\mathbf{Q}$ to be along the x-direction. Because the number of collisions is so large, the particle undergoes a Brownian motion, and we may approximate $V_x(t)$ as a Gaussian random process. The following general theorem developed by R. Kubo for any stationary Gaussian random process, $g(t)$, then applies:

$$\left\langle e^{i\int_0^t g(t')dt'}\right\rangle = e^{-\int_0^t d\tau\,(t-\tau)\langle g(0)g(\tau)\rangle}. \tag{12.56}$$

Taking $g(t) = QV_x(t)$, we get

$$F_s^{\mathrm{cl}}(\mathbf{Q},t) = e^{-Q^2\int_0^t d\tau\,(t-\tau)\langle V_x(0)V_x(\tau)\rangle} \tag{12.57}$$

or

$$F_s^{\mathrm{cl}}(\mathbf{Q},t) = e^{-\frac{1}{2}Q^2W(t)}, \tag{12.58}$$

where the width function is

$$W(t) = 2\int_0^t d\tau\,(t-\tau)\,\langle V_x(0)\,V_x(\tau)\rangle = \int_0^t dt_1 \int_0^t dt_2\,\langle V_x(t_1)\,V_x(t_2)\rangle \cdot$$
$$= \left\langle [R_x(t) - R_x(0)]^2 \right\rangle = \left\langle [\Delta R_x(t)]^2 \right\rangle . \qquad (12.59)$$

We see that, classically, the width function corresponds to the mean-square displacement of the test particle. In the long time limit, the diffusion coefficient, D, of the Brownian motion enters through

$$W(t) = \left\langle [\Delta R_x(t)]^2 \right\rangle = 2Dt. \qquad (12.60)$$

Thus, the intermediate scattering function and dynamic structure factor are

$$F_s^{\mathrm{cl}}(\mathbf{Q},t) = e^{-DQ^2 t} \qquad (12.61)$$

and

$$S_s^{\mathrm{cl}}(\mathbf{Q},\omega) = \frac{1}{\pi}\left[\frac{DQ^2}{\omega^2 + (DQ^2)^2}\right]. \qquad (12.62)$$

The space-time correlation function is

$$G_s^{\mathrm{cl}}(\mathbf{r},t) = [2\pi W(t)]^{-3/2}\, e^{-r^2/2W(t)}, \qquad (12.63)$$

which is the same Gaussian function of r that was found in the short time limit, but with a width function that is proportional to t, rather than t^2.

Based on the results obtained for both short and long times, it is reasonable to assume that a width function, $W(t)$, is definable for a fluid at all times such that

$$F_s^{\mathrm{cl}}(\mathbf{Q},t) = e^{-\frac{1}{2}Q^2 W(t)}. \qquad (12.64)$$

It turns out that this form for the scattering function is only an approximation in the intermediate time regime, and is known as the *Gaussian approximation* [109]. In general, the argument of the exponential contains terms involving Q^2, Q^4, Q^6, etc. However, in fluids at ordinary temperatures, the higher-order corrections to the Gaussian form given by Eq. 12.64 have less than a 10% effect. Figure 12.3 shows the evolution of the width function for an ideal gas, a diffusing particle, and a typical real liquid. Also shown is the width function for a harmonic oscillator, which will be derived in Section 12.2.

12.1.2 The Full Dynamic Structure Factor

The full dynamic structure factor and its associated intermediate scattering function (see Eq. 6.95) are defined, respectively, as

$$S(\mathbf{Q},\omega) = \frac{1}{2\pi}\int_{-\infty}^{+\infty} dt\, e^{-i\omega t} F(\mathbf{Q},t) \qquad (12.65)$$

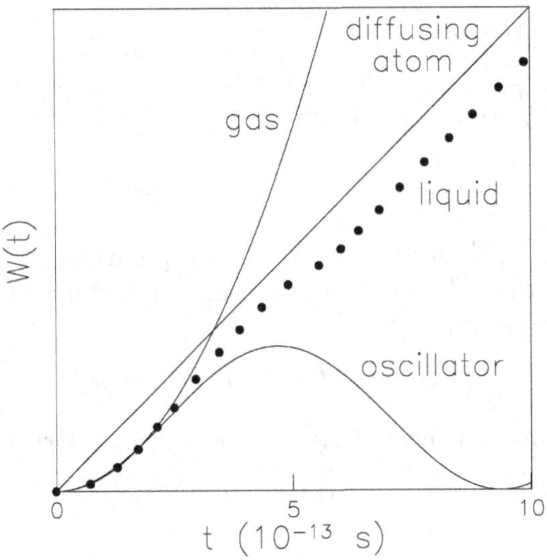

Figure 12.3 Mean-square displacement function (i.e., width function) for a gas, a diffusing particle, an oscillator, and a typical liquid.

$$F(\mathbf{Q},t)=\left\langle \frac{1}{N}\sum_{\ell\ell'} e^{-i\mathbf{Q}\cdot\mathbf{R}_\ell(0)} e^{i\mathbf{Q}\cdot\mathbf{R}_{\ell'}(t)} \right\rangle . \tag{12.66}$$

Let us introduce the real-space particle density by

$$n(\mathbf{r},t)=\sum_{\ell=1}^{N}\delta\left[\mathbf{r}-\mathbf{R}_\ell(t)\right]. \tag{12.67}$$

Its $\mathbf{Q}$-space analog, or the density fluctuation corresponding to wavevector $\mathbf{Q}$ is then

$$n(\mathbf{Q},t)=\int d^3r\, e^{i\mathbf{Q}\cdot\mathbf{r}} n(\mathbf{r},t)=\sum_{\ell=1}^{N} e^{i\mathbf{Q}\cdot\mathbf{R}_\ell(t)}. \tag{12.68}$$

Substitution into Eq. 12.66 gives the intermediate scattering function in the following form:

$$F(\mathbf{Q},t)=\frac{1}{N}\left\langle n(-\mathbf{Q},0)\, n(\mathbf{Q},t)\right\rangle . \tag{12.69}$$

Thus, $F(\mathbf{Q},t)$ gives the correlation between density fluctuations in a many-particle system and $S(\mathbf{Q},\omega)$ is the corresponding spectral density.

In neutron and x-ray diffraction experiments, the important quantity to consider is the angular differential cross-section, which is proportional to the dynamic structure factor integrated over all possible energy transfers. This quantity is denoted by $S(\mathbf{Q})$, and is called the *static structure factor*. Using Eq. 12.65, notice that

$$\begin{aligned} S(\mathbf{Q}) &\equiv \int_{-\infty}^{+\infty} d\omega\, S(\mathbf{Q},\omega) \\ &= \int_{-\infty}^{+\infty} dt\, F(\mathbf{Q},t) \left(\frac{1}{2\pi} \int_{-\infty}^{+\infty} d\omega\, e^{-i\omega t} \right) \\ &= \int_{-\infty}^{+\infty} dt\, F(\mathbf{Q},t)\, \delta(t) \\ &= F(\mathbf{Q},0). \end{aligned} \tag{12.70}$$

Thus, the static structure factor is just the $t = 0$ limit of the intermediate scattering function, which can also be expressed as

$$S(\mathbf{Q}) = \frac{1}{N} \langle n(-\mathbf{Q},0)\, n(\mathbf{Q},0) \rangle = \frac{1}{N} \left\langle \sum_{\ell\ell'} e^{-i\mathbf{Q}\cdot(\mathbf{R}_\ell - \mathbf{R}_{\ell'})} \right\rangle. \tag{12.71}$$

$S(\mathbf{Q})$ provides information about the spatial correlations between pairs of particles in the scattering system. In the case of an isotropic, homogeneous fluid, the structure factor is solely a function of Q (the magnitude of the scattering vector), so that $S(\mathbf{Q}) = S(Q)$. It is the central quantity under consideration in liquid-state theory [11][110][41, pp. 254 - 325]. $S(Q)$ can be interpreted as an equilibrium correlation function of the density fluctuation propagating with wavenumber Q.

The intermediate scattering function, $F(\mathbf{Q},t)$, is expressible in terms of a van-Hove correlation function

$$G(\mathbf{r},t) = \frac{1}{N} \int d^3r' \langle n(\mathbf{r}',0)\, n(\mathbf{r}'+\mathbf{r},t) \rangle \tag{12.72}$$

as

$$F(\mathbf{Q},t) = \int d^3r\, e^{i\mathbf{Q}\cdot\mathbf{r}} G(\mathbf{r},t). \tag{12.73}$$

In a homogeneous system, $G(\mathbf{r},t)$ is independent of $\mathbf{r}'$, so the integration can be carried out, giving the result

$$G(\mathbf{r},t) = \frac{1}{n_0} \langle n(0,0)\, n(\mathbf{r},t) \rangle, \tag{12.74}$$

where $n_0 = N/V$ is the overall number-density of particles in the system (V is the volume of the system). $G(\mathbf{r},t)$ can be identified as a space-time density-density correlation function. To ascertain the physical significance of this function, first combine Eqs. 12.67 and 12.72 to write

$$G(\mathbf{r},t) = \frac{1}{N} \sum_{\ell\ell'} \int d^3r' \langle \delta[\mathbf{r}' - \mathbf{R}_\ell(0)]\, \delta[\mathbf{r}' + \mathbf{r} - \mathbf{R}_{\ell'}(t)] \rangle. \tag{12.75}$$

In the classical limit, this reduces to

$$G^{\mathrm{cl}}(\mathbf{r},t) = \frac{1}{N}\sum_{\ell\ell'}\langle\delta\left[\mathbf{r}+\mathbf{R}_{\ell}(0)-\mathbf{R}_{\ell'}(t)\right]\rangle . \tag{12.76}$$

The physical meaning is as follows:

> $G^{\mathrm{cl}}(\mathbf{r},t)\,d^3r$ is the average density of particles at position $\mathbf{r}$ (within volume d^3r) at time t, given that a particle was located at the origin at $t=0$.

In the limit of $t=0$, the correlation function can always be decomposed as

$$G(\mathbf{r},0) = \delta(\mathbf{r}) + n_0 g(r), \tag{12.77}$$

where

$$n_0 g(r) = \frac{1}{N}\sum_{\ell\neq\ell'}\langle\delta(\mathbf{r}+\mathbf{R}_{\ell}-\mathbf{R}_{\ell'})\rangle . \tag{12.78}$$

$g(r)$ is called the *pair correlation function* or *radial distribution function* [11, p. 15]. $S(\mathbf{Q})$, the static structure factor of the system, can be calculated directly when the radial distribution function is known since

$$\begin{aligned} S(\mathbf{Q}) &= F(\mathbf{Q},0) = \int d^3r\, e^{i\mathbf{Q}\cdot\mathbf{r}} G(\mathbf{r},0) \\ &= 1 + n_0\int d^3r\, e^{i\mathbf{Q}\cdot\mathbf{r}} g(r) \\ &= (2\pi)^3 n_0\delta(\mathbf{Q}) + 1 + n_0\int d^3r\, e^{i\mathbf{Q}\cdot\mathbf{r}}\left[g(r)-1\right]. \end{aligned} \tag{12.79}$$

The delta-function is unimportant since it only contributes to scattering in the forward direction. Thus we write

$$S(\mathbf{Q}) = 1 + n_0\int d^3r\, e^{i\mathbf{Q}\cdot\mathbf{r}}\left[g(r)-1\right]. \tag{12.80}$$

In general, the various scattering/correlation functions that have been discussed are complicated quantities. However, let us consider their behavior in two limiting cases:

Large-Q (Single-Particle) Limit

We can always decompose the intermediate scattering function into two terms:

$$F(\mathbf{Q},t) = F_s(\mathbf{Q},t) + \frac{1}{N}\sum_{\ell\neq\ell'}\left\langle e^{-i\mathbf{Q}\cdot\mathbf{R}_{\ell}(0)} e^{i\mathbf{Q}\cdot\mathbf{R}_{\ell'}(t)}\right\rangle . \tag{12.81}$$

In the limit when $Ql \gg 1$, where l is the typical mean free path of a particle in the system, the second term tends to be small compared to the first. Thus, $F(\mathbf{Q},t)$ simply becomes the self intermediate scattering function, $F_s(\mathbf{Q},t)$, and both the van-Hove correlation function and dynamic structure factor reduce to the single-particle expressions derived in the last section.

Small-Q (Hydrodynamic) Limit

In the limit $Ql \ll 1$, the density fluctuation $n(\mathbf{Q}, t)$ is describable by hydrodynamic equations. Kadanoff and Martin [111] have calculated the classical dynamic structure factor to be

$$\begin{aligned} S^{\text{cl}}(\mathbf{Q},\omega) = & \frac{1}{\pi} n_0 k_B T \chi_T \left[\left(1 - \frac{1}{\gamma}\right) \frac{D_T Q^2}{\omega^2 + (D_T Q^2)^2} \right. \\ & + \frac{1}{\gamma} \frac{c^2 Q^4 \Gamma}{(\omega^2 - c^2 Q^2)^2 + (\omega Q^2 \Gamma)^2} \\ & \left. - \left(1 - \frac{1}{\gamma}\right) \frac{D_T Q^2 (\omega^2 - c^2 Q^2)}{(\omega^2 - c^2 Q^2)^2 + (\omega Q^2 \Gamma)^2} \right], \end{aligned} \tag{12.82}$$

where χ_T is the isothermal compressibility of the system. The first term gives the contribution to density fluctuations arising from heat diffusion (D_T and $\gamma = c_p/c_v$ are the thermal diffusivity and specific heat ratio of the system, respectively). The second term represents the contribution due to sound-wave propagation (c is the adiabatic sound speed). The factor Γ is the sound-wave damping constant, and is given by

$$\Gamma = \frac{1}{M n_0} \left(\frac{4}{3}\eta + \zeta \right) + D_T (\gamma - 1), \tag{12.83}$$

where η and ζ are the shear viscosity and bulk viscosity of the fluid, respectively. The third term is a cross-term produced by the other two.

12.2 Inelastic Neutron Scattering from a Harmonic Oscillator

This section focuses on the fundamental problem of scattering from a one-dimensional simple harmonic oscillator. Since, as was previously learned in Example 5.1, each longitudinal vibration mode of a one-dimenional lattice is entirely equivalent to a single harmonic oscillator, the present problem is basically akin to that of inelastic neutron scattering from phonons in a crystalline solid.† Scattering from a real crystal can then be understood by a further extension of the results to three dimensions.‡

Start out by considering the basic Hamiltonian for an oscillator of mass M having fundamental frequency ω_0:

$$\hat{H} = \frac{\hat{P}^2}{2M} + \frac{1}{2} M \omega_0^2 \hat{X}^2. \tag{12.84}$$

Recall from Example 3.1 that the oscillator is quantized by transforming from the position and momentum operators, $\hat{X}$ and $\hat{P}$, to the lowering and raising operators, $\hat{a}$ and $\hat{a}^\dagger$, via

$$\hat{X} = \sqrt{\frac{\hbar}{2M\omega_0}} \left(\hat{a} + \hat{a}^\dagger \right) \tag{12.85}$$

†See Example 5.2 for a short discussion of phonons in a three-dimensional lattice.
‡See, for example, Ref. [112].

and

$$\hat{P} = -i\sqrt{\frac{M\hbar\omega_0}{2}}\left(\hat{a} - \hat{a}^\dagger\right). \tag{12.86}$$

From the basic commutator property $\left[\hat{X}, \hat{P}\right] = i\hbar$, it follows that

$$\left[\hat{a}, \hat{a}^\dagger\right] = 1, \tag{12.87}$$

and the Hamiltonian takes the form

$$\hat{H} = \hbar\omega_0\left(\hat{a}^\dagger\hat{a} + \frac{1}{2}\right). \tag{12.88}$$

We now consider the intermediate scattering function $F(Q,t) = F_s(Q,t)$ for the oscillator, which, by definition, is

$$F_s(Q,t) = \left\langle e^{-iQX(0)}e^{iQX(t)}\right\rangle. \tag{12.89}$$

First, we use the Baker-Hausdorff theorem (Eq. 12.16) to write

$$e^{-iQ\hat{X}(0)}e^{iQ\hat{X}(t)} = e^{iQ\left[\hat{X}(t)-\hat{X}(0)\right]}e^{\frac{1}{2}Q^2\left[\hat{X}(0),\hat{X}(t)\right]}. \tag{12.90}$$

A second operator theorem [112, p. 95], known as the *Bloch identity*, states that if $\hat{A}$ is any linear combination of oscillator coordinates and momenta, then

$$\left\langle e^A\right\rangle = e^{\frac{1}{2}\left\langle A^2\right\rangle}. \tag{12.91}$$

Taking $\hat{A} = iQ\left[\hat{X}(t) - \hat{X}(0)\right]$, we get

$$\left\langle e^{iQ[X(t)-X(0)]}\right\rangle = e^{-\frac{1}{2}Q^2\left\langle[X(t)-X(0)]^2\right\rangle} = e^{\left\{-Q^2\left\langle X^2\right\rangle+Q^2\langle X(0)X(t)\rangle-\frac{1}{2}Q^2[X(0),X(t)]\right\}}. \tag{12.92}$$

So, using Eqs. 12.90 and 12.92 in Eq. 12.89, we find that

$$F_s(Q,t) = e^{-Q^2\left\langle X^2\right\rangle}e^{Q^2\langle X(0)X(t)\rangle}. \tag{12.93}$$

Thus, the intermediate scattering function consists of the product of two exponential terms: The first term, $\exp\left(-Q^2\left\langle X^2\right\rangle\right)$, is called the *Debye-Waller factor*, and is an attenuation factor arising from thermal vibrations. The second term, which involves the time-correlation function $\langle X(0)\,X(t)\rangle$, gives rise to inelastic scattering.

We now consider two cases:

1. Initial state of the oscillator is a pure state $|m\rangle$:

In this case we have

$$\begin{aligned}\langle X^2\rangle &= \langle m \mid \hat{X}^2 \mid m\rangle \\ &= \frac{\hbar}{2M\omega_0}\langle m \mid \left(\hat{a}+\hat{a}^\dagger\right)^2 \mid m\rangle \\ &= \frac{\hbar}{2M\omega_0}\left(2m+1\right), \qquad (12.94)\end{aligned}$$

where the last step follows from Eq. 3.91. The operator $\hat{X}(t)$ is given by

$$\hat{X}(t) = e^{i\hat{H}t/\hbar}\hat{X}(0)\,e^{-i\hat{H}t/\hbar} = \sqrt{\frac{\hbar}{2M\omega_0}}\left(\hat{a}e^{-i\omega_0 t} + \hat{a}^\dagger e^{i\omega_0 t}\right), \qquad (12.95)$$

so the displacement-displacement correlation function becomes

$$\begin{aligned}\langle X(0)\,X(t)\rangle &= \langle m \mid \hat{X}(0)\,\hat{X}(t) \mid m\rangle \\ &= \frac{\hbar}{2M\omega_0}\langle m \mid \left(\hat{a}+\hat{a}^\dagger\right)\left(\hat{a}e^{-i\omega_0 t}+\hat{a}^\dagger e^{i\omega_0 t}\right) \mid m\rangle \\ &= \frac{\hbar}{2M\omega_0}\left[me^{-i\omega_0 t} + (m+1)\,e^{i\omega_0 t}\right]. \qquad (12.96)\end{aligned}$$

The result for the intermediate scattering function in this case is

$$F_s(Q,t) = \exp\left[-\frac{\hbar Q^2}{2M\omega_0}\left(2m+1\right)\right]\exp\left\{\frac{\hbar Q^2}{2M\omega_0}\left[me^{-i\omega_0 t} + (m+1)\,e^{i\omega_0 t}\right]\right\}. \qquad (12.97)$$

2. The oscillator is in thermal equilibrium at temperature $T = 1/k_B\beta$:

Here, the oscillator is in a mixed state characterized by the density operator[§]

$$\hat{\rho} = \frac{e^{-\beta\hat{H}}}{\mathrm{Tr}\left(e^{-\beta\hat{H}}\right)} = \frac{e^{-\beta\hbar\omega_0\hat{a}^\dagger\hat{a}}}{\sum_m \langle m \mid e^{-\beta\hbar\omega_0\hat{a}^\dagger\hat{a}} \mid m\rangle}. \qquad (12.98)$$

Then,

$$\langle X^2\rangle = \mathrm{Tr}\left(\hat{\rho}\hat{X}^2\right) = \frac{\sum_m \langle m \mid e^{-\beta\hbar\omega_0\hat{a}^\dagger\hat{a}}X^2 \mid m\rangle}{\sum_m \langle m \mid e^{-\beta\hbar\omega_0\hat{a}^\dagger\hat{a}} \mid m\rangle}$$

[§]See the discussion of the density operator in Chapter 7.

$$= \frac{\sum_m e^{-\beta\hbar\omega_0 m} \left\langle m \mid \hat{X}^2 \mid m \right\rangle}{\sum_m e^{-\beta\hbar\omega_0 m}}$$

$$= \frac{\hbar}{2M\omega_0} \left(2 \langle m \rangle + 1\right), \tag{12.99}$$

where the average occupation number is

$$\langle m \rangle = \frac{\sum_m e^{-\beta\hbar\omega_0 m} m}{\sum_m e^{-\beta\hbar\omega_0 m}} = \frac{1}{e^{\beta\hbar\omega_0} - 1}. \tag{12.100}$$

The mean-square displacement now reduces to

$$\left\langle X^2 \right\rangle = \frac{\hbar}{2M\omega_0} \coth(z), \tag{12.101}$$

with $z = \beta\hbar\omega_0/2$. Similarly, the time-correlation function becomes

$$\begin{aligned} \langle X(0)\, X(t) \rangle &= \frac{\hbar}{2M\omega_0} \left[\langle m \rangle e^{-i\omega_0 t} + \langle m+1 \rangle e^{i\omega_0 t} \right] \\ &= \frac{\hbar}{2M\omega_0} \left(\frac{e^z e^{i\omega_0 t} + e^{-z} e^{-i\omega_0 t}}{e^z - e^{-z}} \right). \end{aligned} \tag{12.102}$$

Consider the limiting results at low and high temperature:

- At low temperature, when $z \gg 1$ (or $\langle m \rangle \to 0$), we have $\langle X^2 \rangle \to \hbar/2M\omega_0$ and $\langle X(0)\, X(t) \rangle \to (\hbar/2M\omega_0) \exp(i\omega_0 t)$. Furthermore, if $(\hbar^2 Q^2/2M)/\hbar\omega_0 < 1$, which means that the recoil energy is smaller than the binding energy of the particle, then we can make an expansion in powers of this quantity as

$$\begin{aligned} e^{Q^2 \langle X(0) X(t) \rangle} &= e^{\left(\hbar Q^2/2M\omega_0\right) \exp(i\omega_0 t)} \\ &= 1 + \frac{\hbar Q^2}{2M\omega_0} e^{i\omega_0 t} \\ &\quad + \frac{1}{2} \left(\frac{\hbar Q^2}{2M\omega_0} \right)^2 e^{i2\omega_0 t} + \cdots. \end{aligned} \tag{12.103}$$

  The result is that the dynamic structure factor can be expressed as the following *phonon expansion*:

$$S_s(Q,\omega) = \frac{1}{2\pi} \int_{-\infty}^{+\infty} dt\, e^{-i\omega t} F_s(Q,t)$$

$$= e^{-\hbar Q^2/2M\omega_0}\left[\delta(\omega) + \left(\frac{\hbar^2 Q^2}{2M\hbar\omega_0}\right)\delta(\omega-\omega_0) + \frac{1}{2}\left(\frac{\hbar^2 Q^2}{2M\hbar\omega_0}\right)^2 \delta(\omega - 2\omega_0) + \cdots\right]. \qquad (12.104)$$

Each term in the expansion corresponds to a particular quantized energy exchange between the incoming neutron and the oscillator. The first term is the zero-phonon contribution and represents no excitation of the oscillator, or elastic scattering. The second term, which has a smaller intensity, is the one-phonon contribution—it indicates that the neutron has transferred an energy of $\hbar\omega_0$ to the oscillator. The third term corresponds to the three-phonon contribution, etc.

- At high temperature, when $z \ll 1$, the mean occupation number becomes very large, i.e., $\langle m\rangle = [\exp(2z) - 1]^{-1} \to k_B T/\hbar\omega_0 \gg 1$. Consequently, from Eq. 12.99, the Debye-Waller factor becomes

$$e^{-Q^2\langle X^2\rangle} = e^{-\left(\hbar Q^2/M\omega_0\right)\langle m\rangle} = e^{-k_B T Q^2/M\omega_0^2}. \qquad (12.105)$$

Using Eq. 12.102, the inelastic factor takes the form

$$e^{Q^2\langle X(0)X(t)\rangle} = e^{\left(k_B T Q^2/M\omega_0^2\right)\cos\omega_0 t}. \qquad (12.106)$$

In the high-temperature limit, the intermediate scattering function should reduce to the classical expression, or

$$F_s^{\mathrm{cl}}(Q,t) = e^{-Q^2\langle X^2\rangle} e^{Q^2\langle X(0)X(t)\rangle} = e^{-\left(k_B T Q^2/M\omega_0^2\right)(1-\cos\omega_0 t)}. \qquad (12.107)$$

According to Eq. 12.64, we can now write $F_s^{\mathrm{cl}}(Q,t) = \exp\left[-\frac{1}{2}Q^2 W(t)\right]$, where the width function is identified as¶

$$W(t) = \frac{2k_B T}{M\omega_0^2}(1-\cos\omega_0 t). \qquad (12.108)$$

A power-series expansion of Eq. 12.106 can be performed only if $k_B T Q^2/M\omega_0^2 < 1$. Thus, if the temperature is too high, a phonon expansion is no longer feasible, and will eventually break down. However, for temperatures where the expansion is valid, one can write

$$e^{Q^2\langle X(0)X(t)\rangle} = 1 + \frac{k_B T Q^2}{M\omega_0^2}\cos\omega_0 t + \frac{1}{2}\left(\frac{k_B T Q^2}{M\omega_0^2}\right)^2 \cos^2\omega_0 t + \cdots. \qquad (12.109)$$

The phonon-expansion terms making up the dynamic structure factor will then involve the Fourier transform of various powers of $\cos\omega_0 t$. Thus, $S_s(Q,\omega)$ is

¶This harmonic oscillator width function is shown back in Fig. 12.3.

built up from n-phonon terms involving the factors $\delta(\omega \pm n\omega_0)$. As in the low-temperature case, the minus sign means that energy $n\hbar\omega_0$ has been transferred from the neutron to the oscillator, whereas the plus sign indicates that an energy transfer of $n\hbar\omega_0$ from the oscillator to the neutron is now possible. Energy loss by the neutron is more probable than energy gain, in keeping with the condition of detailed balance.

Example 12.2 Total Cross-Section for Scattering from an Oscillator. Let us calculate the total cross-section, $\sigma(E)$, for the scattering of neutrons of energy E from a harmonic oscillator of mass M in the limit of low temperature. The dynamic structure factor in this case is given by the phonon expansion of Eq. 12.104, so the total cross-section is the sum of an elastic (zero-phonon) term and a series of inelastic (multi-phonon) terms, i.e.,

$$\sigma(E) = \sum_{n=0}^{\infty} \sigma_n(E). \tag{12.110}$$

The nth-phonon term is given by the integral

$$\sigma_n(E) = \frac{\sigma_b}{4\pi}\frac{1}{n!}\int_0^{E/\hbar} d\omega \frac{k'}{k}\int d\Omega\, e^{-\hbar Q^2/2M\omega_0}\left(\frac{\hbar Q^2}{2M\omega_0}\right)^n \delta(\omega - n\omega_0). \tag{12.111}$$

The integration limits for ω reflect the fact that neutrons can only lose energy when scattering at $T = 0$. The solid angle integration is linked to the square of wavevector transfer Q by the following kinematic relation:

$$\frac{\hbar^2 Q^2}{2m_n} = 2E - \hbar\omega - 2\sqrt{E(E-\hbar\omega)}\cos\theta. \tag{12.112}$$

We can therefore introduce a new variable x to replace the solid angle variable $\cos\theta$ by

$$Ax = \frac{A\hbar^2 Q^2}{2M\hbar\omega_0} = 2\frac{E}{\hbar\omega_0} - \frac{\omega}{\omega_0} - 2\sqrt{\frac{E}{\hbar\omega_0}\left(\frac{E}{\hbar\omega_0} - \frac{\omega}{\omega_0}\right)}\cos\theta. \tag{12.113}$$

Denoting the ratio of the incident energy to the binding energy by

$$y = \frac{E}{\hbar\omega_0}, \tag{12.114}$$

we can write

$$d\Omega = 2\pi\sin\theta\, d\theta = \frac{A\pi}{\sqrt{y\left(y - \frac{\omega}{\omega_0}\right)}}dx. \tag{12.115}$$

The integration limits for x can be written as

$$X_{\pm} = \frac{\hbar}{2M\omega_0}(k \pm k')^2, \tag{12.116}$$

and we also note that

$$\frac{k'}{k} = \sqrt{\frac{y - \dfrac{\omega}{\omega_0}}{y}}. \tag{12.117}$$

Putting these relations together in the integral, the total cross-section can be written as

$$\sigma_n(E) = \frac{\sigma_b}{4n!}\frac{A}{y}\int_0^{E/\hbar} \delta(\omega - n\omega_0)\, d\omega \int_{X_-}^{X_+} x^n e^{-x} dx. \tag{12.118}$$

If $E < n\hbar\omega_0$, then the integral over the delta-function vanishes; if $E \geq n\hbar\omega_0$, then the delta-function integral forces the integration limits on the x-integral to be

$$X_{\pm} = \frac{1}{A}\left(\sqrt{y} \pm \sqrt{y-n}\right)^2. \tag{12.119}$$

The final n-phonon cross-section is

$$\sigma_n(E) = \begin{cases} \dfrac{\sigma_b}{4n!}\dfrac{A}{y}\displaystyle\int_{X_-}^{X_+} x^n e^{-x} dx, & \text{for } y \geq n \\ 0, & \text{for } y < n. \end{cases} \tag{12.120}$$

The most interesting feature of this formula is the way it illustrates the effect of chemical binding on the total neutron scattering cross-section:

- In the low-energy limit where $y = E/\hbar\omega_0 < 1$, only the zero-phonon elastic term survives, and the cross-section becomes

$$\sigma(E) = \sigma_0(E) = \frac{\sigma_b}{4}\frac{A}{y}\left(1 - e^{-4y/A}\right). \tag{12.121}$$

  In the limit of zero energy, the cross-section reduces to the bound-atom value, σ_b.

- In the intermediate-energy range where $1 < y < 4$, the one, two, and three-phonon terms also contribute. The behavior of the individual terms, as well as the total cross-section, are illustrated in Fig. 12.4.‖

‖This figure is similar to one first published by E. Fermi [113].

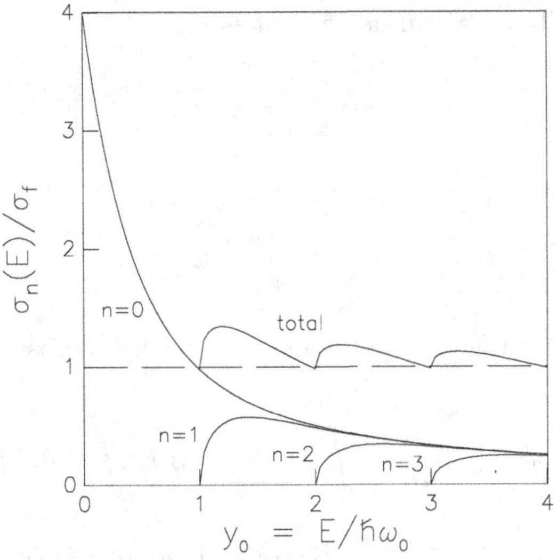

Figure 12.4 Energy-dependence of the cross-section for neutron scattering from a harmonic oscillator ($A = 1$) in the limit of low temperature. The individual n-phonon contributions are shown, along with the total cross-section.

- In the high-energy limit, one has to be careful when evaluating X_- and X_+. Since the phonon expansion was originally made under the assumption that the recoil energy is smaller than the binding energy, we may not take $y \to \infty$ in the latter integration limits. Instead, it is best to just evaluate the successive terms for finite values of the parameter y. As can be inferred from Fig. 12.4, for $y > 4$ the cross-section falls within a few percent of the free-atom value, σ_f. Thus, chemical binding effects are almost negligible over this range of incident energies.

12.3 General Properties of the Dynamic Structure Factor

In this section, we shall derive some rigorous relations involving the dynamic structure factor for a system in thermal equilibrium.

Consider the intermediate scattering function of an N-particle system characterized by a Hamiltonian $\hat{H}$:

$$\begin{aligned} F(\mathbf{Q},t) &= \frac{1}{N}\langle n(-\mathbf{Q},0)\,n(\mathbf{Q},t)\rangle \\ &= \frac{1}{N}\sum_i P(E_i)\,\langle E_i \mid \hat{n}(-\mathbf{Q},0)\,\hat{n}(\mathbf{Q},t) \mid E_i\rangle. \end{aligned} \tag{12.122}$$

$|E_i\rangle$ is a system eigenvector with corresponding eigenvalue E_i, and we assume that the set $\{|E_i\rangle\}$ is complete. $P(E_i)$ is the probability that the system is in state $|E_i\rangle$.

For a system in thermal equilibrium at temperature $T = 1/k_B\beta$, the distribution of energy states follows the Maxwell-Boltzmann distribution with $P(E_i) = e^{-\beta E_i}/Z$, where Z is the system's partition function. Thus, we can write the density correlation function as

$$\begin{aligned}
\langle n(-\mathbf{Q},0)\, n(\mathbf{Q},t)\rangle &= \sum_{i,j} \frac{e^{-\beta E_i}}{Z} \langle E_i \mid \hat{n}(-\mathbf{Q},0) \mid E_j\rangle \langle E_j \mid e^{i\hat{H}t/\hbar} \hat{n}(\mathbf{Q},0)\, e^{-i\hat{H}t/\hbar} \mid E_i\rangle \\
&= \sum_{i,j} \frac{e^{-\beta E_i}}{Z} e^{i(E_j-E_i)t/\hbar} \langle E_i \mid \hat{n}(-\mathbf{Q},0) \mid E_j\rangle \langle E_j \mid \hat{n}(\mathbf{Q},0) \mid E_i\rangle \\
&= \sum_{i,j} \frac{e^{-\beta E j}}{Z} e^{i(E_j-E_i)(t-i\hbar\beta)/\hbar} \langle E_i \mid \hat{n}(-\mathbf{Q},0) \mid E_j\rangle \\
&\qquad \times \langle E_j \mid \hat{n}(\mathbf{Q},0) \mid E_i\rangle \\
&= \sum_{i,j} \frac{e^{-\beta E j}}{Z} \langle E_j \mid \hat{n}(\mathbf{Q},0) \mid E_i\rangle \\
&\qquad \times \langle E_i \mid e^{-i\hat{H}(t-i\hbar\beta)/\hbar} \hat{n}(-\mathbf{Q},0)\, e^{i\hat{H}(t-i\hbar\beta)/\hbar} \mid E_j\rangle \\
&= \sum_{j} \frac{e^{-\beta E j}}{Z} \langle E_j \mid \hat{n}(\mathbf{Q},0)\, e^{i\hat{H}(-t+i\hbar\beta)/\hbar} \\
&\qquad \hat{n}(-\mathbf{Q},0)\, e^{-i\hat{H}(-t+i\hbar\beta)/\hbar} \mid E_j\rangle\,. \qquad (12.123)
\end{aligned}$$

Expressing as simply a thermal average, we arrive at the relation

$$\langle n(-\mathbf{Q},0)\, n(\mathbf{Q},t)\rangle = \langle n(\mathbf{Q},0)\, n(-\mathbf{Q},-t+i\hbar\beta)\rangle \qquad (12.124)$$

or

$$F(\mathbf{Q},t) = F(-\mathbf{Q},-t+i\hbar\beta)\,. \qquad (12.125)$$

By taking the Fourier transform of both sides, we establish the general detailed balance relation valid for a system in thermal equilibrium:

$$S(\mathbf{Q},\omega) = e^{\beta\hbar\omega} S(-\mathbf{Q},-\omega)\,. \qquad (12.126)$$

From our derivation, it should be obvious that Eqs. 12.125 and 12.126 hold true for the self intermediate scattering function and self dynamic structure factor as well.** Equation 12.126 shows that the dynamic structure factor for an isotropic system is an even function of $\mathbf{Q}$, but an asymmetry appears as a function of ω, and hence in the intensities of the Stokes and anti-Stokes lines.†† In the classical limit, i.e., as

**For example, this is illustrated explicitly by the results of Eqs. 12.25 and 12.33.

††Stokes and anti-Stokes lines were discussed previously in the context of the Raman scattering of photons (see Chapter 9).

$\hbar \to 0$, the intermediate scattering function is an even function of t and the dynamic structure factor is even in ω.

Information about a system can sometimes be deduced from the various frequency moments of the dynamic structure factor. The nth moments are defined as

$$\langle \omega_s^n \rangle \equiv \int_{-\infty}^{+\infty} d\omega\, S_s(\mathbf{Q}, \omega)\, \omega^n \tag{12.127}$$

$$\langle \omega^n \rangle \equiv \int_{-\infty}^{+\infty} d\omega\, S(\mathbf{Q}, \omega)\, \omega^n . \tag{12.128}$$

Since the dynamic structure factor is the time Fourier transform of the corresponding intermediate scattering function, the moments are given by time derivatives of the intermediate scattering function evaluated at $t = 0$, i.e.,

$$\langle \omega_s^n \rangle \equiv (-i)^n \frac{\partial^n}{\partial t^n} F_s(\mathbf{Q}, t)\bigg|_{t=0} \tag{12.129}$$

$$\langle \omega^n \rangle \equiv (-i)^n \frac{\partial^n}{\partial t^n} F(\mathbf{Q}, t)\bigg|_{t=0} . \tag{12.130}$$

From the basic definitions for the intermediate scattering functions, it is immediately seen that the zeroth moments are given by

$$\langle \omega_s^0 \rangle = F_s(\mathbf{Q}, 0) = 1 \qquad \text{and} \qquad \langle \omega^0 \rangle = F(\mathbf{Q}, 0) = S(Q) . \tag{12.131}$$

Classically, the self or full dynamic structure factor of a simple fluids mainly consists of a symmetric peak centered about $\omega = 0$. While the zeroth frequency moment gives the area under the peak, the corresponding second moment is related to the width of the peak. This moment is calculated by considering the intermediate scattering functions in the limit of short time, and expanding to second order in t. In the classical limit, it is straightforward to show that

$$\langle \omega_s^2 \rangle = \langle \omega^2 \rangle = Q^2 V_0^2 . \tag{12.132}$$

Thus, the width of the peak depends only on the thermal speed of the particles in the fluid, and is completely independent of interparticle forces. Information on interparticle interactions is encoded in the detailed shape of the peak, especially the region in the wings, as reflected by the higher-order frequency moments. It is possible, but laborious, to obtain the analytical form for the fourth moments in terms of the interparticle potential, $V(r)$, and the pair correlation function, $g(r)$. The results [114] are

$$\langle \omega_s^4 \rangle = \frac{Q^4}{M^2\beta}\left[\frac{3}{\beta} + \frac{n}{Q^4}\int d^3r\, g(r)\, \nabla^2 V(r)\right] \tag{12.133}$$

$$\langle \omega^4 \rangle = \frac{Q^4}{M^2\beta}\left[\frac{3}{\beta} + \frac{n}{Q^4}\int d^3r\, g(r)\,(1 - \cos \mathbf{Q}\cdot\mathbf{r})\,(\mathbf{Q}\cdot\boldsymbol{\nabla})^2\, \nabla^2 V(r)\right] . \tag{12.134}$$

One last general formula we would like to derive is the relationship between the dynamic structure factor and the Fourier transform of the system *response function*. The content of this relation embodies a special case of the so-called *fluctuation-dissipation theorem* [115], which states that, for a system in thermal equilibrium, the linear response of the system to an external probe coupled to the local density is directly related to the spectrum of thermal density fluctuations. The latter quantity is just the dynamic structure factor. The most general form of this theorem will be discussed in Chapter 13.

In the present context, the relevant response function is the density-density response function

$$\phi_Q(t) = \frac{1}{i\hbar N}\left\langle\left[n(\mathbf{Q},t), n(-\mathbf{Q},0)\right]\right\rangle. \tag{12.135}$$

By expanding the commutator in the bracket and using a variation of Eq. 12.124, we can write

$$\begin{aligned}\phi_Q(t) &= \frac{1}{i\hbar N}\left\{\langle n(\mathbf{Q},t)\, n(-\mathbf{Q},0)\rangle - \langle n(-\mathbf{Q},0)\, n(\mathbf{Q},t)\rangle\right\} \\ &= \frac{1}{i\hbar N}\left\{\langle n(-\mathbf{Q},0)\, n(\mathbf{Q},t+i\hbar\beta)\rangle - \langle n(-\mathbf{Q},0)\, n(\mathbf{Q},t)\rangle\right\},\end{aligned} \tag{12.136}$$

or

$$\phi_Q(t) = \frac{1}{i\hbar}\left[F(\mathbf{Q},t+i\hbar\beta) - F(\mathbf{Q},t)\right]. \tag{12.137}$$

The time Fourier transform of both sides then gives

$$\frac{1}{2\pi}\int_{-\infty}^{+\infty} dt\, e^{-i\omega t}\phi_Q(t) = \frac{1}{i\hbar}\left[e^{-\beta\hbar\omega}S(\mathbf{Q},\omega) - S(\mathbf{Q},\omega)\right], \tag{12.138}$$

and a rearrangement of terms produces the desired result:

$$\frac{\hbar}{2\pi i}\int_{-\infty}^{+\infty} dt\, e^{-i\omega t}\phi_Q(t) = \left(1 - e^{-\beta\hbar\omega}\right)S(\mathbf{Q},\omega). \tag{12.139}$$

The fact that the right-hand side of this equation is a real quantity implies that

$$\phi_Q^*(t) = -\phi_Q(-t). \tag{12.140}$$

This property guarantees that the real part of the response function is an odd function of t and that the imaginary part is even. Using these properties, we may rewrite Eq. 12.139 as

$$\frac{\hbar}{\pi}\,\mathrm{Im}\int_0^{\infty} dt\, e^{-i\omega t}\phi_Q(t) = \left(1 - e^{-\beta\hbar\omega}\right)S(\mathbf{Q},\omega). \tag{12.141}$$

In Chapter 13, $\int_0^\infty dt\, e^{-i\omega t}\phi_Q(t)$ will be identified as the density-density *admittance* or *susceptibility function*, $\chi_{nn}(\mathbf{Q},\omega)$. This function reflects the coupling of the external radiation field to the density fluctuations of the system. More specifically, it connects the $\mathbf{Q}$ and ω-components of the scattered field to the spatial and temporal Fourier components of the density-density response function. In the classical limit ($\hbar \to 0$), Eq. 12.141 reduces to the form

$$\mathrm{Im}\int_0^\infty dt\, e^{-i\omega t}\phi_Q(t) = \pi\beta\omega S(\mathbf{Q},\omega)\,. \tag{12.142}$$

The imaginary part of the susceptibility function is a measure of the energy dissipation by the external field and Eq. 12.142 is a specific example of the classical fluctuation-dissipation theorem.

Example 12.3 First Frequency Moment for a System of Non-Interacting Particles. In this example, we make use of Eq. 12.139 to perform the explicit quantum-mechanical calculation of the first frequency moment

$$\langle\omega\rangle \equiv \int_{-\infty}^{+\infty} d\omega\, S(\mathbf{Q},\omega)\,\omega \tag{12.143}$$

for a system of N non-interacting particles of mass M. The physical significance of the first moment is that $\hbar\langle\omega\rangle$ represents the average energy transferred from the incident radiation (i.e., neutron) to the scattering medium.

Begin by Fourier inverting Eq. 12.139 to obtain

$$\phi_Q(t) = \frac{i}{\hbar}\int_{-\infty}^{+\infty} d\omega\, e^{i\omega t}\left(1 - e^{-\beta\hbar\omega}\right) S(\mathbf{Q},\omega)\,. \tag{12.144}$$

Then differentiate both sides with respect to time, and set $t = 0$. By applying the detailed-balance relation (Eq. 12.126), we come up with the following identity:

$$\dot{\phi}_Q(t)\Big|_{t=0} = -\frac{2}{\hbar}\int_{-\infty}^{+\infty} \omega\, S(\mathbf{Q},\omega)\,. \tag{12.145}$$

$\dot{\phi}_Q(t)\Big|_{t=0}$ can also be calculated directly from the definition of the response function given by Eq. 12.135:

$$\dot{\phi}_Q(t)\Big|_{t=0} = \frac{1}{i\hbar N}\left\langle\left[\dot{n}(\mathbf{Q},t)|_{t=0}\,, n(-\mathbf{Q},0)\right]\right\rangle. \tag{12.146}$$

Using the Heisenberg equation of motion for $n(\mathbf{Q}, t)$, namely,

$$\dot{n}(\mathbf{Q}, t)|_{t=0} = \frac{1}{i\hbar}\left[n(\mathbf{Q}, 0), H\right], \tag{12.147}$$

we have

$$\dot{\phi}_Q(t)\Big|_{t=0} = -\frac{1}{\hbar^2 N}\left\langle\left[\left[n(\mathbf{Q}, 0), H\right], n(-\mathbf{Q}, 0)\right]\right\rangle, \tag{12.148}$$

where H is the Hamiltonian for the system of N free particles. It is left as an exercise for the reader to show that the nested commutator reduces to

$$\left\langle\left[\left[n(\mathbf{Q}, 0), H\right], n(-\mathbf{Q}, 0)\right]\right\rangle = N\frac{\hbar^2 Q^2}{M}. \tag{12.149}$$

Combining Eqs. 12.145, 12.148, and 12.149, we find the expression for the first frequency moment:

$$\langle\omega\rangle \equiv \int_{-\infty}^{+\infty} d\omega\, S(\mathbf{Q}, \omega)\,\omega = \frac{\hbar Q^2}{2M}. \tag{12.150}$$

Thus, the average energy transfer is simply $\hbar\langle\omega\rangle = \hbar^2 Q^2/2M$, which is just the recoil energy of the struck particle. This result can be verified directly by substituting the dynamic structure factor for a free particle into Eq. 12.143. The appropriate form to use is the one given by Eq. 12.27.

Suggested References

References on dynamic structure factors, especially as applied to thermal neutron scattering, are

[a] J. P. Boon and S. Yip, *Molecular Hydrodynamics* (McGraw-Hill, New York, 1980).

[b] S. W. Lovesey, *Theory of Neutron Scattering from Condensed Matter*, Vol. 1 (Oxford University Press, New York, 1984).

[c] D. E. Parks, M. S. Nelkin, J. R. Beyster, and N. F. Wikner, *Slow Neutron Scattering and Thermalization* (W. A. Benjamin, New York, 1970).

[d] G. L. Squires, *Introduction to the Theory of Thermal Neutron Scattering* (Cambridge University Press, London, 1978).

Problems

1. The simplest way of taking into account the excluded volume effect on the structure factor of a fluid composed of hard spheres of diameter σ is to take the following approximate form for the pair correlation function:

$$g(r) = \begin{cases} 0, & r < \sigma \\ 1, & r \geq \sigma. \end{cases}$$

This is sometimes referred to as a *"correlation hole approximation."*

(a) What exact features of $g(r)$ does the correlation hole approximation capture?

(b) Under this approximation, calculate the analytical form of the static structure factor, $S(Q)$.

(c) Show that in the limit $Q \to 0$ limit,

$$S(0) = 1 - 8\phi,$$

where ϕ is the volume fraction of the hard spheres. This is an exact result to first order in ϕ. Give a plausible reason why the drastic approximation used for $g(r)$ produces the correct first-order result for $S(0)$.

2. In this chapter, we derived the self intermediate scattering function, $F_s(Q,t)$, for a simple harmonic oscillator (frequency ω_0) in equilibrium at temperature T (see Eqs. 12.93, 12.101, and 12.102). Show that in the limit $\omega_0 \to 0$, the expression for $F_s(Q,t)$ reduces to the ideal gas result (i.e., Eqs. 12.25 and 12.26).

3. In this problem, we examine the calculation of the classical width function, $W(t)$, for a test particle (mass M) in a liquid. The width function must correctly capture both the short-time and long-time behavior of the self intermediate scattering function $F_s^{\text{cl}}(Q,t) = e^{-\frac{1}{2}Q^2 W(t)}$ in the Gaussian approximation (see Eq. 12.64). From Eq. 12.59, we know that the width function can be expressed in terms of the velocity autocorrelation function of the particle, i.e.,

$$W(t) = 2\int_0^t d\tau\,(t-\tau)\langle V_x(0)\,V_x(\tau)\rangle.$$

(a) Let us define the normalized autocorrelation function

$$\psi(t) = \frac{\langle V_x(0)\,V_x(t)\rangle}{\langle V_x^2\rangle},$$

where $\langle V_x^2 \rangle = V_0^2 = k_B T/M$, and its Fourier transform (the density of states function)

$$\Psi(\omega) = \frac{1}{2\pi} \int_{-\infty}^{+\infty} dt\, e^{i\omega t} \psi(t) = \frac{1}{\pi} \int_0^{\infty} dt\, \psi(t) \cos \omega t.$$

Show that the width function can be written as

$$W(t) = 4V_0^2 \int_0^{\infty} d\omega \frac{1 - \cos \omega t}{\omega^2} \Psi(\omega).$$

(b) In the case when the test particle undergoes random, or Brownian, motion, its velocity $V_x(t)$ is governed by the *Langevin equation* (see, for example, reference [116])

$$M \frac{dV_x}{dt} = -bV_x + \zeta(t).$$

The net force on the particle due to its interaction with the surrounding fluid is represented by the sum of the two terms on the right-hand side. The first term, $-bV_x$, has the form of a frictional force, and corresponds to the slowly-varying part of the net force. b is the *friction constant.* The second term, $\zeta(t)$, represents the rapidly-varying, stochastic part of the force that comes about because of thermal fluctuations in the system. Due to its truly random nature, the average value of $\zeta(t)$ vanishes. Show that if the test particle indeed satisfies the Langevin equation, then the velocity autocorrelation function is given by

$$\psi(t) = e^{-|t|/\tau} \tag{12.151}$$

and relate the relaxation time τ to the friction constant b. Calculate the width function $W(t)$ and give its short-time and long-time limits.

(c) By examining the long-time behavior of the width function for the Brownian particle, derive the general relationship between the diffusion coefficient, D, of the test particle and the autocorrelation function $\psi(t)$. Also give the relation between D and the density of states function $\Psi(\omega)$.

(d) The Brownian particle model given above is based on a single parameter, namely, the relaxation time, τ. It correctly produces the classical short-time and long-time limits of the self intermediate scattering function. A more advanced model, known as the *relaxing cage model*, was developed by Desai and Yip [117]. This is a two-parameter model in which the density of states function is modeled as

$$\Psi(\omega) = \frac{2}{\pi} \frac{\omega_0^2/\tau_0}{\left(\omega^2 - \omega_0^2\right)^2 + \left(\omega/\tau_0\right)^2}.$$

with parameters ω_0 and τ_0. Show that

$$\psi(t) = \int_0^\infty d\omega\, \Psi(\omega) \cos \omega t = e^{-t/2\tau_0} \left(\cos \Omega t + \frac{1}{2\tau_0 \Omega} \sin \Omega t \right),$$

where $\Omega^2 = \omega_0^2 - \frac{1}{4\tau_0^2}$.

(e) Determine the expression for the width function in the relaxing cage model. Check its short-time and long-time behavior. You should find that this model describes the motion of the test particle to be a vibrational one with a characteristic frequency ω_0 at short time, and a diffusional one with a frictional constant $M\omega_0^2\tau_0$ at long time. In terms of a typical atom in a liquid, the picture corresponds to putting the atom in an external parabolic potential well which relaxes in time—hence the name "relaxing cage model". Eventually the atom diffuses away and experiences only the frictional force.

4. For a simple homogeneous fluid, where the various particles are treated as statistically equivalent, the self intermediate scattering function of a test particle is given by Eq. 12.14, i.e.,

$$F_s(\mathbf{Q}, t) = \int d^3r\, e^{i\mathbf{Q}\cdot\mathbf{r}} G_s(\mathbf{r}, t),$$

where $G_s(\mathbf{r}, t)$ is the van-Hove space-time self correlation function. However, for a liquid under confinement, the various particles of the fluid are no longer statistically identical, and we need to make an extension as follows:

$$F_s(\mathbf{Q}, t) = \int\int d^3r\, d^3r_0\, e^{i\mathbf{Q}\cdot(\mathbf{r}-\mathbf{r}_0)} G_s(\mathbf{r}, t \mid \mathbf{r}_0)\, p(\mathbf{r}_0).$$

Here, $G_s(\mathbf{r}, t \mid \mathbf{r}_0)$ is a *conditional* space-time self correlation function, which is subject to the initial equilibrium distribution, $p(\mathbf{r}_0) = e^{-V(\mathbf{r}_0)/k_BT}/Z$, for the test particle under the influence of the confining potential, $V(\mathbf{r})$ (and Z is the usual normalization factor).

(a) Using the new definition for $F_s(\mathbf{Q}, t)$, show the following four important properties:

i. Conservation of Probability:
$F_s(\mathbf{Q} = 0, t) = \int\int d^3r\, d^3r_0\, G_s(\mathbf{r}, t \mid \mathbf{r}_0)\, p(\mathbf{r}_0) = 1$

ii. Initial Condition:
$G_s(\mathbf{r}, t = 0 \mid \mathbf{r}_0) = \delta(\mathbf{r} - \mathbf{r}_0)$ leads to $F_s(\mathbf{Q}, t = 0) = 1$

iii. Equilibrium Distribution:
$G_s(\mathbf{r}, t \to \infty \mid \mathbf{r}_0) = p(\mathbf{r}) = \frac{1}{Z} e^{-V(\mathbf{r})/k_BT}$

iv. Elastic Incoherent Structure Factor (*EISF*):
$EISF = F_s(\mathbf{Q}, t \to \infty) = \left| \int p(\mathbf{r}) e^{i\mathbf{Q}\cdot\mathbf{r}} d^3r \right|^2$

(b) In particular, use the last relation to calculate the $EISF$ for the following two cases:

i. The test particle is confined within a sphere of radius a:

$$EISF = \left[\frac{3j_1(Qa)}{Qa} \right]^2,$$

where $j_1(x)$ is the first-order spherical Bessel function.

ii. The test particle is confined within a harmonic potential well:

$$EISF = e^{-Q^2 \langle r^2 \rangle /3},$$

where $\langle r^2 \rangle$ is the mean square vibrational amplitude of the particle in the well. Calculate this quantity in terms of the parameters of the potential well.

5. The *Mössbauer effect* is the phenomenon of resonance absorption of a gamma ray by a nucleus embedded in a condensed medium (a gas, liquid, or crystalline solid). A gamma ray is a photon emitted in the process of de-excitation of a nuclear excited state to a lower energy level. Let us denote this transition energy by E_0. Then, if one sends in a photon of energy E close to E_0, the resonance absorption cross-section for the nucleus embedded in the medium can be written as

$$\sigma_a(E) = \sigma_0 \frac{\Gamma^2}{4} \sum_{n,n_0} g_{n_0} \frac{\left| \langle n | e^{i\mathbf{p}\cdot\mathbf{R}/\hbar} | n_0 \rangle \right|^2}{\left(E_0 - E + \varepsilon_n - \varepsilon_{n_0}\right)^2 / \hbar^2 + \Gamma^2/4},$$

where $\Gamma = \Delta E/\hbar$ is the natural linewidth of the excited nuclear state, $\mathbf{p}$ is the momentum of the incident gamma-ray photon, and $\mathbf{R}$ is the position-vector for the center-of-mass of the absorbing nucleus. $|n_0\rangle$ and $|n\rangle$ are the respective initial and final states of the condensed medium, with corresponding energy eigenvalues ε_{n_0} and ε_n, and g_{n_0} is the statistical weight-factor for the initial state.

(a) Show that the cross-section formula can be transformed into the form

$$\sigma_a(E) = \sigma_0 \frac{\Gamma}{4} \int_{-\infty}^{+\infty} dt \, e^{-i\omega t - \Gamma|t|/2} F_s(p,t),$$

where

$$F_s(p,t) = \left\langle e^{-i\mathbf{p}\cdot\mathbf{R}(0)/\hbar} e^{i\mathbf{p}\cdot\mathbf{R}(t)/\hbar} \right\rangle$$

and $\hbar\omega = E - E_0$. In order to proceed, first write

$$\frac{1}{\left(E_0 - E + \varepsilon_n - \varepsilon_{n0}\right)^2/\hbar^2 + \Gamma^2/4} = \int_{-\infty}^{+\infty} d\rho \frac{\delta\left[\rho - \left(\varepsilon_n - \varepsilon_{n0}\right)\right]}{\left(E - E_0 - \rho\right)^2/\hbar^2 + \Gamma^2/4}$$

and then use the Fourier transform formula

$$\int_{-\infty}^{+\infty} d\rho \frac{e^{-it\rho/\hbar}}{\left(E - E_0 - \rho\right)^2/\hbar^2 + \Gamma^2/4} = \frac{2\pi\hbar}{\Gamma} e^{-it[(E-E_0)/\hbar] - \Gamma|t|/2}.$$

One now sees that the resonance absorption cross-section for gamma rays takes on nearly the same form as a neutron inelastic scattering cross-section. As a result, the Mössbauer effect can be used to study dynamics in the condensed medium.

(b) In particular, one can study the dynamics of liquids by measuring the time Fourier transform of the self intermediate scattering function, $F_s(p,t)$. If we let $\mathbf{p} = \hbar\mathbf{k}$, then for a diffusing nucleus in a liquid, we know that the classical self intermediate scattering function is

$$F_s^{\mathrm{cl}}(k,t) = e^{-Dk^2|t|}.$$

Show that substitution into the expression for the absorption cross-section leads to

$$\sigma_a(E) = \sigma_0 \frac{\Gamma}{4} e^{\hbar\omega/2k_BT} e^{-\hbar^2k^2/8Mk_BT} \frac{\Gamma + 2k^2D}{\left(E - E_0\right)^2/\hbar^2 + \left(\Gamma + 2k^2D\right)^2/4},$$

where the two exponential factors correspond to the quantum-mechanical correction terms, i.e., the detailed balance factor and the recoil factor, discussed in the text (see Eq. 12.33). The last factor shows that the effect of single-particle diffusion is to broaden the resonance absorption line by the amount $\Delta\Gamma = 2k^2D = 2E^2D/\hbar^2c^2$. As an example consider iron in a liquid state with a typical diffusion constant on the order of $D \approx 10^{-5}$ cm^2/s. The Mössbauer gamma rays for the radioactive isotope Fe^{57} have an energy of 14.4 keV. Estimate the numerical value of the energy-broadening $\hbar\Delta\Gamma$, and compare to the natural broadening. (For more complete details on the theory, refer to the paper by Singwi and Sjölander [118].)

Chapter 13
LINEAR RESPONSE THEORY

In a system which is weakly coupled to an external perturbation field, the response of the system is proportional to the perturbation. In other words, one can speak of a "linear" response to the perturbation. In this, the final chapter, we present both the classical and quantum-mechanical approach to what is known as *linear response theory* [115]. It will be found that either approach leads to the calculation of a so-called *susceptibility* or *admittance function*, $\chi(\omega)$, which embodies the system response to the perturbation. The examples presented will emphasize the calculation of NMR susceptibilities.

13.1 Classical Treatment of Linear Response Theory

Imagine the following experiment: We apply a perturbation, $F(t')$, at $t = t'$ and follow the system response by monitoring some observable, $N(t)$, at a later time t. We can write a linear response relation as

$$\Delta N(t) = n(t - t')\, F(t')\Delta t', \tag{13.1}$$

where $n(t - t')$ is the response function which should be a function of $t - t'$, the duration of the applied perturbation. If the perturbation is switched on at $t' = -\infty$, the system response at time t will be

$$N(t) = \int_{-\infty}^{t} n(t - t')F(t')dt'. \tag{13.2}$$

The causality relationship demands that $n(t - t') = 0$ for $t - t' < 0$. To find the function $n(t)$, we can apply a delta-function perturbation, $F(t') = \delta(t')$. Then,

$$N(t) = \int_{-\infty}^{t} n(t - t')\delta(t')dt' = n(t). \tag{13.3}$$

So $n(t)$ is the response function of the system to a sharp pulse-excitation at $t = 0$. Knowledge of $n(t)$ enables one to find, from Eq. 13.2, the observable resulting from an arbitrary perturbation.

Consider the case of a step-function excitation of unit strength starting at time zero. Then Eq. 13.2 can be written in this case as

$$N_{\text{step}}(t) = \int_0^t n(t-t')dt'. \tag{13.4}$$

Let $\tau = t - t'$, $d\tau = -dt'$. Then,

$$N_{\text{step}}(t) = \int_0^t n(\tau)d\tau, \tag{13.5}$$

and

$$\frac{d}{dt}N_{\text{step}}(t) = n(t). \tag{13.6}$$

Thus, $n(t)$ can also be obtained from a time-derivative of the response function to a step excitation.

An arbitrary perturbation can always be decomposed into a set of sinusoidal perturbations, or Fourier components. It is therefore important to consider the response of a system to a single sinusoidal perturbation at frequency ω. Let us write such a perturbation using the complex form

$$F^c(t) = F_0 e^{-i\omega t}. \tag{13.7}$$

Then the response of the system is

$$\begin{aligned} N^c(t) &= \int_{-\infty}^t n(t-t')F_0 e^{-i\omega t'}dt' \\ &= F_0 e^{-i\omega t}\int_{-\infty}^t n(t-t')e^{i\omega(t-t')}dt' \\ &= F_0 e^{-i\omega t}\int_0^\infty n(\tau)e^{i\omega\tau}d\tau. \end{aligned} \tag{13.8}$$

We now define a complex *susceptibility* (or *admittance*), $\chi(\omega)$, as the one-sided Fourier transform of the response function, namely,

$$\chi(\omega) \equiv \chi'(\omega) - i\chi''(\omega) = \int_0^{+\infty} n(\tau)e^{i\omega\tau}d\tau. \tag{13.9}$$

Eq. 13.8 can then be written as

$$N^c(t) = \chi(\omega)\, F^c(t). \tag{13.10}$$

This shows that the response of the system is proportional to the external perturbation and the proportionality constant is the complex susceptibility. It is important to notice that $\chi(\omega)$ is a Fourier transform of the delta-function response, $n(t)$. In particular,

$$\chi'(\omega) = \int_0^\infty n(t) \cos \omega t \, dt \qquad \text{and} \qquad \chi''(\omega) = -\int_0^\infty n(t) \sin \omega t \, dt. \tag{13.11}$$

An important relationship exists between the real and imaginary parts of the susceptibility. It is derived by considering the susceptibility to be a function of the complex variable $z = x + iy$, i.e.,

$$\chi(z) = \int_0^\infty n(\tau) e^{iz\tau} \, d\tau = \int_0^\infty n(\tau) e^{-y\tau} e^{ix\tau} \, d\tau = u(x,y) + iv(x,y). \tag{13.12}$$

Equating the real and imaginary parts gives

$$u(x,y) = \int_0^\infty n(\tau) e^{-y\tau} \cos x\tau \, d\tau \qquad \text{and} \qquad v(x,y) = \int_0^\infty n(\tau) e^{-y\tau} \sin x\tau \, d\tau. \tag{13.13}$$

We readily see that the Cauchy-Riemann relations,

$$\frac{\partial u}{\partial x} = \frac{\partial v}{\partial y} \qquad \text{and} \qquad \frac{\partial u}{\partial y} = -\frac{\partial v}{\partial x}, \tag{13.14}$$

are satisfied. Therefore, $\chi(z) = u + iv$ is an analytic function of z. In order that u and v be finite over the entire z-plane, we require that $y \geq 0$, making $\chi(z)$ analytic on the real axis and in the upper half of the complex plane. Now examine a new function, $\chi(z)/(z - \omega)$, where ω is real, and integrate this function over the closed contour C drawn in Fig. 13.1. Since C avoids the point $z = \omega$ on the real axis, the contour contains no poles, hence

$$\oint_C \frac{\chi(z)}{z - \omega} dz = 0. \tag{13.15}$$

Now let the radius of the large semicircle tend to infinity. Since $|\chi(z)| \to 0$ as $y \to +\infty$, there is no contribution to the integral from the semicircle itself. The remaining portion of the integral can be decomposed as

$$\int_{-\infty}^{\omega - R} \frac{\chi(\omega')}{\omega' - \omega} d\omega' + \int_{C_R} \frac{\chi(z)}{z - \omega} dz + \int_{\omega + R}^{\infty} \frac{\chi(\omega')}{\omega' - \omega} d\omega' = 0, \tag{13.16}$$

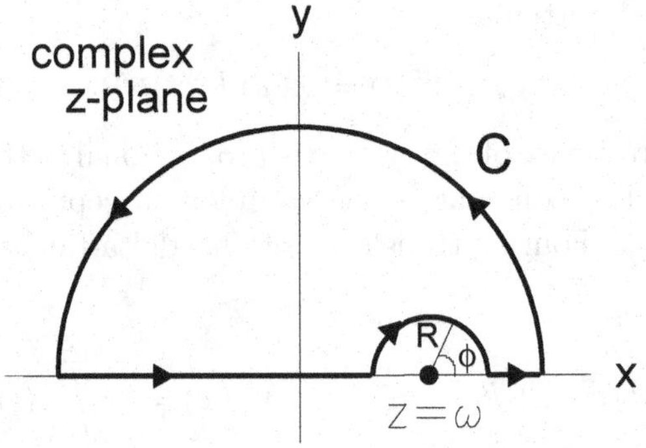

Figure 13.1 Contour used to integrate the function $\chi(z)/(z-\omega)$.

where the curve C_R corresponds to the small semicircle (radius R) circumventing the pole at $z=\omega$. The second integral can be evaluated by writing $z=\omega+Re^{i\phi}$, so that

$$\int_{C_R} \frac{\chi(z)}{z-\omega}dz = \int_{\pi}^{0} \frac{\chi\left(\omega+Re^{i\phi}\right)}{Re^{i\phi}} iRe^{i\phi}d\phi = -i\int_{0}^{\pi} \chi\left(\omega+Re^{i\phi}\right) d\phi. \tag{13.17}$$

In the limit of vanishingly small R, the result is

$$\int_{C_R} \frac{\chi(z)}{z-\omega}dz = -\pi i\chi(\omega), \tag{13.18}$$

and Eq. 13.16 becomes

$$P\int_{-\infty}^{+\infty} \frac{\chi(\omega')}{\omega'-\omega}d\omega' - \pi i\chi(\omega) = 0. \tag{13.19}$$

Since $\chi(\omega)=\chi'(\omega)-i\chi''(\omega)$, we identify

$$\chi'(\omega) = -\frac{1}{\pi}P\int_{-\infty}^{+\infty} \frac{\chi''(\omega')}{\omega'-\omega}d\omega' \tag{13.20}$$

$$\chi''(\omega) = \frac{1}{\pi}P\int_{-\infty}^{+\infty} \frac{\chi'(\omega')}{\omega'-\omega}d\omega'. \tag{13.21}$$

These two relations between $\chi'(\omega)$ and $\chi''(\omega)$ are the famous *Kramers-Kronig relations*. Knowledge of either the real or imaginary part of the susceptibility for all frequencies, allows one to calculate the other part.

Example 13.1 Steady-State Solution of the Bloch Equations. In a nuclear magnetic resonance experiment (see Chapter 10), a spin system is immersed in a strong static magnetic field, $\mathbf{B}_0 = B_0\mathbf{e}_z$, along with a superimposed weak transverse field oscillating at an RF frequency, ω. The purpose of this example is to show that the response of the spin system to the sinusoidal perturbation presented by the RF field obeys the linear response relationship of Eq. 13.10. This is accomplished by solving the classical Bloch equations for the spin magnetization components (previously introduced as Eqs. 10.103–10.105). In Sect. 10.5, the Bloch equations were used to examine the transient response to an RF pulse. In the present context, we are seeking the steady-state solution resulting from the application of a steady RF field.

It was previously shown (see Sect. 10.4) that the alternating transverse field can be decomposed into two counter-rotating fields. However, at resonance, only the field rotating in a clockwise sense is of importance, and the total magnetic field is written as

$$\mathbf{B} = B_0\mathbf{e}_z + B_1\left(\mathbf{e}_x\cos\omega t - \mathbf{e}_y\sin\omega t\right). \tag{13.22}$$

Applying this field to the Bloch equations gives

$$\frac{dM_x}{dt} - \left(\omega_0 M_y + \omega_1 M_z \sin\omega t\right) + \frac{M_x}{T_2} = 0 \tag{13.23}$$

$$\frac{dM_y}{dt} + \left(\omega_0 M_x - \omega_1 M_z \cos\omega t\right) + \frac{M_y}{T_2} = 0 \tag{13.24}$$

$$\frac{dM_z}{dt} + \left(\omega_1 M_x \sin\omega t + \omega_1 M_y \cos\omega t\right) + \frac{M_z - M_0}{T_1} = 0, \tag{13.25}$$

where $\omega_0 = \gamma B_0 \gg \omega_1 = \gamma B_1$.

In order to solve Eqs. 13.23–13.25, it becomes simpler to transform to the rotating coordinate frame previously discussed in Sect. 10.4. Recall that the z-axes of the rotating and laboratory coordinate systems coincide, and are aligned with the main magnetic field, $\mathbf{B}_0$. The rotating frame turns (in a clockwise sense) about the z-axis at frequency ω such that its x-axis always points in the direction of $\mathbf{B}_1$, the rotating magnetic field. Let the components of the magnetization along the x- and y-axes of the rotating frame be denoted by u and v, respectively. Then

$$\begin{aligned} M_x &= u\cos\omega t + v\sin\omega t \\ M_y &= -u\sin\omega t + v\cos\omega t \end{aligned} \tag{13.26}$$

or

$$\begin{aligned} u &= M_x\cos\omega t - M_y\sin\omega t \\ v &= M_x\sin\omega t + M_y\cos\omega t. \end{aligned} \tag{13.27}$$

By forming $\cos\omega t \times$ (Eq. 13.23) $-\sin\omega t \times$ (Eq. 13.24), we get

$$\frac{du}{dt} - \left(\omega_0 - \omega\right)v + \frac{u}{T_2} = 0. \tag{13.28}$$

By forming $\sin\omega t \times$ (Eq. 13.23) $+ \cos\omega t \times$ (Eq. 13.24), we get

$$\frac{dv}{dt} + (\omega_0 - \omega)\,u - \omega_1 M_z + \frac{v}{T_2} = 0. \tag{13.29}$$

From Eq. 13.25, upon using the second of Eqs. 13.27, we obtain

$$\frac{dM_z}{dt} + \omega_1 v + \frac{M_z - M_0}{T_1} = 0. \tag{13.30}$$

Equations 13.28–13.30 are the Bloch equations in the rotating frame. The steady-state solution to these equations is obtained by setting $du/dt = dv/dt = dM_z/dt = 0$. This is because u and v are slowly varying functions in the rotating frame. The resulting steady-state equations are then

$$u = (\omega_0 - \omega)\,vT_2 \tag{13.31}$$

$$v = -(\omega_0 - \omega)\,uT_2 + \omega_1 M_z T_2 \tag{13.32}$$

$$M_z = M_0 - \omega_1 T_1 v. \tag{13.33}$$

Solving these equations for the magnetization components in the rotating frame produces

$$\frac{u}{M_0} = \frac{\omega_1 (\omega_0 - \omega)\,T_2^2}{1 + (\omega_0 - \omega)^2 T_2^2 + \omega_1^2 T_1 T_2} \tag{13.34}$$

$$\frac{v}{M_0} = \frac{\omega_1 T_2}{1 + (\omega_0 - \omega)^2 T_2^2 + \omega_1^2 T_1 T_2} \tag{13.35}$$

$$\frac{M_z}{M_0} = \frac{1 + (\omega_0 - \omega)^2 T_2^2}{1 + (\omega_0 - \omega)^2 T_2^2 + \omega_1^2 T_1 T_2}. \tag{13.36}$$

Then using Eqs. 13.26 to transform back to the laboratory frame, we have the transverse magnetizations

$$\begin{aligned} \frac{M_x}{M_0} &= \frac{\omega_1 T_2\left[(\omega_0 - \omega)\,T_2 \cos\omega t + \sin\omega t\right]}{1 + (\omega_0 - \omega)^2 T_2^2 + \omega_1^2 T_1 T_2} \\ \frac{M_y}{M_0} &= \frac{\omega_1 T_2\left[-(\omega_0 - \omega)\,T_2 \sin\omega t + \cos\omega t\right]}{1 + (\omega_0 - \omega)^2 T_2^2 + \omega_1^2 T_1 T_2}. \end{aligned} \tag{13.37}$$

Recalling that $M_0 = \chi_0 B_0$ (see Eq. 10.39), along with the relations $\omega_0 = \gamma B_0$ and $\omega_1 = \gamma B_1$, Eqs. 13.37 can be written as

$$\begin{aligned} M_x &= B_1\chi'(\omega)\cos\omega t - B_1\chi''(\omega)\sin\omega t \\ M_y &= -B_1\chi''(\omega)\cos\omega t - B_1\chi'(\omega)\sin\omega t, \end{aligned} \tag{13.38}$$

where we define $\chi'(\omega)$ and $\chi''(\omega)$ as

$$\chi'(\omega) = \frac{\chi_0\omega_0(\omega_0 - \omega)T_2^2}{1 + (\omega_0 - \omega)^2 T_2^2 + \omega_1^2 T_1 T_2} \tag{13.39}$$

and

$$\chi''(\omega) = \frac{-\chi_0\omega_0 T_2}{1 + (\omega_0 - \omega)^2 T_2^2 + \omega_1^2 T_1 T_2}. \tag{13.40}$$

Let us introduce a complex transverse magnetization

$$M_+(t) = M_x(t) + iM_y(t). \tag{13.41}$$

It then follows that

$$M_+(t) = \chi(\omega) B_1 e^{-i\omega t}, \tag{13.42}$$

where, as before, $\chi(\omega)$ is the complex susceptibility given by $\chi(\omega) = \chi'(\omega) - i\chi''(\omega)$. In other words, $M_+(t)$ is the system response to the sinusoidal perturbation presented by the complex RF field, $B_1 e^{-i\omega t}$. Equation 13.42 follows the general form of the linear response relationship given by Eq. 13.10.

Using $\mathbf{B}_1 = B_1(\mathbf{e}_x \cos\omega t - \mathbf{e}_y \sin\omega t)$, it is left as an exercise for the reader to show that the power absorbed by the sample, $P(\omega)$, is given by

$$P(\omega) = \mathbf{B}_1 \cdot \frac{d\mathbf{M}}{dt} = \omega B_1 v. \tag{13.43}$$

Upon using Eq. 13.35 for v, one can see that the power absorption is related to $\chi''(\omega)$ by

$$P(\omega) \approx -\omega_0 B_1^2 \chi''(\omega) \tag{13.44}$$

(where it has been assumed that $\omega \approx \omega_0$). Since $\chi''(\omega)$ is related to the dissipation of RF power in the sample, we call it the dissipative part of the susceptibility.

13.2 Quantum-Mechanical Treatment of Linear Response Theory

13.2.1 Response to a Time-Dependent Perturbation

Consider a situation where a physical system is subjected to an external field. Examples include atom-EM field and neutron-nucleus interactions. Let

$$\hat{H} = \hat{H}_0 + \hat{A}F(t), \tag{13.45}$$

where $\hat{A}$ is the physical quantity through which the external field couples to the system and $F(t)$ is the time dependence of the external field. An illustration of this will occur in Example 13.2, where we will consider an atom subjected to an incoming light wave. For that case (see Eq. 9.149),

$$\hat{A}F(t) = -\hat{\boldsymbol{\mu}} \cdot \boldsymbol{\epsilon} E_0 e^{-i\omega t}, \tag{13.46}$$

where the electric field $\mathbf{E} = \boldsymbol{\epsilon} E_0 e^{-i\omega t}$ couples to the dipole moment $\widehat{\boldsymbol{\mu}}$ of the atom.

The Schrödinger equation governing the evolution of the perturbed system is

$$i\hbar \frac{\partial}{\partial t} |\psi_s(t)\rangle = \hat{H} |\psi_s(t)\rangle . \tag{13.47}$$

By making a transformation to the interaction picture (see Sect. 6.1),

$$|\psi_I(t)\rangle = e^{i\hat{H}_0 t/\hbar} |\psi_s(t)\rangle , \tag{13.48}$$

we can write

$$i\hbar \frac{\partial}{\partial t} |\psi_I(t)\rangle = -\hat{H}_0 e^{i\hat{H}_0 t/\hbar} |\psi_s(t)\rangle + i\hbar e^{i\hat{H}_0 t/\hbar} \frac{\partial}{\partial t} |\psi_s(t)\rangle . \tag{13.49}$$

Substituting Eq. 13.47 into Eq. 13.49 produces

$$i\hbar \frac{\partial}{\partial t} |\psi_I(t)\rangle = \hat{A}_I(t) F(t) |\psi_I(t)\rangle , \tag{13.50}$$

where

$$\hat{A}_I(t) = e^{i\hat{H}_0 t/\hbar} \hat{A} e^{-i\hat{H}_0 t/\hbar} . \tag{13.51}$$

We now consider a system which, at $t = -\infty$, is in a ground state of the Hamiltonian, $\hat{H}_0$, i.e.,

$$|\psi_s(t = -\infty)\rangle = |\psi_I(t = -\infty)\rangle = |\varphi_0\rangle , \tag{13.52}$$

and wish to calculate the system's wavefunction at time t assuming that the external perturbation is switched on gradually beginning at $t = -\infty$. Integrating Eq. 13.50, we have

$$|\psi_I(t)\rangle = |\varphi_0\rangle + \frac{1}{i\hbar} \int_{-\infty}^{t} dt' \hat{A}_I(t') F(t') |\psi_I(t')\rangle . \tag{13.53}$$

When the perturbation is weak, to first order in $\hat{A}$ we have

$$|\psi_I(t)\rangle = |\varphi_0\rangle + \frac{1}{i\hbar} \int_{-\infty}^{t} dt' \hat{A}_I(t') F(t') |\varphi_0\rangle . \tag{13.54}$$

This equation is good enough if we are interested only in the "linear" response of the system to the perturbation. The expectation value of any arbitrary observable C of the system is obtained from

$$\langle C(t)\rangle \equiv \langle \psi_I(t) \mid \hat{C}_I(t) \mid \psi_I(t)\rangle . \tag{13.55}$$

Therefore, inserting Eq. 13.54 into Eq. 13.55, we have

$$\begin{aligned}\langle C(t)\rangle &= \langle\varphi_0 \mid \hat{C}_I(t) \mid \varphi_0\rangle + \frac{1}{i\hbar}\int_{-\infty}^{t} dt' \langle\varphi_0 \mid \hat{C}_I(t)\,\hat{A}_I(t') \mid \varphi_0\rangle F(t') \\ &\quad -\frac{1}{i\hbar}\int_{-\infty}^{t} dt' \langle\varphi_0 \mid \hat{A}_I(t')\,\hat{C}_I(t) \mid \varphi_0\rangle F(t') \\ &= C_0 + \frac{1}{i\hbar}\int_{-\infty}^{t} dt' \langle\varphi_0 \mid \left[\hat{C}_I(t), \hat{A}_I(t')\right] \mid \varphi_0\rangle F(t'), \end{aligned} \tag{13.56}$$

where C_0 is the unperturbed expectation value of observable C at $t = -\infty$. Thus, the perturbation causes the expectation value of C to change from C_0 to $\langle C(t)\rangle$.

Two cases of interest to consider are the response to an impulse and the response to a periodic perturbation:

Impulse response function

If the applied external perturbation is a delta-function, i.e., $F(t) = \delta(t)$, one obtains a so-called *impulse response function*, $h_{CA}(t) \equiv \langle C(t)\rangle - C_0$, for the system:

$$h_{CA}(t) \equiv \langle C(t)\rangle - C_0 = \begin{cases} 0, & \text{for } t < 0 \\ \frac{1}{i\hbar}\langle\varphi_0 \mid \left[\hat{C}_I(t), \hat{A}\right] \mid \varphi_0\rangle, & \text{for } t \geq 0. \end{cases} \tag{13.57}$$

Response to a periodic excitation

Consider a periodic perturbation given as

$$F(t) = e^{-i\omega t + \eta t}; \qquad \eta > 0 \text{ and small.} \tag{13.58}$$

Here, η is introduced to ensure that the perturbation is zero at $t = -\infty$. The system response is then

$$\langle C(t)\rangle - C_0 = \frac{1}{i\hbar}\int_{-\infty}^{t} dt' \langle\varphi_0 \mid \left[\hat{C}_I(t), \hat{A}_I(t')\right] \mid \varphi_0\rangle e^{-i\omega t' + \eta t'}. \tag{13.59}$$

Noticing that

$$\langle\varphi_0 \mid \left[\hat{C}_I(t), \hat{A}_I(t')\right] \mid \varphi_0\rangle = \langle\varphi_0 \mid \left[\hat{C}_I(t - t'), \hat{A}\right] \mid \varphi_0\rangle, \tag{13.60}$$

and setting $\tau = t - t'$, Eq. 13.59 can be written as

$$\langle C(t)\rangle - C_0 = \chi_{CA}(\omega)\, e^{-i\omega t + \eta t}, \tag{13.61}$$

where

$$\chi_{CA}(\omega) = \frac{1}{i\hbar}\int_0^\infty d\tau \langle\varphi_0 \mid \left[\hat{C}_I(\tau), \hat{A}\right] \mid \varphi_0\rangle e^{i\omega\tau - \eta\tau} \tag{13.62}$$

is the susceptibility of the system. In terms of the impulse response function, $h_{CA}(t)$, the susceptibility is simply

$$\chi_{CA}(\omega) = \int_0^\infty h_{CA}(t) e^{i\omega t - \eta t} dt. \tag{13.63}$$

$\chi_{CA}(\omega)$ can be evaluated by noting that

$$\begin{aligned}\langle\varphi_0 \left| \left[\hat{C}_I(\tau), \hat{A}\right] \right| \varphi_0\rangle &= \sum_n \left[\left\langle\varphi_0 \mid \hat{C}_I(\tau) \mid \varphi_n\right\rangle \left\langle\varphi_n \mid \hat{A} \mid \varphi_0\right\rangle \right. \\ &\quad \left. - \left\langle\varphi_0 \mid \hat{A} \mid \varphi_n\right\rangle \left\langle\varphi_n \mid \hat{C}_I(\tau) \mid \varphi_0\right\rangle\right] \\ &= \sum_n \left[e^{-i\omega_{n0}\tau} C_{0n} A_{n0} - e^{i\omega_{n0}\tau} A_{0n} C_{n0}\right],\end{aligned} \tag{13.64}$$

where $\omega_{n0} = (E_n - E_0)/\hbar$. Substituting this into Eq.13.62, the susceptibility becomes

$$\chi_{CA}(\omega) = \frac{1}{\hbar}\sum_n \left[\frac{C_{0n}A_{n0}}{\omega - \omega_{n0} + i\eta} - \frac{A_{0n}C_{n0}}{\omega + \omega_{n0} + i\eta}\right]. \tag{13.65}$$

$\chi_{CA}(\omega)$ has two important properties: First, as $\omega \to \infty$,

$$\chi_{CA}(\omega) \longrightarrow \begin{cases} \dfrac{1}{\hbar\omega}\langle\varphi_0 \mid \left[\hat{C}, \hat{A}\right] \mid \varphi_0\rangle, & \text{for } \left[\hat{C}, \hat{A}\right] \neq 0 \\ \dfrac{1}{\hbar^2\omega^2}\langle\varphi_0 \mid \left[\left[\hat{H}_0, \hat{A}\right], \hat{C}\right] \mid \varphi_0\rangle, & \text{for } \left[\hat{C}, \hat{A}\right] = 0. \end{cases} \tag{13.66}$$

The second property is the following sum rule:

$$\int_{-\infty}^{\infty} \chi_{CA}(\omega)\, d\omega = -i\pi\langle\varphi_0 \mid \left[\hat{C}, \hat{A}\right] \mid \varphi_0\rangle. \tag{13.67}$$

Example 13.2 Response of an Atom to a Periodic EM Field. For an atom subjected to an incoming light wave having wavevector $\mathbf{k}$ and polarization λ, we have

$$\hat{A}F(t) = -\hat{\boldsymbol{\mu}}\cdot\boldsymbol{\epsilon}_{\mathbf{k}\lambda} E_0 e^{-i\omega t + \eta t}. \tag{13.68}$$

In other words,

$$\hat{A} = -\hat{\boldsymbol{\mu}}\cdot\boldsymbol{\epsilon}_{\mathbf{k}\lambda} E_0 = -\hat{\mu}_k E_0, \tag{13.69}$$

where we are using the shorthand notation $\hat{\mu}_k \equiv \hat{\boldsymbol{\mu}}\cdot\boldsymbol{\epsilon}_{\mathbf{k}\lambda}$. If $\hat{\mu}_{k'} \equiv \hat{\boldsymbol{\mu}}\cdot\boldsymbol{\epsilon}_{\mathbf{k}'\lambda'}$ takes on the role of our operator $\hat{C}$, then, from Eqs. 13.61 and 13.65, we are able to calculate the primed component of the induced dipole moment as

$$\langle \mu_{k'}(t)\rangle = \chi_{k'k}(\omega)\, e^{-i\omega t+\eta t}, \tag{13.70}$$

with susceptibility

$$\begin{aligned}
\chi_{k'k}(\omega) &= \frac{1}{\hbar}\sum_n \left[-\frac{\langle 0|\hat{\mu}_{k'}|n\rangle\langle n|\hat{\mu}_k|0\rangle}{\omega-\omega_{n0}+i\eta} + \frac{\langle 0|\hat{\mu}_k|n\rangle\langle n|\hat{\mu}_{k'}|0\rangle}{\omega+\omega_{n0}+i\eta}\right] E_0 \\
&= \frac{1}{\hbar}\sum_n \left[\frac{\langle 0|\hat{\mu}_{k'}|n\rangle\langle n|\hat{\mu}_k|0\rangle}{\omega_{n0}-\omega-i\eta} + \frac{\langle 0|\hat{\mu}_k|n\rangle\langle n|\hat{\mu}_{k'}|0\rangle}{\omega_{n0}+\omega+i\eta}\right] E_0 \\
&= \alpha_{k'k}(\omega)\, E_0.
\end{aligned} \tag{13.71}$$

Thus,

$$\langle \mu_{k'}(t)\rangle = \alpha_{k'k}(\omega)\, E_0 e^{-i\omega t+\eta t}, \tag{13.72}$$

where $\alpha_{k'k}(\omega)$ is the atomic polarizability tensor

$$\alpha_{k'k}(\omega) = \frac{1}{\hbar}\sum_n \left[\frac{\langle 0|\hat{\mu}_{k'}|n\rangle\langle n|\hat{\mu}_k|0\rangle}{\omega_{n0}-\omega-i\eta} + \frac{\langle 0|\hat{\mu}_k|n\rangle\langle n|\hat{\mu}_{k'}|0\rangle}{\omega_{n0}+\omega+i\eta}\right]. \tag{13.73}$$

Let us go back to the cross-section for coherent Rayleigh scattering derived in Chapter 9 (i.e., Eq. 9.146). Using our present notation (i.e., $|\mathcal{A}\rangle \to |0\rangle$, $|\mathcal{I}\rangle \to |n\rangle$, $E_{\mathcal{I}} - E_{\mathcal{A}} \to \hbar\omega_{n0}$, $\omega_k \to \omega$),

$$\left(\frac{d\sigma}{d\Omega}\right)_{\text{Rayleigh}} = k^4 \left|\frac{1}{\hbar}\sum_I \frac{\langle 0|\hat{\mu}_{k'}|n\rangle\langle n|\hat{\mu}_k|0\rangle}{\omega_{n0}-\omega} + \frac{\langle 0|\hat{\mu}_k|n\rangle\langle n|\hat{\mu}_{k'}|0\rangle}{\omega_{n0}+\omega}\right|^2. \tag{13.74}$$

Comparing with Eq. 13.73, we see that

$$\left(\frac{d\sigma}{d\Omega}\right)_{\text{Rayleigh}} = k^4 \left|\boldsymbol{\epsilon}_{\mathbf{k}'\lambda'}\cdot\widetilde{\widetilde{\boldsymbol{\alpha}}}(\omega)\cdot\boldsymbol{\epsilon}_{\mathbf{k}\lambda}\right|^2. \tag{13.75}$$

$\widetilde{\widetilde{\boldsymbol{\alpha}}}(\omega)$, the polarizability tensor of the atom in a field of frequency ω, is also the Rayleigh scattering tensor previously identified in Eq. 9.148.

13.2.2 Response of a System at Temperature T

The linear response theory can be extended to handle a system at a specified temperature. Such a system is in a mixed state and, as discussed in Chapter 7, is characterized by a density operator, $\hat{\rho}$.

As before, let us assume that a perturbation is switched on gradually from $t = -\infty$ to time t. Then the density operator can be written as

$$\hat{\rho}(t) = \hat{\rho}_0 + \hat{\rho}'(t). \tag{13.76}$$

$\hat{\rho}_0 = e^{-\beta\hat{H}_0}/Z$ is the density operator of the system in the unperturbed, thermal equilibrium state at $t = -\infty$, with $\beta = 1/k_BT$. Recall that $Z = \sum_n e^{-\beta E_n}$ is the partition function of the system, where the E_n's are the eigenvalues of the unperturbed Hamiltonian, $\hat{H}_0$. For a mixed state, the expectation value of an observable C is

$$\langle C \rangle \equiv \mathrm{Tr}\left(\hat{\rho}\hat{C}\right), \tag{13.77}$$

so, using Eq. 13.76, we have

$$\langle C \rangle = \mathrm{Tr}\left(\hat{\rho}_0\hat{C}\right) + \mathrm{Tr}\left(\hat{\rho}'\hat{C}\right) = C_0 + \mathrm{Tr}\left(\hat{\rho}'\hat{C}\right). \tag{13.78}$$

At this point it will prove useful to work in the interaction picture where

$$\hat{\rho}_I(t) = \hat{\rho}_0 + e^{i\hat{H}_0t/\hbar}\,\hat{\rho}'(t)\; e^{-i\hat{H}_0t/\hbar}. \tag{13.79}$$

Using this relation, along with the fact that $\mathrm{Tr}\left[\hat{A}\hat{B}\right] = \mathrm{Tr}\left[\hat{B}\hat{A}\right]$, Eq. 13.78 becomes

$$\begin{aligned}
\langle C \rangle &= \mathrm{Tr}\left[e^{-i\hat{H}_0t/\hbar}\hat{\rho}_I(t)\, e^{i\hat{H}_0t/\hbar}\hat{C}\right] \\
&= \mathrm{Tr}\left[e^{i\hat{H}_0t/\hbar}\hat{C}e^{-i\hat{H}_0t/\hbar}\hat{\rho}_I(t)\right] \\
&= \mathrm{Tr}\left[\hat{C}_I(t)\,\hat{\rho}_I(t)\right].
\end{aligned} \tag{13.80}$$

A perturbation expansion of the density operator was previously performed in Sect. 7.3. For the linear response to a perturbation, one need only consider the first-order expansion (see Eq. 7.57), namely,

$$\hat{\rho}_I(t) = \hat{\rho}_0 + \frac{1}{i\hbar}\int_{-\infty}^{t} dt'\left[\hat{V}_I(t'), \hat{\rho}_0\right]. \tag{13.81}$$

Substituting into Eq. 13.80 produces

$$\begin{aligned}
\langle C(t) \rangle &= \mathrm{Tr}\left[\hat{C}_I(t)\,\hat{\rho}_0\right] + \frac{1}{i\hbar}\int_{-\infty}^{t} dt'\mathrm{Tr}\left\{\hat{C}_I(t)\left[\hat{V}_I(t'), \hat{\rho}_0\right]\right\} \\
&= C_0 + \frac{1}{i\hbar}\int_{-\infty}^{t} dt'\mathrm{Tr}\left\{\hat{C}_I(t)\,\hat{V}_I(t')\,\hat{\rho}_0 - \hat{C}_I(t)\,\hat{\rho}_0\hat{V}_I(t')\right\}
\end{aligned}$$

$$= C_0 + \frac{1}{i\hbar}\int_{-\infty}^{t} dt'\mathrm{Tr}\left\{\hat{C}_I(t)\,\hat{V}_I(t')\,\hat{\rho}_0 - \hat{V}_I(t')\,\hat{C}_I(t)\,\hat{\rho}_0\right\}$$

$$= C_0 + \frac{1}{i\hbar}\int_{-\infty}^{t} dt'\mathrm{Tr}\left\{\left[\hat{C}_I(t)\,,\hat{V}_I(t')\right]\hat{\rho}_0\right\}. \quad (13.82)$$

The trace appearing on the right-hand side represents the expectation value associated with the commutator $\left[\hat{C}_I(t)\,,\hat{V}_I(t')\right]$. We shall denote this expectation value simply as $\langle[C_I(t)\,,V_I(t')]\rangle$.* Then the response of the system to the perturbation is written as

$$\langle C(t)\rangle - C_0 = \frac{1}{i\hbar}\int_{-\infty}^{t} dt'\,\langle[C_I(t)\,,V_I(t')]\rangle\,. \quad (13.83)$$

As before, assume that

$$\hat{H} = \hat{H}_0 + \hat{V}(t) = \hat{H}_0 + \hat{A}F(t)\,, \quad (13.84)$$

so that

$$\hat{V}_I(t) = \hat{A}_I(t)\,F(t) \quad (13.85)$$

and the response then becomes

$$\langle C(t)\rangle - C_0 = \frac{1}{i\hbar}\int_{-\infty}^{t} dt'\,\langle[C_I(t)\,,A_I(t')]\rangle\,F(t')\,. \quad (13.86)$$

In the case of a periodic excitation, we have

$$\langle C(t)\rangle - C_0 = \frac{1}{i\hbar}\int_{-\infty}^{t} dt'\,\langle[C_I(t)\,,A_I(t')]\rangle\,e^{-i\omega t'+\eta t'}\,. \quad (13.87)$$

Note that

$$\langle[C_I(t)\,,A_I(t')]\rangle = \langle[C_I(t-t')\,,A_I(0)]\rangle = \langle[C_I(t-t')\,,A]\rangle\,. \quad (13.88)$$

Then, by setting $\tau = t - t'$, we obtain the result

$$\langle C(t)\rangle - C_0 = \chi_{CA}(\omega)\,e^{-i\omega t+\eta t} \quad (13.89)$$

with susceptibility

$$\chi_{CA}(\omega) = \frac{1}{i\hbar}\int_{0}^{\infty} d\tau\,\langle[C_I(\tau)\,,A]\rangle\,e^{i\omega\tau-\eta\tau}\,. \quad (13.90)$$

The use of Eq. 13.90 is illustrated in our final example:

*The hats (^) have been intentionally dropped, in keeping with our previous notation used for expectation values, namely, $\mathrm{Tr}\left(\hat{A}\hat{\rho}\right) = \langle A\rangle$, not $\left\langle\hat{A}\right\rangle$.

Example 13.3 Microscopic Theory of NMR Susceptibility. As in the case of Example 13.1, consider again a spin system placed in a static longitudinal field, $\mathbf{B}_0$. The system is in thermal equilibrium at temperature T. The response of the spin system to a rotating transverse magnetic field, $\mathbf{B}_1$, can be obtained by observing the transverse components of the system's magnetization vector. Specifically, let us consider the complex transverse magnetization, $M_+(t) = M_x(t) + iM_y(t)$, introduced previously in Example 13.1. Then, using Eq. 13.89, and remembering that the transverse component of the magnetization at thermal equilibrium is zero, the expectation value of M_+ due to the RF transverse field can be written as

$$\langle M_+(t)\rangle = \chi(\omega)e^{-i\omega t+\eta t}, \tag{13.91}$$

where the susceptibility is given by Eq. 13.90. The interaction Hamiltonian between the spin system and the (complex) alternating field is given by

$$\hat{V}(t) = \hat{A}F(t) = -\hat{M}_+B_1e^{-i\omega t}, \tag{13.92}$$

which assumes a system of unit volume. If the amplitude of the RF field is also assumed to be unity, then $\hat{A} = -\hat{M}_+$, and the susceptibility is

$$\chi(\omega) = -\frac{1}{i\hbar}\int_0^\infty d\tau \,\langle[M_+(\tau), M_+(0)]\rangle\, e^{i\omega\tau}, \tag{13.93}$$

where η has been set to zero. The imaginary part of the susceptibility is then

$$\begin{aligned}\chi''(\omega) &= -\operatorname{Im}[\chi(\omega)]\\ &= -\frac{1}{\hbar}\int_0^\infty d\tau\,\langle[M_+(\tau), M_+(0)]\rangle \cos\omega\tau.\end{aligned} \tag{13.94}$$

Assuming that the thermal average is an even function of τ, we can rewrite $\chi''(\omega)$ as

$$\begin{aligned}\chi''(\omega) &= -\frac{1}{2\hbar}\int_{-\infty}^\infty d\tau\,\langle[M_+(\tau), M_+(0)]\rangle \cos\omega\tau\\ &= -\frac{1}{2\hbar}\int_{-\infty}^\infty d\tau\,\langle[M_+(\tau), M_+(0)]\rangle\, e^{i\omega\tau}.\end{aligned} \tag{13.95}$$

The last step follows because the $\sin\omega\tau$-component of $e^{i\omega\tau}$ does not contribute to the integration. It is now left to the reader to show that one of the terms making up the commutator can be rewritten as the following variation on the form of Eq. 12.124:

$$\langle M_+(0)\, M_+(\tau)\rangle = \langle M_+(\tau - i\hbar\beta)\, M_+(0)\rangle\,. \tag{13.96}$$

Then Eq. 13.95 becomes

$$\chi''(\omega) = -\frac{1}{2\hbar}\left\{\int_{-\infty}^{\infty} d\tau \left\langle M_+(\tau)\, M_+(0)\right\rangle e^{i\omega\tau} - \int_{-\infty}^{\infty} d\tau \left\langle M_+(\tau - i\hbar\beta)\, M_+(0)\right\rangle e^{i\omega\tau}\right\}. \tag{13.97}$$

In the second integration let $\tau - i\hbar\beta = t$, so that

$$\begin{aligned}\chi''(\omega) &= -\frac{1}{2\hbar}\left[\int_{-\infty}^{\infty} d\tau \left\langle M_+(\tau)\, M_+(0)\right\rangle e^{i\omega\tau} - e^{-\beta\hbar\omega}\int_{-\infty}^{\infty} dt \left\langle M_+(t)\, M_+(0)\right\rangle e^{i\omega t}\right] \\ &= -\frac{1}{2\hbar}\left(1 - e^{-\beta\hbar\omega}\right)\int_{-\infty}^{\infty} d\tau \left\langle M_+(\tau)\, M_+(0)\right\rangle e^{i\omega\tau}. \end{aligned}\tag{13.98}$$

Finally, since $\beta\hbar\omega \ll 1$ in NMR experiments, we can approximate $e^{-\beta\hbar\omega}$ as $1 - \beta\hbar\omega$, and the imaginary part of the susceptibility is of the form

$$\chi''(\omega) = \frac{-\omega}{2k_BT}\int_{-\infty}^{\infty} d\tau \left\langle M_+(\tau)\, M_+(0)\right\rangle e^{i\omega\tau}. \tag{13.99}$$

$\chi''(\omega)$ in liquids has a sharp peak centered around ω_0, so the factor ω outside the integral can be safely replaced with ω_0 without any loss of generality. Equation 13.99 shows that $\chi''(\omega)$ is proportional to the Fourier transform of the time-autocorrelation function of the transverse component of the magnetization. This is an example of the general fluctuation-dissipation theorem introduced in Section 12.3. The right-hand side of Eq. 13.99 represents the fluctuations in the transverse component of the system's magnetization. The left-hand side is the power dissipation of the RF field in the sample, as we found in Eq. 13.44.

We learned in Chapter 10 that the transverse magnetization undergoes an exponential relaxation with time-constant T_2. This suggests that the correlation function decays according to

$$\left\langle M_+(\tau)\, M_+(0)\right\rangle = \left\langle |M_+|^2\right\rangle e^{-|\tau|/T_2} e^{-i\omega_0\tau}. \tag{13.100}$$

The Fourier transform of this correlation function is

$$\frac{1}{2\pi}\int_{-\infty}^{+\infty} d\tau \left\langle |M_+|^2\right\rangle e^{-|\tau|/T_2} e^{-i(\omega_0-\omega)\tau} = \frac{1}{\pi}\left\langle |M_+|^2\right\rangle \left[\frac{T_2}{1+(\omega_0-\omega)^2 T_2^2}\right], \tag{13.101}$$

so, from Eq. 13.99, the susceptibility function is

$$\chi''(\omega) = -\left\langle |M_+|^2 \right\rangle \frac{\omega_0}{k_B T} \left[\frac{T_2}{1 + (\omega_0 - \omega)^2 T_2^2} \right]. \tag{13.102}$$

Compare this expression to the one derived from the classical Bloch equations, namely, Eq. 13.40. In the limit of zero longitudinal relaxation time ($T_1 \to 0$), that result was

$$\chi''(\omega) = \frac{-\chi_0 \omega_0 T_2}{1 + (\omega_0 - \omega)^2 T_2^2}. \tag{13.103}$$

A direct comparison of terms in Eqs. 13.102 and 13.103 results in

$$\left\langle |M_+|^2 \right\rangle = k_B T \chi_0. \tag{13.104}$$

Since $\left\langle |M_+|^2 \right\rangle = \langle M_x^2 \rangle + \langle M_y^2 \rangle$, and since $\langle M_x^2 \rangle = \langle M_y^2 \rangle$, we can identify

$$\langle M_x^2 \rangle = \langle M_y^2 \rangle = \frac{1}{2} k_B T \chi_0. \tag{13.105}$$

From Eq. 10.40, we have $\chi_0 = N \langle \mu^2 \rangle / 3 k_B T$, where N is the number of spins per unit volume and $\langle \mu^2 \rangle = \gamma^2 \hbar^2 I (I+1)$. Therefore,

$$\langle M_x^2 \rangle = \langle M_y^2 \rangle = \frac{1}{6} N \langle \mu^2 \rangle. \tag{13.106}$$

More importantly, both the quantum-mechanical and the classical treatment of NMR susceptibility predict a Lorentzian form for $\chi''(\omega)$, the dissipative part of the susceptibility.

Suggested References

The following texts contain sections on general linear response theory and correlation functions:

[a] D. Forster, *Hydrodynamics, Fluctuations, Broken Symmetry, and Correlation Functions* (Perseus Books Group, Cambridge, MA, 1990).

[b] R. Kubo, M. Toda, and N. Hashitsume, *Statistical Physics II: Nonequilibrium Statistical Mechanics*, 2nd ed. (Springer-Verlag, Berlin, 1991).

[c] L. E. Reichl, *A Modern Course in Statistical Physics*, 2nd ed. (John Wiley and Sons, New York, 1998).

[d] P. M. Chaikin and T. C. Lubensky, *Principles of Condensed Matter Physics* (Cambridge University Press, Cambridge, 1995).

For a comprehensive coverage of NMR susceptibility, refer to

[e] C. P. Slichter, *Principles of Magnetic Resonance*, 3rd ed. (Springer-Verlag, Berlin, 1996).

Problems

1. The magnetization of a particular material responds to a step of applied magnetic field, turned on at $t = 0$, according to

$$M(t) = M_0 \left(1 - e^{-t/T}\right).$$

   Determine the real and imaginary parts of the susceptibility function, $\chi'(\omega)$ and $\chi''(\omega)$, and show that they satisfy the Kramers-Kronig relations.

2. (a) Provide an argument for why $\chi'(\omega)$ and $\chi''(\omega)$, the real and imaginary parts of the susceptibility function, must be even and odd functions of ω, respectively.

   (b) A system exhibits a narrow absorption peak at frequency $\omega = \omega_0$. A naive student claims that the *absorption spectrum*, which is the dissipative part of the susceptibility, is given by

$$\chi''(\omega) = \alpha\delta(\omega - \omega_0),$$

   where α is a real constant. However, because $\chi''(\omega)$ must be an odd function, the expression given is not acceptable. Fix the incorrect expression by adding in a second delta-function term. From the correct expression, determine $\chi'(\omega)$, the corresponding real part of the susceptibility (also referred to as the *dispersion spectrum*). Sketch both $\chi'(\omega)$ and $\chi''(\omega)$.

(c) For a narrow absorption peak, show that the static susceptibility, $\chi_0 \equiv \chi(\omega = 0)$, is proportional to the area under the peak.

3. The aim of this problem is to determine the response function and related susceptibility for a classical one-dimensional damped harmonic oscillator (mass m) subject to an applied external force, $f(t)$. The displacement $x(t)$ of the oscillator obeys the equation

$$m\ddot{x} + m\gamma\dot{x} + m\omega_0^2 x = f(t), \tag{13.107}$$

where γ is the damping constant and ω_0 is the natural frequency of the undamped oscillator. The linear response function $n(t)$ is then defined by (see Eq. 13.2)

$$x(t) = \int_{-\infty}^{t} n(t - t') f(t') \, dt'.$$

If one considers an impulse force, $f(t) = F_0\delta(t)$, then we have $x(t) = F_0 n(t)$, and the equation for the response function is

$$m\ddot{n} + m\gamma\dot{n} + m\omega_0^2 n = \delta(t). \tag{13.108}$$

(a) From Eq. 13.9, the susceptibility function can be written as a Laplace transform

$$\chi(\omega) = \int_{0}^{\infty} n(t) e^{-st} dt$$

setting $s = -i\omega + \varepsilon$, in the limit $\varepsilon \to 0^+$. Taking the Laplace transform of Eq. 13.108, subject to the initial conditions $x(0) = \dot{x}(0) = 0$, show that the complex susceptibility becomes

$$\chi(\omega) = \frac{1}{m(\omega^2 - \omega_0^2 - i\gamma\omega)}$$

and that the dissipative part of the susceptibility is

$$\chi''(\omega) = -\frac{\gamma\omega}{m\left[(\omega^2 - \omega_0^2)^2 + (\gamma\omega)^2\right]}.$$

(b) For $t > 0$, the response function can be calculated from the susceptibility function according to

$$n(t) = \frac{1}{2\pi} \int_{-\infty}^{+\infty} \chi(\omega) e^{-i\omega t} d\omega.$$

For the damped oscillator, show that

$$n(t) = \frac{2}{m\Omega} e^{-\gamma t/2} \sin(\Omega t/2), \quad \text{for } t > 0$$

where $\Omega^2 = 4\omega_0^2 - \gamma^2$. Sketch the behavior of the response function for the two cases of underdamping ($\gamma < 2\omega_0$) and overdamping ($\gamma > 2\omega_0$).

(c) As an example, consider the response of the oscillator to an isolated force-pulse of duration T, i.e., determine $x(t)$ for the case

$$f(t) = \begin{cases} 0, & \text{for } t < 0 \\ F_0, & \text{for } 0 < t < T \\ 0, & \text{for } t > T. \end{cases}$$

4. This problem investigates a classical harmonic oscillator of mass m immersed in a viscous fluid at temperature T. The oscillator motion is again governed by Eq. 13.107 of the previous problem, where the average viscous force from the fluid gives rise to the damping term. In this case, however, the external force $f(t)$ is replaced by $\zeta(t)$, a fluctuating, stochastic force that comes about because of random collisions of the oscillator with the surrounding fluid molecules (with $\langle \zeta(t) \rangle = 0$).

(a) Let $X(\omega)$ be the Fourier transform of the displacement function $x(t)$. Starting with the susceptibility function $\chi(\omega)$ derived in the last problem, show that the power spectral density of the displacement is given by

$$|X(\omega)|^2 = \frac{2\gamma T}{m} \frac{1}{(\omega^2 - \omega_0^2)^2 + (\gamma\omega)^2}.$$

(**Hint:** You will need to obtain the power spectrum of the random force, $\zeta(t)$. Since the nature of this force does not depend on the oscillator frequency ω_0, its power spectral density, which is a constant (why?), can be obtained by setting $\omega_0 = 0$ in the equation of motion and taking its Fourier transform. With the help of the Wiener-Khintchine theorem try to relate the power spectrum of the random force signal to $\langle v^2 \rangle$, the mean-square velocity.)

(b) Show, using the Wiener-Khintchine theorem, that the displacement auto-correlation function is

$$\langle x(0)\, x(t) \rangle = \frac{k_B T}{m\omega_0^2} e^{-\gamma t/2} \left[\cos(\Omega t/2) + \frac{\gamma}{\Omega} \sin(\Omega t/2) \right], \quad \text{for } t > 0 \tag{13.109}$$

where, as before, $\Omega^2 = 4\omega_0^2 - \gamma^2$.

(c) Explain the connection of the latter results to the fluctuation-dissipation theorem.

(d) If $x(t)$ is a stationary, random process, show that the velocity autocorrelation function can be calculated from

$$\langle v(0)\, v(t)\rangle = -\frac{d^2}{dt^2}\langle x(0)\, x(t)\rangle .$$

Use this result to calculate $\langle v(0)\, v(t)\rangle$ from the displacement corrrelation function of Eq. 13.109.

(e) By letting $\omega_0 \to 0$, the situation corresponds to a free particle diffusing through a liquid. For this case, show that the normalized velocity autocorrelation function, in fact, reduces to the previous result for a diffusing particle obtained at the end of Chapter 12, i.e., Eq. 12.151.

Some Constants and Conversion Factors

Basic Physical Constants

speed of light in vacuum, $c = 2.998 \times 10^{10}\text{cm/s}$

Planck's constant, $\hbar = 1.055 \times 10^{-27}\text{erg}\cdot\text{s} = 6.582 \times 10^{-16}\text{eV}\cdot\text{s}$

Boltzmann's constant, $k_B = 1.381 \times 10^{-16}\text{erg/K} = 8.618 \times 10^{-5}\text{eV/K}$

Energy Conversion

$1\ \text{eV} = 1.602 \times 10^{-12}\text{erg}$

The Photon

$$\text{wavelength, } \lambda_\text{p}\,[\text{Å}] = \frac{12.40}{E\,[\text{keV}]}$$

The Neutron

mass, $m_n = 1.675 \times 10^{-24}\text{g}$

$$\text{wavelength, } \lambda_\text{n}\,[\text{Å}] = \frac{0.2861}{\sqrt{E\,[\text{eV}]}} = \frac{9.047}{\sqrt{E\,[\text{meV}]}}$$

References

1. G. L. Baker and J. P. Golub, *Chaotic Dynamics: An Introduction* (Cambridge University Press, Cambridge, 1990).
2. H. L. Anderson, ed., *A Physicist's Desk Reference*—Physics Vade Mecum, 2nd ed. (AIP, New York, 1989) p. 241.
3. E. Merzbacher, *Quantum Mechanics*, 2nd ed. (John Wiley and Sons, New York, 1970) p. 159.
4. G. Herzberg, *Molecular Spectra and Structure* (Van Nostrand and Reinhold, New York, 1950).
5. M. Born and J. Oppenheimer, *Ann. Phys.* **84** (1927) 457.
6. A. Goswami, *Quantum Mechanics* (Wm. C. Brown Publishers, Dubuque, Iowa, 1992) p. 433.
7. P. M. Morse and H. Feshbach, *Methods of Theoretical Physics* (McGraw-Hill, New York, 1953) p. 1781.
8. M. Kerker, *The Scattering of Light and Other Electromagnetic Radiation* (Academic Press, New York, 1969) pp. 83–91.
9. I. L. Fabelinskii, *Molecular Scattering of Light* (Plenum Press, New York, 1968).
10. S. H. Chen, in *Physical Chemistry: An Advanced Treatise*, Vol. 8A, ed. D. Henderson (Academic Press, New York, 1971) pp. 85–156.
11. P. A. Egelstaff, *An Introduction to the Liquid State* (Academic Press, New York, 1967) pp. 19–22, 73.
12. A. Einstein, *Ann. Phys.* **33** (1910) 1275.
13. G. Mie, *Ann. Physik.* **25** (1908) 377.
14. Lord Rayleigh, *Phil. Mag.* **36** (1918) 365.
15. J. R. Wait, *Can. J. Phys.* **33** (1955) 189.
16. S. Asano and G. Yamamoto, *Appl. Opt.* **14** (1975) 29.
17. J. A. Stratton, *Electromagnetic Theory* (McGraw-Hill, New York, 1941).
18. H. C. van de Hulst, *Light Scattering by Small Particles* (John Wiley and Sons, New York, 1957) pp. 114–130.
19. A. N. Lowan, *Tables of Scattering Functions for Spherical Particles*, Natl. Bur. Standards, (U.S.), Appl. Math. Series 4 (Govt. Printing Office, Washington, 1948) pp. 4–19.

20. S. H. Chen, M. Holz, and P. Tartaglia, *Appl. Opt.* **16** (1977) 187.
21. W. H. Louisell, *Quantum Statistical Properties of Radiation* (John Wiley and Sons, New York, 1973) p. 496.
22. M. Planck, *Verh. dt. phys. Ges.* **2** (1900) 202 and 237.
23. A. Einstein, *Annln Phys.* **17** (1905) 132.
24. G. N. Lewis, *Nature* **118** (1926) 874.
25. D. Marcuse, *Engineering Quantum Electrodynamics* (Harcourt, Brace and World, New York, 1970) pp. 97–99.
26. T. Uyematsu, K. Yamazaki, S. Bun, H. Koyano, H. Tsushima, K. Inagaki, Y. Yoshida, M. Osaki, and K. Kasai, in *Squeezed Light*, ed. O. Hirota (Elsevier Science Publishers, Amsterdam, 1992) p. 101.
27. M. J. Collett and R. Loudon, *J. Opt. Soc. Am.* **B4** (1987) 1525.
28. F. A. M. de Oliveira and P. L. Knight, *Phys. Rev. A* **39** (1989) 3417.
29. E. Fermi, *Nuclear Physics* (University of Chicago Press, Chicago, 1950) p. 142.
30. E. Fermi, *Ric. Sci.* **7** (1936) 13.
31. A. Foderaro, *The Elements of Neutron Interaction Theory* (M.I.T. Press, Cambridge, Mass., 1971) pp. 529–530.
32. R. G. Sachs, *Nuclear Theory* (Addison-Wesley, Reading, Mass., 1953) pp. 89–91.
33. N. F. Mott and H. S. W. Massey, *The Theory of Atomic Collisions*, 2nd ed. (Oxford University Press, Oxford, 1949) pp. 116–131.
34. L. Koester, H. Rauch, M. Herkens, and K. Schröder, *Summary of Neutron Scattering Lengths (*KFA-Report, Jül-1755, Kernforschungsanlage, Jülich, GmbH, 1981).
35. W. Heitler, *The Quantum Theory of Radiation* (Oxford University Press, London, 1954) pp. 256–275.
36. R. G. Gordon, in *Advances in Magnetic Resonance*, Vol. 3, ed. J. S. Waugh (Academic Press, New York, 1968) pp. 4–10.
37. R. G. Gordon, *J. Chem. Phys.* **43** (1965) 1307.
38. M. O. Bulanin and N. D. Orlova, *Opt. Spectry.* **15** (1963) 112.
39. V. G. Weisskopf and E. Wigner, *Z. Phys.* **63** (1930) 54.
40. J. H. Van Vleck and D. L. Huber, *Rev. Mod. Phys.* **49** (1977) 939.
41. D. A. McQuarrie, *Statistical Mechanics* (Harper and Row, New York, 1976) pp. 357–378.
42. M. P. Auger, *J. Phys. Radium* **6** (1925) 205.
43. M. P. Auger, *Ann. Phys., Paris* **6** (1926) 183.
44. R. W. Fink, R. C. Jopson, H. Mark, and C. D. Swift, *Rev. Mod. Phys.* **38** (1966) 513.
45. E. H. S. Burhop, *The Auger Effect and Other Radiationless Transitions* (Cambridge University Press, London, 1952) pp. 44–57.
46. H. Lay, *Z. Phys.* **91** (1934) 533.

47. E. J. Callan, *Bull. Am. Phys. Soc.* **7** (1962) N416.
48. E. J. Callan, *Rev. Mod. Phys.* **35** (1963) 524.
49. M. A. Listengarten, *Izv. Akad. Nauk SSSR, Ser. Fiz.* **25** (1961) 792 [*Bull. Acad. Sci. USSR, Phys., Columbia Tech. Transl.* **25** (1961) 803].
50. M. A. Listengarten, *Izv. Akad. Nauk SSSR, Ser. Fiz.* **26** (1962) 182 [*Bull. Acad. Sci. USSR, Phys., Columbia Tech. Transl.* **26** (1962) 182].
51. A. A. Jaffe, *Bull. Res. Council (Israel)* **3**, No. 4, (March 1954) 316.
52. R. C. Jopson, H. Mark, C. D. Swift, and M. A. Williamson, *Phys. Rev.* **137** (1965) A1353.
53. M. Stobbe, *Ann. d. Phys.* **7** (1930) 661.
54. F. Sauter, *Ann. de Phys.* **9** (1931) 217.
55. F. Sauter, *Ann. de Phys.* **11** (1931) 454.
56. H. R. Hulme, J. McDougall, R. A. Buckingham, and R. H. Fowler, *Proc. Roy. Soc.* **149** (1935) 131.
57. H. Hall, *Rev. Mod. Phys.* **8** (1936) 358.
58. R. D. Evans, *The Atomic Nucleus* (McGraw-Hill, New York, 1955) pp. 330–337.
59. J. D. Jackson, *Classical Electrodynamics*, 2nd ed. (John Wiley and Sons, New York, 1975) pp. 658–659.
60. P. Eisenberger and P. M. Platzman, *Phys. Rev. A* **2** (1970) 415.
61. C. H. Macgillavry, G. D. Rieck, and K. Lonsdale, eds., *International Tables for X-Ray Crystallography*, Vol. III (D. Reidel Publishing Co., Dordrecht, Holland, 1985) pp. 201–216.
62. P. Eisenberger, *Phys. Rev. A* **2** (1970) 1678.
63. P. Eisenberger, *Phys. Rev. A* **5** (1972) 628.
64. P. Eisenberger, W. H. Henneker, and P. E. Cade, *J. Chem. Phys.* **56** (1972) 1207.
65. P. Eisenberger, L. Lam, P. M. Platzman, and P. Schmidt, *Phys. Rev. B* **6** (1972) 3671.
66. R. M. Eisberg, *Fundamentals of Modern Physics* (John Wiley and Sons, New York, 1961) pp. 393–396.
67. P. Bosi, F. Dupré, F. Menzinger, F. Sacchetti, and M. C. Spinelli, *Nuovo Cimento Lett.* **21** (1978) 436.
68. J. Teixeira, M. C. Bellissent-Funel, S. H. Chen, and B. Dorner, *Phys. Rev. Lett.* **54** (1985) 2681.
69. C. Y. Liao, S. H. Chen, and F. Sette, *Phys. Rev. E* **61** (2000) 1518.
70. G. Ruocco and F. Sette, *J. Phys.: Condens. Matter* **11** (1999) R259.
71. I. M. de Schepper, E. G. D. Cohen, C. Bruin, J. C. van Rijs, W. Montfrooij, and L. A. de Graaf, *Phys. Rev. A* **38** (1988) 271.
72. B. Kamgar-Parsi, E. G. D. Cohen, and I. M. de Schepper, *Phys. Rev. A* **35** (1987) 4781.

73. W. Montfrooij, E. C. Svensson, I. M. de Schepper, and E. G. D. Cohen, *J. Low Temp. Phys.* **105** (1996) 149.
74. E. G. D. Cohen and I. M. de Schepper, *Nuovo Cimento* **12** (1990) 521.
75. J. P. Boon and S. Yip, *Molecular Hydrodynamics* (McGraw-Hill, New York, 1980).
76. F. Sciortino and S. Sastry, *J. Chem. Phys.* **100** (1994) 3881.
77. E. A. Power and T. Thirunamachandran, *Am. J. Phys.* **46** (1978) 370.
78. E. M. Purcell, H. C. Torrey, and R. V. Pound, *Phys. Rev.* **69** (1946) 37.
79. F. Bloch, W. W. Hansen, and M. E. Packard, *Phys. Rev.* **69** (1946) 127.
80. W. G. Proctor and F. C. Yu, *Phys. Rev.* **77** (1950) 717.
81. W. C. Dickinson, *Phys. Rev.* **77** (1950) 736.
82. J. P. Hornak, in Kirk-Othmer *Encyclopedia of Chemical Technology*, 4th ed.,Vol. 16 (John Wiley and Sons, New York, 1995) pp. 107–134.
83. R. C. Weast, ed., *CRC Handbook of Chemistry and Physics*, 67th ed. (CRC Press, Cleveland, 1986) pp. E78–E80.
84. F. Bloch, *Phys. Rev.* **70** (1946) 460.
85. E. L. Hahn, *Phys. Rev.* **80** (1950) 580.
86. H. Y. Carr and E. M. Purcell, *Phys. Rev.* **94** (1954) 630.
87. E. O. Stejskal, *J. Chem. Phys.* **43** (1965) 3597.
88. B. Chu, *Laser Light Scattering: Basic Principles and Practice*, 2nd ed. (Academic Press, San Diego, 1991) pp. 201–214.
89. L. Mandel, E. C. G. Sudarshan, and E. Wolf, *Proc. Phys. Soc.* **84** (1964) 435.
90. R. J. Glauber, in *Quantum Optics and Electronics*, eds. C. de Witt, A. Blandin, and C. Cohen-Tannoudji (Gordon and Breach Publishers, New York, 1965) p. 65.
91. R. Loudon, *The Quantum Theory of Light*, 2nd ed. (Oxford University Press, Oxford, 1983) pp. 101–105.
92. L. Mandel and E. Wolf, *J. Opt. Soc. Am.* **53** (1963) 1315.
93. L. Mandel, in *Progress in Optics*, Vol. 2, ed. E. Wolf (John Wiley and Sons, New York, 1963) p. 181.
94. R. J. Glauber, in *Physics of Quantum Electronics*, eds. P. L. Kelley, B. Lax, and P. E. Tannenwald (McGraw-Hill, New York, 1966) p. 788.
95. T. Uyematsu, K. Yamazaki, S. Bun, H. Koyano, H. Tsushima, K. Inagaki, Y. Yoshida, M. Osaki, and K. Kasai, in *Squeezed Light*, ed. O. Hirota (Elsevier Science Publishers, Amsterdam, 1992) p. 153.
96. R. Nossal, S. H. Chen, and C. C. Lai, *Opt. Comm.* **4** (1971) 35.
97. B. J. Berne and R. Pecora, *Dynamic Light Scattering: With Applications to Chemistry, Biology, and Physics* (John Wiley and Sons, New York, 1976) pp. 67–73.
98. S. H. Chen and A. V. Nurmikko, in *Spectroscopy in Biology and Chemistry: Neutron, X-Ray, Laser*, eds. S. H. Chen and S. Yip (Academic Press, New York, 1974) p. 377.
99. S. H. Chen, W. B. Veldkamp, and C. C. Lai, *Rev. Sci. Instrum.* **46** (1975) 1356.

100. S. H. Chen and P. Tartaglia, *Opt. Comm.* **6** (1972) 119.
101. L. Mandel, *Phys. Rev.* **181** (1969) 75.
102. P. Tartaglia and S. H. Chen, *Opt. Comm.* **7** (1973) 379.
103. P. Tartaglia and S. H. Chen, *J. Chem. Phys.* **58** (1973) 4389.
104. E. Jakeman and E. R. Pike, *J. Phys.* **A2** (1969) 411.
105. M. Kotlarchyk, S. H. Chen, and S. Asano, *Appl. Opt.* **18** (1979) 2470.
106. R. Nossal and S. H. Chen, *J. de Physique*, Suppl. 2–3, **33** (1972) C1-171.
107. M. Kotlarchyk and S. H. Chen, *J. Chem. Phys.* **79** (1983) 2461.
108. L. van Hove, *Phys. Rev.* **95** (1954) 249.
109. G. H. Vineyard, *Phys. Rev.* **110** (1958) 999.
110. J. P. Hansen, *Theory of Simple Liquids* (Academic Press, London, 1976).
111. L.P. Kadanoff and P. C. Martin, *Ann. Phys.* **24** (1963) 419.
112. S. W. Lovesey, *Theory of Neutron Scattering from Condensed Matter*, Vol. 1 (Oxford University Press, New York, 1984) pp. 98–169.
113. E. Fermi, *Ric. Sci.* **7** (1936) 13 [English translation in *Enrico Fermi: Collected Papers*, Vol. 1 (University of Chicago Press, Chicago, 1962) p. 980].
114. J. P. Boon and S. Yip, *Molecular Hydrodynamics* (McGraw-Hill, New York, 1980), pp. 35–45.
115. R. Kubo, *J. Phys. Soc. Japan* **12** (1957) 570.
116. F. Reif, *Fundamentals of Statistical and Thermal Physics* (McGraw-Hill, New York, 1965) pp. 560–577.
117. R. C. Desai and S. Yip, *Phys. Rev.* **166** (1968) 129.
118. K. S. Singwi and A. Sjölander, *Phys. Rev.* **120** (1960) 1093.

Index